D1555721

Edited by Hagen Klauk

Organic Electronics II

Related Titles

Ramm, P., Lu, J. J.-Q., Taklo, M. M. V. (eds.)

Handbook of Wafer Bonding

2011

ISBN: 978-3-527-32646-4

Samori, P., Cacialli, F. (eds.)

Functional Supramolecular Architectures

For Organic Electronics and Nanotechnology

2011

ISBN: 978-3-527-32611-2

Chujo, Y. (ed.)

Conjugated Polymer Synthesis

Methods and Reactions

2011

ISBN: 978-3-527-32267-1

Leclerc, M., Morin, J.-F. (eds.)

Design and Synthesis of Conjugated Polymers

2010

ISBN: 978-3-527-32474-3

Perepichka, I. F., Perepichka, D. F. (eds.)

Handbook of Thiophene-Based Materials

Applications in Organic Electronics and Photonics, 2 Volume Set

2009

ISBN: 978-0-470-05732-2

Garrou, P., Bower, C., Ramm, P. (eds.)

Handbook of 3D Integration

Technology and Applications of 3D Integrated Circuits

2008

ISBN: 978-3-527-32034-9

Brabec, C., Scherf, U., Dyakonov, V. (eds.)

Organic Photovoltaics

Materials, Device Physics, and Manufacturing Technologies

2008

ISBN: 978-3-527-31675-5

Klauk, H. (ed.)

Organic Electronics

Materials, Manufacturing and Applications

2006

ISBN: 978-3-527-31264-1

Edited by Hagen Klauk

Organic Electronics II

More Materials and Applications

WILEY-VCH Verlag GmbH & Co. KGaA

The Editor

Dr. Hagen Klauk
Max Planck Institute for Solid
State Research
Heisenbergstr. 1
70569 Stuttgart
Germany

Cover
The Cover was provided by Huai-Yuan Tseng working in the group of Vivek Subramanian at the University of California, Berkeley.
It shows an optical micrograph of a self-aligned organic transistor manufactured entirely by inkjet printing.

Back Cover
The structure on the back cover was kindly provided by John Anthony.

Library of Congress Card No.: applied for

British Library Cataloguing-in-Publication Data
A catalogue record for this book is available from the British Library.

Bibliographic information published by the Deutsche Nationalbibliothek
The Deutsche Nationalbibliothek lists this publication in the Deutsche Nationalbibliografie; detailed bibliographic data are available on the Internet at <http://dnb.d-nb.de>.

Cover Design Adam-Design, Weinheim
Typesetting Laserwords Private Limited, Chennai, India
Printing and Binding Markono Print Media Pte Ltd, Singapore

Printed in Singapore
Printed on acid-free paper

Print ISBN: 978-3-527-32647-1
ePDF ISBN: 978-3-527-64023-2
oBook ISBN: 978-3-527-64021-8
ePub ISBN: 978-3-527-64022-5
Mobi ISBN: 978-3-527-64024-9

Contents

Preface

Organic Electronics is not only a fascinating field of research and development, but also one that continues to move ahead at a swift pace. And thus six years after the first installment of this book series (*Organic Electronics – Materials, Manufacturing, and Applications*, Wiley-VCH, 2006) it was time for an update. The 12 chapters of the sequel provide a detailed look at some important new developments and advancements in the field of organic thin-film transistors, including novel semiconductors for p-channel and n-channel organic transistors, advanced thin-film characterization techniques, new insights into charge transport in organic semiconductors, research into low-voltage electrolyte-gated transistors, solution-processing techniques and high-resolution printing approaches for the manufacture of organic transistors, the science and technology of organic light-emitting transistors, and advanced design strategies for large-scale organic integrated circuits.

Once again, I am deeply indebted to the many professionals who have contributed to this book. First and foremost I would like to extend my sincere gratitude to the 42 authors who have taken time out of their busy schedules to share their wisdom and knowledge. Second I want to thank Martin Preuss, Bernadette Gmeiner, and Bente Flier at Wiley-VCH for the encouragement and organizational oversight to make this book happen. And finally my thanks go out to the readers of both the first and second book of this series for their interest.

Stuttgart, 2011 *Hagen Klauk*

List of Contributors

John E. Anthony
Department of Chemistry
University of Kentucky
Lexington
KY 40506-0055
USA

Thomas D. Anthopoulos
Centre for Plastic Electronics
Imperial College
Department of Physics
South Kensington Campus
London SW7 2AZ
UK

Monique J. Beenhakkers
Polymer Vision
Kastanjelaan 1000
5616 LZ Eindhoven
The Netherlands

Mario Caironi
Center for Nano Science and
Technology @PoliMi
Istituto Italiano di Tecnologia
Via Pascoli, 70/3
20133 Milano
Italy

Eugenio Cantatore
Eindhoven University of
Technology
Department of Electrical
Engineering
PO Box 513
5600 Eindhoven
The Netherlands

Wim Dehaene
Imec
Kapeldreef 75
3001 Leuven
Belgium

and

Katholieke Universiteit Leuven
ESAT department
Kasteelpark Arenberg 10
PO Box 2440
3001 Leuven
Belgium

Alejandro de la Fuente Vornbrock
University of California
Department of Electrical
Engineering and Computer
Sciences
Berkeley
CA 94720-1770
USA

Dean M. DeLongchamp
Polymers Division
National Institute of Standards and Technology
100 Bureau Dr. MS 8541
Gaithersburg
MD 20899-8541
USA

Antonio Facchetti
Northwestern University
Department of Chemistry
2145 Sheridan Road
Evanston
IL 60208-3113
USA

and

Polyera Corporation
8045 Lamon Avenue
Skokie
IL 60077
USA

C. Daniel Frisbie
University of Minnesota–Twin Cities
Department of Chemical Engineering and Materials Science
421 Washington Ave SE
Minneapolis
MN 55455
USA

Gerwin H. Gelinck
Holst Centre TNO
High Tech Campus 31
5656 AE Eindhoven
The Netherlands

Jan Genoe
Imec
Kapeldreef 75
3001 Leuven
Belgium

and

Katholieke Hogeschool Limburg
IWT department
Universitaire Campus
3590 Diepenbeek
Belgium

Enrico Gili
University of Cambridge
Cavendish Laboratory
J J Thomson Avenue
Cambridge CB3 0HE
UK

David J. Gundlach
National Institute of Standards and Technology
Electronics and Electrical Engineering Laboratory
100 Bureau Drive Gaithersburg
MD 20899
USA

Martin Heeney
Centre for Plastic Electronics
Imperial College
Department of Chemistry
South Kensington Campus
London SW7 2AZ
UK

Paul Heremans
Imec
Kapeldreef 75
3001 Leuven
Belgium

and

Katholieke Universiteit Leuven
ESAT department
Kasteelpark Arenberg 10
PO Box 2440
3001 Leuven
Belgium

Thomas N. Jackson
Penn State University
Department of Electrical Engineering
121 Elect Engineering East
University Park
PA 16802
USA

David Ian James
Centre for Plastic Electronics
Imperial College
Department of Chemistry
South Kensington Campus
London SW7 2AZ
UK

and

Centre for Plastic Electronics
Imperial College
Department of Physics
South Kensington Campus
London SW7 2AZ
UK

Adolphus G. Jones
Department of Chemistry
University of Kentucky
Lexington
KY 40506-0055
USA

R. Joseph Kline
Polymers Division
National Institute of Standards and Technology
100 Bureau Dr. MS 8541
Gaithersburg
MD 20899-8541
USA

Oana D. Jurchescu
Wake Forest University
Department of Physics
1834 Wake Forest Road
Winston-Salem
NC 27109
USA

Yuanyuan Li
Penn State University
Department of Electrical Engineering
121 Elect Engineering East
University Park
PA 16802
USA

Hagen Marien
Imec
Kapeldreef 75
3001 Leuven
Belgium

and

Katholieke Universiteit Leuven
ESAT department
Kasteelpark Arenberg 10
PO Box 2440
3001 Leuven
Belgium

Iain McCulloch
Centre for Plastic Electronics
Imperial College
Department of Chemistry
South Kensington Campus
London SW7 2AZ
UK

Steve Molesa
University of California
Department of Electrical
Engineering and Computer
Sciences
Berkeley
CA 94720-1770
USA

Devin A. Mourey
Penn State University
Department of Electrical
Engineering
121 Elect Engineering East
University Park
PA 16802
USA

Kris Myny
Imec
Kapeldreef 75
3001 Leuven
Belgium

and

Katholieke Universiteit Leuven
ESAT department
Kasteelpark Arenberg 10
PO Box 2440
3001 Leuven
Belgium

and

Katholieke Hogeschool Limburg
IWT department
Universitaire Campus
3590 Diepenbeek
Belgium

Rodrigo Noriega
Stanford University
Department of Applied Physics
476 Lomita Mall
Stanford
CA 94305
USA

Alberto Salleo
Stanford University
Department of Materials Science
and Engineering
476 Lomita Mall
Stanford
CA 94305
USA

Tsuyoshi Sekitani
The University of Tokyo
Department of Electrical and
Electronic Engineering and
Information Systems
7-3-1 Hongo
Bunkyo-ku
Tokyo 113-8656
Japan

Henning Sirringhaus
University of Cambridge
Cavendish Laboratory
J J Thomson Avenue
Cambridge CB3 0HE
UK

Jeremy Smith
Centre for Plastic Electronics
Imperial College
Department of Physics
South Kensington Campus
London SW7 2AZ
UK

Takao Someya
The University of Tokyo
Department of Electrical and
Electronic Engineering and
Information Systems
7-3-1 Hongo
Bunkyo-ku
Tokyo 113-8656
Japan

and

The University of Tokyo
Institute for Nano Quantum
Information Electronics (INQIE)
4-6-1 Komaba
Meguro-ku
Tokyo 153-8505
Japan

Daniel Soltman
University of California
Department of Electrical
Engineering and Computer
Sciences
Berkeley
CA 94720-1770
USA

Soeren Steudel
Imec
Kapeldreef 75
3001 Leuven
Belgium

Michiel Steyaert
Katholieke Universiteit Leuven
ESAT department
Kasteelpark Arenberg 10
PO Box 2440
3001 Leuven
Belgium

Vivek Subramanian
University of California
Department of Electrical
Engineering and Computational
Sciences
Berkeley
CA 94720-1770
USA

Huai-Yuan Tseng
University of California
Department of Electrical
Engineering and Computer
Sciences
Berkeley
CA 94720-1770
USA

Peter Vicca
Imec
Kapeldreef 75
3001 Leuven
Belgium

Nick A.J.M. van Aerle
Polymer Vision
Kastanjelaan 1000
5616 LZ Eindhoven
The Netherlands

and

ASML
De Run 6501
5504 DR Veldhoven
The Netherlands

Yu Xia
University of Minnesota–
Twin Cities
Department of Chemical
Engineering and Materials
Science
421 Washington Ave SE
Minneapolis
MN 55455
USA

Jana Zaumseil
Institute of Polymer Materials
University Erlangen-Nürnberg
Martensstraße 7
D-91058 Erlangen
Germany

Part I
Materials

1
Organic Semiconductor Materials for Transistors

David Ian James, Jeremy Smith, Martin Heeney, Thomas D. Anthopoulos, Alberto Salleo, and Iain McCulloch

1.1
General Considerations

Recent advances in the electrical performance of organic semiconductor materials position organic electronics as a viable alternative to technologies based on amorphous silicon (a-Si). Traditionally a-Si-based transistors, which are used as the switching and amplifying components in modern electronics [1], require energy intensive batch manufacturing techniques. These include material deposition and patterning using a number of high-vacuum and high-temperature processing steps in addition to several subtractive lithographic patterning and mask steps, limiting throughput. Although this allows for the cost of individual transistors to be extremely low because of the high circuit density that can be obtained, the actual cost per unit area is very high. Alternatively, organic semiconductors can be formulated into inks and processed using solution-based printing processes [2–5]. This allows for large-area, high-throughput, low-temperature fabrication of organic field-effect transistors (OFETs), enabling not only a reduction in cost but also the migration to flexible circuitry, as lower temperatures enable the use of plastic substrates. The potential applications for these OFETs are numerous, ranging from flexible backplanes in active matrix displays to item-level radiofrequency identification tags.

OFETs are typically p-type (hole transporting) devices that are composed of a source and drain electrode connected by an organic semiconductor, with a gate electrode, insulated from the organic semiconductor via a dielectric material, as shown in Figure 1.1b. Holes are injected into the highest occupied molecular orbital (HOMO) of the organic semiconductor upon application of a negative gate voltage. The holes migrate to the accumulation layer, which forms at the semiconductor interface with the dielectric, and are transported between the source and drain upon application of an electric field between the two. Modulation of the gate voltage is used to turn the transistor ON and OFF, with the ON current and voltage required to turn the device on being figures of merit for the electrical performance of the device. The performance of the transistor is also governed by the charge carrier mobility of the semiconductor, which should be high to ensure fast charging speeds.

Organic Electronics II: More Materials and Applications, First Edition. Edited by Hagen Klauk.

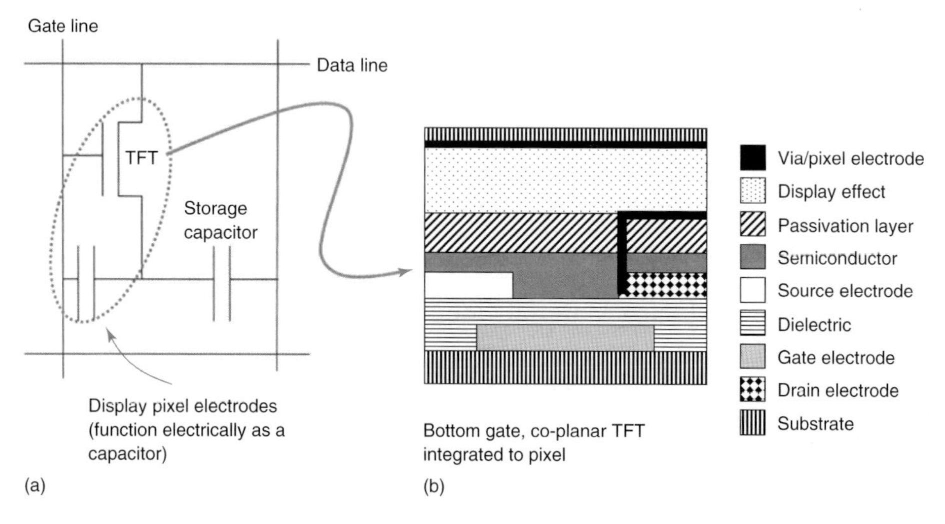

Figure 1.1 (a) Simple diagram of active matrix backplane circuitry and (b) cross section of corresponding TFT and pixel architecture.

In displays, OFETs can act as individual pixel switches in the backplane active matrix circuitry, as shown in Figure 1.1a. This technology is currently being used commercially in small-sized electrophoretic displays (EPDs), marketed as e-paper [6], to charge both the pixel and the storage capacitor. Active matrix backplanes are found in both liquid crystal displays (LCDs) and organic light emitting diode (OLED) displays, where a transistor also provides current to the emitting diode element. An advantage of the EPD effect is that the pixels are reflective to ambient light, which allows the pixel transistor to occupy the majority of the area underneath the pixel. This maximizes the transistor width, enabling more current to be delivered to the pixel, resulting in lower mobility specifications being required from the semiconductor. For small-sized devices ($\lesssim$10 cm diagonally) with low resolutions and low refresh rates, the mobility required is in the region of 0.01 cm^2 V^{-1} s^{-1}, which is well within the capabilities of both polymer and small molecule semiconductors. In comparison, medium- to large-sized LCDs commonly used for monitor and television displays require semiconductor mobilities in excess of 0.5 cm^2 V^{-1} s^{-1}, and currently employ a-Si or polysilicon for higher-resolution displays. EPDs are also bistable, as once the pixel and the storage capacitor are charged, no additional power is needed to maintain the image. This minimizes the duty cycle load of the transistor, thus extending the lifetime. One problem with EPDs is that it is possible for ionic impurities within the liquid EPD cell to facilitate current leakage from the capacitor, which means that higher charge carrier mobilities are required than would be expected and thus high-purity electrophoretic inks are required to reduce the current demands of the display effect.

The function of the transistor in a LCD is to apply an electric field across the pixel, thus switching the direction of the optical axis of the liquid crystals, which therefore generates the image. As the display operates in the transmissive mode, the transistor is directly in the path of the light source, and so must be small to maximize the aperture ratio of the pixel and thus increase the efficiency of light output. However, the use of smaller transistors means that the organic semiconducting material within the transistor needs to have a higher charge carrier mobility than that of the materials used in EPDs.

OLED displays have the potential to be fabricated using high-throughput printing techniques such as gravure or ink jet. Using organic transistors will allow for the complete integration of both front- and backplane fabrication processes. Top emitting devices, in which the OLED frontplane cathode is transparent, can be fabricated, allowing the OFET to be positioned underneath the emissive layer. Thus, the OFETs can be larger per pixel than the transistors used in LCDs with equivalent-sized pixels. However, as the current output from the transistor dictates the brightness of the pixel, the transistor to transistor uniformity must be very tight. Additionally, multiple OFETs are needed per OLED pixel, requiring OLED OFETs to be smaller than the OFETs used in EPDs, where only one OFET is required per pixel. This leads to the need for higher-mobility organic semiconductors as well as reduced transistor to transistor anisotropy to avoid issues of color shifts from differential pixel aging effects and nonuniform pixel brightness.

Higher-mobility semiconductor materials are also required, as future demand for larger screen sizes, better resolution, and faster refresh rates for video rate displays will lead to the need for higher ON currents as a result of the larger number of rows and columns, as well as the requirement for faster pixel charging speeds. The development of these materials is discussed in the next section.

1.2 Materials Properties of Organic Semiconductors

Organic semiconductors are based on the fact that the sp^2 hybridization of carbon in a double bond leaves a p_z orbital available for π bonding. The electrons in the π bond can be delocalized via conjugation with neighboring π bonds, thus giving rise to charge carrier mobility. For this reason, the majority of organic semiconductors are composed of aromatic units linked together, allowing π orbital conjugation along the length of the molecule. Charge transport within both small molecule and polymeric organic semiconductors generally occurs via a thermally activated hopping mechanism, and in an OFET, this occurs along the plane of the substrate, propagating within a thin layer of semiconductor only a few molecules thick at the dielectric interface. Thus, the semiconductor at this interface must be highly ordered into closely packed organized π stacks with correctly oriented and interconnected domains as illustrated in Figure 1.2. This can be achieved by utilizing coplanar aromatic molecules, which form a highly crystalline thin film microstructure domain leading to high charge carrier mobility.

Most high-performing semiconducting polymers exhibit a crystalline phase, melt transition, and amorphous phase on heating. The temperature at which the phase

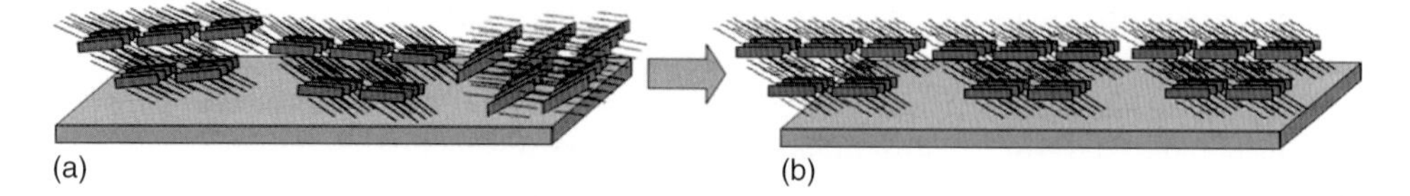

Figure 1.2 Schematic representation of (a) misaligned and poorly connected lamellar domains and (b) coaligned domains. Morphology in (b) leads to more optimal charge transport.

transition appears is dependent on the Gibbs free energy of each phase with respect to temperature, with the lowest free energy phase prevailing. Aromatic planar extended rigid-rod-type polymers have a predisposition to exhibit a liquid crystalline phase, due in part to their calamitic conformation. This phase is often masked by the lower free energy of the crystalline or amorphous phase. However, for some polymers, a liquid crystalline phase occurs between the crystalline and the isotropic melt phases (Figure 1.3). Annealing of the polymer within the liquid crystalline phase produces highly ordered and aligned crystalline thin films, which is desirable for high charge carrier mobilities. So in order to design polymers that incorporate a liquid crystalline phase, the entropy (dg/dT) of the isotropic phase must be decreased. This can be achieved by increasing the stiffness of the polymer backbone by the use of coplanar aromatic molecules, which decreases the disorder of the melt (decreased slope of dg/dT), allowing the liquid crystalline phase to appear.

The molecular weight of polymers also has an influence on the charge carrier mobility [7]. Increasing the molecular weight has been shown [8, 9] to be beneficial up to a plateau region, and so average molecular weights above 20 kDa are typically desired. The reason for this is that high polymer molecular weights enable the crystalline domains within the transistor thin film to be better defined and more interlinked. On the other hand, low-molecular-weight polymer films, although they

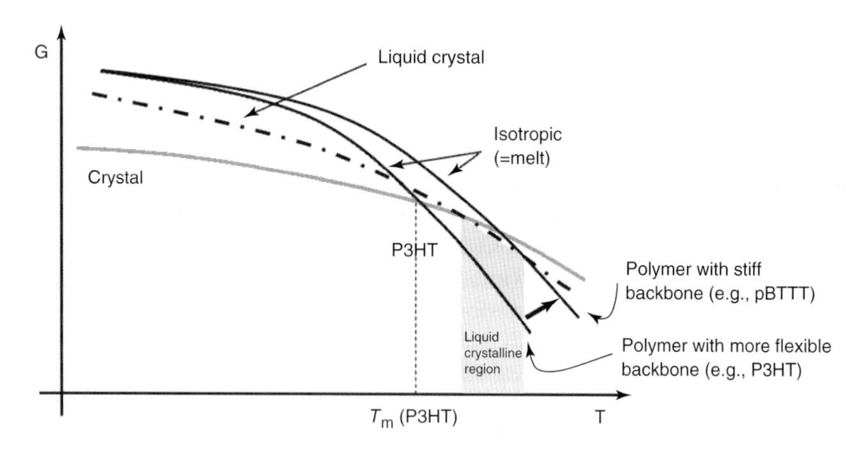

Figure 1.3 Effect of stiffening polymer backbone on the polymer-phase free energy.

exhibit higher crystallinity, they have more defined grain boundaries, thus leading to lower mobilities [10]. However, at very high molecular weights (Mn $>$ 150 kDa), the polymer has a higher viscosity, which hinders the crystallization during annealing. This leads to reduced mobilities. Having a high polydispersity index, that is, a broad range of molecular weights is also expected to lead to reduced mobilities because of poor crystallization, although no systematic studies have investigated this so far [11].

Another consideration, in addition to the charge carrier mobility, is the stability of the semiconductor under ambient conditions. This has an impact on both the device lifetime as well as the device ON and OFF currents, as the introduction of bulk charge carriers through doping leads to an increase in the conductivity of the semiconductor when the gate voltage is off. The electrochemical stability of the organic semiconductor is dictated by the HOMO energy level of the molecule or polymer as redox electrochemistry with oxygen in the presence of water leads to the loss of electrons from the HOMO, provided that the HOMO is $<$4.9 eV from the vacuum energy level [12]. Thus, it is important to lower the HOMO energy level below this value through judicious molecular design of the molecule or polymer.

One distinction between small molecule semiconductors and polymeric semiconductors is in the way they can be processed. Small molecule semiconductors can be processed either by evaporation or using solution-based techniques. Polymeric aromatic semiconductors, however, cannot usually be processed by evaporation, so they must be functionalized to ensure that they are solution processable. As polymeric semiconductors designed for charge transport typically have closely packed stiff aromatic backbones, are highly crystalline, and have low polarity, they are usually insoluble in the majority of commonly used organic solvents. Thus, aliphatic side chains must be employed to enhance their solubility. The length, degree of branching, and density of the alkyl side chains influence both the vertically separated chains d spacing, and π stacking distances, which in turn affect the thin film morphology and consequently the transistor field-effect mobility. Both the formulation rheology and the thin film thermal properties can be controlled by tuning the polymer molecular weight and polydispersity, allowing compatibility with printing techniques such as flexography and gravure, which have high viscosity requirements. This allows control over the thickness and morphology of the conformal and cohesive thin films, minimizing thin film reticulation, which often occurs as the solution dries on a low-energy surface substrate. Solution deposition of multilayer device stacks is possible with polymeric semiconductors, as their limited solubility (narrow solubility parameter profile) and their high bulk viscosity enable solvent orthogonality. This is a prerequisite, as sequential solution deposition of one polymer layer on top of another requires that each deposited layer is inert to the solvents and temperature conditions that they are subsequently exposed to. Additionally, the negligible vapor pressure of polymers means that they are not susceptible to interlayer diffusion during the thermal cycles undertaken during device fabrication. The robust mechanical properties of polymeric thin films make them ideal for flexible processing or flexible substrate operation. As polymers crystallize forming crystalline domains, which are small relative to the transistor channel length, fairly isotropic in-plane transport can be obtained. This leads to low device to device performance variability, which is important for applications where large numbers of transistors are integrated.

1.3
Small Molecule Semiconductors

1.3.1
Sexithiophene

Thiophene is an electron-rich planar aromatic heterocycle and its 2,5-coupled oligomers form well-ordered structures in thin films. One of the first promising thiophene-containing organic semiconductors was sexithiophene (6T) [13]. This oligomer showed mobilities in field-effect transistors of 10^{-3} cm^2 V^{-1} s^{-1}, which was the highest of any organic semiconductor at the time [14]. The rigid rod molecules of 6T were able to adopt an all-transplanar configuration because of the sulfur atoms arranging spatially along the length of the molecule to maximize their separation from each other across the short axis. As the angle formed between adjacent thiophene units linked at the positions 2 and 5 is $<180^\circ$, rotation of an individual thiophene unit around the long axis of the polymer is restricted due to the energetically unfavorable requirement that neighboring molecules will also need to rotate in a cooperative manner. Thus, the coupled thiophene units can adopt a coplanar conformation, allowing close intermolecular approach between neighboring backbones. For 6T, this results in a herringbone packing arrangement with relatively close $\pi-\pi$ distances, which in turn allows efficient intermolecular (or interchain) hopping of charges, leading to high charge mobility. However, one drawback of 6T is that it requires deposition from the vapor phase under vacuum in order for it to be processed [15] as it is scarcely soluble [16] and so is unsuitable for processing by high-throughput solution-based printing techniques.

1.3.2
Pentacene and Derivatives

Another small molecule that has been frequently employed in transistors is pentacene [17] with mobilities of up to 5 cm^2 V^{-1} s^{-1} achieved for thin film devices whose processing conditions have been extensively optimised [18]. As pentacene is a planar acene molecule, it is also able to pack in a herringbone arrangement, thus giving rise to ordered, crystalline films, with high mobilities. However, it is poorly soluble in most solvents, except for hot chlorinated aromatics, and so also requires vapor deposition to form thin films. Thus to make pentacene more amenable to solution processing, Anthony and coworkers [19–22] have produced a number of pentacene derivatives. These involve the introduction of bulky alkynyl silyl groups at the positions central 6 and 13 of the pentacene ring, which have shown a marked improvement not only in solubility but also in oxidative stability. Additionally, the crystal packing arrangement has been altered by the substituents, resulting in highly ordered two-dimensional slipped stack arrays being formed for certain derivatives of substituted pentacenes and related dithienoanthracenes. Recent work [23] has shown pronounced grain boundary effects on the charge transport of slip stacked crystal motifs in comparison to the more isotropic herringbone arrangements of pentacene. However, on optimizing processing conditions to form continuous polycrystalline thin films, these materials have given rise to impressive mobilities in excess of $1\,cm^2\,V^{-1}\,s^{-1}$ [21, 24], which exceeds requirements for most initial organic electronics applications. Unfortunately, problems with these materials still persist, including the issues of anisotropic in-plane transport, which causes device to device nonuniformity, interlayer mixing upon solution deposition of multilayer stacks, and the difficulty in controlling the process of crystallization over large substrate areas. One method to overcome this is the use of small-molecule-polymer blends, which is discussed later in this chapter.

1.4 Polymer Semiconductors

1.4.1 Thiophene-Based Polymers

1.4.1.1 Poly(3-Alkylthiophenes)

Thiophene-containing polymers have emerged as leading examples of high-performing semiconductors. These polymers are typically electron rich, with lamellar microstructures exhibiting closely packed π-stacked backbones, optimal for charge transport. In this chapter, we review three examples of this class of polymer, shown in Figure 1.4.

Regioregular (RR) poly(3-hexylthiophene) (P3HT) is the most extensively studied semiconducting polymer. This is not only due to its ease of processing from solution and the fact that it is widely available but also due to its highly crystalline microstructure that gives rise to its promising electrical properties. It has been shown that a systematic improvement in charge carrier mobility can be obtained

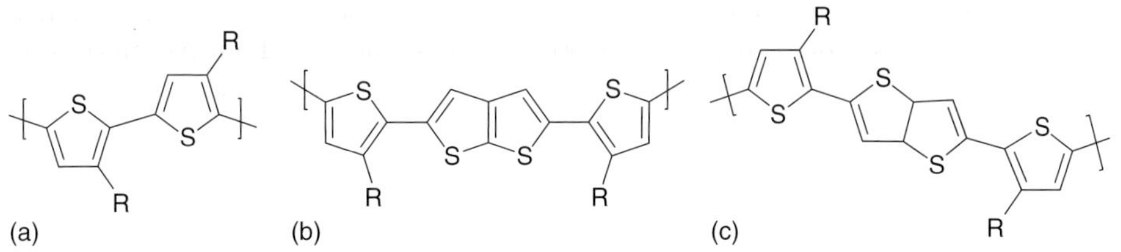

(a) (b) (c)

Figure 1.4 Structure of thiophene polymers (a) poly(3-hexythiophene) (P3HT), (b) poly(2,5-bis(3-alkylthiophen-2-yl)thieno[2,3-*b*]thiophene (pBTCT), and (c) poly(2,5-bis(3-alkylthiophen-2-yl)thieno[3,2-*b*]thiophene (pBTTT).

by maximizing the regioregularity and the molecular weight of the polymer. The effect of increased regioregularity is to avoid out-of-plane twists along the backbone of the polymer, as monomers in a head to head orientation experience steric repulsions between adjacent alkyl groups. This disrupts the planarity of the molecule, decreasing the effective conjugation length of the polymer, as can be seen in the hypsochromatic shifts in optical absorbance as the regioregularity decreases [25]. This in turn reduces the efficiency of charge hopping. P3HT with a head to tail regioregularity in excess of 96% has been shown to exhibit charge carrier mobilities of up to 0.1 $cm^2 V^{-1} s^{-1}$ under inert atmospheric conditions [26]. Indeed, there are still efforts to further push the measured transistor mobilities through improvements in processing conditions, device architecture, and electrodes [27–30]. On changing the molecular weight of the polymer, atomic force microscopy (AFM) measurements show that low-molecular-weight high-RR P3HT ($\sim$5 kDa) forms highly crystalline rodlike structures in which the width of the rod does not exceed the length of the individual polymer chains. RR P3HT with higher molecular weights ($>$30 kDa) exhibits less crystallinity with small nodule like crystallites interconnected with amorphous regions. The higher-molecular-weight RR P3HT shows higher mobilities than the more crystalline low-molecular-weight RR P3HT films. One explanation for this is that the higher-molecular-weight P3HT has better defined and more connected grains allowing better intergrain charge transport, whereas the low-molecular-weight RR P3HT has more defined grain boundaries [10]. Another explanation for the reduced mobility in low-molecular-weight films cites enhanced out-of-plane twisting in the polymer chains, leading to shorter conjugation lengths and thus reduced charge hopping rates [31]. Studies have also correlated increasing molecular weight with increasing crystalline quality within domains, in the high-mobility regime, as there are fewer chain ends per domain, as well as the possibility for individual polymer chains to interconnect domains at high molecular weight [9]. However, there is a limit to the improvements on increasing the molecular weight, as at over $\sim$50 kDa, crystalline disorder increases, a phenomenon that can be attributed to slower crystallization kinetics.

Studies regarding the alkyl side chain influence on the electrical performance of polyalkylthiophenes have been undertaken by several groups. These studies indicate that hexyl side chains are the optimum length, as the charge carrier mobility decreases as the chain length increases [32–34]. As the face to face ($\pi - \pi$) distance is similar across the series of polymers fabricated [35], it is apparent that the decrease in charge carrier mobility may be due to the increase in the fraction of insulating side chains in the polymer. If the polymer lamellae are misaligned in the plane of the substrate and the direction of charge flow, then hopping or tunneling between the insulating alkyl side chains will be necessary for charge transport. However, as hopping rates are dependent on the distance between neighboring polymer chains, longer alkyl side chains would be expected to be detrimental to charge mobility.

The effect of bulky or highly polar substituted end groups on the side chains has been studied by Bao and Lovinger [35]. The series of RR polythiophenes prepared showed poor field-effect mobilities of around 10^{-6} cm^2 V^{-1} s^{-1} because of the low degrees of crystallinity and ordering exhibited. The incorporation of chiral alkyl side chains was shown to maintain crystallinity at the expense of the $\pi-\pi$ stacking distance – the distance was increased to 4.3 Å as opposed to 3.8 Å for P3HT. This lead to a decrease in charge carrier mobility of around one order of magnitude. The overall findings of the study showed that nonpolar linear alkyl chains were optimal for charge transport because of the efficient backbone stacking.

The HOMO energy level of P3HT is about −4.6 eV from the vacuum energy level as a consequence of the electron-rich, π-conjugated, highly planar aromatic backbone that has electron-donating alkyl side chains. This HOMO level renders P3HT potentially unstable to ambient air and humidity, as electrochemical oxidation occurs at potentials above −4.9 eV [12]. Even though the sensitivity to water and oxygen redox electrochemistry is not the only contributing factor to the instability of P3HT, as well as for other π-conjugated aromatics, it is crucial to ensure that the electrochemical oxidation of the organic semiconductor is not thermodynamically favorable. There are a number of reports in the literature that have observed instabilities in OFET performance under ambient air conditions [36, 37] and have credited this to an interaction with molecular oxygen [38]. For instance, charge-transfer complexes between oxygen and thiophene have been proposed, which can generate reversible charged states and show a doping effect on transistor performance.

On photoexcitation, singlet oxygen can be generated by energy transfer from excited electronic states, which causes irreversible chemical degradation to the polymer. The mechanism of this photooxidation is that the singlet oxygen undergoes a 1,4-Diels-Alder addition reaction to the thienyl double bonds of the thiophene ring, breaking the conjugation of the backbone. However, in the absence of light, oxygen is not a strong oxidant for thiophene polymers. Instead, ozone, and possibly other pollutants such as NO_x and SO_x found in ambient air, have high electron affinities and are likely to participate in doping [39]. If the ozone molecule stays intact during the complexation with the polymer, then the doping is reversible. Irreversible doping occurs when, on dissociation, an exothermic reaction between the polymer and the ozone molecule occurs, cleaving the carbon-carbon bonds in the backbone, leading to reduced conjugation. This rationalization for organic semiconductor instability is consistent with the evidence that top-gate devices usually display enhanced stability in comparison with bottom-gate devices [40]. In a top-gate device architecture, the organic semiconducting layer is protected from the environment by the dielectric and gate layers, which potentially act as sacrificial layers for reactions with highly reactive dopants such as ozone.

There are many synthetic design parameters that affect the performance of the semiconductor. Polydispersity [41], molecular weight [10, 31, 42–46], levels of impurities [47], end groups, and chemical defects in the polymer backbone [48–50] are all important, as they influence both the morphology and electrical properties of the organic semiconductor within the devices. All these parameters are affected by the choice of the polymerization conditions, with the final purification steps being able to control the polydispersity, molecular weight, and impurities

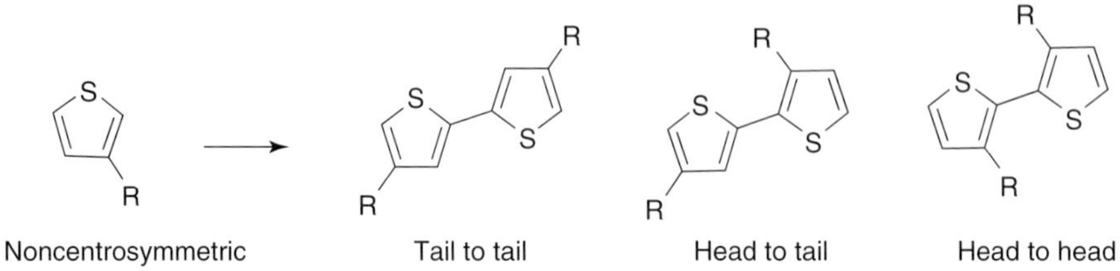

Noncentrosymmetric Tail to tail Head to tail Head to head

Figure 1.5 Possible regiochemistries from the dimerization of a noncentrosymmetric thiophene monomer.

to some extent. However, it is usually not possible to remove chemical defects in the polymer backbone, and so these are best avoided by carefully choosing and optimizing an appropriate synthetic route. As the monomers used for the synthesis of poly(3-alkylthiophenes) are noncentrosymmetric, the regiochemistry of the solubilizing side chains is difficult to control. One of the most studied examples is the polymerization of halogenated 3-hexylthiophene monomer to produce P3HT. Here, there are three possible products that can be formed, tail to tail, head to tail, or head to head, as shown in Figure 1.5. For poly(3-alkylthiophenes), great progress in optimizing the regioregularity and polymer molecular weight has been made over the years by employing a number of different synthetic routes [51–57].

1.4.1.2 **Thienothiophene Copolymers**

The majority of design strategies to reduce the susceptibility of thiophene polymers to degradation or oxidative doping involves decreasing the HOMO energy level below the electrochemical oxidation threshold [58]. As the number of conjugated units along the backbone increases, the HOMO energy level increases up to a critical conjugation length, at which point it remains fairly constant. However, the conjugation length of the backbone can be controlled either by changing the coplanarity (more coplanar π orbitals promote increased delocalization) and therefore the π orbital overlap between neighboring thiophene rings or electronically by introducing a repeat unit into the backbone, which inhibits or prevents π orbital delocalization. The HOMO energy level also increases when the electron density of the conjugated π system is increased.

In the alternating copolymer thieno[2,3-*b*]thiophene and 4,4′-dialkyl-2,2′-bithiophene referred to as poly(bithiophene-cross-conjugated thiophene (pBTCT), the thieno[2,3-*b*]thiophene is a π orbital conjugation blocker, as the central cross-conjugated double bond prevents conjugation between the substituents at the positions 2 and 5. Thus, full conjugation along the backbone is not permitted [59], leading to a lower lying HOMO (larger ionization potential). Additionally, the reduced number of electron-donating chains per aromatic group also contributes to the lower lying HOMO. Ultraviolet photoelectron spectroscopy (UPS) has confirmed that the pBTCT polymer series, composed of pBTCT with alkyl chain lengths from C8 to C12, exhibit a 0.4 eV lowering of the HOMO energy level in comparison to P3HT [60]. The strategy employed for pBTCT synthesis used a regiosymmetrical backbone repeat unit, which was polymerized by Stille coupling. This prevents regioisomerism, which can produce conformational irregularities and in turn reduce crystallinity. The backbone conformation that is most energetically favorable for pBTCT is a "crankshaft" conformation in which the sulfurs in neighboring monomer units arrange in an "anti" configuration because of their large size, which requires them to maximize their spatial separation. The alkyl side chains have two different separation distances on the backbone, which are wide enough to enable side chain interdigitation between the neighboring polymer chains, unlike in P3HT where the side chain packing density is too high to allow interdigitation. As the thieno[2,3-*b*]thiophene monomer unit is planar, pBTCT is able to adopt a coplanar conformation, with the alkyl side chains arranged in a tail-to-tail regiosymmetrical arrangement along the backbone. This regiopositioning of the alkyl groups ensures that there are no steric interactions between adjacent alkyl chains, enabling pBTCT to have a highly planar backbone conformation with optimal π orbital overlap and delocalization within the bithiophene. The backbone planarity is evident from high-resolution grazing X-ray scattering measurements that show interchain $\pi-\pi$ stacking distance of 3.67 Å, which is 0.13 Å less than that of P3HT, as well as charge carrier mobilities of $0.04\,cm^2\,V^{-1}\,s^{-1}$. ON/OFF ratios of around 10^6 have been obtained in air, with devices showing only very minor changes in transfer characteristics when measured over a two-month period [60]. This oxidative stability can be attributed to the lowering of the HOMO energy level as previously mentioned.

The thieno[2,3-*b*]thiophene monomer unit has been synthesized in a variety of ways [61–63]. One example is the elegant synthesis reported by Otsubo *et al.*, in which 1-trimethylsilylpentadiyne is lithiated with *n*-BuLi/BuOK, followed by trapping of the created anion with carbon disulfide [62], producing 2-trimethylsilylthieno[2,3-*b*]thiophene, as a result of the ring closure reaction that occurs during the workup, and is further desilylated upon treatment with tetrabutylammonium fluoride to give thieno[2,3-*b*]thiophene [63]. A more convenient route (Scheme 1.1) has been developed for larger-scale reactions in which commercially available 2-thiophenethiol is used as the starting material (Scheme 1.1) [64]. Alkylation using bromoacetaldehyde dimethyl acetal under Williamson ether conditions yields the protected aldehyde, which on deprotection can be ring closed under

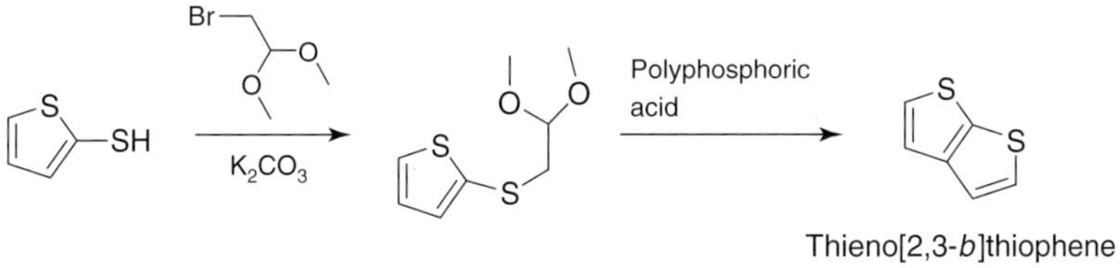

SH Br O O K_2CO_3 O O S S Polyphosphoric acid S S Thieno[2,3-*b*]thiophene

Scheme 1.1 Synthesis of thieno[2,3-*b*]thiophene.

reflux with phosphoric acid in chlorobenzene. This affords thieno[2,3-*b*]thiophene as a colorless oil.

There are generally two approaches to the synthesis of regiosymmetric polymers, the homopolymerization of a suitable centrosymmetric monomer and polymerization of two difunctional symmetric monomers (a so-called AA + BB approach), to produce an alternating copolymer. The benefit of the AA + BB approach is that a variety of copolymers can be synthesized purely by changing one of the comonomers in the polymerization. However, the disadvantage of this approach is that the molecular weight of the polymer is governed by the Carothers equation, so in order to achieve high molecular weights, a strict 1 : 1 stoichiometry needs to be maintained to achieve a high degree of reaction conversion. In practice, this means that only high yielding cross-coupling chemistries are suitable for this approach, and very-high-purity crystalline monomers are necessary to obtain high molecular weights [65].

pBTCT can be synthesized according to the AA + BB methodology using Stille cross-coupling (Scheme 1.2). 2,5-Trimethyl(stannyl)thieno[2,3-*b*]thiophene is readily prepared by lithiation of thieno[2,3-*b*]thiophene with 2 equiv. of *n*-butyllithium, followed by quenching of the resulting dianion with trimethylstannyl chloride. The choice of using trimethyltin as the organometallic group, in spite of its high toxicity, is because it affords a highly crystalline product that can be readily purified. The trimethylstannyl monomer has been polymerized with a range of 5,5-dibromo-4,4-dialkyl-2,2-bithiophenes in the presence of a palladium catalyst to afford polymers with typical weight average molecular weights of around 50–60 000 g mol^{-1} and polydispersities of around two [11].

1.4.1.3 pBTTT

An analogous copolymer to pBTCT is poly(2,5-bis(3-alkylthiophen-2-yl)thieno[3,2-*b*]thiophene (pBTTT) composed of alternating thieno[3,2-*b*]thiophene and 4,4-dialkyl 2,2-bithiophene monomer units [66, 67]. The difference between the two copolymers is that the thieno[3,2-*b*]thiophene monomer has a different arrangement of double bonds in comparison to thieno[2,3-*b*]thiophene, as the sulfur atoms are arranged in an "anti" orientation as opposed to "syn." This allows conjugation between the neighboring thiophenes at the positions 2 and 5, thus enabling extended π

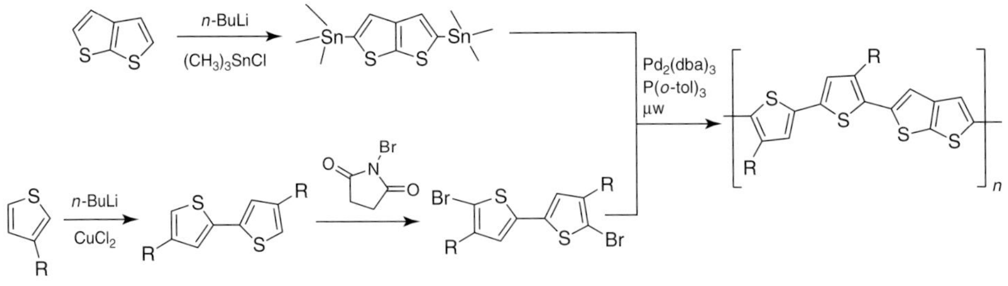

Scheme 1.2 Polymerization of pBTCT.

orbital delocalization along the polymer backbone. In turn, this leads to a lower bandgap and ionization potential than pBTCT. As is the case with pBTCT, pBTTT has alkyl groups solely on the bithiophene units. These alkyl groups inductively donate σ electron density into the π electron system of the polymer backbone, raising the HOMO energy level. However, the fact that there are fewer alkyl groups per unit length in the polymer backbone compared to P3HT, because of the unsubstituted thieno[3,2-*b*]thiophene units, means that the HOMO energy level is lower than P3HT. In addition, the aromatic thieno[3,2-**b**]thiophene ring has a larger resonance stabilization energy than a thiophene ring, as the quinoidal form of the fused ring has a higher energy and is thus less favorable. The consequence of this is to decrease the delocalization of electron density from the aromatic thieno[3,2-**b**]thiophene ring, reducing delocalization along the polymer backbone and thus lowering the HOMO energy level, giving rise to improved ambient operational stability. The backbone of pBTTT adopts a planar conformation as the planar monomers, both of which are centrosymmetric, have an all "anti" sulfur arrangement across the short axis. This enables main chain extension as a "rigid rod" shape, with the monomer units alternately bending to accommodate the nonlinear bond angle between neighboring thiophene units. As the side chain attachment density along the polymer chain is low and the symmetry of the repeat units allows them to rotate around the backbone axis, the alkyl side chains are able to interdigitate with the alkyl groups of neighboring polymer chains. Additionally, the long-range linearity of the polymer backbone facilitates backbone π stacking, enabling neighboring polymer backbones to assemble in a closely packed, tilted face-to-face arrangement [68]. This creates an extended order microstructure, denoted as π-stacked lamella, in which the interdigitated side chains facilitate "registration" between the vertically π-stacked adjacent lamella, thus promoting the formation of large three-dimensional ordered domains.

pBTTT polymers exhibit a thermotropic liquid crystalline phase, which originates on heating when the side chains melt. Transitions in differential scanning calorimetry (DSC) experiments show that the mesophase persists until a further main chain melting thermal transition occurs. As fairly high melting enthalpies are observed, it is evident that the polymers exhibit a high level of crystallinity of both side and main chains, as is consistent with the model of an interdigitated, closely packed polymer conformation. Thermal annealing within the mesophase can further develop the lamella microstructure to create well-connected three-dimensional polycrystalline thin films. Combining thermal annealing with low-energy surface treatments such as with hexamethyldisilazane (HMDS), which forms a self-assembled monolayer (SAM) on the substrate surface, promotes the edge on orientation of the polymer backbone along the surface [69, 70]. This allows the growth of highly ordered crystalline domains, and the coalignment of the domains by the low-energy surface promotes good intergrain connectivity. The effect of annealing on the ordering and orientation of the polymer has been shown by two-dimensional grazing incident X-ray diffraction studies [71] to improve both the crystallinity of the film by forming larger grain sizes, enhance in-plane orientation, and reduce the lamellar spacing from 19.5 to 19.2 Å. This has the effect of increasing the charge carrier mobility and has led to devices with high hole mobilities of up to $1\,cm^2\,V^{-1}\,s^{-1}$ in OFET devices under N_2 [72].

The thieno[3,2-*b*]thiophene monomer can be synthesized from commercially available 3-bromothiophene in four steps. Lithiation of the 3-bromothiophene at the position 2 with diisopropylamine (LDA), a nonnucleophilic base, followed by quenching of the resulting anion with *N*-formylpiperidine or dimethylformamide produces *o*-bromoaldehyde. Reacting this thiophene aldehyde with ethyl 2-sulfanylacetate in conjunction with a base produces a good yield of substituted thieno[3,2-*b*]thiophene. The ester group is converted to a carboxylic acid by hydrolysis, which is subsequently removed by thermal decarboxylation with quinoline in the presence of copper. This produces the unsubstituted thieno[3,2-*b*]thiophene with an overall yield of 60% over the course of the four steps (Scheme 1.3) [73, 74].

Like pBTCT, pBTTT can be synthesized using the AA + BB strategy. Lithiation of thieno[3,2-*b*]thiophene at the positions 2 and 5 with 2 equiv. of *n*-butyllithium, followed by quenching with trimethyltin chloride affords the difunctional tin monomer. Stille coupling of the trimethylstannyl monomer with a range of 5,5-dibromo-4,4-dialkyl-2,2-bithiophenes using a palladium catalyst produced a crude pBTTT polymer, which was purified. Purification involved the precipitation of the polymer from the reaction solvent, followed by solvent extraction to remove low-molecular-weight oligomers and the catalyst. For pBTTT, it was possible to remove traces of metal catalyst via chromatographic purification by filtration over a plug of silica, a process that is unusual for the purification of most polymers. Reprecipitation into acetone, a nonsolvent in this case typically gave the polymer in 90% yields, with molecular weights in the range of 20–30 000 $g\,mol^{-1}$, depending on the length of the alkyl side chains, with polydispersities of around two.

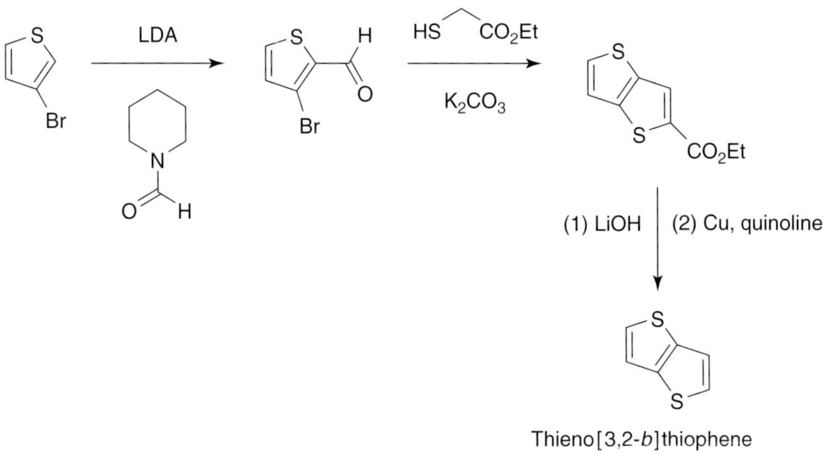

Scheme 1.3 Synthesis of thieno[3,2-*b*]thiophene.

1.5 Semiconductor Blends

Two or more organic semiconductors within a blend film can be used to combine the advantageous properties of each component. As previously mentioned, the processability of small molecular semiconductors is generally lower than that of polymers; however, control of crystallization and increased film uniformity can be achieved using a polymer-small-molecule blend while retaining high charge carrier mobilities. Other benefits to using blends include the ability to combine n- and p-type materials within the same film and the possibility for self-assembly from solution of the device constituents. Although the range of possible film microstructures increases with the addition of more components, this added complexity is often outweighed by the ease with which certain features such as film uniformity, processability, ambipolarity, solubility, and environmental stability can be combined.

There have been several examples of blend OFETs designed for high mobility and ease of processing. First, the use of crystalline-crystalline polymer systems, where RR P3HT and common bulk polymers such as poly(styrene) are blended, allows low concentrations of the semiconducting component and improved mechanical properties [75]. Second, polymer-small-molecule systems, based on both oligothiophenes and acenes, have been solution processed for OFETs. In the case of 2,8-difluoro-5,11-bis(triethylsilylethynyl) anthradithiophene blended with poly(triarylamine), mobilities of well over $2\,cm^2\,V^{-1}\,s^{-1}$ have been demonstrated [76]. The key to understanding these systems and the behavior of many blends is the phase separation of the components. Owing to the low entropy of mixing of polymer–polymer systems and even small-molecule-polymer systems, phase separation in many organic systems is thermodynamically favorable. However, during solution processing, such as spin coating, the system is often far from equilibrium and the solvent itself will strongly affect the final film morphology. Thus, solvent evaporation rates, substrate–solution interactions, solution viscosities, and solute–solvent interactions (or solubility) are all important factors. Phase separation can occur both laterally and/or vertically within a thin film because of the presence of interfaces. Thus, it is possible to control the location of a component within the blend film and, in particular, increase the concentration of high-mobility material at the semiconductor–dielectric interface. This has been achieved by vertical phase separation in, for example, acene–polymer blend OFETs in a top-gate architecture. Similarly, in crystalline polymer blends, the process of crystallization-induced phase separation can lead to semiconductor accretion at the interfaces and therefore allows electrical percolation even at very low semiconductor concentrations. In this case, the order of solidification of the components during processing is critical to obtaining the correct morphology. Controlling crystallization by annealing after film deposition is also possible using a glass-forming additive. This has been demonstrated in a rubrene blend [77] and allows for improved processability in the amorphous state and high mobilities after annealing.

1.6 Device Physics and Architecture

The design of an OFET can be one of the four different architectures depending on whether the gate electrode is deposited prior or after the semiconductor layer and whether the source and drain electrodes are in the same plane as the dielectric (coplanar) or not (staggered) (Figure 1.6). These can affect the ease of device manufacture and the final performance of the device. A bottom-gate, bottom-contact geometry is commonly employed because it allows simple screening of new materials, is well established for display backplane applications, and is easy to fabricate. However, coplanar architectures do not usually give the optimal performance since improved gate-field-enhanced charge injection is present in staggered electrode devices. For all geometries, a simple thin film model that was first developed for inorganic transistors can be applied [78]. Despite the differences between inorganic and organic electronics, the model has been widely employed and allows the estimation of field-effect mobilities. Deviations from this ideal model arise in OFETs because of contact resistances [79] and electric-field-dependent mobilities [80].

If we consider a three terminal device with source, drain, and gate, the gate is separated from the semiconductor by a dielectric layer, and charge carriers are injected into the semiconductor from the source. The source is normally grounded

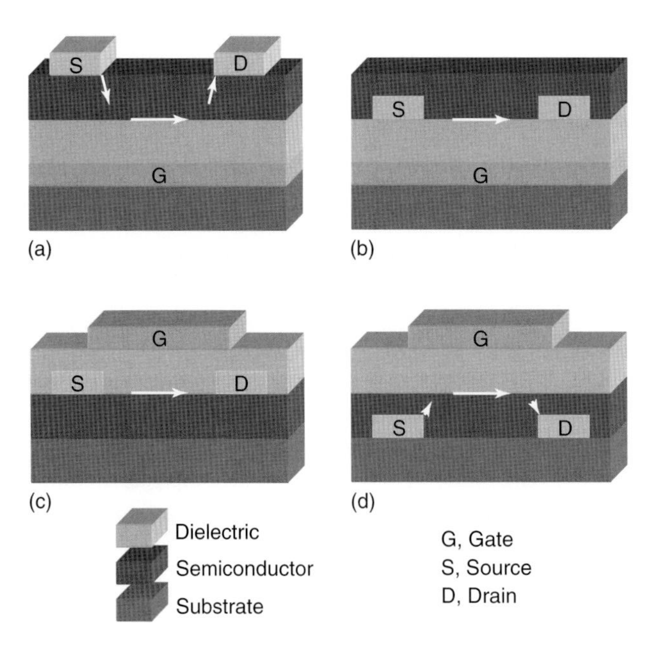

Figure 1.6 OFET device architectures: (a) top contact, bottom gate; (b) bottom contact, bottom gate; (c) top contact, top gate; and (d) bottom contact, top gate. Arrows represent the charge carrier pathways in the device and show the difference between staggered and coplanar arrangements.

and the drain voltage, V_D, is used to induce a source-drain current, I_D, which in turn is modulated by the gate field caused by V_G. It is well established that OFET conduction occurs within a few monolayers of material, essentially in a two-dimensional fashion at the semiconductor–dielectric interface [81]. If we first consider the resistance, dR, of a thin element of the channel, dx, we obtain,

$$dR = \frac{dx}{WQ(x)\mu} \tag{1.1}$$

where W is the channel width, μ is the charge carrier mobility, and $Q(x)$ is the surface charge density at point x (Figure 1.7).

The magnitude of this charge depends mainly on the applied gate voltage and the voltage at x due to V_D; however, in addition, there will be a threshold voltage, V_T, that accounts for the flat-band potential, for the charge donor or acceptor state present, and/or for charge trapping of injected carriers. Hence, we can express $Q(x)$ in terms of these voltages and the geometric capacitance of the dielectric layer, C_i,

$$Q(x) = C_i[V_G - V_T - V(x)] \tag{1.2}$$

Several assumptions are needed in order to calculate the current through the channel. Firstly, we take the mobility to be independent of voltage and thus x, and secondly, we use the gradual channel approximation, that is, the channel length is much greater than the film thickness so that the perpendicular electric field is greater than that in the x-direction. Substituting into $dV = I_D dR$ and integrating over the channel length, L, gives

$$I_D = \frac{W}{L}\mu C_i\left[(V_G - V_T)V_D - \frac{V_D^2}{2}\right] \tag{1.3}$$

This is the general form of the equation for I_D, but we can apply it to the two regimes of the OFET, namely, linear and saturation. In the linear regime $V_G - V_T \gg V_D$, the accumulation region is uniform along the channel, and hence,

$$I_{D\,lin} = \frac{W}{L}\mu C_i (V_G - V_T) V_D \tag{1.4}$$

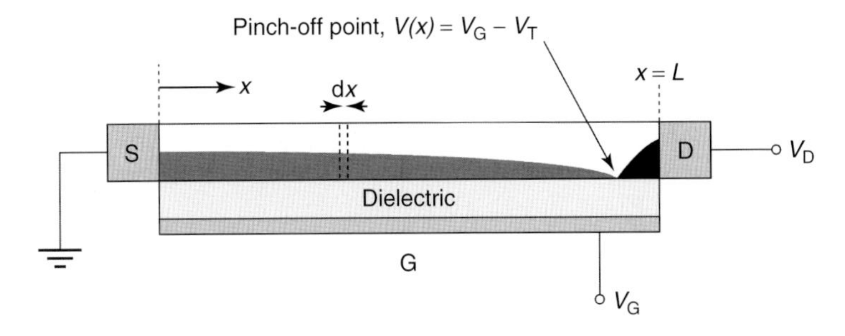

Figure 1.7 Schematic diagram of an OFET channel of length, L, showing source (S), drain (D), and gate (G) electrodes. The accumulation layer is represented in dark grey, and saturation occurs when pinch-off of this region leads to a region next to the drain that is depleted of carriers. At this point, $V_{D\,sat} = V_G - V_T$.

In the saturation regime where pinch-off of the accumulation region occurs and $V_{D\,sat} = V_G - V_T$, we have

$$I_{D\,sat} = \frac{W}{2L}\mu C_i (V_G - V_T)^2 \tag{1.5}$$

From these equations, we can therefore calculate the field-effect mobility within the OFET as follows

$$\mu_{lin} = \frac{L}{WC_i V_D}\left(\frac{\partial I_{D\,lin}}{\partial V_G}\right) \tag{1.6}$$

$$\mu_{sat} = \frac{L}{WC_i}\left(\frac{\partial I_{D\,sat}}{\partial V_G}\right) = \frac{2L}{WC_i}\left(\frac{\partial \sqrt{I_{D\,sat}}}{\partial V_G}\right)^2 \tag{1.7}$$

Although these expressions are widely used, they contain an error if the intrinsic mobility is a function of the gate voltage [82] (or equivalently the charge density), which is common for organic semiconductors. Therefore, several other estimates for mobility can be used that, for example, do not incorporate the fitting parameter V_T but use the more physically meaningful onset voltage, V_{on}. One example would be the effective or average mobility, μ_{EFF}, calculated from Eq. (1.4) but considering all charges rather than just those when $V_G > V_T$.

$$\mu_{EFF}(V_G) = \frac{L}{WC_i V_D}\frac{I_{D\,lin}}{(V_G - V_{on})} \tag{1.8}$$

When optimizing the electrical performance of OFETs, there are several factors to be considered. High charge carrier mobility is often important since higher currents are possible for fixed transistor dimensions. Also, the switching speed of integrated circuits increases with increased mobility [83]. These are key features for the driving circuitry in organic active matrix displays where high currents are needed to operate the OLEDs and small transistors are preferable compared to the pixel size in the case of bottom-emission devices. An often-quoted aim for OFETs is to achieve mobilities equal to or greater than that of a-Si, which is widely used for display backplanes. Owing to the weaker interactions between organic molecules than the covalent inorganic semiconductors, charge transport is rarely bandlike in nature. Instead, charges, which may well be polaronic [84], that is, distort their own molecular environment, are limited by hopping from one spatial region to another or by escaping from trap states. This leads, in most cases, to a thermally activated μ with only a few exceptions [85, 86]. Combined with the problems of charge injection from metallic electrodes and the nature of the semiconductor–dielectric interface, this makes the design of high-mobility OFETs challenging. The other main parameters that have an influence on the use of OFETs in circuits are the threshold voltage, V_T, and the *ON/OFF ratio* (defined as the ratio of currents between the on and off states of the device). Threshold voltages close to zero indicate negligible charge trapping or doping and are generally advantageous since large gate voltages are not required to switch the device on or off. A high ON/OFF ratio is also useful, as the leakage through the transistor, when it is in the off state, is minimized. If being used to charge a capacitor in, for example, a display application, this is particularly important.

The ON/OFF ratio can be approximated when there is negligible charge depletion by

$$\frac{I_{\mathrm{ON}}}{I_{\mathrm{OFF}}} = 1 + \frac{\mu C_{\mathrm{i}} V_{\mathrm{D}}}{2\sigma_{\mathrm{bulk}} t} \tag{1.9}$$

where σ_{bulk} is the conductivity of the bulk film and t is the film thickness [87]. Thus, by reducing conduction through the bulk of the device, the off current can be lowered. In this respect, high mobilities, thin films, and low defect/impurity concentrations are essential for high ON/OFF ratios. Typical device characteristics for a high-mobility OFET are shown in Figure 1.8, demonstrating both linear and saturation regimes of the device as well as highlighting ON and OFF currents, threshold voltage, and onset voltage.

The performance of an OFET depends not only on the semiconducting material employed but also critically on the interfaces that this material makes with both the source and drain electrodes and the dielectric material. First, we must be able to inject and extract electrons or holes to and from the metallic contacts. When a metal and an organic semiconductor are placed in contact, there is an equalization of fermi levels in a similar fashion to inorganic materials. Electrons (holes) must then overcome any energetic barrier created for injection from the metal into the LUMO (HOMO) of the organic material. However, unlike inorganics, the magnitude of this barrier depends not only on the work function of the metal, ϕ_{m}, but also on the strength of the interaction between organic and metal and the possible creation of interface dipoles [88], leading to a shift in vacuum levels, Δ. Thus, instead of the

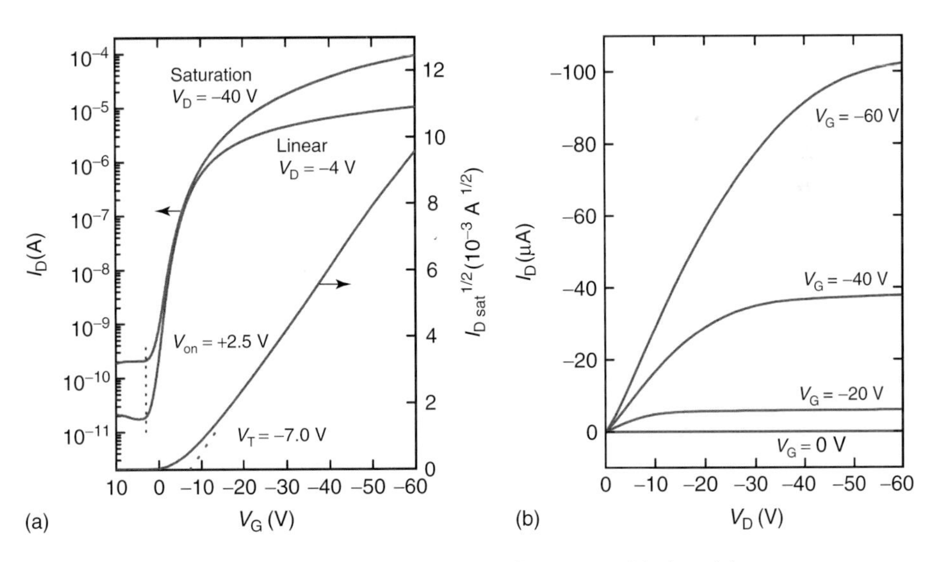

Figure 1.8 (a) Transfer and (b) output characteristics from a typical high-mobility p-type blend OFET. The channel length and width are 70 and 1000 μm, respectively, and mobility in the saturation regime can be calculated to be 2.5 cm^2 V^{-1} s^{-1}.

Mott–Schottky limit for electron injection, which gives a barrier height, ϕ_B, of

$$\varphi_B = \varphi_m - \chi \tag{1.10}$$

with χ being the electron affinity of the semiconductor, that is, the position of the LUMO, we obtain

$$\varphi_B = \varphi_m + \Delta - \chi \tag{1.11}$$

By selecting a suitable metal with a work function close to either the HOMO or LUMO energies of the organic material and/or by controlling interface dipole formation, the device performance can be made nonlimited by charge injection rates. An often-used technique to modify this interface is the application of SAMs to the metal. This involves the use of small organic molecules, such as thiols for metals [89], that can chemically bind to surfaces in such a way that a single, uniform layer forms. If the molecule has a dipole moment, it can alter Δ depending on its orientation to the surface or it can modify the surface energy and thus the interaction of the subsequent organic layer with the substrate.

The second important interface in an OFET is between the dielectric and semiconductor, where the accumulation layer forms and the majority of charge conduction occurs. Again SAMs are often employed to optimize this interface, especially with regard to minimizing charge trapping and modifying film morphology. The choice of dielectric itself, its surface properties, and its dielectric constant will play a key role in device operation. Some examples include the use of low permittivity polymeric materials to reduce energetic disorder at the interface [90] and the use of silane SAMs to passivate oxide dielectrics [91]. Layers with a large value of C_i are also potentially important for reducing the operating voltage of the OFET, key to realizing portable, low-power organic electronics. In this case, the SAM itself can be used as the dielectric such as long chain alkyl phosphonic acids on thin aluminum oxide [92].

1.7 Summary

One of the main advantages of organic electronics is the ability to produce large-scale circuits relatively cheaply and on a wide variety of surfaces such as flexible substrates. Fabrication using solution-based deposition including printing is one way to achieve this. However, there are several constraints when using organic materials in this way. Firstly, the majority of the processing must occur at low temperatures (typically $\lesssim 200\,^{\circ}\mathrm{C}$) to prevent chemical decomposition. Secondly, designing materials that are easily processable can compromise their electrical performance. Thirdly, patterning of organics requires new techniques such as nanoimprinting, self-aligned printing, self-assembly, or soft lithography to make suitable device structures while keeping fabrication costs low. Finally, deposition of several materials is needed for even the most simple of devices, and thus one layer must not, for example, dissolve previous layers. The process of material deposition will often strongly affect the microstructure of the solid film produced and therefore also the electrical properties of the OFET. Most solution deposition methods will be far from equilibrium, and the rates of solvent evaporation, material surface energies, and solution viscosities will influence film morphology. Over the past several years, many materials systems have been developed that aim to improve ease of processing while forming highly controllable or ordered thin films on a microstructural and/or molecular level. When employing these materials in organic devices, other factors such as interface effects, environmental stability, and device uniformity must then be considered and incorporated. Therefore, a combination of good electrical performance, device architecture, and material properties is needed in order to fabricate high-performance organic electronic devices suitable for possible commercial applications.

References

1. Horowitz, P. and Hill, W. (1990) *The Art of Electronics*, 2nd edn, Cambridge University Press.
2. Gelinck, G.H., Huitema, H.E.A., Van Veenendaal, E., Cantatore, E., Schrijnemakers, L., Van der Putten, J., Geuns, T.C.T., Beenhakkers, M., Giesbers, J.B., Huisman, B.H., Meijer, E.J., Benito, E.M., Touwslager, F.J., Marsman, A.W., Van Rens, B.J.E., and De Leeuw, D.M. (2004) *Nat. Mater.*, **3**, 106–110.
3. Loo, Y.-L. (2007) *AIChE J.*, **53**, 1066–1074.
4. Sirringhaus, H. (2005) *Adv. Mater.*, **17**, 2411–2425.
5. Gundlach, D.J., Royer, J.E., Park, S.K., Subramanian, S., Jurchescu, O.D., Hamadani, B.H., Moad, A.J., Kline, R.J., Teague, L.C., Kirillov, O., Richter, C.A., Kushmerick, J.G., Richter, L.J., Parkin, S.R., Jackson, T.N., and Anthony, J.E. (2008) *Nat. Mater.*, **7**, 216–221.
6. Gelinck, G.H., Huitema, H.E.A., van Veenendaal, E., Cantatore, E., Schrijnemakers, L., van der Putten, J.B.P.H., Geuns, T.C.T., Beenhakkers, M., Giesbers, J.B., Huisman, B.-H., Meijer, E.J., Benito, E.M., Touwslager, F.J., Marsman, A.W., van Rens, B.J.E., and de Leeuw, D.M. (2004) *Nat. Mater.*, **3**, 106–110.
7. Fumagalli, L., Binda, M., Natali, D., Sampietro, M., Salmoiraghi, E., and Di Gianvincenzo, P. (2008) *J. Appl. Phys.*, **104**, 084513/1–084513/8.
8. Chang, J.F., Sirringhaus, H., Giles, M., Heeney, M., and McCulloch, I. (2007) *Phys. Rev. B*, **76**, 205204–205215.
9. Chang, J.-F., Clark, J., Zhao, N., Sirringhaus, H., Breiby, D.W., Andreasen, J.W., Nielsen, M.M., Giles, M., Heeney, M., and McCulloch, I. (2006) *Phys. Rev. B: Condens. Matter Mater. Phys.*, **74**, 115318/1–115318/12.
10. Kline, R.J., McGehee, M.D., Kadnikova, E.N., Liu, J., Frechet, J.M.J., and Toney, M.F. (2005) *Macromolecules*, **38**, 3312–3319.
11. Iain, M., Martin, H., Michael, L.C., Dean, D., Kline, R.J., Michael, C., Warren, D., Daniel, F., David, G., Behrang, H., Rick, H., Lee, R., Alberto, S., Maxim, S., David, S., Steven, T., and Weimin, Z. (2009) *Adv. Mater.*, **21**, 1091–1109.
12. de Leeuw, D.M., Simenon, M.M.J., Brown, A.R., and Einerhand, R.E.F. (1997) *Synth. Met.*, **87**, 53–59.
13. Horowitz, G., Fichou, D., Peng, X., Xu, Z., and Garnier, F. (1989) *Solid State Commun.*, **72**, 381–384.
14. Horowitz, G., Fichou, D., Peng, X.Z., Xu, Z.G., and Garnier, F. (1989) *Solid State Commun.*, **72**, 381.
15. Wittmann, J.C., Straupé, C., Meyer, S., Lotz, B., Lang, P., Horowitz, G., and Garnier, F. (1997) *Thin Solid Films*, **311**, 317–322.

16. Sotgiu, G., Zambianchi, M., Barbarella, G., and Botta, C. (2002) *Tetrahedron*, **58**, 2245–2251.
17. Horowitz, G., Peng, X., Fichou, D., and Garnier, E. (1991) *J. Mol. Electron.*, **7**, 85.
18. Kelley, T.W., Muyres, D.V., Paul, F.B., Smith, T.P., and Jones, T.D. (2003) *MRS Symp. Proc.*, **771**, 169–179.
19. Sheraw, C.D., Jackson, T.N., Eaton, D.L., and Anthony, J.E. (2003) *Adv. Mater. (Weinheim, Ger.)*, **15**, 2009–2011.
20. Anthony, J.E., Eaton, D.L., and Parkin, S.R. (2002) *Org. Lett.*, **4**, 15–18.
21. Payne, M.M., Parkin, S.R., Anthony, J.E., Kuo, C.C., and Jackson, T.N. (2005) *J. Am. Chem. Soc.*, **127**, 4986–4987.
22. Anthony, J.E., Brooks, J.S., Eaton, D.L., and Parkin, S.R. (2001) *J. Am. Chem. Soc.*, **123**, 9482–9483.
23. Rivnay, J., Jimison, L.H., Northrup, J.E., Toney, M.F., Noriega, R., Lu, S., Marks, T.J., Facchetti, A., and Salleo, A. (2009) *Nat. Mater.*, **8**, 952–958.
24. Subramanian, S., Park, S.K., Parkin, S.R., Podzorov, V., Jackson, T.N., and Anthony, J.E. (2008) *J. Am. Chem. Soc.*, **130**, 2706–2707.
25. Kim, Y., Cook, S., Tuladhar, S.M., Choulis, S.A., Nelson, J., Durrant, J.R., Bradley, D.D.C., Giles, M., McCulloch, I., Ha, C.-S., and Ree, M. (2006) *Nat. Mater.*, **5**, 197–203.
26. Sirringhaus, H., Brown, P.J., Friend, R.H., Nielsen, M.M., Bechgaard, K., Langeveld-Voss, B.M.W., Spiering, A.J.H., Janssen, R.A.J., Meijer, E.W., Herwig, P., and de Leeuw, D.M. (1999) *Nature*, **401**, 685–688.
27. Singh, K.A., Sauve, G., Zhang, R., Kowalewski, T., McCullough, R.D., and Porter, L.M. (2008) *Appl. Phys. Lett.*, **92**, 263303/1–263303/3.
28. Wang, C., Jimison, L.H., Goris, L., McCulloch, I., Heeney, M., Ziegler, A., and Salleo, A. (2010) *Adv. Mater.*, **22**, 697–701.
29. Chan, C.K., Richter, L.J., Dinardo, B., Jaye, C., Conrad, B.R., Ro, H.W., Germack, D.S., Fischer, D.A., De Longchamp, D.M., and Gundlach, D.J. (2010) *Appl. Phys. Lett.*, **96**, 133304/1–133304/3.
30. Baeg, K.-J., Khim, D., Kim, D.-Y., Koo, J.B., You, I.-K., Choi, W.S., and Noh, Y.-Y. (2010) *Thin Solid Films*, **518**, 4024–4029.
31. Zen, A., Pflaum, J., Hirschmann, S., Zhuang, W., Jaiser, F., Asawapirom, U., Rabe, J.P., Scherf, U., and Neher, D. (2004) *Adv. Funct. Mater.*, **14**, 757–764.
32. Babel, A. and Jenekhe, S.A. (2005) *Synth. Met.*, **148**, 169–173.
33. Zen, A., Saphiannikova, M., Neher, D., Asawapirom, U., and Scherf, U. (2005) *Chem. Mater.*, **17**, 781–786.
34. Sauve, G., Javier, A.E., Zhang, R., Liu, J., Sydlik, S.A., Kowalewski, T., and McCullough, R.D. (2010) *J. Mater. Chem.*, **20**, 3195–3201.
35. Bao, Z. and Lovinger, A.J. (1999) *Chem. Mater.*, **11**, 2607–2612.
36. Meijer, E.J., Detcheverry, C., Baesjou, P.J., van Veenendaal, E., de Leeuw, D.M., and Klapwijk, T.M. (2003) *J. Appl. Phys.*, **93**, 4831–4835.
37. McCulloch, I., Bailey, C., Giles, M., Heeney, M., Love, I., Shkunov, M., Sparrowe, D., and Tierney, S. (2005) *Chem. Mater*, **17**, 1381–1385.
38. Abdou, M.S.A., Orfino, F.P., Son, Y., and Holdcroft, S. (1997) *J. Am. Chem. Soc.*, **119**, 4518–4524.
39. Michael, L.C., Robert, A.S., and John, E.N. (2007) *Appl. Phys. Lett.*, **90**, 123508.
40. Rost, H., Ficker, J., Alonso, J.S., Leenders, L., and McCulloch, I. (2004) *Synth. Met.*, **145**, 83–85.
41. Olsen, B.D., Jang, S.-Y., Luning, J.M., and Segalman, R.A. (2006) *Macromolecules*, **39**, 4469–4479.
42. Kline, R.J., McGehee, M.D., Kadnikova, E.N., Liu, J., and Fréchet, J.M.J. (2003) *Adv. Mater.*, **15**, 1519–1522.
43. Pokrop, R., Verilhac, J.-M., Gasior, A., Wielgus, I., Zagorska, M., Travers, J.-P., and Pron, A. (2006) *J. Mater. Chem.*, **16**, 3099–3106.
44. Verilhac, J.-M., Pokrop, R., LeBlevennec, G., Kulszewicz-Bajer, I., Buga, K., Zagorska, M., Sadki, S., and Pron, A. (2006) *J. Phys. Chem. B*, **110**, 13305–13309.
45. Jui-Fen, C., Jenny, C., Ni, Z., Henning, S., Dag, W.B., Jens, W.A., Martin, M.N., Mark, G., Martin, H., and Iain, M.

(2006) *Phys. Rev. B (Condens. Matter Mater. Phys.)*, **74**, 115318.
46. Brinkmann, M. and Rannou, P. (2007) *Adv. Funct. Mater.*, **17**, 101–108.
47. Urien, M., Wantz, G., Cloutet, E., Hirsch, L., Tardy, P., Vignau, L., Cramail, H., and Parneix, J.-P. (2007) *Org. Electron.*, **8**, 727–734.
48. Kulkarni, A.P., Kong, X., and Jenekhe, S.A. (2004) *J. Phys. Chem. B*, **108**, 8689–8701.
49. Chunyan, C., Chan, I., Volker, E., Andreas, Z., Günter, L., and Gerhard, W. (2005) *Chem. A Eur. J.*, **11**, 6833–6845.
50. List, E.J.W., Guentner, R., de Freitas, P.S., and Scherf, U. (2002) *Adv. Mater.*, **14**, 374–378.
51. Waltman, R.J., Bargon, J., and Diaz, A.F. (2002) *J. Phys. Chem.*, **87**, 1459–1463.
52. Sugimoto, R.I., Takeda, S., Gu, H.B., and Yoshino, K. (1986) *Chem. Express*, **1**, 635–638.
53. McCullough, R.D. and Lowe, R.D. (1992) *J. Chem. Soc., Chem. Commun.*, **1**, 70–72.
54. Iraqi, A. and Barker, G.W. (1998) *J. Mater. Chem.*, **8**, 25–29.
55. Guillerez, S. and Bidan, G. (1998) *Synth. Met.*, **93**, 123–126.
56. Wu, X., Chen, T.-A., and Rieke, R.D. (2002) *Macromolecules*, **28**, 2101–2102.
57. Heuze, K. and McCullough, R.D. (1999) *Polym. Prepr.*, **40**, 854.
58. Yoon, M.H., DiBenedetto, S.A., Facchetti, A., and Marks, T.J. (2005) *J. Am. Chem. Soc.*, **127**, 1348–1349.
59. Milian Medina, B., Van Vooren, A., Brocorens, P., Gierschner, J., Shkunov, M., Heeney, M., McCulloch, I., Lazzaroni, R., and Cornil, J. (2007) *Chem. Mater.*, **19**, 4949–4956.
60. Heeney, M., Bailey, C., Genevicius, K., Shkunov, M., Sparrowe, D., Tierney, S., and McCulloch, I. (2005) *J. Am. Chem. Soc.*, **127**, 1078–1079.
61. Archer, W.J. and Taylor, R. (1982) *J. Chem. Soc. Perkin Trans. 2*, 295–299.
62. Otsubo, T., Kono, Y., Hozo, N., Miyamoto, H., Aso, Y., Ogura, F., Tanaka, T., and Sawada, M. (1993) *Bull. Chem. Soc. Jpn.*, **66**, 2033–2041.
63. Yasuike, S., Kurita, J., and Tsuchiya, T. (1997) *Heterocycles*, **45**, 1891–1894.
64. Heeney, M., Bailey, C., Genevicius, K., Giles, M., Shkunov, M., Sparrowe, D., Tierney, S., Zhang, W., and McCulloch, I. (2005) *Proc. SPIE-Int. Soc. Opt. Eng.*, **5940**, 594007/1.
65. Tsao, H.N., Cho, D., Andreasen, J.W., Rouhanipour, A., Breiby, D.W., Pisula, W., and Müllen, K. (2009) *Adv. Mater.*, **21**, 209–212.
66. McCulloch, I., Heeney, M., Chabinyc, M.L., DeLongchamp, D., Kline, R.J., Cölle, M., Duffy, W., Fischer, D., Gundlach, D., Hamadani, B., Hamilton, R., Richter, L., Salleo, A., Shkunov, M., Sparrowe, D., Tierney, S., and Zhang, W. (2009) *Adv. Mater.*, **21**, 1091–1109.
67. McCulloch, I., Heeney, M., Bailey, C., Genevicius, K., MacDonald, I., Shkunov, M., Sparrowe, D., Tierney, S., Wagner, R., Zhang, W., Chabinyc, M.L., Kline, R.J., McGehee, M.D., and Toney, M.F. (2006) *Nat. Mater.*, **5**, 328–333.
68. Anthony, J.E. (2006) *Chem. Rev.*, **106**, 5028–5048.
69. Salleo, A., Chabinyc, M.L., Yang, M.S., and Street, R.A. (2002) *Appl. Phys. Lett.*, **81**, 4383–4385.
70. Kline, R.J., Dean, M.D., Daniel, A.F., Eric, K.L., Martin, H., Iain, M., and Michael, F.T. (2007) *Appl. Phys. Lett.*, **90**, 062117.
71. Chabinyc, M.L., Toney, M.F., Kline, R.J., McCulloch, I., and Heeney, M. (2007) *J. Am. Chem. Soc.*, **129**, 3226–3237.
72. Hamadani, B.H., Gundlach, D.J., McCulloch, I., and Heeney, M. (2007) *Appl. Phys. Lett.*, **91**, 243512.
73. Hawkins, D.W., Iddon, B., Longthorne, D.S., and Rosyk, P.J. (1994) *J. Chem. Soc., Perkin Trans. 1*, **19**, 2735–2743.
74. Fuller, L.S., Iddon, B., and Smith, K.A. (1997) *J. Chem. Soc. Perkin Trans. 1*, **22**, 3465–3470.
75. Goffri, S., Muller, C., Stingelin-Stutzmann, N., Breiby, D.W., Radano, C.P., Andreasen, J.W., Thompson, R., Janssen, R.A.J., Nielsen, M.M., Smith, P., and Sirringhaus, H. (2006) *Nat. Mater.*, **5**, 950–956.
76. Hamilton, R., Smith, J., Ogier, S., Heeney, M., Anthony, J.E., McCulloch, I., Bradley, D.D.C., Veres, J., and Anthopoulos, T.D. (2009) *Adv. Mater.*, **21**, 1166–1171.

77. Stingelin-Stutzmann, N., Smits, E., Wondergem, H., Tanase, C., Blom, P., Smith, P., and de Leeuw, D. (2005) *Nat. Mater.*, **4**, 601–606.
78. Horowitz, G. and Delannoy, P. (1991) *J. Appl. Phys.*, **70**, 469–475.
79. Meijer, E.J., Gelinck, G.H., van Veenendaal, E., Huisman, B.-H., de Leeuw, D.M., and Klapwijk, T.M. (2003) *Appl. Phys. Lett.*, **82**, 4576–4578.
80. Horowitz, G., Hajlaoui, R., Fichou, D., and Kassmi, A.E. (1999) *J. Appl. Phys.*, **85**, 3202–3206.
81. Dodabalapur, A., Torsi, L., and Katz, H.E. (1995) *Science*, **268**, 270–271.
82. Hoffman, R.L. (2004) *J. Appl. Phys.*, **95**, 5813–5819.
83. Fix, W., Ullmann, A., Ficker, J., and Clemens, W. (2002) *Appl. Phys. Lett.*, **81**, 1735–1737.
84. Fishchuk, I.I., Kadashchuk, A., Bässler, H., and Nešpurek, S. (2003) *Phys. Rev. B*, **67**, 224303.
85. Horowitz, G., Hajlaoui, M.E., and Hajlaoui, R. (2000) *J. Appl. Phys.*, **87**, 4456–4463.
86. Podzorov, V., Menard, E., Borissov, A., Kiryukhin, V., Rogers, J.A., and Gershenson, M.E. (2004) *Phys. Rev. Lett.*, **93**, 086602.
87. Brown, A.R., Jarrett, C.P., de Leeuw, D.M., and Matters, M. (1997) *Synth. Met.*, **88**, 37–55.
88. Peisert, H., Knupfer, M., and Fink, J. (2002) *Appl. Phys. Lett.*, **81**, 2400–2402.
89. Hong, J.-P., Park, A.-Y., Lee, S., Kang, J., Shin, N., and Yoon, D.Y. (2008) *Appl. Phys. Lett.*, **92**, 143311.
90. Veres, J., Ogier, S., Leeming, S.W., Cupertino, D.C., and Khaffaf, S.M. (2003) *Adv. Funct. Mater.*, **13**, 199–204.
91. Lim, S.C., Kim, S.H., Lee, J.H., Kim, M.K., Kim, D.J., and Zyung, T. (2005) *Synth. Met.*, **148**, 75–79.
92. Klauk, H., Zschieschang, U., Pflaum, J., and Halik, M. (2007) *Nature*, **445**, 745–748.

2
Characterization of Order and Orientation in Semiconducting Polymers

Dean M. DeLongchamp and R. Joseph Kline

2.1
Introduction

The electrical properties of an organic electronics-based device typically depend greatly on the extent of structural order in its semiconducting layer. Regular structural order (e.g., crystallinity) creates regular overlap of π orbitals involved in the transport of holes and electrons, ensuring that charges can move throughout the film. Furthermore, the conformational regularity afforded by high levels of long-range order often permits the greater delocalization of π orbitals, facilitating intramolecular charge transport. The precise measurement of structural ordering, to determine the type and quality of molecular packing, is therefore critical both to the development of new organic semiconductors and to the development of better processing routes for those known today.

Organic semiconductors typically feature a common conjugated plane composed of either a single fused-ring system or many conformationally independent rings that become coplanar by adoption of a low-free-energy trans configuration. Thus, the molecules are typically regarded as two-dimensional sheets, and facile π overlap is typically only observed in the sheet-stacking direction, nominally perpendicular to the molecular conjugated plane. These features of organic semiconductors lead naturally to an inherent anisotropy in π overlap in three-dimensional crystals of organic semiconductors, and hence, the expectation of orientation-dependent transport. The precise measurement of orientation is then critical to understand how the intermolecular interactions are arranged with respect to a given device geometry to ensure that transport is facilitated in the directions in which it is desired for a given function.

It is important to emphasize that order and orientation in organic semiconductor films develop dynamically during a drying process, and thus, they can depend greatly on the processing conditions employed. Although one of the greatest merits of organic semiconductors is their ability to be processed from solution using simple graphics art techniques, variations in drying are responsible for much of the variability that is observed in electrical performance. These variations are often manifested in order and orientation differences in the semiconductor device layer.

Organic Electronics II: More Materials and Applications, First Edition. Edited by Hagen Klauk.

2.2
X-Ray Diffraction

X-ray diffraction (XRD) is a technique commonly used in materials science to measure the atomic and molecular structures of solids and liquids. The most common implementations of XRD in conventional materials science are used as a single technique to determine the crystal structure [1]. These methods are typically practiced on either a powder, as is often done in laboratory-scale XRD instruments for inorganic crystals such as ceramics or metals, or a single crystal, as is often done on synchrotron XRD beamlines for proteins [2–4]. In both cases, a large ensemble of diffraction peaks are collected over a large range of crystal orientations, sufficient to identify the unit cell from a large number of peaks with mixed index. The variations in the peak intensities are used to determine the structure factor and the positions of all atoms in the unit cell. For protein crystallography involving complex molecules with a large number of atoms, the detailed arrangement of the molecule within the unit cell is typically determined by molecular simulation, comparing proposed molecular packing structures with the locations and intensities of the peaks that they would produce, and iterating until a match is found [2]. Often times the diffraction is repeated near atomic absorption edges to selectively enhance the scattering from specific parts of the molecule to aid the determination of a unique solution [5].

These conventional implementations of XRD are often of limited practical use for organic semiconductors. Powder diffraction of organic semiconductor crystals often fails to yield sufficient peaks of mixed index to identify a unique unit cell, and the complex shape of semiconducting molecules with fused cores and multiple flexible side chains makes refinement of all the atomic positions impractical because of the large parameter space. Furthermore, it is often difficult to grow single crystals of sufficient size to practice single-crystal diffraction techniques, again especially for polymer semiconductors. However, it is important to mention that some approaches like these have been successful. Most notably, single-crystal diffraction is often practiced on small molecule organic semiconductors, such as the soluble oligoacenes (Figure 2.1) [6]. Some researchers have used molecular simulations to reduce the parameter space and allow fitting of diffraction patterns of organic semiconductors, even in the absence of a rich collection of peaks, with results that agree well with detailed structure measurements by other methods [7, 8]. Figure 2.2 shows an example by Brocorens *et al.*, where molecular mechanics modeling of the molecular packing was used to determine the likely crystal structure of the common semiconducting polymer pBTTT.

Organic semiconductor devices normally consist of a thin film of the organic semiconductor as the active layer of the device (a few devices use single crystals). Since many organic semiconductors have been shown to have thin film phases that are different from the bulk phase, a thin film structure measurement is required to characterize the device-relevant structure [9, 10]. The thin film requirement limits the utility of single-crystal and powder diffraction methods for characterizing organic semiconductors, and most publications primarily use thin-film

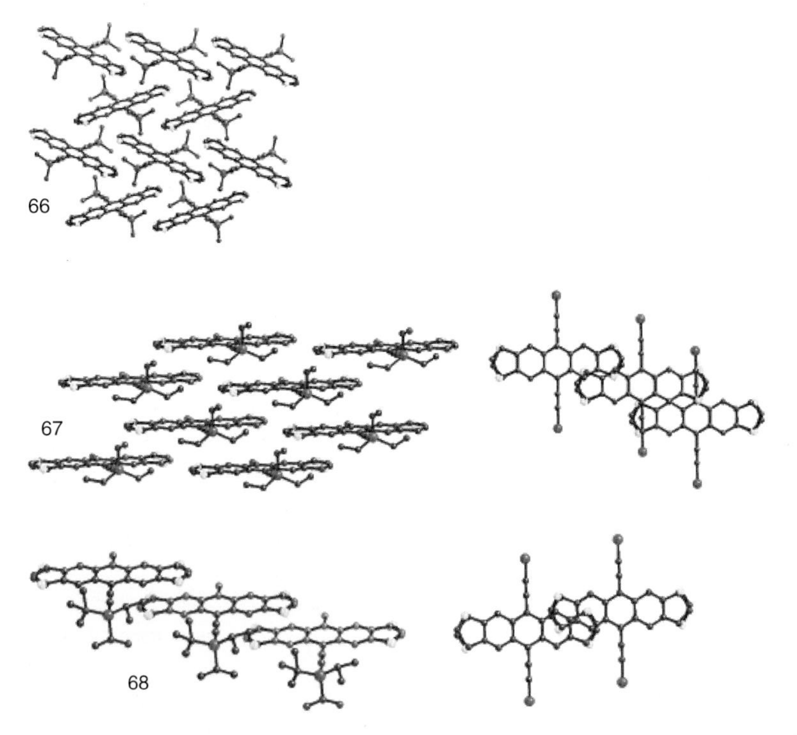

Figure 2.1 Crystal structures of various anthradiothiophenes solved from single-crystal diffraction (some trialkylsilyl groups omitted for clarity). Views normal to the aromatic plane for 67 and 68 show details of π-overlap with nearest neighboring molecule(s) (Cambridge structural database (CSD) RefCodes MAMNUK, MAMPAS, and FANGUX) [6].

diffraction methods. These measurements are particularly important because of the anisotropic properties of most organic semiconductors and the importance of packing defects on the electronic properties. The anisotropic optical and electronic properties make the measurement of the thin film crystal orientation distribution among the most important structural measurements [11]. Additionally, most semiconducting polymers and some small molecule organic semiconductors are semicrystalline or amorphous, making measurements of the crystallinity key to the determination of the electronic properties [12–16]. The most common thin-film diffraction methods are thin-film XRD and grazing-incidence X-ray diffraction (GIXD). It should be noted that electron diffraction has also been used to elucidate some of the microstructural features discussed in this section, but is not be covered in this chapter [17, 18].

2.2.1 Thin-Film XRD

Thin-film XRD is the most common method applied to organic semiconductors in the literature and also usually the most misinterpreted. The measurements can

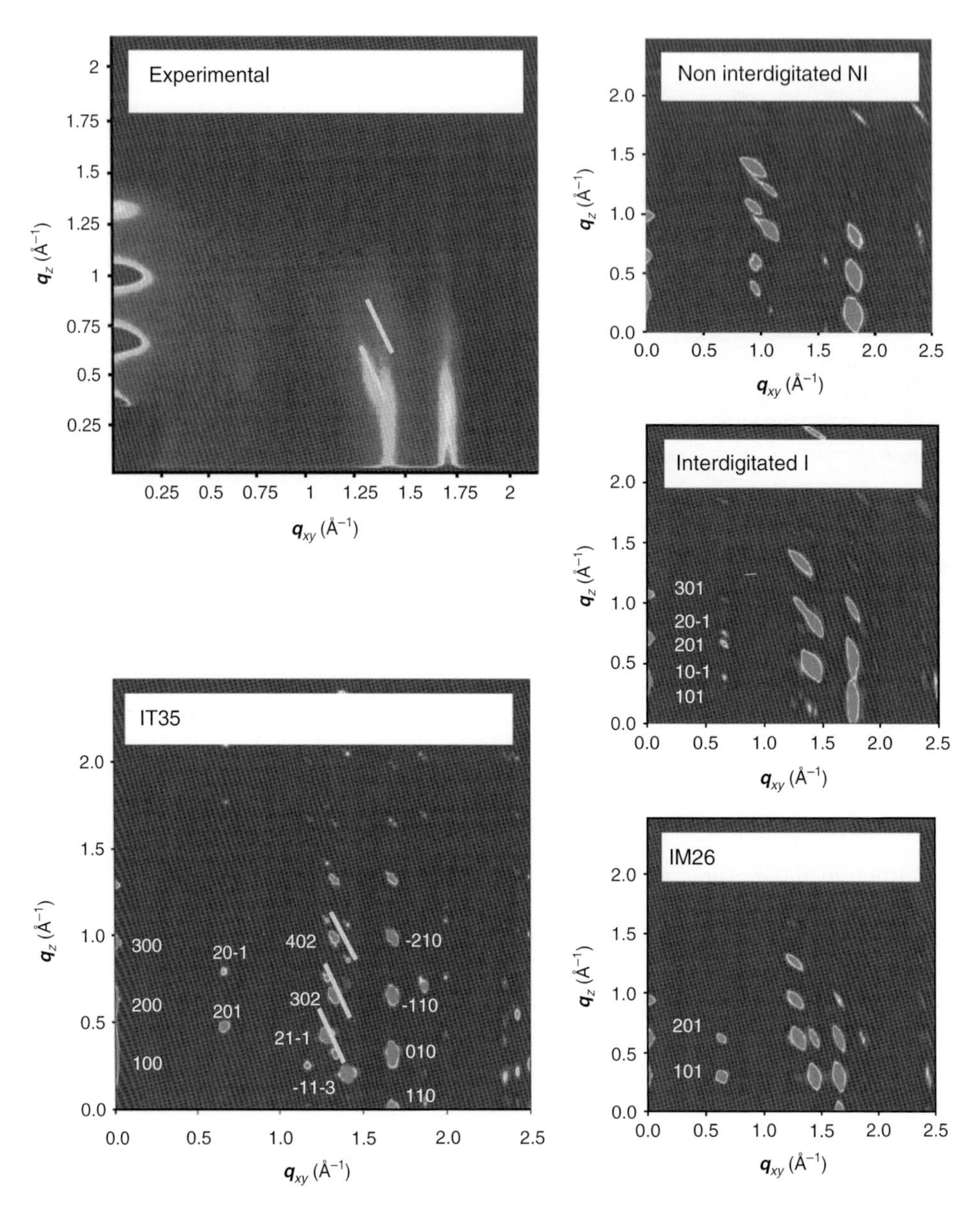

Figure 2.2 Experimental and simulated 2D XRD patterns of pBTTT for the noninterdigitated configuration NI and interdigitated configurations I, IM26, and IT35. For I, NI, and IM26, a low level of disorder of the crystallites in the film (with a standard deviation s = 2°) and a small peak broadening (0.2°) are considered. For IT35, smaller standard deviation (s = 0.75°) and peak broadening (0.1°) are used to better localize the individual diffraction peaks [7].

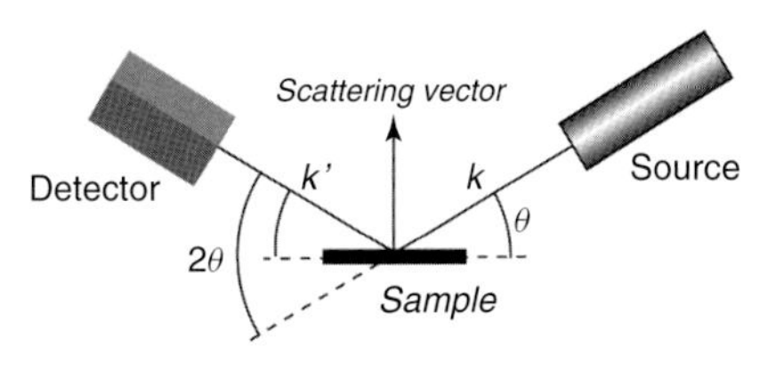

Figure 2.3 Specular thin-film XRD geometry.

be made either on a laboratory source or at a synchrotron. The accessibility of the measurement allows researchers not familiar with XRD to easily make the measurement. The geometry of thin-film XRD is nominally identical to that used for powder diffraction, as is shown in Figure 2.3. An X-ray beam from the source is directed at the sample plane with an incidence angle θ_i and diffracting angle θ_d. For the specular condition, θ_i and θ_d are equal. As for all elastic scattering, the magnitude and orientation of the scattering vector $\boldsymbol{q}$ can defined by the directional vectors formed between source and sample ($\boldsymbol{k}$) and the vector formed between sample and detector ($\boldsymbol{k}'$). These vectors have magnitude

$$|k| = |k'| = \frac{2\pi}{\lambda} \tag{2.1}$$

where λ is the wavelength of incident X-ray radiation. The scattering vector $\boldsymbol{q}$ is defined as

$$q = k - k' \tag{2.2}$$

and therefore $\boldsymbol{q}$ has magnitude

$$|q| = \frac{4\pi \sin\theta}{\lambda} \tag{2.3}$$

where θ is more generally defined as one half of the scattering angle 2θ formed between $\boldsymbol{k}$ and $\boldsymbol{k}'$. The orientation of $\boldsymbol{q}$ is given by the geometry of the source and detector and is perpendicular to the sample plane in specular thin-film XRD and X-ray reflectivity.

When the magnitude and orientation of $\boldsymbol{q}$ match the Bragg condition, diffraction peaks occur, indicating the spacing and orientation of crystal lattice planes. Bragg's law describes the condition of constructive interference due to regularly spaced scattering objects as

$$n\lambda = 2d \sin\theta \tag{2.4}$$

or equivalently as

$$d = \frac{2\pi n}{q} \tag{2.5}$$

Bragg's law thus requires the scattering vector orientation and magnitude to have specific values that correspond to Bragg reflections. The magnitude of $\boldsymbol{q}$ is adjusted in specular thin-film XRD by changing θ, but its orientation remains perpendicular to the substrate normal. A practical consequence of a fixed $\boldsymbol{q}$ orientation is that Bragg reflections will only be observed if the vector normal to a real-space crystallographic plane within the measured thin film is normal to the substrate in specular XRD.

It is, therefore, possible to observe a complete absence of peaks in specular XRD even if a sample is highly crystalline, if there is no crystallographic plane parallel to the substrate plane. Figure 2.4 shows an example from Rivnay *et al.* of a polymer that was initially reported as amorphous from out-of-plane thin-film diffraction measurements but was later shown to be highly crystalline, but oriented with all the primary diffraction peaks in plane [19]. Most importantly, it is not possible to assess *relative or absolute crystallinity* from peak intensity/area in specular XRD of thin films if the film is not completely isotropic. If the sample has a preferred crystal orientation, specular XRD measures only a narrow fraction of the possible crystal orientation distribution depending on the angular resolution of the measurement. The importance of this concept occurs in the common situation where the degree of orientation changes but the crystallinity does not (Figure 2.5). In this case, the intensity of the specular diffraction would increase as the film becomes more oriented in the specular direction but is not a result of increased crystallinity. Similarly, an oriented film that becomes isotropic or reoriented away from the specular condition would appear to have reduced crystallinity. To accurately access crystallinity, a full rocking curve must be done [11] and is discussed in detail in later paragraphs. It is most typical in specular XRD of crystalline organic semiconductors to observe reflections due to a single set of Miller indices that denotes a set of planes parallel to a preferred contact face; for example, only (h00) reflections when the real-space b–c plane is parallel to the substrate.

Specular XRD is practiced on both conventional laboratory diffractometers and synchrotron beamlines. The required resolution of the measurement depends on the broadness of the diffraction peaks and the degree of sample orientation. The X-ray beam at a synchrotron is typically considerably more collimated than a laboratory source. For a laboratory source, the measurements are usually done at lower resolution to maximize the X-ray flux. When measuring a highly oriented sample with high-resolution optics, it is important to properly align the sample to achieve the specular condition, as misalignment would cause the reflected beam to be blocked by the receiving slits and to not reach the detector. This is particularly important for *in situ* temperature measurements where thermal drift of the sample could cause apparent drops in the intensity of the scattering peaks that are actually a result of misalignment and not due to changes in the sample.

Rocking curves are used to measure the orientation distribution of diffraction peaks. In a rocking curve, the source and detector geometry are maintained at the Bragg condition for the crystal plane of interest, and the sample is "rocked" such that the substrate normal is tilted away from the scattering vector. The resulting

Figure 2.4 X-ray characterization of P(NDI2OD-T2) structure. (a–c) Two-dimensional grazing-incidence diffraction pattern from spun cast, isotropic film (a); dip coated, aligned film with scattering vector $\boldsymbol{q}$ nominally perpendicular to the fiber direction (b); and $\boldsymbol{q}$ nominally parallel to the fiber direction (c). (d) High-resolution specular (top, gray) and grazing (second from top, black) scans of isotropic film, and grazing scans of aligned films with scattering perpendicular (third from top, dark gray) and parallel (bottom, light gray) to the fiber direction [19].

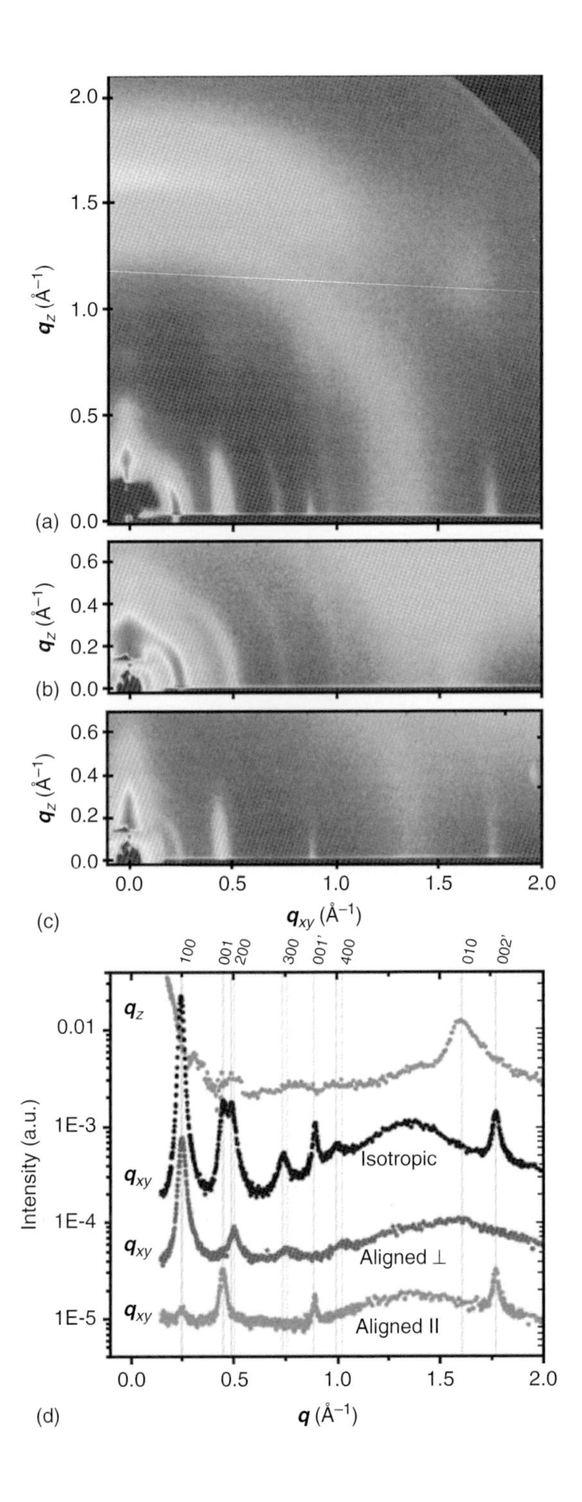

(a)
(b)
(c)
(d)
q_z (Å^{-1})
q_{xy} (Å^{-1})
100
001
200
300
001'
400
010
002'
Intensity (a.u.)
Isotropic
Aligned ⊥
Aligned II
q (Å^{-1})

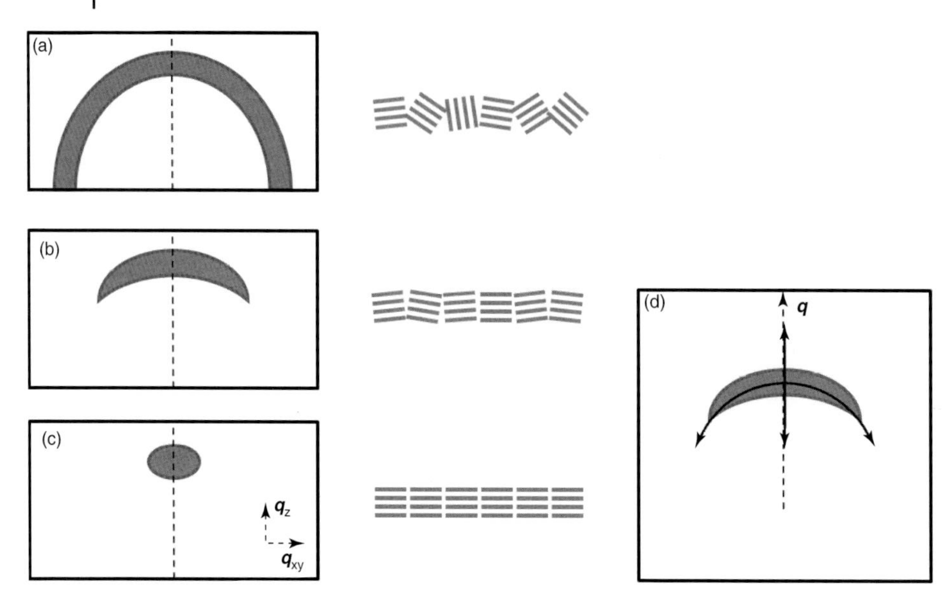

Figure 2.5 Illustrations of example diffraction patterns for different microstructures. (a) Randomly oriented films produce a sharp ring, (b) oriented films with broad orientation distribution produce an arc, and (c) highly oriented films produce an ellipse. The inset shows the crystal orientation distribution that produced the pattern. (d) Diagram of the measurement scan for $\boldsymbol{q}_z$ scan for measuring plane spacing and a rocking curve for measuring orientation distribution [20].

curve of intensity versus the rocking angle (typically called ω or χ) describes the orientation distribution of crystals on the substrate. Rocking curves can be performed such that the surface normal vector is kept within the plane formed by $\boldsymbol{k}$ and $\boldsymbol{k}'$ (e.g., "rocking back and forth" toward and away from the detector by changing θ) or such that the surface normal vector is kept within the plane that is both orthogonal to the $\boldsymbol{k} - \boldsymbol{k}'$ plane and includes $\boldsymbol{q}$ (e.g., "side to side" or by changing χ). It is important to note that the rocking curve at small scattering angles is a combination of Bragg scattering caused by the crystal orientation distribution with a background of the specular and diffuse (e.g., off-specular) reflectivity, and thus without proper background subtraction, its shape does not define a crystal orientation distribution. For example, a rocking curve on the reflectivity background is often used for aligning the sample to the specular condition. This sharp, resolution-limited peak from the reflectivity will be superimposed on top of any neighboring diffraction peaks and must be accounted for when quantifying the presence of highly oriented crystals. Rocking curves have proven to be of great use in discriminating fine differences in crystal orientation for organic semiconductors such as poly(3-hexylthiophene) (P3HT) [21] and pentacene [22]. Figure 2.6 shows examples of rocking curves from these systems. For P3HT, it was found that the presence of highly oriented crystals with resolution-limited rocking curves was correlated with the charge carrier mobility. This correlation

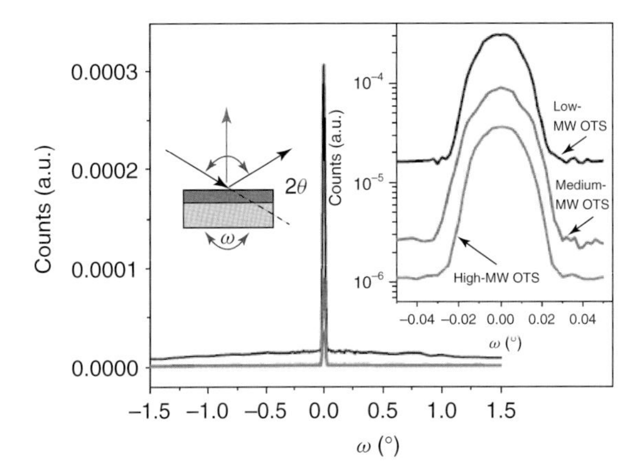

Figure 2.6 Rocking curves of the (100) peak for films of low-, medium-, and high-molecular-weight (MW) P3HT on octyltrichlorosilane (OTS)-treated substrates showing the presence of highly oriented crystals. Left inset: the rocking curve geometry showing the angle relative to the sample normal ω. Right inset: a magnification of the $\omega = 0$ region plotted on a logarithmic scale [21].

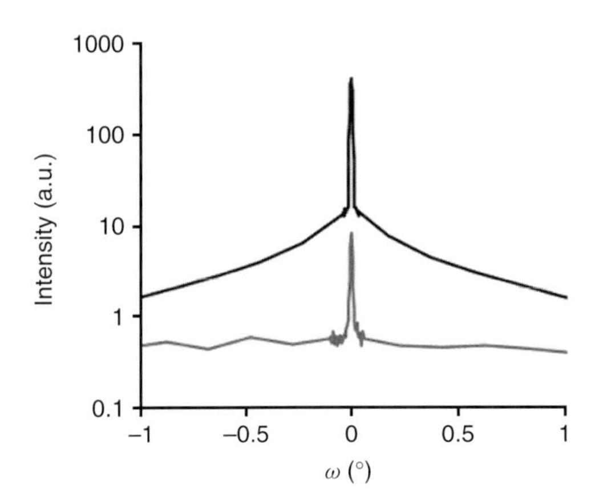

Figure 2.7 High-resolution rocking curve of the (100) peak from a film of pBTTT-C12 on OTS/SiO_2 as-spun (gray line) and annealed at 180 °C (black line) [23].

was further extrapolated to pBTTT, which was shown to be dominated by highly oriented crystals (Figure 2.7) [23]. Figure 2.8 shows results for pentacene crystals where the lateral width of the rocking curve was used to estimate the dislocation density on a variety of substrates [22]. It should also be noted that a high-resolution measurement is required to observe these highly oriented crystals. The common configuration of laboratory sources to optimize the X-ray flux utilizes a beam with a large divergence, substantially broadening the scattering from highly oriented crystals. Additionally, it should be noted that geometric corrections are required to account for the change in beam footprint on the sample as the sample is tilted. Wide-angle rocking curves can also be extracted from two-dimensional area

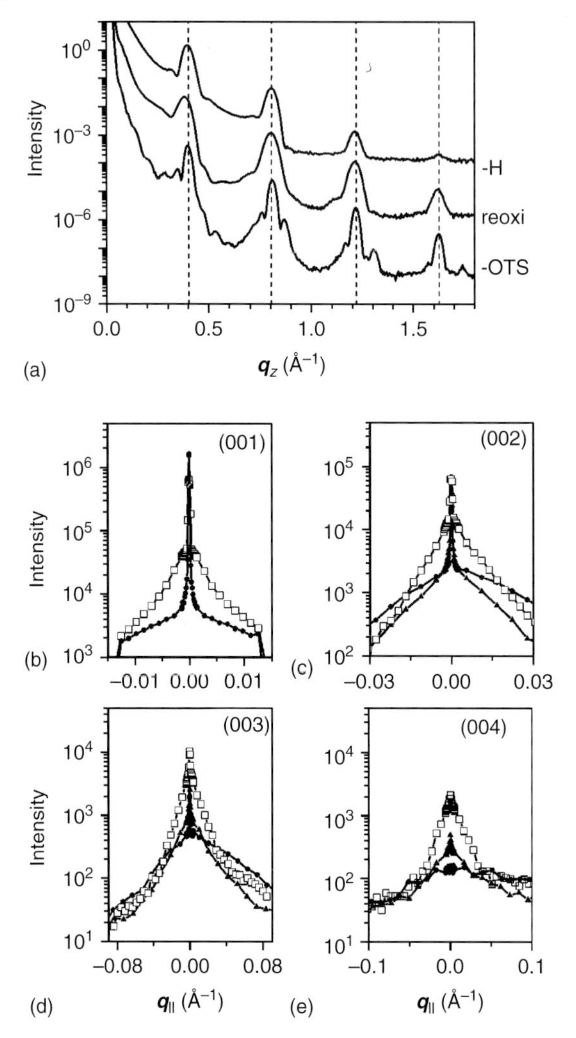

Figure 2.8 X-ray intensities for a nominal coverage of 8 ml of pentacene. (a) Reflectivity data recorded under specular conditions for different surface terminations (from top to bottom: H terminated, reoxidized, and OTS-grafted surface). (b–e) X-ray rocking scan intensities. The scans are performed at the center of the 00L Bragg reflections (dashed lines in (a)). (b) $L = 1$, (c) $L = 2$, (d) $L = 3$, and (e) $L = 4$. The OTS, H termination, and reoxidized surface are indicated by squares, disks, and triangles, respectively [22].

detectors as described earlier. Figure 2.9 shows an example from Baker *et al.* of a method for extracting full rocking curve images from two-dimensional area detectors [11]. They combine images taken near the Bragg condition with those at grazing angle to produce the full rocking curve. These full rocking curves can been used to estimate the relative crystallinity and full orientation distribution of thin films after correcting for geometry, polarization, and intensity.

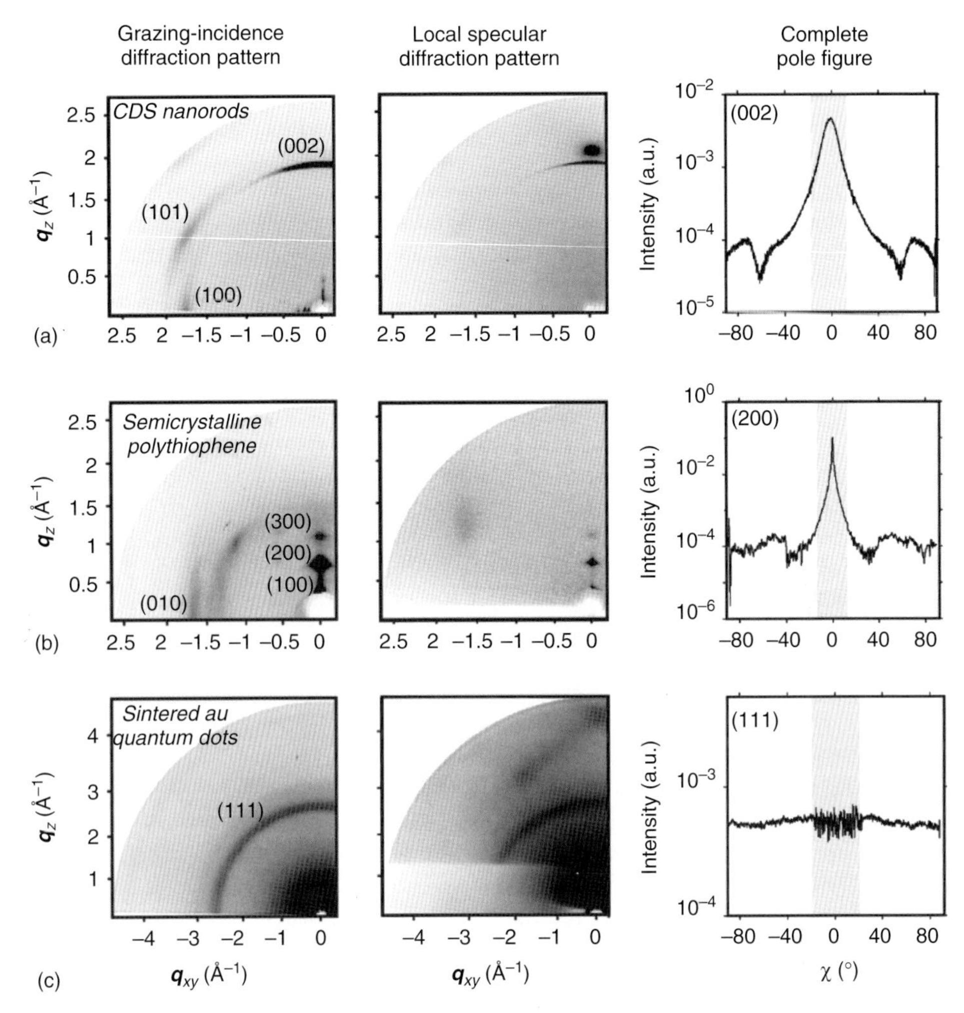

Figure 2.9 Example data for three different materials: (a) CdS nanorods deposited by slow evaporation of nanorod solution, (b) semicrystalline polymer poly[5,50-bis(3-dodecyl-2-thienyl)-2,20-bithiophene] ((poly(3,3''''-didodecyl quaterthiophene) PQT-12) deposited by spin casting, and (c) sintered Au quantum dots deposited by spin casting. The column on the left shows diffraction patterns collected in grazing-incidence geometry. The middle column shows diffraction patterns taken with the specular condition satisfied for the desired Bragg reflection, referred to as *local specular*. The column on the right shows the complete pole figure compiled by appropriately combining data from the grazing-incidence diffraction patterns (unshaded regions) and the specular diffraction patterns (shaded regions). The specular diffraction patterns are necessary to complete the pole figures: the intensity highlighted in gray is not probed in the grazing-incidence geometry and is instead collected in the local specular curves. Note that the intensity shown along $q_{xy} = 0$ in the grazing-incidence images is distorted as described above (e.g., are not truly at $q_{xy} = 0$) but images are shown in this fashion for simplicity [11].

The Bragg peaks are typically analyzed using the Scherrer equation (Eq. (2.6)).

$$t = \frac{0.9\lambda}{B\cos\theta} \tag{2.6}$$

where B is the full-width at half max (FWHM) peak width in radians and t is the size dimension [24]. The Scherrer equation represents the coherence length parallel to the scattering vector, which in specular XRD can be regarded as the crystal thickness. For many organic semiconductors, the crystals thickness and the film thickness are the same, indicating crystals that cover the entire film. It is important to note that the coherence length extracted from specular diffraction is not directly related to the crystal directions important in thin-film transistors (the lateral crystal size). The Scherrer equation involves several key assumptions that are often not valid for organic semiconductors, such as a constant *d*-spacing within the crystal. In organic semiconductors, defects and disorder result in a large distribution of lattice plane spacings that limit the long-range order. Importantly, results from a Scherrer analysis must be interpreted only as a lower bound of domain size because many other factors that also increase the width of a Bragg peak. The most significant of these factors for polymer semiconductors is the presence of nonuniform strain, which is manifested as variations in *d*-spacing within the film due to defects and disorder. The peak broadening resulting from nonuniform strain and coherence length have different dependences on the peak order [25, 26]. Peak analysis of the peak broadening as a function of the peak order has been used to show that nonuniform strain is present in films of polythiophene-based semiconductors [27]. Prosa *et al.* used a Warren–Averbach method that utilizes Fourier analysis to separate the contributions of coherence length and nonuniform strain as a function of temperature. They found that the peak width varied as a function of peak order, with the peaks broadening as the order increases. Figure 2.10 shows the values for plane spacing, domain size, and disorder extracted from the diffraction peak width series. They found that the polymer underwent a phase change on heating. The domain size and the disorder parameter increased at the phase change. The domain sizes extracted by the Scherrer equation overestimated the domain size.

2.2.2
Grazing-Incidence X-Ray Diffraction (GIXD)

Specular XRD measures crystals that are oriented perpendicular to the substrate and covers only a small region of reciprocal space. For applications such as TFTs, the in-plane crystal structure, not the out-of-plane crystals, is most directly correlated with charge transport and device performance. The out-of-plane diffraction measured by specular XRD is only related to TFT transport through inference because the symmetry of the crystals results in the measured scattering peaks usually being approximately orthogonal to the in-plane scattering. The diffracted intensity for organic semiconductor films away from the specular condition (off-specular) decreases substantially. Similarly, the diffracted intensity for transmission measurements of weakly scattering thin films such as organic semiconductors is also very small. The intensity can be increased by increasing the film thickness, but

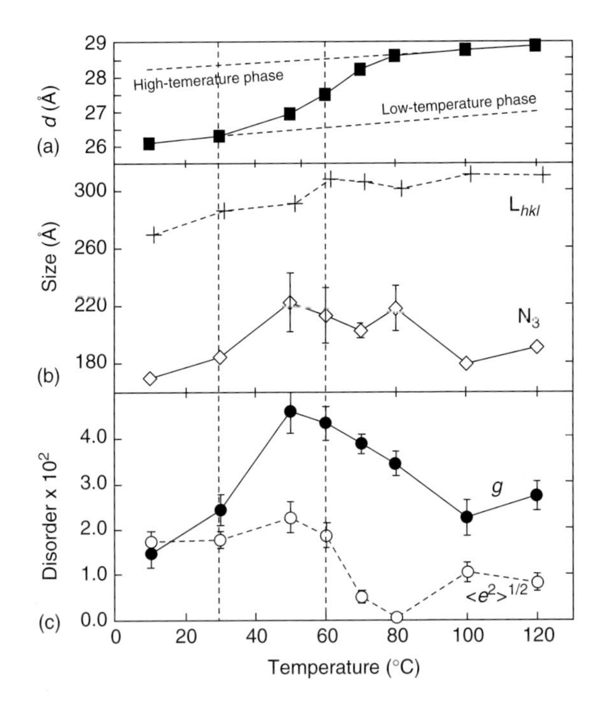

Figure 2.10 Plots of various parameters as a function of temperature for (a) interlamellar *d*-spacing (■), (b) crystal size according to Scherrer formula (+) and WA analysis (◇), and (c) displacement disorder *g* (•) and lattice parameter fluctuation *e* (○). Error bars represent 95% confidence levels based on a Monte Carlo bootstrap procedure [27].

this creates the concern that the film morphology is dependent on the film thickness and that the measured thicker films are not device relevant. GIXD allows the measurement of the in-plane packing by enhancing the scattering intensity through a combination of increased beam footprint and electric field enhancements due to the grazing geometry [28–30]. GIXD requires a highly collimated beam to tightly define the incidence angle α (Figure 2.11) and thus is typically practiced at synchrotrons or specialized laboratory sources [31, 32].

GIXD involves irradiating the sample with an X-ray beam with an incidence angle α near that of the critical angle α_c, such that the X-rays travel in an evanescent wave across the sample surface. The critical angle is defined by the wavelength-dependent index of refraction of the solid film (Eq. (2.7)):

$$n = 1 - \frac{\lambda^2 e^2 \rho_a}{2\pi m_e c^2} f \tag{2.7}$$

where λ is the X-ray wavelength, e is the electron charge, ρ_a is the atomic density, m_e is the electron mass, c is the speed of light, and f is the atomic scattering factor. The critical angle is then

$$\alpha_c = \cos^{-1} \frac{n_{solid}}{n_{air}} \tag{2.8}$$

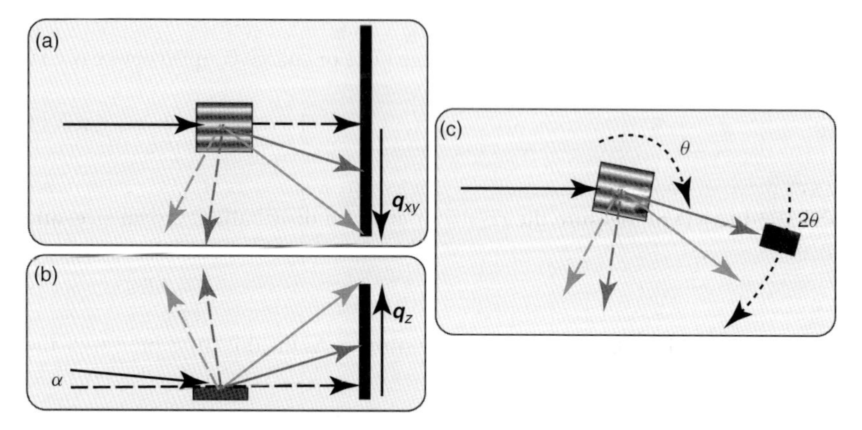

Figure 2.11 Common geometries for GIXD assuming a horizontal sample. (a) Top and (b) side views of 2D detector geometry illustrating how the orientation of the scattering vector $\boldsymbol{q}$ (dashed arrow) varies with the scattering angle. (c) Top view of the point detector geometry illustrating how the sample is rotated at half the angular velocity of the detector to maintain the scattering vector at a constant in-plane sample orientation [20].

The α_c of an organic semiconductor film is typically $\approx 0.2^\circ$ for X-rays at 8 keV. In GIXD, the propagation of the X-ray wave front is identical to that of X-ray reflectivity. Precise control of the incidence angle allows the depth penetration of the X-ray intensity to be varied and for the formation of standing waves to adjust the X-ray intensity distribution through the film [33–36]. If α is less than α_c, then the diffraction will represent the structure of the near surface of the film; if α is greater than the α_c of the film but less than that of the substrate, then the diffraction will represent the film bulk with limited contribution from the substrate. It is therefore possible to vary α such that a comparison of surface and bulk diffraction is possible, although small changes in α near α_c can dramatically change the measured depth and position of standing waves within the film. Determining the exact profile distribution of X-ray intensity within the film requires process knowledge of the incidence angle, the film density distribution, surface roughness, and extremely low beam divergence and accurate sample alignment. Accurate standing-wave measurements require the ability to do X-ray reflectivity measurements on the films before GIXD characterization. The reflectivity measurement can be modeled to determine the correct incidence angle for the desired standing-wave distribution. For normal X-ray energies, incidence angle accuracies of less than one hundredth of a degree and similar beam divergences are required to accurately control the depth penetration of the X-rays. As a practical consequence, most GIXD measurements on organic semiconductors are performed in a bulk-sensitive mode, just above α_c for the organic film; and if α is purposefully varied, it is typically for the purpose of performing a surface versus bulk comparison.

In either the bulk- or surface-sensitive modes of GIXD, the X-ray beam will diffract if it meets the Bragg condition. Since the incident beam vector $\boldsymbol{k}$ is fixed, the orientation and magnitude of the scattering vector $\boldsymbol{q}$ depends solely on the position of the detector (e.g., on $\boldsymbol{k}'$). The scattered radiation is typically measured either by a point detector, just as for specular XRD, or by an area detector, which is a two-dimensional detector such as an image plate or charge-coupled device (CCD) with a fixed pixel resolution.

The literature contains numerous examples of using GIXD to characterize organic electronic materials [13, 14, 37–65]. Below we discuss a few examples selected from the large collection of published studies. In Figure 2.12, Joshi *et al.* used GIXD to measure the crystal orientation distribution, crystal size, and qualitatively the crystallinity and degree of ordering for a series of P3HT films of varying thicknesses [61]. They found that as the film thickness increases, the degree of orientation decreases. For very thin films, the crystals were highly oriented with their (100) axis normal to the surface. As the thickness increased, the films contained more randomly oriented crystals, showing that for thicker films (~200 nm), the crystals nucleate in the bulk of the film and not at the substrate surface. For very thin films (~10 nm), the crystal size decreased to less than the film thickness. Interestingly, despite substantial variation in microstructure, the charge carrier mobility in TFTs was unaffected by the film thickness, indicating that the thin layer of polymer at the dielectric–polymer interface did not contribute substantially to the bulk-dominated diffraction pattern.

In Figure 2.13, Donley *et al.* used GIXD to investigate the microstructure of poly(9,9′-dioctylfluorene-co-benzothiadiazole) (F8BT) films of various molecular weights and annealing conditions [41]. They found that spin-cast as-deposited (pristine) films had a disordered structure with a significant torsion on the backbone between the F8 and BT units and BT units on neighboring molecules adjacent to each other. Annealing above the glass transition temperature (T_g) allowed neighboring molecules to rearrange such that the BT units were adjacent to an F8 unit of the neighboring molecule. This new structure (referred to as the *alternating structure*) reduced the torsion along the backbone between the comonomer units and resulted in the appearance of strong diffraction peaks from oriented crystals. The two structures had considerably different optical and electrical properties, with the alternating structure having a significantly lower electron mobility.

In Figure 2.14, Germack *et al.* used GIXD to investigate the microstructure of a blend of P3HT and [6,6]-phenyl-C61-butyric acid methyl ester (PCBM) for bulk-heterojunction (BHJ) organic photovoltaics [66]. The BHJ films were deposited on a variety of substrates with varying surface energy. Other measurements showed a vertical stratification of the P3HT and PCBM, with an enrichment of P3HT on the film surface and an enrichment of PCBM on silicon oxide substrates. The films are relatively thick (~200 nm), and the thin interfacial layers do not contribute strongly to the diffraction. The GIXD patterns show a mixture of a broad distribution of (100) oriented P3HT with a diffuse ring from the amorphous PCBM. No evidence of PCBM crystals was observed in these films. Additionally, the lattice spacing and peak positions of P3HT are unchanged with the addition of PCBM, indicating that PCBM does not intercalate the P3HT crystals.

In Figure 2.15, Jimison *et al.* used GIXD to measure the structure of PQT as a function of processing [65]. They found multiple lattice spacings for PQT, indicating the presence of two different polymorphs. Annealing changed the relative amounts

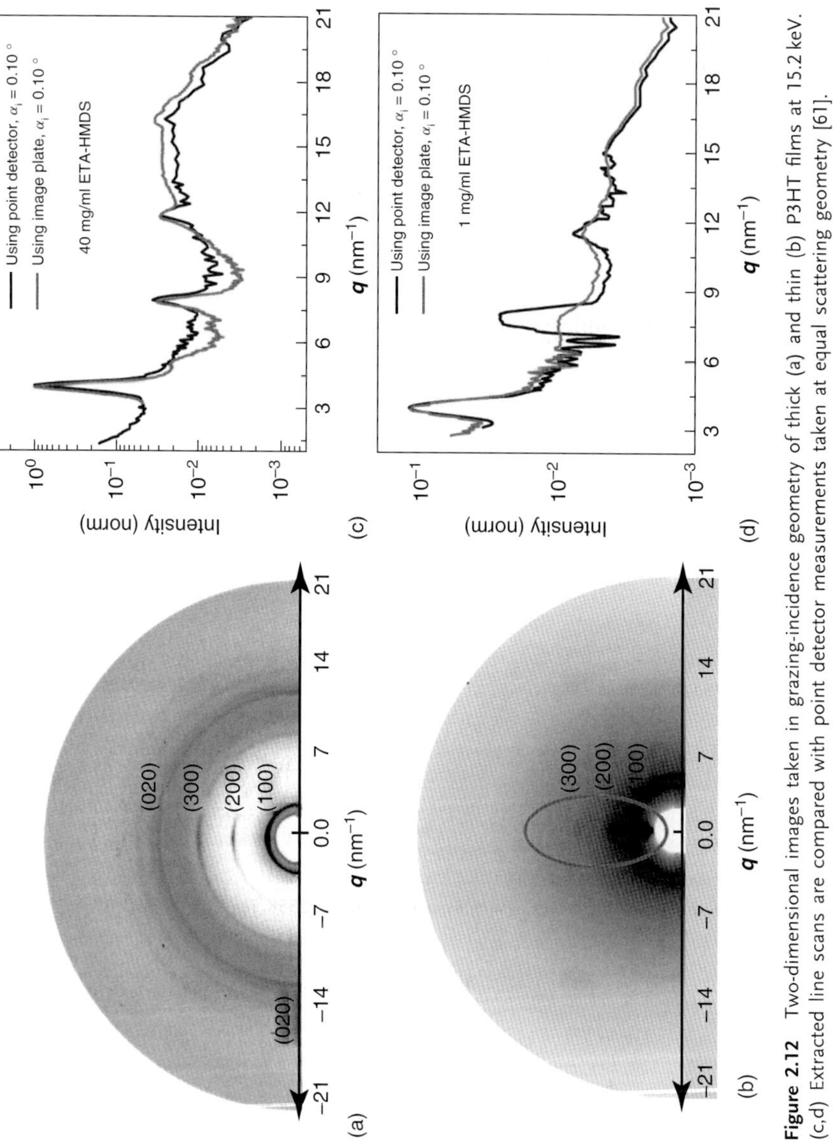

Figure 2.12 Two-dimensional images taken in grazing-incidence geometry of thick (a) and thin (b) P3HT films at 15.2 keV. (c,d) Extracted line scans are compared with point detector measurements taken at equal scattering geometry [61].

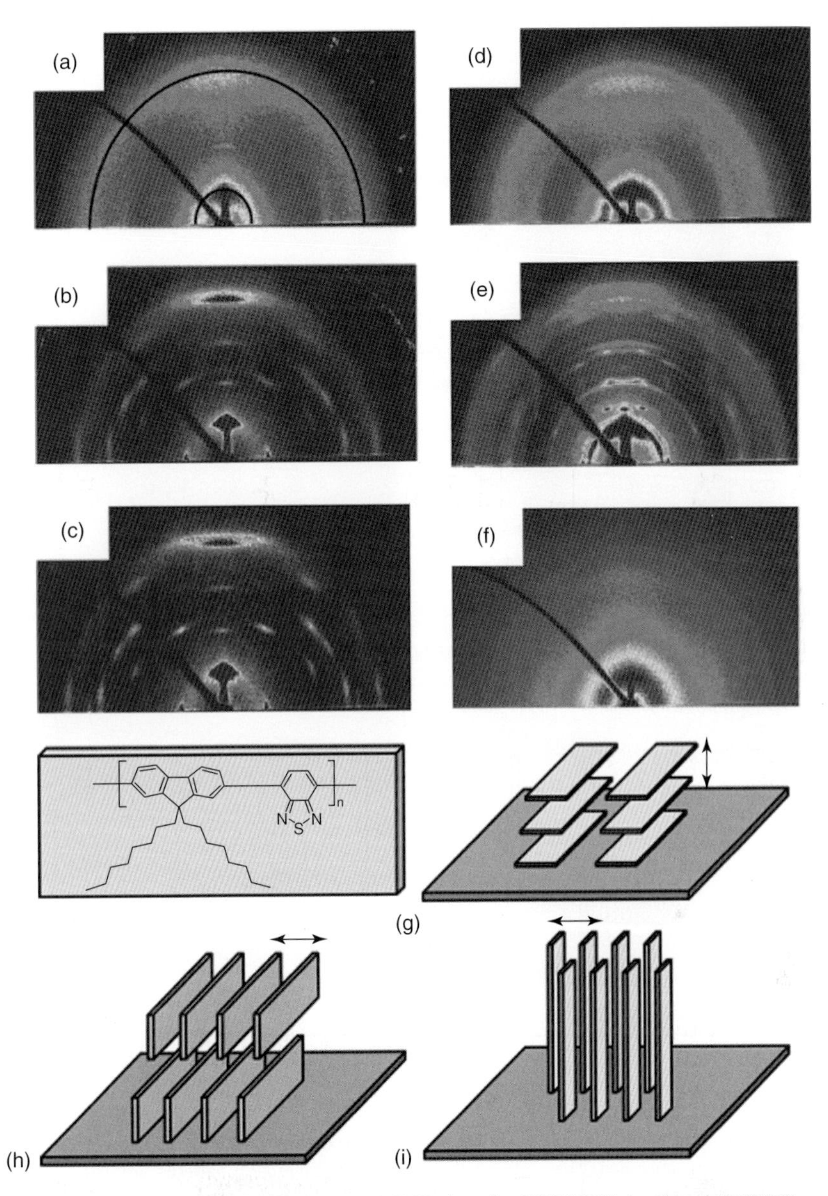

Figure 2.13 Wide-angle x-ray scattering (WAXS) data for F8BT/255K (a–c) and F8BT/9K (d–f). (a,d) Pristine, (b,e) annealed to T_g and slowly cooled, and (c,f) annealed to T_m and slowly cooled. The inner and outer rings in (a) correspond to the (001) and (004) reflections, respectively. (g,i) Possible orientations of the polymers with respect to the substrate, with the ð-stacking direction indicated by arrows. Analysis of the WAXS and ellipsometry data indicates that in all samples the predominant orientation is similar to that in (g). An edge-on orientation (h) is not observed [41].

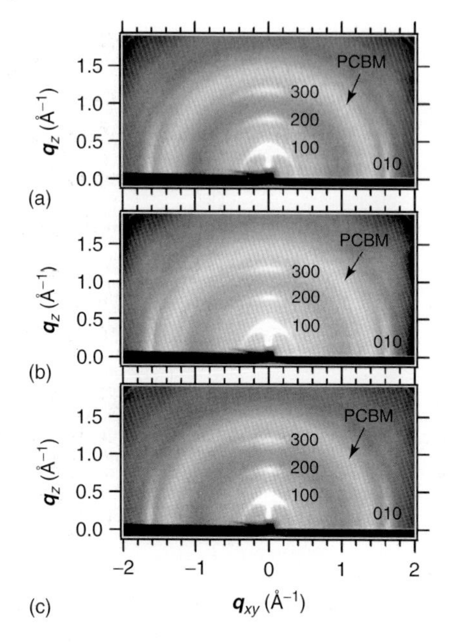

Figure 2.14 GIXD images of P3HT–PCBM blend films on (a) SiO_2, (b) (PEDOT/PSS) Poly(3,4-ethylenedioxythiophene)/poly(styrenesulfonate), and (c) (poly(thienothiophene)) PTT/Nafion showing oriented P3HT domains [66].

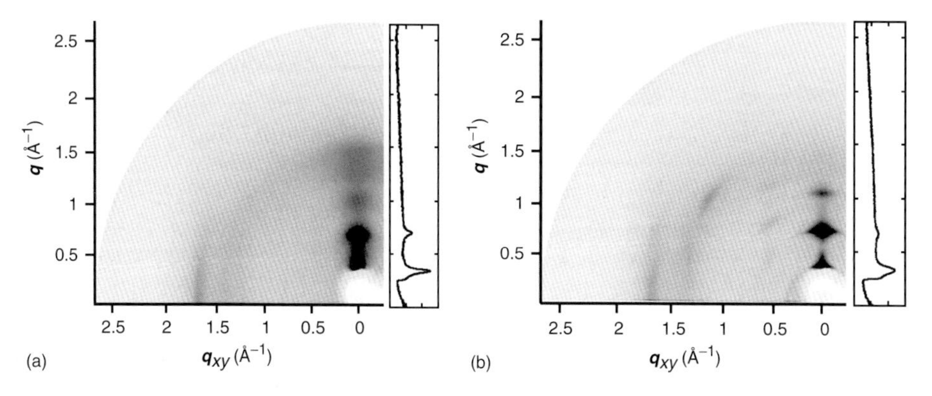

Figure 2.15 Two-dimensional survey diffraction patterns of PQT-12 films on bare SiO_2 (a) as-spun and (b) annealed. The plots on the right of the diffraction patterns are vertical line scans intensity versus q_z extracted at $q_{xy} = 0$ [65].

of the two polymorphs and increased the crystal size as shown by the sharpening of the peaks and addition of new mixed index peaks. The annealing also corresponded to an increase in charge carrier mobility.

In Figure 2.16, Chabinyc *et al.* used GIXD to measure the structure of pBTTT as a function of annealing [23]. Earlier in Figure 2.7, it was shown that the orientation distribution of pBTTT changes with annealing to the mesophase at 180 °C. As-spun

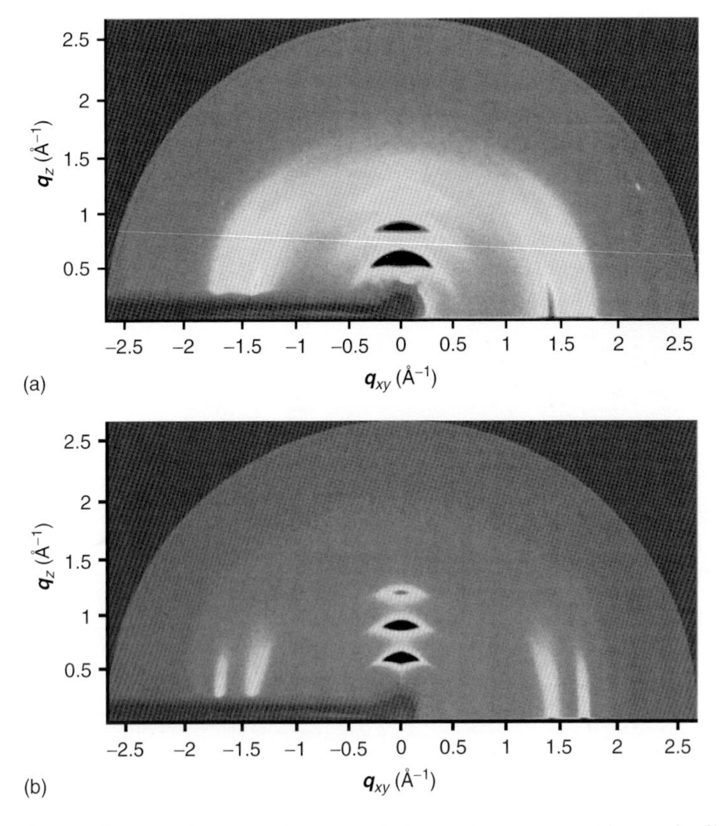

Figure 2.16 Two-dimensional X-ray scattering at the grazing incidence of a film of pBTTT-C14 (a) as-spun and (b) annealed at 180 °C OTS/SiO_2 [23].

films tend to have a broad orientation distribution, annealing causes a transition to a film with a large fraction of highly oriented crystals. This transition is easily seen in the 2D GIXD patterns. The as-spun film has broad arcs from the broad orientation distribution, while the annealed film has ellipses from the highly oriented crystals. The ellipse shape is due to a combination of the instrumental resolution and disorder in the crystals (Figure 2.5). This is similar to what was discussed in the earlier section on rocking curves and the extraction of dislocation density from the lateral width of peaks [22]. Also, the area detector geometry usually has a much lower resolution than the point detector geometry.

In Figure 2.17, Knaapila *et al.* used GIXD to investigate the molecular packing in fiber-oriented films of the semiconducting polymer poly[9,9-bis (2-ethylhexyl)fluorene-2,7-diyl] (PF2/6) [45]. The fiber orientation allows the separation of peaks from different directions in the unit cell. They found that the films consisted of two different crystal orientations (types I and II) with a similar hexatic phase, with the molecular backbone having a helical structure. They were able to estimate the concentration of the two phases and determined that the films were primarily type I.

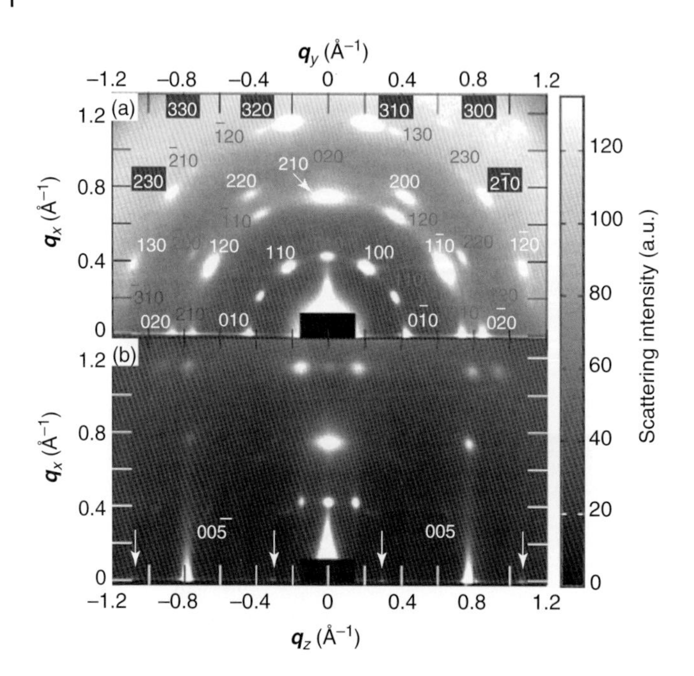

Figure 2.17 GIXD patterns of PF2/6 (number-average molecular mass (Mn) = 29 kg mol^{-1}) and schematics of the orientation types (below). The GIXD patterns were measured with the incident beam along the z- and y-axes, respectively. (a) ($xy0$) plane. White and gray indices show the primary reflections of the types I and II, respectively. (b) ($x0z$) plane. The reflections close to 002 and 007 are shown by arrows. Percent values are the fractions of the orientation types I and II, respectively [45].

In Figure 2.18, Osaka *et al.* used GIXD to study the microstructure of poly (2,5-bis(3-dodecyl-5-(3-dodecylthiophen-2-yl)thiophen-2-yl)thiazolo[5,4-d]thiazole) (PTzQT) as a function of alkane side chain length [50]. They found for short side chain lengths that the lamellae are somewhat disordered and only result in two orders of the ($h00$) diffraction series. This was not surprising because the attachment of the side chains on the molecule is such that the takeoff angles and spacing are asymmetric, making efficient packing of short chains problematic. For longer side chains, diffraction peaks up to the fifth order were seen for the ($h00$) series, indicating relatively good lamellar packing and an increase in the degree of ordering. The longer chains have enough flexibility to allow the side chains to pack well, whereas the short chains do not have enough freedom to accommodate the asymmetric attachment.

In conclusion, XRD is an invaluable tool for measuring the packing of ordered material. In particular, XRD has had a substantial effect on the organic electronic field for determining the film microstructure. In particular, XRD is used to measure the crystal type, orientation, and size, as well as the degree of ordering. More advanced measurements are under development to determine the structure of disordered regions and the quantitative degree of crystallinity.

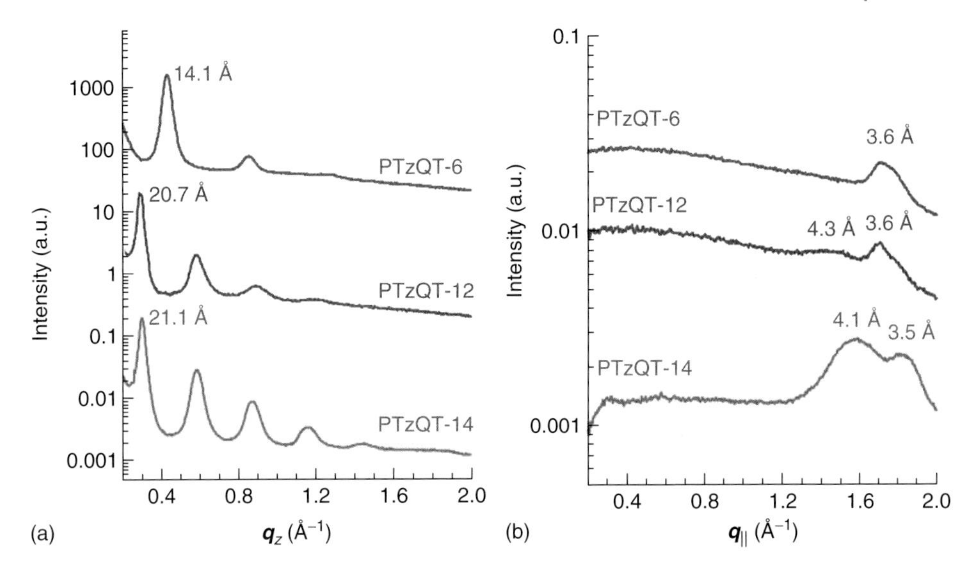

Figure 2.18 One-dimensional GIXD patterns of the polymer thin film cast on OTS-8 treated SiO_2 surfaces. (a) Out-of-plane and (b) in-plane patterns of PTzQT-6, -12, and -14 (after annealing) [50].

2.3 Near-Edge X-Ray Absorption Fine Structure (NEXAFS) Spectroscopy

2.3.1 Background and General Features of NEXAFS

The near-edge X-ray absorption fine structure (NEXAFS) spectroscopic technique is generally regarded as a powerful synchrotron-based surface science tool for the measurement of composition and orientation, especially in organic media. The technique was first published in the early 1980s, with an emphasis on monolayers or submonolayers of molecules adsorbed on catalytic surfaces from the gas phase [67, 68]. Owing to experimental complications in thick organic films, the NEXAFS technique was rather narrowly applied to molecularly thin organic layers adsorbed on inorganic substrates for many years. Methods to properly ground or otherwise eliminate charging artifacts enabled the application of NEXAFS to a much wider variety of films and powders in the 1990s [69]. In the last decade, it has found widespread use in the measurement of organic semiconductors [14, 20, 70, 71].

NEXAFS spectroscopy is the measurement of photon absorption features near an elemental edge where core-shell electrons are promoted to low-lying unfilled orbitals. The well-defined energy levels of the low-lying orbitals give rise to distinct peaks in the absorption spectra within 30 eV of the elemental edge. NEXAFS spectroscopy can be performed at a variety of elemental edges, but it is most commonly performed for organic materials at the K-edge of low-atomic-number elements such as carbon (285 eV), nitrogen (400 eV), oxygen (535 eV), and fluorine

(685 eV). The technique is elementally selective by the core shell that is accessed (e.g., the energy range over which the spectrum is collected), it delivers elemental composition information by the height of the elemental edge (e.g., intensity $\approx$30 eV above the edge in the postresonant part of a spectrum), and it delivers bond information by the energy and orientation of the final state of the excitation (e.g., the peaks in the spectrum). It should be noted that NEXAFS spectroscopy can also be named X-ray absorption near-edge structure (XANES) spectroscopy, which is typically the term used when the fine structure or elemental edge intensity is measured for higher-atomic-number elements.

The low-lying unfilled orbitals that are the final state of an NEXAFS excitation are typically molecular antibonding orbitals of π^* or σ^* character. These excitations can occur both above and below the edge, where it becomes possible for core-shell electrons to be excited directly into the continuum. The edge is manifested in an NEXAFS spectrum as "step," as in a step function, but typically has a nonzero width that may be described by a Gauss error function (Figure 2.19). A 1s $\rightarrow \pi^*$ transition typically occurs at energies less than those at the edge, and a 1s $\rightarrow \pi^*$ transition typically occurs at energies greater than those at the edge. For period 2 elements such as carbon and oxygen, the initial state is typically understood to be the K-shell, and resonance description is often simplified to the final state alone, for example, a 1s $\rightarrow \sigma^*$ resonance is referred to as a σ^* *resonance*.

2.3.1.1 NEXAFS Experimental Considerations

Owing to its requirement for a high-intensity monochromatic beam in the soft X-ray range, NEXAFS is practiced almost exclusively at synchrotron light sources. The beam may originate with a single bending magnet or an insertion device. Signal-to-noise ratio is typically not a significant constraint in routine NEXAFS experiments, so the added intensity of insertion devices is not required, but the greater control over polarization afforded by insertion devices can occasionally simplify orientation measurements. In the soft X-ray range required for NEXAFS measurements of low-atomic-number elements, the beam is highly attenuated through air and most X-ray window materials. The measurement must be performed in a vacuum environment, most typically on a differentially pumped beamline open to the ring vacuum at the beam source.

The most common NEXAFS measurement scheme is shown in Figure 2.20. The sample is affixed to a translator and adjusted to define an incident angle Θ. The sample holder is typically grounded to prevent sample charging that can alter the shape of the spectrum. Chemical composition measurements are typically performed at the orientation-insensitive "magic angle" of 54.7° incidence. The sample may be translated so that a fresh spot is used for each scan. The requirement that the samples be pumped to a high-vacuum environment favors high-throughput sample arrangements where the translator can accept a large number of samples that can be loaded simultaneously.

Spectra are collected by varying the photon energy of the X-rays incident on the sample, typically using a grating monochromator, and measuring their absorption. It is possible to directly measure the attenuation of the X-ray, as is done in scanning

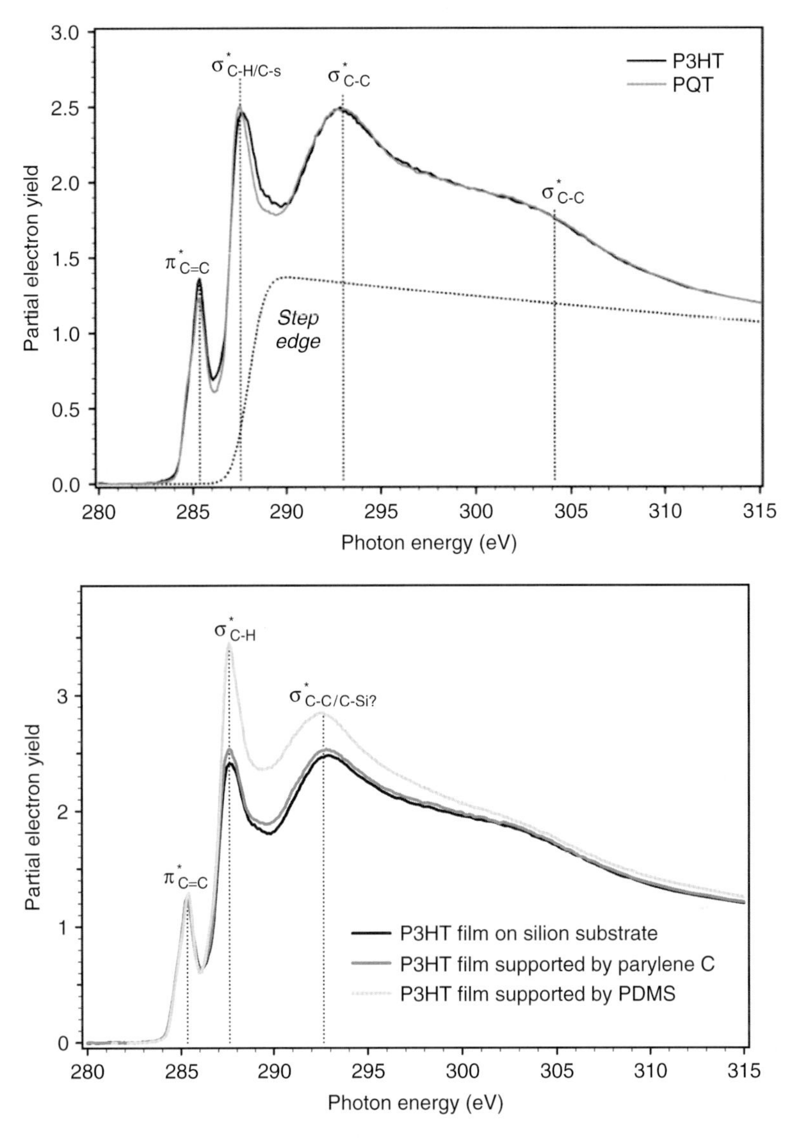

Figure 2.19 NEXAFS carbon K-edge spectra of P3HT and PQT films collected in partial electron-yield mode at a grid bias of −50 V, with a beam incident angle of 54.7°. The step edge is shown for illustrative purpose; it is not fit to this spectrum.

transmission X-ray microscopy (STXM). Generally, however, direct measurement of absorption is not common because it requires a thin film on a substrate that does not itself absorb significant photons, perhaps a grid or thin nitride membrane as is typically used for transmission electron microscopy. The absorbance of X-rays is, therefore, more commonly measured indirectly via decay processes of the excited state. The excited state may decay via Auger electron emission or fluorescent

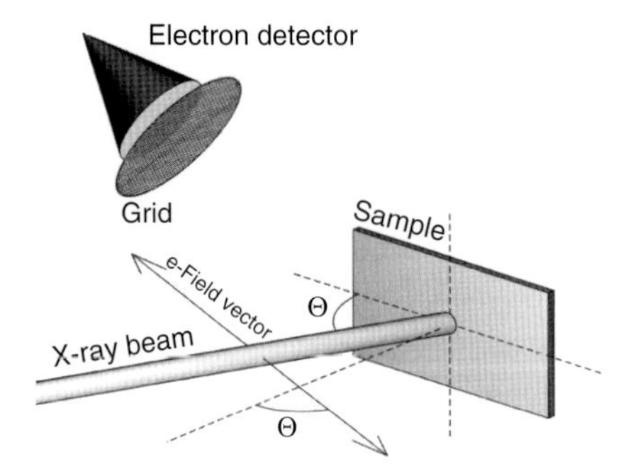

Figure 2.20 Schematic of NEXAFS experiment. The substrate is tilted with an angle of incidence Θ. Signal is collected via an electron detector with a wire mesh grid that acts as a low-pass kinetic energy filter.

photon emission; Auger electron emission is the predominant decay route for low-atomic-number elements. The number of Auger electrons is proportional to the number of absorbed photons, and therefore, electron detection schemes are a convenient route to measure absorbance. The simplest electron yield measurement is to measure the compensation current that must flow into the sample to balance Auger emission. This mode, called total electron yield (TEY), requires a conducting substrate grounded via a low-resistance path. The emitted Auger electrons can also be detected in the high-vacuum measurement chamber using an electron detector. An unshielded electron detector will typically receive a large number of electrons of various energies emitted by many processes, especially photoelectron processes (it is important to recognize that these processes are also represented in the TEY signal). For this reason, the electron detector is shielded with a polarized grid that acts as a high-pass kinetic energy filter, as shown in Figure 2.20. If only Auger electrons are desired to be collected, the grid may be polarized with a negative bias near the Auger energy. For example, the carbon KLL Auger at $\approx$263 eV can be collected with a grid bias of -250 to -260 eV. This collection mode is termed Auger electron yield (AEY). Adjustment of the grid bias to a less negative potential permits electrons that originated as KLL Auger electrons but have been inelastically scattered within the sample [72]. Manipulation of the grid bias in this manner, therefore, provides a means for near-surface depth profiling [73–75]; this collection mode is typically called partial electron yield (PEY). The depth profile near the film surface can alternatively be judged by comparing the AEY signal to the TEY signal, which represents a greater depth into the sample because it includes contributions from *all* KLL Auger electrons that have lost kinetic energy via scattering processes yet have not been recaptured by the film. Finally, fluorescence yield (FY) measurement, the measurement of photons emitted via the fluorescence decay of the excited state, can be used to gain a bulk-sensitive NEXAFS measurement [76]. Comparison of electron yield and FY NEXAFS spectra can reveal surface segregation and heterostructure formation [77, 78].

Regardless of the yield mode, NEXAFS data must be processed to provide spectra that are useful for determination of chemistry or molecular orientation

[79]. Most importantly, the yield measurement must first be divided by a reference to beam intensity, which can vary considerably over a spectrum because of the characteristics of the beamline or synchrotron ring. Most often, the reference to beam intensity is provided by a gold-covered wire or wire mesh, which provides a photoelectron current proportional to beam intensity. NEXAFS spectra are then typically normalized. There are two common normalization schemes. The first, pre-edge normalization is almost always performed. In pre-edge normalization, a constant value is typically subtracted from the spectrum to render the yield measurement zero in the energy range just below the onset of spectral features. In carbon, for example, pre-edge normalization is typically performed by subtracting the nonnormalized value of $\approx$280 eV from the nonnormalized spectrum. This process generally eliminates background contributions from other elements and isolates the carbon portion of the soft X-ray absorption for comparison. In the second common normalization scheme, postedge normalization, a pre-edge normalized spectrum is divided by the intensity in the postresonance region. In carbon, postedge normalization can be performed by dividing a pre-edge normalized spectrum by the pre-edge normalized value of $\approx$330 eV. This process in essence normalizes the spectral intensity to the total amount of carbon atoms present. Postedge normalization is essential for comparison of measurements collected with different illumination areas, which is especially common for orientation determination measurements, where the intersection of a nominally cylindrical beam with the planar sample at different angles can result in a greatly varying spot size. Postedge normalization can also be used to compare chemistry across samples with different amounts of material present, such as self-assembled monolayers with different packing densities.

A common source of experimental artifacts in NEXAFS measurements is sample charging [80]. Owing to Auger processes, a sample can collect significant positive charge over the course of an NEXAFS measurement if it is not well grounded or otherwise charge compensated. The positive charge can attenuate electron emission by electrostatic binding. The result is a lowered electron yield signal. This artifact can interact significantly with normalization procedures in ways that lead to an incorrect interpretation of data. Since the postedge region is typically collected last, it is often influenced the most by sample charging, and its intensity can therefore be depressed relative to other parts of the spectrum. After postedge normalization, a depressed postedge can exaggerate the intensity of spectral features. It is therefore possible for NEXAFS spectra of identical materials to exhibit what falsely appear to be significant chemical differences if grounding schemes or substrate conductivities are different. Importantly, there is a generally observed dependence of charging on the beam spot size; it tends to be greatest for larger spots. This dependence of charging on beam spot size leads to the tendency of the charging artifact to influence the interpretation of NEXAFS orientation measurements. Because the beam spot size varies systematically with angle, if charging is present, it too will vary systematically with angle. If the charging depresses the postedge intensity, then the postedge will be artificially depressed in a manner that depends on incident angle. Postedge normalization, which is essential for orientation measurements to accommodate the different spot sizes, will result in spectral features that falsely appear to change in intensity with incident angle. In some cases, the intensity variation can even fit into the orientation models discussed below.

Severe charging is obvious when it results in a postedge intensity lower than the pre-edge intensity, but it is certainly possible for subtle charging to go undiagnosed, and strategies should be used to check whether charging-induced artifacts are present. The most straightforward approach is to reverse the scan direction; if the spectra still appear identical, then charging is less likely. If charging is present, a collection of postedge normalized NEXAFS data collected for orientation measurement will generally exhibit a lack of isosbestic points, and all resonances will appear greatest for the shallowest incident angle spectra, where charging is the greatest.

2.3.1.2 Chemistry Determination by NEXAFS

The fine sensitivity of NEXAFS to the chemical structure of organic compounds makes it an excellent probe of composition. For example, NEXAFS has been used to monitor chemical changes in oligothiophenes for organic transistor applications, as they are converted from soluble precursors to insoluble organic semiconductors [81–84]. The interface selectivity of NEXAFS in electron-yield modes makes it valuable in systems where the interface composition is expected to vary from the bulk. For example, the composition of a blend at interfaces can be determined by fitting a linear combination of reference spectra to a spectrum of the film interface collected under similar conditions [85]. This method was used to show that the interfaces of polymer-fullerene BHJ films used in photovoltaic devices obey segregation behavior to miscible polymer blends, with interface composition controlled by the surface energy of the adjacent interface [66, 86].

In many cases, interface composition determination by NEXAFS has lower uncertainty than composition determination by X-ray photoelectron spectroscopy (XPS) because it combines aspects of elemental analysis (because NEXAFS post-edge intensity reports elemental composition) with bond detection in the energy and intensity of specific resonances. NEXAFS peaks specific to different carbon chemistries are typically well separated in energy, such that less peak deconvolution is required than that might be typical for XPS. Finally, a key strength of NEXAFS is its ability to determine molecular orientation. A disadvantage of NEXAFS, however, is that it requires a synchrotron source, whereas XPS can be practiced with readily available laboratory sources. Furthermore, the elemental range of most NEXAFS beamlines is typically constrained – those designed for organic films typically span boron to fluorine – such that XPS can be a useful complement for evaluating the presence of higher-atomic-number species such as metals.

Like XPS, NEXAFS electron-yield modes have inherent interfacial sensitivity due to the limited escape depth of electrons. The evaluation of buried interface composition requires some means of freeing the interface such that its surface can then be exposed to the measurement chamber. A common approach is to apply a flexible support to the film top surface, and then lift the film off of the substrate. This approach requires that the film fails adhesively at the preferred interface and not at the film–support interface or cohesively within the film. We have evaluated several support materials for this purpose. Elastomeric poly(dimethylsiloxane) (PDMS) stamps can be used as supports [87]; they are typically cast and cured on a flat mold, and then simply pressed onto the film top. Adhesion of the PDMS to the hydrophobic organic semiconductor is typically greater than the adhesion of the semiconductor to the substrate, provided the substrate is also relatively hydrophobic. The film can thus be removed. This approach has been shown to have a minimal influence on the charge-transport properties of organic semiconductor films on relamination. However, PDMS presents significant constraints to NEXAFS measurements. Typical commercially available PDMS formulations will continually outgas in a high-vacuum environment, providing difficulties in evacuating the measurement chamber and maintaining a low pressure. The outgassing species, which are presumably dimethylchlorosilane or some analog thereof, appear in the NEXAFS measurement, specifically as a strong enhancement in the C–H σ^* positioned at $\approx$288 eV, consistent with a contribution that is predominantly from methyl groups. Because the spectral contribution is relatively isolated to this region, we often use PDMS as a delamination support for evaluation of π^* orientation at the buried interface [88]. This measurement is of high relevance to OTFT materials because the region measured by NEXAFS PEY at a grid bias of -50 V, where $\approx$50% of the signal originates within the topmost $\approx$4 nm of the film, closely corresponds to the $\approx$6 nm mobile channel region of OTFT devices [89]. An alternative support material is gas-phase deposited parylene C. Parylene does not outgas in a vacuum environment, but because it is typically applied in a very thin layer, it is not as rigid as a typical PDMS support and is therefore of less utility in orientation measurements that require a rigorously flat substrate. Parylene is preferred in measurements where composition measurement is critical because it does not influence the shape of the spectrum.

To illustrate the correspondence of primary chemical structure to NEXAFS resonances, spectra of P3HT and poly[3,3''''-bis(3-didodecyl quaterthiophene) (PQT) [90] are shown in Figure 2.19. The lowest energy resonance at $\approx$285 eV is the carbon–carbon π^* of the backbone thiophene rings. It should be noted that even for P3HT, this resonance appears to be the combination of several (at least two) peaks, possibly because of the inequivalent carbon-carbon double bonds that arise from a single β-substituted thiophene ring. The shape of this π^* region is subtly different for PQT potentially because of the presence of unsubstituted thiophene rings. The $\approx$287.5 eV resonance is likely a combination of carbon–hydrogen and carbon–sulfur σ^*, with some Rydberg contribution. It can be quite difficult to separate and interpret peaks in this region, as they typically lie atop the step edge transition region. The carbon–carbon σ^* resonance at $\approx$293 eV can be attributed primarily to the methylene groups of the side chains, and the 303 eV σ^* resonance can be attributed to carbon-carbon bonds along the backbone. The appearance of a higher-energy carbon σ^* is typical for conjugated systems, but it should be noted that a $\approx$293 eV carbon-carbon σ^* is typically observed even in systems that are entirely conjugated, such as benzene. The $\approx$293 eV peak in Figure 2.19 must therefore be regarded as having some contribution from the backbone rings; this consideration becomes important in analyses of bond orientation.

2.3.1.3 Orientation Analysis in Organic Semiconductors

NEXAFS resonances have a characteristic transition dipole direction that is determined largely by the orientation of the final state orbital because the X-ray wavelength at the carbon K-edge is sufficiently large that the dipole approximation remains valid. For conjugated systems, the π^* orbitals extend as lobes nominally perpendicular to the plane of conjugation. Thus, for a single fused-ring system or a series of conformationally free yet coplanar rings, the orientation of the π^* director can be a definitive descriptor of the molecular orientation with respect to substrate normal. Often the π^* orientation can be further supported by the orientation of a σ^* director, which lies parallel to a bond, although σ^* orientation can be more difficult to quantify. The orientation of the conjugated plane can be determined by varying the electric field vector of incident light with respect to the sample, which is typically achieved by changing the angle of beam incidence. NEXAFS orientation analysis for several conjugated organics is shown in Figure 2.21.

Because the pentacene molecule consists of a single fused ring, its surface-relative orientation in a thin film can be partially determined by the π^* director normal to the fused-ring plane. Note that the pentacene spectra shown in Figure 2.21 exhibit multiple strong π^* resonances, with major peaks at $\approx$284 and $\approx$286 eV. Each major peak has a minor shoulder, indicating at least four distinct contributions to the pentacene π^*. All these π^* contributions have the same intensity trend with incident angle. Separation and analysis of each peak produces a similar quantified orientation, proving that the orientations of these many π^* contributions are parallel. The π^* intensity for pentacene is greatest near normal incidence, where the electric field vector of incident light is parallel to the substrate plane. This result confirms that the pentacene molecular orientation is "edge on" upon the substrate; for example, the π^* director is preferentially parallel to the substrate. Orientation can be quantified by fitting the trend of resonance intensity with respect to incident light; a plot of intensity versus the squared sine of incident angle should be linear by Eq. (2.9).

$$I = A \cdot \left\{ \frac{P}{3} \left[1 + \frac{1}{2}(3\cos^2\Theta - 1)(3\cos^2\alpha - 1) \right] + \frac{(1-P)}{2}\sin^2\alpha \right\} \tag{2.9}$$

where A is a constant; P is the beamline polarization factor, $\approx$0.83 for the data shown here; Θ is the incident angle of the beam; and α is the average angle of a vector orbital relative to substrate normal.

For pentacene, as shown in Figure 2.21, the fit results in $\alpha \approx 73^\circ$, where $\alpha = 90^\circ$ would indicate "perfect" edge-on orientation. It is important to note that the π^* director does not necessarily describe the pentacene long-axis orientation. Long-axis orientation, either parallel or perpendicular to the substrate, could be consistent with an "edge-on" conjugated plane orientation. The σ^* trend in Figure 2.21 does show significant orientation, such that the net σ^* orientation appears to be perpendicular to the substrate. Because the sum of carbon-carbon bond vectors in

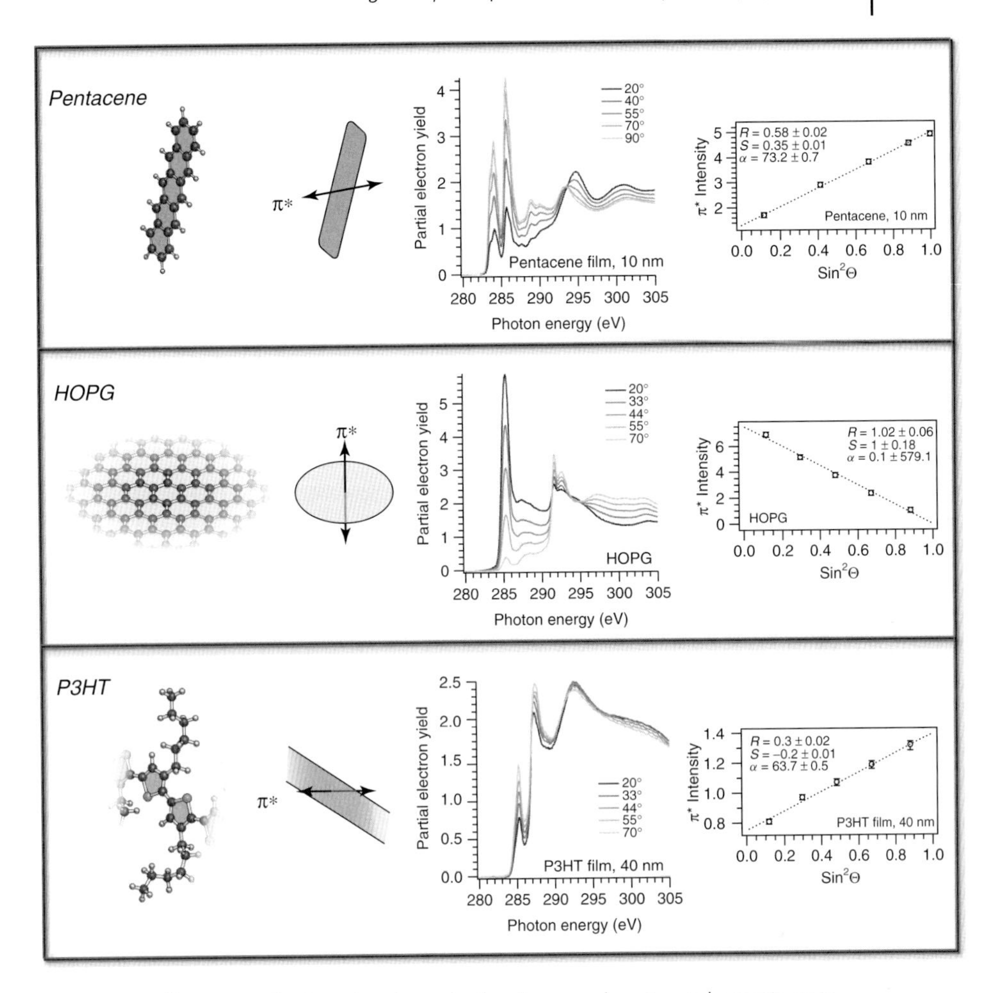

Figure 2.21 Illustration of orientation determination for several conjugated organic materials by NEXAFS. From left to right: an illustration of the molecule, the π^* director with respect to the common conjugated plane, the NEXAFS spectra as they vary with angle of beam incidence, and the π^* intensity trend with the squared sine of incident angle. [14]

pentacene has a greater projection on its long axis than its short axis, this trend indicates that the molecules are roughly "vertical"; for example, the long axis is preferentially oriented toward the substrate normal. If we assume that all the tilt in the π^* director is due to the tilting of the long axis, then the long axis would be tilted by $\approx 17^\circ$. This is a greater tilt than that would be expected if all the pentacene molecules were perfectly (001) oriented in their "thin film" phase [9]; this result could be due to some tilt distribution in the short axis or more generally due to some distribution in crystal orientation on the substrate.

If a film is completely crystalline, and is composed only of a single-crystal type with a common substrate contact plane, it is then appropriate to interpret the orientation determined from NEXAFS spectroscopy as the orientation of a chemical moiety within the unit cell. In all other cases, the orientation must be understood to be dependent on the molecular orientation distribution. The orientation α determined from NEXAFS spectroscopy is an azimuthal average, described as arc $\cos(\langle\cos^2\alpha\rangle 0.5)$, for samples that are azimuthally isotropic (e.g., films lacking in-plane orientation). For this reason, it can be preferred to express orientation from NEXAFS spectroscopy by another quantity that does not express a particular angle but still permits the relative comparison of orientations. For this purpose, we prefer a dichroic ratio R, which is calculated via extrapolation of the linear relationship shown in Eq. (2.10) to the extremes of 0° and 90° incidence; R is thus defined as

$$R = \frac{I(90^\circ) - I(0^\circ)}{I(90^\circ) + I(0^\circ)} = \frac{P(1-3\langle\cos^2\alpha\rangle)}{2(1 - \langle\cos^2\alpha\rangle) - P(1-3\langle\cos^2\alpha\rangle)} \tag{2.10}$$

For pentacene, $R \approx 0.58$, indicating a strong preference for the π^* director to be parallel to the substrate. "Perfect" orientation of the π^* director parallel to the substrate would produce $R \approx 0.7$, which we have never observed in organic semiconductors.

Although we are not aware of a reference material that produces a π^* aligned perfectly parallel to the substrate, a reference material that produces orientation of the opposite extreme is readily available in graphitic carbons. Freshly exfoliated, highly oriented pyrolytic graphic (HOPG), with NEXAFS spectra shown in Figure 2.21, has extremely large π^* intensity at shallow incident angles and nearly no intensity at normal incidence. The π^* director is thus essentially orthogonal to the sample plane. Analysis of the data in Figure 2.19 delivers $\alpha \approx 0^\circ$ and $R \approx -1$. It is interesting to note that HOPG presents what appears to be only a single π^* peak, whereas pentacene appears to exhibit four and P3HT appears to exhibit at least two. This result is consistent with the equivalency of bond lengths in HOPG and the greater diversity of bond lengths expected in pentacene and P3HT.

The orientation of P3HT is significantly more modest than that of pentacene or HOPG. The π^* orientation is $R \approx 0.3$, which indicates a modest edge-on orientation to the conjugated plane. The side chain orientation is also revealed in the orientation of the σ^* at $\approx$293 eV, which shows a modest orientation preference normal to the substrate. It is important to note, however, that the P3HT backbone may also contribute to this resonance, and the backbone is expected to be significantly parallel to the substrate. Therefore, the "true" side chain orientation is somewhat more vertical than what is revealed by the angular dependence of the 293 eV resonance intensity.

We have so far discussed orientation primarily as it is measured from the orientation of the π^* orbital, which has a transition dipole moment that extends as a vector normal to the conjugated plane of a molecule. Orientation can also be determined from the orientation of the C$-$C σ^* resonance, but such analyses are typically only qualitative or comparative because the σ^* resonance intensity can only be determined after subtraction of the edge background, as illustrated in Figure 2.19a. If the height or decay of the absorption edge is incorrectly modeled, then the measured intensity change of the σ^* intensity with respect to incident angle can be in error, and any quantification can also be in error. This challenge is highlighted in early attempts to determine the orientation of all-alkane arachidate monolayers formed by the Langmuir–Blodgett technique [91–93]. These reports produced extremely different interpretations of the monolayer molecular orientation based on different strategies for modeling the absorption edge. In practice, we have found that it is challenging to produce unique simultaneous fits for the absorption edge and C–C σ^* peaks, and therefore, orientation analysis for organic semiconductors from the variation of σ^* intensity with incident angle is best left quantitative.

Although quantification of molecular orientation from the σ^* can be challenging, the qualitative information it provides can prove quite useful. For example, the pentacene spectrum in Figure 2.21 exhibits a significant trend with incident angle for the σ^* resonance at $\approx$294 eV, which indicates that the net σ^* vector orientation is preferentially perpendicular to the substrate. Since the fused-ring structure contains more σ^* bonds parallel to its long axis than to its short axis, and the transition dipole moment of a σ^* resonance is parallel to the σ^* bond, this trend therefore indicates that the molecule's long axis is preferentially perpendicular to the substrate. This molecular orientation is therefore quite consistent with the pentacene 001 orientation that was deduced from the highly edge-on conjugated plane and provides a reassuring self-consistency check on the overall molecular orientation determination. Similarly, the HOPG spectra exhibit σ^* resonances at $\approx$293 and $\approx$295 eV that are preferentially parallel to the substrate, which is certainly consistent with its molecular structure. In P3HT, the small observed σ^* dichroism appears to suggest a modest preference for the alkane side chains to be perpendicular to the substrate, although this orientation is by no means pervasive, consistent with the known conformational disorder in the side chains at room temperature [94].

The determinations of molecular orientation for the model systems described above illustrate the potential for NEXAFS analysis as an essential part of the organic semiconductor measurement toolkit. It has shown a great deal of practical use in comparing the extents of orientation that occur from processing variations. For example, NEXAFS has been used to determine the sensitivity of P3HT molecular orientation to the solidification rate during spin coating. For fast spin rates and fast solidification, P3HT adopts a modest "plane-on" orientation, whereas for slow spin rates and slow solidification, it adopts the "edge-on" orientation described above [95]. NEXAFS was also used to determine the effects of varying solvent volatility on the P3HT molecular orientation [96]. Other studies have been performed that exploit the capabilities of NEXAFS to determine the orientation behaviors of semiconductors including phthalocyanines [97, 98], oligofluorenes [8, 99], oligoacenes [10,100–105], oligophenylenes [106], oligothiophenes [107], and perylene [108].

2.3.2
Horizons for NEXAFS

Although the typical implementation of NEXAFS uses a beam spot that is no smaller than ≈200 μm, greater spatial resolution can be achieved using STXM. This technique relies on contrast in the low-Z K-edge spectra to generate a microscopic image [109]. The general concept of STXM is similar to scanning transmission electron microscopy (STEM); the beam is focused onto a sample using zone plate optics and then rastered across the surface. The beam is transmitted through the sample, and absorption is evaluated by measuring the beam before and after it is transmitted through the sample. Films for STXM must therefore be relatively thin and are typically supported on a copper or graphite grid, such as that used for STEM. A complete spectrum can be collected at every point, or the raster can be collected much faster at a single energy where there is known chemical contrast between the components. STXM images superimpose all the vertical layers in the sample, although it may be possible to determine a true three-dimensional image using tomography. STXM has been used in studies of polymer–polymer BHJ photovoltaic systems to exploit chemical contrast and image the segregation of nanoscale domains, as shown in Figure 2.22 [110]. More recently, STXM has been applied to study fullerene crystals that form in polymer/fullerene solar cells [111]. The chemical contrast between the two components was sufficient to measure the PCBM diffusion coefficient. A very recent study has demonstrated that in-plane orientation can make an excellent contrast mechanism, such that STXM could be used to collect domain images of compositionally homogeneous, polycrystalline

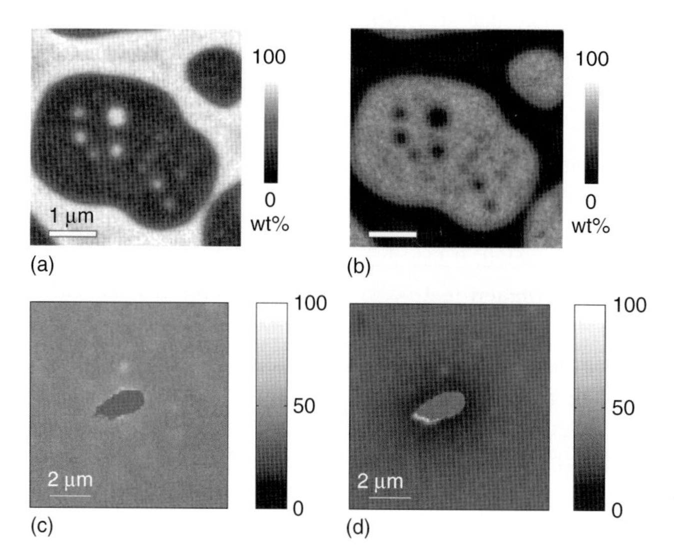

Figure 2.22 STXM images of (a,b) polymer–polymer and (c,d) polymer–fullerene solar cell blends. In (d), note the polymer depletion surrounding the fullerene crystallite (colored gray) [109, 110].

pentacene films, with clear resolution of individual grains from different azimuthal orientations [112].

Other methods in which the chemical contrast of NEXAFS is exploited to gain information about nanoscale morphology are resonant soft X-ray scattering (RSoXS) [113, 114] and resonant soft X-ray reflectivity (RSoXR) [115–117]. Both techniques exploit the significant inflection in X-ray scattering length density that occurs near an NEXAFS absorption. For systems with multiple components featuring different NEXAFS resonances, energy of maximum contrast in the scattering length density can typically be found near a spectral peak unique to one of the materials. The techniques are then practiced at that energy essentially as conventional versions of small-angle X-ray scattering and X-ray reflectivity, albeit with more stringent constraints on sample dimensions and accessible range in reciprocal space. RSoXS can provide detailed information about nanoscale domain characteristics such as shape, size, and positional arrangement, whereas RSoXR is a sensitive probe of the interfacial width between laminar film layers. Because these techniques do not require expensive or time-consuming labeling strategies and can be practiced on materials combinations that typically lack contrast in hard X-rays and neutrons, they may soon prove quite valuable to investigations of phase behavior in multicomponent soft matter systems such as those employed in the active layers of organic photovoltaic devices [114].

References

1. Cullity, B.D. and Stock, S.R. (2001) *Elements of X-Ray Diffraction*, Prentice Hall, New York.
2. Sali, A., Glaeser, R., Earnest, T., and Baumeister, W. (2003) From words to literature in structural proteomics. *Nature*, **422** (6928), 216–225.
3. Kornberg, R.D. (1974) Chromatin structure – repeating unit of histones and DNA. *Science*, **184** (4139), 868–871.
4. Kornberg, R.D. and Lorch, Y.L. (1999) Twenty-five years of the nucleosome, fundamental particle of the eukaryote chromosome. *Cell*, **98** (3), 285–294.
5. Hendrickson, W.A. (1991) Determination of macromolecular structures from anomalous diffraction of synchrotron radiation. *Science*, **254** (5028), 51–58.
6. Anthony, J.E. (2006) Functionalized acenes and heteroacenes for organic electronics. *Chem. Rev.*, **106** (12), 5028–5048.
7. Brocorens, P., Van Vooren, A., Chabinyc, M.L., Toney, M.F., Shkunov, M., Heeney, M., Mcculloch, I., Cornil, J., and Lazzoroni, R. (2009) Solid-state supramolecular organization of polythiophene chains containing thienothiophene units. *Adv. Mater.*, **21** (10–11), 1193–1198.
8. Quan, Y., Mannsfeld, S.C.B., Tang, M.L., Roberts, M., Toney, M.F., Delongchamp, D.M., and Bao, Z. (2008) Microstructure of oligofluorene asymmetric derivatives in organic thin film transistors. *Chem. Mater.*, **20** (8), 2763–2772.
9. Fritz, S.E., Martin, S.M., Frisbie, C.D., Ward, M.D., and Toney, M.F. (2004) Structural characterization of a pentacene monolayer on an amorphous SiO2 substrate with grazing incidence X-ray diffraction. *J. Am. Chem. Soc.*, **126** (13), 4084–4085.
10. Kafer, D. and Witte, G. (2007) Evolution of pentacene films on Ag(111): Growth beyond the first monolayer. *Chem. Phys. Lett.*, **442** (4–6), 376–383.
11. Baker, J.L., Jimison, L.H., Mannsfeld, S., Volkman, S., Yin, S., Subramanian, V., Salleo, A., Alivisatos, A.P., and

Toney, M.F. (2010) Quantification of thin film crystallographic orientation using X-ray diffraction with an area detector. *Langmuir*, **26** (11), 9146–9151.

12. Chabinyc, M.L. (2008) X-ray scattering from films of semiconducting polymers. *Polym. Rev.*, **48** (3), 463–492.
13. Kline, R.J. and Mcgehee, M.D. (2006) Morphology and charge transport in conjugated polymer. *Polym. Rev.*, **46** (1), 27–45.
14. Salleo, A., Kline, R.J., Delongchamp, D.M., and Chabinyc, M.L. (2010) Microstructural characterization and charge transport in thin films of conjugated polymers. *Adv. Mater.*, **22** (34), 3812–3838.
15. Winokur, M.J. and Chunwachirasiri, W. (2003) Nanoscale structure-property relationships in conjugated polymers: implications for present and future device applications. *J. Polym. Sci. B, Polym. Phys. (U.S.A.)*, **41** (21), 2630–2648.
16. Hugger, S., Thomann, R., Heinzel, T., and Thurn-Albrecht, T. (2004) Semicrystalline morphology in thin films of poly(3-hexylthiophene). *Colloid Polym. Sci.*, **282** (8), 932–938.
17. Kayunkid, N., Uttiya, S., and Brinkmann, M. (2010) Structural model of regioregular poly(3-hexylthiophene) obtained by electron diffraction analysis. *Macromolecules*, **43** (11), 4961–4967.
18. Bao, Z.N., Feng, Y., Dodabalapur, A., Raju, V.R., and Lovinger, A.J. (1997) High-performance plastic transistors fabricated by printing techniques. *Chem. Mater.*, **9** (6), 1299.
19. Rivnay, J., Toney, M.F., Zheng, Y., Kauvar, I.V., Chen, Z., Wagner, V., Facchetti, A., and Salleo, A. (2010) Unconventional face-on texture and exceptional in-plane order of a high mobility n-type polymer. *Adv. Mater.* **22** (39), 4359–4363.
20. Delongchamp, D.M., Kline, R.J., Fischer, D.A., Richter, L.J., and Toney, M.F. (2011) Molecular characterization of organic electronic films. *Adv. Mater.*, **23** (3), 319–337.
21. Kline, R.J., Mcgehee, M.D., and Toney, M.F. (2006) Highly oriented crystals at the buried interface in polythiophene thin-film transistors. *Nat. Mater.*, **5** (3), 222–228.
22. Nickel, B., Barabash, R., Ruiz, R., Koch, N., Kahn, A., Feldman, L.C., Haglund, R.F., and Scoles, G. (2004) Dislocation arrangements in pentacene thin films. *Phys. Rev. B*, **70** (12), 125401.
23. Chabinyc, M.L., Toney, M.F., Kline, R.J., Mcculloch, I., and Heeney, M. (2007) X-ray scattering study of thin films of poly(2,5-bis(3-alkylthiophen-2-yl) thieno[3,2-b]thiophene). *J. Am. Chem. Soc.*, **129** (11), 3226–3237.
24. Scherrer, P. (1918) Zsigmondy's Kolloidchemie. *Nachr. chaft.*, **98**, (3), 394.
25. Warren, B. and Averbach, B. (1952) The separation of cold-work distortion and particle size broadening in X-ray patterns. *J. Appl. Phys.*, **23** (4), 497.
26. Crist, B. and Cohen, J. (1979) Fourier-analysis of polymer X-ray diffraction patterns. *J. Polym. Sci. B, Polym. Phys.*, **17** (6), 1001–1010.
27. Prosa, T.J., Moulton, J., Heeger, A.J., and Winokur, M.J. (1999) Diffraction Line-shape analysis of poly(3-dodecylthiophene): A study of layer disorder through the liquid crystalline polymer transition. *Macromolecules*, **32** (12), 4000–4009.
28. Fuoss, P.H. and Brennan, S. (1990) Surface sensitive X-ray scattering. *Annu. Rev. Mater. Sci.*, **20**, 365–390.
29. Vineyard, G.H. (1982) Grazing-incidence diffraction and the distorted-wave approximation for the study of surfaces. *Phys. Rev. B*, **26** (8), 4146–4159.
30. Sinha, S.K., Sirota, E.B., Garoff, S., and Stanley, H.B. (1988) X-ray and neutron-scattering from rough surfaces. *Phys. Rev. B*, **38** (4), 2297–2311.
31. Breiby, D.W., Bunk, O., Andreasen, J.W., Lemke, H.T., and Nielsen, M.M. (2008) Simulating X-ray diffraction of textured films. *J. Appl. Crystallogr.*, **41**, 262–271.

32. Gundlach, D.J., Royer, J.E., Park, S.K., Subramanian, S., Jurchescu, O.D., Hamadani, B.H., Moad, A.J., Kline, R.J., Teague, L.C., Kirillov, O., Richter, C.A., Kushmerick, J.G., Richter, L.J., Parkin, S.R., Jackson, T.N., and Anthony, J.E. (2008) Contact-induced crystallinity for high-performance soluble acene-based transistors and circuits. *Nat. Mater.*, **7** (3), 216–221.
33. Factor, B.J., Russell, T.P., and Toney, M.F. (1993) Grazing incidence X-ray scattering studies of thin films of an aromatic polyimide. *Macromolecules*, **26** (11), 2847–2859.
34. Lee, B., Lo, C.T., Thiyagarajan, P., Lee, D.R., Niu, Z., and Wang, Q. (2008) Structural characterization using the multiple scattering effects in grazing-incidence small-angle X-ray scattering. *J. Appl. Crystallogr.*, **41** (1), 134–142.
35. Bedzyk, M.J., Bommarito, G.M., and Schildkraut, J.S. (1989) X-ray standing waves at a reflecting mirror surface. *Phys. Rev. Lett.*, **62** (12), 1376.
36. Wang, J., Bedzyk, M.J., Penner, T.L., and Caffrey, M. (1991) Structural studies of membranes and surface layers up to 1,000 A thick using X-ray standing waves. *Nature*, **354** (6352), 377–380.
37. Aasmundtveit, K.E., Samuelsen, E.J., Guldstein, M., Steinsland, C., Flornes, O., Fagermo, C., Seeberg, T.M., Pettersson, L.A.A., Inganas, O., Feidenhans'l, R., and Ferrer, S. (2000) Structural anisotropy of poly(alkylthiophene) films. *Macromolecules*, **33** (8), 3120–3127.
38. Aasmundtveit, K.E., Samuelsen, E.J., Mardalen, J., Bakken, E., Carlsen, P.H.J., and Lienert, U. (1997) Orientation effect in thin layers of poly(octylthiophene) on glass studied by synchrotron X-ray diffraction. *Synth. Met.*, **89** (3), 203–208.
39. Sirringhaus, H., Brown, P.J., Friend, R.H., Nielsen, M.M., Bechgaard, K., Langeveld-Voss, B.M.W., Spiering, A.J.H., Janssen, R.A.J., Meijer, E.W., Herwig, P., and De Leeuw, D.M. (1999) Two-dimensional charge transport in self-organized, high-mobility conjugated polymers. *Nature*, **401** (6754), 685–688.
40. Chang, J.F., Sun, B.Q., Breiby, D.W., Nielsen, M.M., Solling, T.I., Giles, M., Mcculloch, I., and Sirringhaus, H. (2004) Enhanced mobility of poly(3-hexylthiophene) transistors by spin-coating from high-boiling-point solvents. *Chem. Mater.*, **16** (23), 4772–4776.
41. Donley, C.L., Zaumseil, J., Andreasen, J.W., Nielsen, M.M., Sirringhaus, H., Friend, R.H., and Kim, J.S. (2005) Effects of packing structure on the optoelectronic and charge transport properties in poly(9,9-di-n-octylfluorene-alt-benzothiadiazole). *J. Am. Chem. Soc.*, **127** (37), 12890.
42. Tsao, H.N., Cho, D., Andreasen, J.W., Rouhanipour, A., Breiby, D.W., Pisula, W., and Müllen, K. (2008) The influence of morphology on high-performance polymer field-effect transistors. *Adv. Mater.*, **21**, 209–212.
43. Breiby, D.W., Samuelsen, E.J., Konovalov, O., and Struth, B. (2004) Ultrathin films of semiconducting polymers on water. *Langmuir*, **20** (10), 4116–4123.
44. Knaapila, M., Lyons, B.P., Hase, T.P.A., Pearson, C., Petty, M.C., Bouchenoire, L., Thompson, P., Serimaa, R., Torkkeli, M., and Monkman, A.P. (2005) Influence of molecular weight on the surface morphology of aligned, branched side-chain polyfluorene. *Adv. Funct. Mater.*, **15** (9), 1517–1522.
45. Knaapila, M., Stepanyan, R., Lyons, B.P., Torkkeli, M., and Monkman, A.P. (2006) Towards general guidelines for aligned, nanoscale assemblies of hairy-rod polyfluorene. *Adv. Funct. Mater.*, **16**, 599–609.
46. Knaapila, M., Hase, T.P.A., Torkkeli, M., Stepanyan, R., Bouchenoire, L., Cheun, H.-S., Winokur, M.J., and Monkman, A.P. (2007) Meridional orientation in biaxially aligned thin films of hairy-rod polyfluorene. *Cryst. Growth Des.*, **7** (9), 1706–1711.
47. Cheun, H., Liu, X., Himpsel, F.J., Knaapila, M., Scherf, U., Torkkeli, M.,

and Winoku, M.J. (2008) Polarized optical absorption spectroscopy, NEXAFS, and GIXRD measurements of chain alignment in polyfluorene thin films. *Macromolecules*, **41** (17), 6463–6472.

48. Yang, H., Shin, T.J., Bao, Z., and Ryu, C.Y. (2007) Structural transitions of nanocrystalline domains in regioregular poly(3-Hexyl thiophene) thin films. *J. Polym. Sci., Part B: Polym. Phys.*, **45**, 1303–1312.

49. Yang, H., Shin, T.J., Yang, L., Cho, K., Ryu, C.Y., and Bao, Z. (2005) Effect of mesoscale crystalline structure on the field-effect mobility of regioregular poly(3-hexyl thiophene) in thin-film transistors. *Adv. Funct. Mater.*, **15** (4), 671–676.

50. Osaka, I., Zhang, R., Sauve, G., Smilgies, D.-M., Kowalewski, T., and Mccullough, R.D. (2009) High-lamellar ordering and amorphous-like pi-network in short-chain thiazolothiazole-thiophene copolymers lead to high mobilities. *J. Am. Chem. Soc.*, **131** (7), 2521–2529.

51. Osaka, I. and Mccullough, R.D. (2008) Advances in molecular design and synthesis of regioregular polythiophenes. *Acc. Chem. Res.*, **41** (9), 1202–1214.

52. Liu, J., Zhang, R., Sauve, G., Kowalewski, T., and Mccullough, R.D. (2008) Highly disordered polymer field effect transistors: N-Alkyl Dithieno[3,2-b:2',3'-d]pyrrole-based copolymers with surprisingly high charge carrier mobilities. *J. Am. Chem. Soc.*, **130** (39), 13167–13176.

53. Zhang, R., Li, B., Iovu, M.C., Jeffries-El, M., Sauve, G., Cooper, J., Jia, S.J., Tristram-Nagle, S., Smilgies, D.M., Lambeth, D.N., Mccullough, R.D., and Kowalewski, T. (2006) Nanostructure dependence of field-effect mobility in regioregular poly(3-hexylthiophene) thin film field effect transistors. *J. Am. Chem. Soc.*, **128** (11), 3480–3481.

54. Mayer, A.C., Ruiz, R., Zhou, H., Headrick, R.L., Kazimirov, A., and Malliaras, G.G. (2006) Growth dynamics of pentacene thin films: real-time synchrotron x-ray scattering study *Phys. Rev. B*, **73** (20), 205307-1–205307-5.

55. Mayer, A.C., Toney, M.F., Scully, S.R., Rivnay, J., Brabec, C.J., Scharber, M., Koppe, M., Heeney, M., Mcculloch, I., and Mcgehee, M.D. (2009) Bimolecular crystals of fullerenes in conjugated polymers and the implications of molecular mixing for solar cells. *Adv. Funct. Mater.*, **19** (8), 1173–1179.

56. Mcculloch, I., Heeney, M., Chabinyc, M.L., Delongchamp, D., Kline, R.J., Coelle, M., Duffy, W., Fischer, D., Gundlach, D., Hamadani, B., Hamilton, R., Richter, L., Salleo, A., Shkunov, M., Sporrowe, D., Tierney, S., and Zhong, W. (2009) Semiconducting thienothiophene copolymers: design, synthesis, morphology, and performance in thin-film organic transistors. *Adv. Mater.*, **21** (10–11), 1091–1109.

57. Dhagat, P., Haverinen, H.M., Kline, R.J., Jung, Y., Fischer, D.A., Delongchamp, D.M., and Jabbour, G.E. (2009) Influence of dielectric surface chemistry on the microstructure and carrier mobility of an n-type organic semiconductor. *Adv. Funct. Mater.*, **19** (15), 2365–2372.

58. Lucas, L.A., Delongchamp, D.M., Richter, L.J., Kline, R.J., Fischer, D.A., Kaafarani, B.R., and Jabbour, G.E. (2008) Thin film microstructure of a solution processable pyrene-based organic semiconductor. *Chem. Mater.*, **20** (18), 5743–5749.

59. Kline, R.J., Mcgehee, M.D., Kadnikova, E.N., Liu, J.S., Frechet, J.M.J., and Toney, M.F. (2005) Dependence of regioregular poly(3-hexylthiophene) film morphology and field-effect mobility on molecular weight. *Macromolecules*, **38** (8), 3312–3319.

60. Joshi, S., Pingel, P., Grigorian, S., Panzner, T., Pietsch, U., Neher, D., Forster, M., and Scherf, U. (2009) Bimodal temperature behavior of structure and mobility in high molecular weight P3HT thin films. *Macromolecules*, **42** (13), 4651–4660.

61. Joshi, S., Grigorian, S., Pietsch, U., Pingel, P., Zen, A., Neher, D., and Scherf, U. (2008) Thickness dependence of the crystalline structure and hole mobility in thin films of low molecular weight poly(3-hexylthiophene). *Macromolecules*, **41**, 6800–6808.

62. Zen, A., Saphiannikova, M., Neher, D., Grenzer, J., Grigorian, S., Pietsch, U., Asawapirom, U., Janietz, S., Scherf, U., Lieberwirth, I., and Wegner, G. (2006) Effect of molecular weight on the structure and crystallinity of poly(3-hexylthiophene). *Macromolecules*, **39** (6), 2162–2171.

63. Rivnay, J., Jimison, L.H., Northrup, J.E., Toney, M.F., Noriega, R., Lu, S.F., Marks, T.J., Facchetti, A., and Salleo, A. (2009) Large modulation of carrier transport by grain-boundary molecular packing and microstructure in organic thin films. *Nat. Mater.*, **8** (12), 952–958.

64. Jimison, L.H., Toney, M.F., Mcculloch, I., Heeney, M., and Salleo, A. (2009) Charge-transport anisotropy due to grain boundaries in directionally crystallized thin films of regioregular poly(3-hexylthiophene). *Adv. Mater.*, **21**, 1568–1572.

65. Jimison, L.H., Salleo, A., Chabinyc, M.L., Bernstein, D.P., and Toney, M.F. (2008) Correlating the microstructure of thin films of poly[5,5-bis(3-dodecyl-2-thienyl)-2,2-bithiophene] with charge transport: Effect of dielectric surface energy and thermal annealing. *Phys. Rev. B.*, **78**, 125319.

66. Germack, D.S., Chan, C.K., Kline, R.J., Fischer, D.A., Gundlach, D.J., Toney, M.F., Richter, L.J., and Delongchamp, D.M. (2010) Interfacial segregation in polymer/fullerene blend films for photovoltaic devices. *Macromolecules*, **43** (8), 3828–3836.

67. Stöhr, J. and Jaeger, R. (1982) Absorption-edge resonances, core-hole screening, and orientation of chemisorbed molecules - Co, No, and N-2 on Ni(100). *Phys. Rev. B*, **26** (8), 4111–4131.

68. Stöhr, J., Gland, J.L., Kollin, E.B., Koestner, R.J., Johnson, A.L., Muetterties, E.L., and Sette, F. (1984) Desulfurization and structural transformation of thiophene on the Pt(111) surface. *Phys. Rev. Lett.*, **53** (22), 2161–2164.

69. Stöhr, J. (1992) *NEXAFS Spectroscopy*, vol. 392, Springer-Verlag, Berlin.

70. Delongchamp, D.M., Lin, E.K., and Fischer, D.A. (2007) in *Organic Field-Effect Transistors* (eds Z. Bao and J. Locklin), CRC Press, New York, pp. 277–300.

71. Delongchamp, D.M., Lin, E.K., and Fischer, D.A. (2005) Organic semiconductor structure and chemistry from near-edge X-ray absorption fine structure (NEXAFS) spectroscopy. *Proc. SPIE*, **5940**, 54–64.

72. Zharnikov, M., Frey, S., Heister, K., and Grunze, M. (2002) An extension of the mean free path approach to X-ray absorption spectroscopy. *J. Electron Spectrosc. Relat. Phenom.*, **124** (1), 15–24.

73. Genzer, J., Kramer, E.J., and Fischer, D.A. (2002) Accounting for Auger yield energy loss for improved determination of molecular orientation using soft x-ray absorption spectroscopy. *J. Appl. Phys.*, **92** (12), 7070–7079.

74. Lenhart, J.L., Jones, R.L., Lin, E.K., Soles, C.L., Wu, W.L., Fischer, D.A., Sambasivan, S., Goldfarb, D.L., and Angelopoulos, M. (2002) Probing surface and bulk chemistry in resist films using near edge X-ray absorption fine structure. *J. Vac. Sci. Technol. B*, **20** (6), 2920–2926.

75. Prabhu, V.M., Sambasivan, S., Fischer, D., Sundberg, L.K., and Allen, R.D. (2006) Quantitative depth profiling of photoacid generators in photoresist materials by near-edge X-ray absorption fine structure spectroscopy. *Appl. Surf. Sci.*, **253** (2), 1010–1014.

76. Fischer, D.A., Colbert, J., and Gland, J.L. (1989) Ultrasoft (C,N,O) X-ray-fluorescence detection – proportional-counters, focusing multilayer mirrors, and scattered-light systematics. *Rev. Sci. Instrum.*, **60** (7), 1596–1602.

77. Wu, W.L., Sambasivan, S., Wang, C.Y., Wallace, W.E., Genzer, J., and Fischer, D.A. (2003) A direct comparison of surface and bulk chain-relaxation in polystyrene. *Eur. Phys. J. E*, **12** (1), 127–132.

78. Wallace, W.E., Fischer, D.A., Efimenko, K., Wu, W.L., and Genzer, J. (2001) Polymer chain relaxation: surface outpaces bulk. *Macromolecules*, **34** (15), 5081–5082.
79. Scholl, A., Zou, Y., Schmidt, T., Fink, R., and Umbach, E. (2003) Energy calibration and intensity normalization in high-resolution NEXAFS spectroscopy. *J. Electron Spectrosc. Relat. Phenom.*, **129** (1), 1–8.
80. Zimmermann, U., Schnitzler, G., Wustenhagen, V., Karl, N., Dudde, R., Koch, E.E., and Umbach, E. (2000) NEXAFS and ARUP spectroscopy of an organic single crystal: alpha-perylene. *Mol. Cryst. Liquid Cryst.*, **339**, 231–259.
81. Delongchamp, D.M., Sambasivan, S., Fischer, D.A., Lin, E.K., Chang, P., Murphy, A.R., Frechet, J.M.J., and Subramanian, V. (2005) Direct correlation of organic semiconductor film structure to field-effect mobility. *Adv. Mater.*, **17** (19), 2340–2344.
82. Murphy, A.R., Chang, P.C., Vandyke, P., Liu, J.S., Frechet, J.M.J., Subramanian, V., Delongchamp, D.M., Sambasivan, S., Fischer, D.A., and Lin, E.K. (2005) Self-assembly, molecular ordering, and charge mobility in solution-processed ultrathin oligothiophene films. *Chem. Mater.*, **17** (24), 6033–6041.
83. Delongchamp, D.M., Jung, Y.S., Fischer, D.A., Lin, E.K., Chang, P., Subramanian, V., Murphy, A.R., and Frechet, J.M.J. (2006) Correlating molecular design to microstructure in thermally convertible oligothiophenes: the effect of branched versus linear end groups. *J. Phys. Chem. B*, **110** (22), 10645–10650.
84. Mauldin, C.E., Puntambekar, K., Murphy, A.R., Liao, F., Subramanian, V., Frechet, J.M.J., Delongchamp, D.M., Fischer, D.A., and Toney, M.F. (2009) Solution-processable alpha, omega-distyryl oligothiophene semiconductors with enhanced environmental stability. *Chem. Mater.*, **21** (9), 1927–1938.
85. Epps, T.H., Delongchamp, D.M., Fasolka, M.J., Fischer, D.A., and Jablonski, E.L. (2007) Substrate surface energy dependent morphology and dewetting in an ABC triblock copolymer film. *Langmuir*, **23** (6), 3355–3362.
86. Germack, D.S., Chan, C.K., Hamadani, B.H., Richter, L.J., Fischer, D.A., Gundlach, D.J., and Delongchamp, D.M. (2009) Substrate-dependent interface composition and charge transport in films for organic photovoltaics. *Appl. Phys. Lett.*, **94**, 233303.
87. Chabinyc, M.L., Salleo, A., Wu, Y.L., Liu, P., Ong, B.S., Heeney, M., and Mcculloch, L. (2004) Lamination method for the study of interfaces in polymeric thin film transistors. *J. Am. Chem. Soc.*, **126** (43), 13928–13929.
88. Delongchamp, D.M., Kline, R.J., Lin, E.K., Fischer, D.A., Richter, L.J., Lucas, L.A., Heeney, M., Mcculloch, I., and Northrup, J.E. (2007) High carrier mobility polythiophene thin films: structure determination by experiment and theory. *Adv. Mater.*, **19** (6), 833–837.
89. Horowitz, G. (2004) Organic thin film transistors: from theory to real devices. *J. Mater. Res.*, **19**, 1946–1962.
90. Ong, B.S., Wu, Y.L., Liu, P., and Gardner, S. (2004) High-performance semiconducting polythiophenes for organic thin-film transistors. *J. Am. Chem. Soc.*, **126** (11), 3378–3379.
91. Outka, D.A., Stöhr, J., Rabe, J.P., and Swalen, J.D. (1988) The orientation of Langmuir-Blodgett monolayers using NEXAFS. *J. Chem. Phys.*, **88** (6), 4076–4087.
92. Outka, D.A., Stohr, J., Rabe, J.P., Swalen, J.D., and Rotermund, H.H. (1987) Orientation of arachidate chains in Langmuir-Blodgett monolayers on Si(111). *Phys. Rev. Lett.*, **59** (12), 1321–1324.
93. Hahner, G., Kinzler, M., Woll, C., Grunze, M., Scheller, M.K., and Cederbaum, L.S. (1991) Near edge X-ray-absorption fine-structure determination of alkyl-chain orientation – breakdown of the building-block

scheme. *Phys. Rev. Lett.*, **67** (7), 851–854.

94. Gurau, M.C., Delongchamp, D.M., Vogel, B.M., Lin, E.K., Fischer, D.A., Sambasivan, S., and Richter, L.J. (2007) Measuring molecular order in poly(3 alkylthiophene) thin films with polarizing spectroscopies. *Langmuir*, **23** (2), 834–842.
95. Delongchamp, D.M., Vogel, B.M., Jung, Y., Gurau, M.C., Richter, C.A., Kirillov, O.A., Obrzut, J., Fischer, D.A., Sambasivan, S., Richter, L.J., and Lin, E.K. (2005) Variations in semiconducting polymer microstructure and hole mobility with spin-coating speed. *Chem. Mater.*, **17** (23), 5610–5612.
96. Ho, P.K.H., Chua, L.L., Dipankar, M., Gao, X.Y., Qi, D.C., Wee, A.T.S., Chang, J.F., and Friend, R.H. (2007) Solvent effects on chain orientation and interchain pi-interaction in conjugated polymer thin films: direct measurements of the air and substrate interfaces by near-edge X-ray absorption spectroscopy. *Adv. Mater.*, **19** (2), 215–221.
97. Huang, Y.L., Chen, W., Chen, S., and Wee, A.T.S. (2009) Low-temperature scanning tunneling microscopy and near-edge X-ray absorption fine structure investigation of epitaxial growth of F16CuPc thin films on graphite. *Appl. Phys. A: Mater. Sci. Process.*, **95** (1), 107–111.
98. Kera, S., Casu, M.B., Bauchspiess, K.R., Batchelor, D., Schmidt, T., and Umbach, E. (2006) Growth mode and molecular orientation of phthalocyanine molecules on metal single crystal substrates: a NEXAFS and XPS study. *Surf. Sci.*, **600** (5), 1077–1084.
99. Yuan, Q., Mannsfeld, S.C.B., Tang, M.L., Toney, M.F., Luening, J., and Bao, Z.A. (2008) Thin film structure of tetraceno[2,3-b]thiophene characterized by grazing incidence X-ray scattering and near-edge X-ray absorption fine structure analysis. *J. Am. Chem. Soc.*, **130** (11), 3502–3508.
100. Zheng, F., Park, B.N., Seo, S., Evans, P.G., and Himpsel, F.J. (2007) Orientation of pentacene molecules on SiO2: From a monolayer to the bulk. *J. Chem. Phys.*, **126** (15), 6.
101. Hu, W.S., Tao, Y.T., Hsu, Y.J., Wei, D.H., and Wu, Y.S. (2005) Molecular orientation of evaporated pentacene films on gold: alignment effect of self-assembled monolayer. *Langmuir*, **21** (6), 2260–2266.
102. Yoshikawa, G., Miyadera, T., Onoki, R., Ueno, K., Nakai, I., Entani, S., Ikeda, S., Guo, D., Kiguchi, M., Kondoh, H., Ohta, T., and Saiki, K. (2006) In-situ measurement of molecular orientation of the pentacene ultrathin films grown on SiO2 substrates. *Surf. Sci.*, **600** (12), 2518–2522.
103. Pedio, M., Doyle, B., Mahne, N., Giglia, A., Borgatti, F., Nannarone, S., Henze, S.K.M., Temirov, R., Tautz, F.S., Casalis, L., Hudej, R., Danisman, M.F., and Nickel, B. (2007) Growth of pentacene on Ag(111) surface: a NEXAFS study. *Appl. Surf. Sci.*, **254** (1), 103–107.
104. Chiodi, M., Gavioli, L., Beccari, M., Di Castro, V., Cossaro, A., Floreano, L., Morgante, A., Kanjilal, A., Mariani, C., and Betti, M.G. (2008) Interaction strength and molecular orientation of a single layer of pentacene in organic-metal interface and organic-organic heterostructure. *Phys. Rev. B*, **77** (11), 7.
105. Tersigni, A., Shi, J., Jiang, D.T., and Qin, X.R. (2006) Structure of tetracene films on hydrogen-passivated Si(001) studied via STM, AFM, and NEXAFS. *Phys. Rev. B*, **74** (20), 9.
106. Hu, W.S., Lin, Y.F., Tao, Y.T., Hsu, Y.J., and Wei, D.H. (2005) Highly oriented growth of p-sexiphenyl molecular nanocrystals on rubbed polymethylene surface. *Macromolecules*, **38** (23), 9617–9624.
107. Oehzelt, M., Berkebile, S., Koller, G., Ivanco, J., Surnev, S., and Ramsey, M.G. (2009) Alpha-sexithiophene on Cu(110) and Cu(110)-(2 x 1)O: an STM and NEXAFS study. *Surf. Sci.*, **603** (2), 412–418.
108. Casu, M.B., Scholl, A., Bauchspiess, K.R., Hubner, D., Schmidt, T., Heske, C., and Umbach, E. (2009) Nucleation in organic thin film growth: perylene

on Al2O3/Ni3Al(111). *J. Phys. Chem. C*, **113** (25), 10990–10996.

109. Ade, H. and Stoll, H. (2009) Near-edge X-ray absorption fine-structure microscopy of organic and magnetic materials. *Nat. Mater.*, **8**, 282.

110. Mcneill, C.R., Watts, B., Thomsen, L., Ade, H., Greenham, N.C., and Dastoor, P.C. (2007) X-ray microscopy of photovoltaic polyfluorene blends: relating nanomorphology to device performance. *Macromolecules*, **40** (9), 3263–3270.

111. Watts, B., Belcher, W.J., Thomsen, L., Ade, H., and Dastoor, P.C. (2009) A quantitative study of PCBM diffusion during annealing of P3HT:PCBM blend films. *Macromolecules.*, **42** (21), 8392–8397.

112. Brauer, B., Virkar, A., Mannsfeld, S.C.B., Bernstein, D.P., Kukreja, R., Chou, K.W., Tyliszczak, T., Bao, Z.A., and Acremann, Y. (2010) X-ray microscopy imaging of the grain orientation in a pentacene field-effect transistor *Chem. Mater.*, **22** (12), 3693–3697.

113. Virgili, J.M., Tao, Y.F., Kortright, J.B., Balsara, N.P., and Segalman, R.A. (2007) Analysis of order formation in block copolymer thin films using resonant soft X-ray scattering. *Macromolecules*, **40** (6), 2092–2099.

114. Swaraj, S., Wang, C., Araki, T., Mitchell, G., Liu, L., Gaynor, S., Deshmukh, B., Yan, H., Mcneill, C.R., and Ade, H. (2009) The utility of resonant soft x-ray scattering and reflectivity for the nanoscale characterization of polymers. *Eur. Phys. J.: Spec. Top.*, **167**, 121–126.

115. Wang, C., Araki, T., Watts, B., Harton, S., Koga, T., Basu, S., and Ade, H. (2007) Resonant soft x-ray reflectivity of organic thin films. *J. Vac. Sci. Technol. A*, **25** (3), 575–586.

116. Ade, H., Wang, C., Garcia, A., Yan, H., Sohn, K.E., Hexemer, A., Bazan, G.C., Nguyen, T.Q., and Kramer, E.J. (2009) Characterization of multicomponent polymer trilayers with resonant soft x-ray reflectivity. *J. Polym. Sci. B, Polym. Phys.*, **47** (13), 1291–1299.

117. Wang, C., Garcia, A., Yan, H.P., Sohn, K.E., Hexemer, A., Nguyen, T.Q., Bazan, G.C., Kramer, E.J., and Ade, H. (2009) Interfacial widths of conjugated polymer bilayers. *J. Am. Chem. Soc.*, **131** (35), 12538.

3
Charge Transport Theories in Organic Semiconductors

Rodrigo Noriega and Alberto Salleo

3.1
Introduction

Development of organic semiconductors has progressed steadily in the past 10 years, assisted by great commercial interest. As we write this chapter, smartphones using organic light-emitting diode (OLED) displays are available in the mass market, with several display manufacturers ramping up their production to meet the demands of larger market segments. Organic transistors and solar cells are widely believed to appear next in consumer products while more futuristic applications such as chemical sensors for point-of-care diagnostics or robotic skin are being envisioned. These developments have been made possible by the constant increase in materials quality. For instance, organic semiconductors with mobilities in excess of 5 $cm^2 V^{-1} s^{-1}$ for crystalline films and 1 $cm^2 V^{-1} s^{-1}$ in polymer films have been reported [1–4]. The promise of organic electronics, that is, to use the power of organic synthesis to rationally design and make materials that exhibit desired properties is, however, only partially fulfilled. For instance, the design of high-mobility semiconductors is still largely empirical and requires a great amount of intuition. This difficulty arises in part due to the lack of a coherent picture of charge transport mechanisms in organic semiconductors. While the variety of microstructures (from amorphous to semi-, poly-, and single-crystalline microstructures) that can be encountered in this materials family is partially responsible, there remains a lack of understanding of the fundamental transport mechanisms. For instance, even in the simplest case of an ultrapure single crystal, the role of charge localization is still hotly debated.

In this chapter, we review the transport theories that have been used to explain the electrical properties of organic semiconductors. Models often simplify reality by reducing it to extremes: at the opposite ends of the spectrum, one would find a perfect single crystal and a perfectly amorphous material. Real thin films, however, are found somewhere in between these extremes. The challenge of understanding charge transport in organic semiconductors lies in the extension of simplified models to intermediate cases.

Organic Electronics II: More Materials and Applications, First Edition. Edited by Hagen Klauk.

3.2
Well-Ordered Systems: Organic Single Crystals

3.2.1
General Conditions for Band Transport

To form an ideal crystal of an organic molecule, one can perform the thought experiment of bringing the molecules from an infinite distance in the gas phase to their final positions in the equilibrium crystal structure. The electronic states of the molecules are well-defined, and it is understood that overlap of molecular orbitals in the condensed phase can give rise to electronic bands. The band structure of organic solids is best understood by starting from its constituents: single molecules. Conjugated molecules are of particular interest since they have arrays of carbon p orbitals occupied by one electron each. Intramolecular overlap of these initially degenerate orbitals gives rise to splitting and delocalization. As a result, a single conjugated molecule already possesses a recognizable "germ" of a bandgap, with the highest occupied molecular orbital (HOMO) playing the role of a valence-band edge and the lowest-unoccupied molecular orbital (LUMO) playing the role of the bottom of the conduction band [5]. As the original p orbitals were occupied by a single electron, the HOMO is filled and the LUMO is empty. Hence, in the previous analogy, a conjugated molecule has the electronic structure of an intrinsic semiconductor. When isolated conjugated molecules are brought together from the gas phase into a crystal, the molecular levels can overlap substantially and, in fact, give rise to electronic bands with energy dispersions of the order of a few hundreds of millielectronvolts [6, 7]. There is ample experimental evidence for the existence of electronic bands in organic solids, further complemented by theoretical calculations of the band structure of some of the most interesting crystalline organic semiconductors, such as rubrene or pentacene [8–10]. Organic solids are held together by weak van der Waals interactions; therefore the formation of the crystal can be viewed as a perturbation of the original molecular electronic levels. For instance, the polarizability of the organic crystal shifts the HOMO and LUMO levels of the isolated molecule by amounts of the order of ~1 eV in such a way as to decrease the bandgap on the formation of the crystal [11, 12]. Because of the electronic structure of the constituent molecules, such an ideal crystal is an intrinsic semiconductor.

The existence of electronic bands is not by itself an indication that charge transport occurs via coherent Bloch waves. In particular, in organic molecules, the molecular relaxation that occurs on introducing charge is large. As a result, the band structure calculated for a crystal made of neutral molecules may not be preserved when charge carriers are present. Since ideal organic crystals are intrinsic, relatively wide bandgap semiconductors, this effect is of great relevance to most electronic devices, which rely on transport of charge. Intuitively, if the transfer integral between molecules is large, leading to a large bandwidth, band transport can still occur in spite of molecular relaxation. This effect can be understood in terms of charge delocalization, which effectively spreads the spatial extent of the charge wavefunction, and hence the resulting molecular relaxation, over several molecules. A semiquantitative criterion that has been used to determine whether the bandwidth W is sufficient to give rise to band transport is [7]:

$$W > \frac{\hbar}{\tau_{\mathrm{vib}}} \tag{3.1}$$

where τ_{vib} is the characteristic vibration time and $\hbar$ is the usual reduced Planck's constant. This criterion ensures that a charge leaves a molecule before geometrical relaxation and self-trapping can take place. For typical molecules, bandwidths W of 0.1–0.2 eV or more are necessary to ensure that the band transport criterion is satisfied.

If charge in a crystal can move, it will be able to drift in an externally applied electric field. The linear response of the charge consists of acquiring a constant velocity caused by the successive acceleration in the field and scattering in the crystal. The ratio of the velocity to the electric field is the mobility μ. The carrier mobility is regarded as a fundamental materials property and is conventionally used as a figure of merit for semiconductors. At high fields, the response may not be linear anymore, which is usually expressed as a field-dependent mobility $\mu(F)$.

As the transfer integral is directly related to the bandwidth (the larger the former the larger is the latter), it is immediately evident that the transfer integral and hence the crystal packing play a central role in coherent transport theories. Uncorrelated thermal vibrations perturb the overlap between orbitals and therefore affect the electronic band structure. As the temperature is lowered, the amplitude of these vibrations is decreased, and as a consequence, the mobility increases. Hence, the observation of increasing mobility as a function of decreasing temperature is often taken as a hallmark of band transport. For example, in a semiconductor in which band transport is limited by scattering from acoustical phonons, the mobility is proportional to $T^{-3/2}$ [13]. Furthermore, in a crystal, the overlap integrals and therefore the bandwidth depend on the crystallographic direction. Organic molecules have extremely anisotropic polarizabilities, and therefore one may expect large variations in mobility with crystallographic direction. The resulting dependence of mobility on crystallographic direction is also often taken as an indicator of band transport.

3.2.2
Experimental Evidence for Band Transport in Organic Crystals

The most common test to determine whether band transport is observed in an organic semiconductor consists of measuring the temperature dependence of the carrier mobility. Because organic solids are held together by weak van der Waals bonds, thermal vibrations at room temperature can acquire amplitudes that prohibit coherent propagation of charges and lead to localization into polarons. Furthermore, impurities and structural defects may constitute electronic traps that prevent charge from propagating freely [9]. In view of these considerations, it is not surprising that the experimental observation of band transport is challenging. Indeed, a necessary condition is the growth of high-quality crystals after repeated purifications steps. Highest quality crystals are usually grown by sublimation and recrystallization methods. Obtaining high-purity crystals is challenging as some of the organic impurities may result from thermally induced dimerization or other simple reactions that involve the starting molecule, making them structurally very similar to the material under study and difficult to extract from the host crystal. In order to freeze out the effects of vibrations and impurities, measurements aimed at verifying the occurrence of band transport are often performed at low temperature. The first evidence of band transport in organic crystals was obtained by studying the mobility measured using the time-of-flight (ToF) technique [9]. Since the density of photogenerated carriers must be maintained low in ToF measurements, even a relatively low density of traps can negatively affect the results of the experiment. Hence, in this case, the availability of high-purity crystals is of paramount importance. Nevertheless, in clean crystals and at temperatures low enough to freeze out the effects of vibrations, a negative temperature dependence of the mobility was observed by Probst and Karl in anthracene (Figure 3.1) [14].

Similar measurements made on naphthalene single crystals illustrate very clearly the typical hallmarks of band transport (Figure 3.2). A clear $T^{-3/2}$ dependence of the mobility in all crystallographic directions is only observed below 100 K. Furthermore, whereas the three principal directions of the mobility tensor are clearly distinguished at low temperature, the mobilities in two of those directions are identical at room temperature.

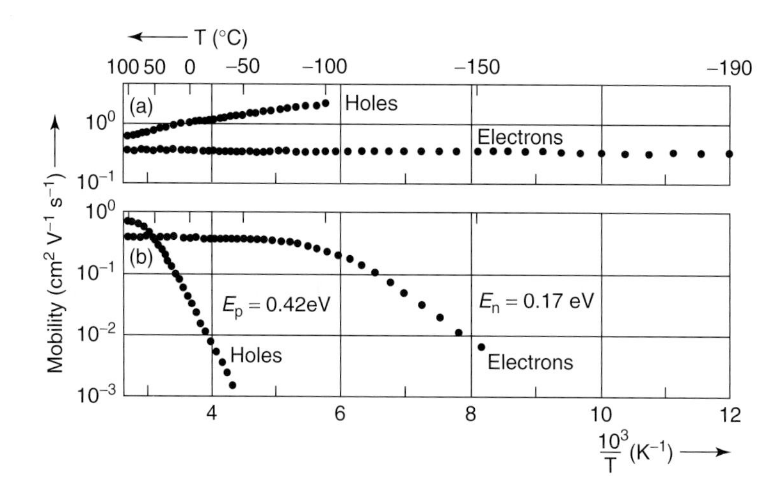

Figure 3.1 Temperature dependence of carrier mobility in pure anthracene (a) and tetracene-doped anthracene (b). The negative temperature dependence is observed only in the pure material. Tetracene, being a larger polyacene than anthracene, constitutes a charge trap in the anthracene host. (Reprinted with permission from Ref. [14]. Copyright (1975) by John Wiley and Sons.)

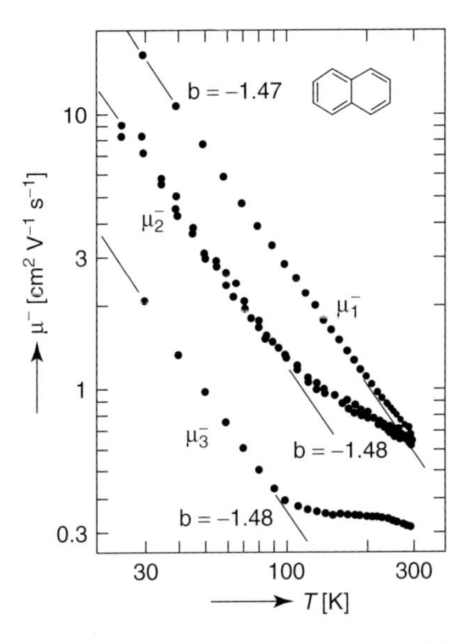

Figure 3.2 Three principal components of the mobility tensor in a naphthalene crystal measured by ToF, and fitted to a T^b dependence. A clear $T^{-3/2}$ dependence is observed only below ~100 K. (Reprinted with permission from Ref. [9]. Copyright (2001) by Springer-Verlag.)

In ToF measurements, charge is generated by using radiation that is well above the bandgap and as a consequence is in an excited state. This situation is not necessarily representative of what happens in an electronic device in which electrons (holes) are injected at the bottom (top) of the conduction (valence) band. Recent experiments pioneered by Podzorov *et al.* using vacuum-gap field-effect transistor (FET) structures fabricated with elastomers showed that a negative temperature dependence of the mobility can be observed in single-crystal electronic devices [15–18]. Rubrene FETs displayed an increasing mobility from room temperature ($\sim$10 cm^2 V^{-1} s^{-1}) down to 150 K ($>$20 cm^2 V^{-1} s^{-1}). The mobility was also found to depend on the crystallographic direction [17–19]. At $T < 150$ K, the mobility was thermally activated with an activation energy of ~ 70 meV. A gated Hall-effect measurement showed that the activated behavior was because of shallow traps near the valence-band edge (the FETs were hole-only devices): the mobility of the free carriers keeps increasing as the temperature is decreased, even at temperatures at which the apparent mobility shows the opposite trend (Figure 3.3).

Owing to constraints in the experimental set-up, the measurements could not be carried out at temperatures low enough to observe the $T^{-3/2}$ dependence of μ, and therefore, a weaker temperature dependence was observed. The use of FET structures also somewhat relaxes the purity and quality requirements of the crystal. Indeed, in FETs, the areal charge density is much higher than in ToF measurements and the gate voltage can be used to populate traps and push the Fermi level close to the valence-band edge. It should be noted, however, that

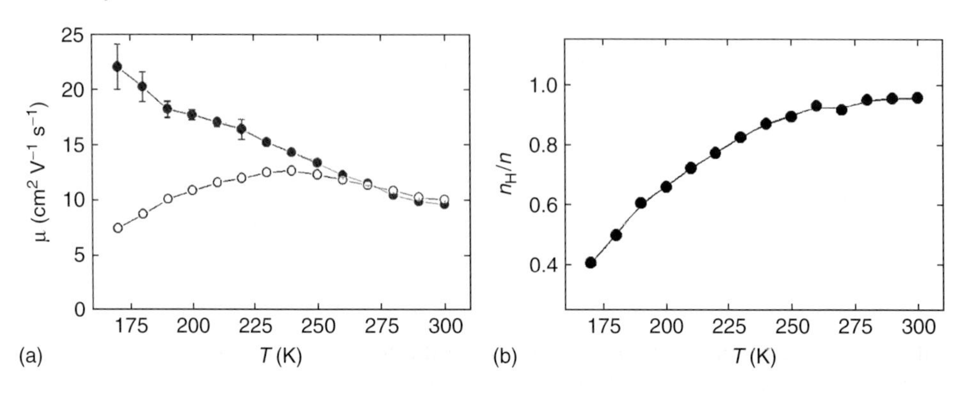

Figure 3.3 (a) Temperature dependence of the Hall mobility (solid circles) and field-effect mobility (open circles) in a rubrene single-crystal air-gap FET. At $T < 225$ K, the apparent FET mobility decreases, while the Hall mobility increases. The decrease in FET mobility is attributed to charge trapping as demonstrated by the ratio of Hall carrier density n_H to total gate-induced carrier density n (b). (Reprinted with permission from Ref. [18]. Copyright (2005) by the American Physical Society.)

poor-quality rubrene crystals (as determined by a room temperature mobility smaller than 10 cm^2 V^{-1} s^{-1}) still do not display negative temperature dependence of the mobility. Results generating similar conclusions have since been obtained in other laboratories with different materials [20–35].

In summary, carrier mobility measurements obtained using different techniques and made in a range of charge density regimes strongly suggest that band transport can occur in high-quality organic single crystals. Room temperature mobilities as high as 20 cm^2 V^{-1} s^{-1} are reproducibly observed in high-quality rubrene crystals.

3.2.3 Band or Bandlike?

As described in the previous section, the experimental evidence for band transport in organic single crystals relies heavily on the negative temperature dependence of carrier mobility. Furthermore, for many large conjugated molecules, the reorganization energy is smaller than the transfer integral, indicating that charge delocalization is possible. It was argued, however, that band transport and delocalization are inconsistent with the estimated carrier mean free path at room temperature (approximately 1 unit cell). Troisi *et al.* recently proposed a different charge transport mechanism in organic single crystals that resolves this apparent paradox [36–38]. Computational studies showed that for many compounds, the fluctuations of the transfer integral are of the same order as the integral itself. Under these conditions, translational symmetry is greatly perturbed and the band description of the electronic structure of the crystal is no longer valid. The transfer integral fluctuations are mostly because of low-frequency modes situated around 40 cm^{-1}. In fact, it was experimentally demonstrated that in pentacene, charge carriers are strongly coupled to low-frequency molecular motions with frequencies centered around ~1.2 THz [39]. The proposed transport model postulates that charge is dynamically localized on molecules by these intermolecular vibrations. By observing the temporal spread of an initially localized electron density, the authors conclude that charge obeys a diffusive behavior and that a mobility can be extracted by applying Einstein's relation. As expected, an increase in the transfer integral leads to an increase in mobility, while an increase in the electron–phonon coupling parameter leads to a decrease in mobility. By using molecular parameters found in the literature, the authors calculate room temperature mobility values compatible with experimental measurements (of the order of 30 $cm^2\ V^{-1}\ s^{-1}$ for rubrene). More importantly, because localization is due to dynamic disorder, reducing the temperature reduces the disorder and consequently increases the mobility. Hence this model predicts a negative temperature dependence of the mobility without introducing band transport. In rubrene, the calculated temperature dependence is $T^{-2.1}$, which is steeper than the experimental value. It should be noted, however, that the experimental functional dependence may be affected by trapping, which would reduce the steepness of the temperature dependence of mobility by leading to an inversion at ~150 K.

Single crystals of organic semiconductors are ideal systems to explore the limits of these materials and study fundamental charge transport physics in the cleanest possible systems. Most materials of practical relevance, however, have some degree of disorder.

3.3 Disordered Materials

3.3.1 Different Types of Disorder

In organic materials, the building blocks that make up the sample are not individual atoms but molecules; hence the states of the charge carriers can be described as a linear combination of their molecular orbitals. In the absence of disorder, the energy levels of such orbitals would all be the same and the position of the molecules would be regular, with an infinite correlation length. In actuality, the energy levels of the molecules follow a random distribution with a breadth that is characteristic of the amount of disorder in the material. This is particularly true in highly disordered or amorphous materials. Furthermore, a nonzero amount of positional and conformational disorder is present, which means that molecules are either not associated with points in a crystalline lattice or deviate significantly from them and that distortions in the orientation or the shape of molecules are present, for example, in the form of twists in the conjugated backbones of polymers or in the conjugated cores of smaller molecules.

A straightforward way to understand these disorder effects is to think of a 1-D chain of sites in which each site represents a molecule and is associated with a single energy level E_i, where i is an index labeling the molecules along the chain [40]. The hopping matrix element V_i between the ith and $(i+1)$th sites can then be defined, which reduces the Schrödinger equation to

$$E\Psi_i = E_i\Psi_i + V_i\Psi_{i+1} + V_{i-1}\Psi_{i-1}, \tag{3.2}$$

where E is the energy of the eigenstate of the equation and Ψ_i is the amplitude of such wavefunction on the ith site. Using a transfer matrix formalism, this equation can be rewritten as:

$$\begin{bmatrix} \Psi_{i+1} \\ \Psi_i \end{bmatrix} = M_i \begin{bmatrix} \Psi_i \\ \Psi_{i-1} \end{bmatrix}, \tag{3.3}$$

where

$$M_i = \begin{bmatrix} \frac{E-E_i}{V_i} & -\frac{V_{i-1}}{V_i} \\ 1 & 0 \end{bmatrix} \tag{3.4}$$

is the transfer matrix.

In this form, the nomenclature for diagonal and off-diagonal disorders becomes evident. In a *diagonal disorder* model, the site energies E_i are disordered, with the transfer integrals V_i remaining position independent. When these hopping integrals are disordered, the transfer matrix has position-dependent terms in its off-diagonal elements as well, which is referred to as *off-diagonal disorder*.

Another important distinction when discussing disorder effects is that between static and dynamic disorder. As seen in the previous section, dynamic disorder may play an important role in transport even in ideal single crystals. Such difference refers to the time evolution of the disorder effects and arises from a disparity in the characteristic timescales of the electronic Hamiltonian compared to the dynamics of charge carriers. When the electronic Hamiltonian changes very slowly compared to the dynamics of the charge carriers, such carriers experience a practically static energy landscape, in which the site energies and transfer integrals vary negligibly during the motion of the charge carriers through the sample. This situation is known as *static disorder*. In contrast, when the electronic Hamiltonian varies rapidly compared to the dynamics of charge carriers, the resulting time-dependent differential equation must be solved. This latter case is called *dynamic disorder*. In general, materials can be in an intermediate regime in which there is a built-in static disorder with an added component of dynamic disorder [41]. Molecular crystals generally have a low degree of static disorder. Because the intermolecular interactions are weak and the transfer integrals are very sensitive to small positional or conformational deformations [7], thermal fluctuations are sufficient to bring these systems into the dynamic (off-diagonal) disorder regime [41]. For noncrystalline materials, the static disorder dominates at most temperatures, and they are studied using the formalism originally developed for amorphous inorganic semiconductors in which charge carriers hop in a time-independent disordered energy landscape. These materials are the object of this section.

It is interesting to consider the intermediate charge transport regimes residing between the static and dynamic disorder cases. Two main approaches have been developed to analyze this problem, each being an extension of one of the limiting cases. The first intends to solve for the time-dependent wavefunction of charge carriers as originally developed for dynamically disordered systems, and the other is a variation of a method initially intended to study statically disordered materials, which aims to track the time-dependent occupation probability density [41].

3.3.2
Effect of Disorder on Charge Transport

Without any loss of generality, the possible electronic states in a semiconductor can be divided into either extended states or localized states. As their names suggest, extended states are those in which the charge carrier is "spread out" (delocalized) over the entire lattice. These states are often thought of as Bloch waves [13]. On the other hand, localized states are those in which the charge carrier's wavefunction has a nonzero amplitude in a finite region about a particular point (localization site). The size of this region determines whether a state is weakly localized (with a larger localization region) or highly localized (with a smaller localization region).

As first suggested by Anderson and since then discussed by many [42, 43], a high degree of disorder can induce localization of the charge carrier wavefunctions in a semiconductor. A mathematical criterion to differentiate between localized and extended states has been presented by Ball [44, 45], in which the electronic Hamiltonian was solved using perturbation theory and the states' wavefunctions are expressed as a linear combination of the individual orbitals of the sites comprising the lattice. Taking into account the fact that if a state is localized, there can only be a finite number of orbitals on which the amplitude is nonnegligible, it was found that when the conditions for localization are met, the resulting states are not weakly localized but instead highly localized, which is in accordance with the methods used to study charge transport in disordered materials.

The mechanism for charge transport is largely determined by the type of states that are involved in the process, and can be divided as follows. When charge transport is due to extended states, the mean free path length of the carriers is very large and a bandlike transport behavior is observed. This is described by the standard Fermi liquid approach and Boltzmann theory, in which the charge carriers are treated in the nearly free approximation and are only weakly scattered by a small amount of disorder, structural defects, or impurities [46]. The other extreme is when charge transport is described by a hopping mechanism between individual localized states, where charge carriers occupy a single site at a time and hop from one to another, with rates determined by the spatial separation and energy difference between the states involved in the hop. A third mechanism that involves both extended and localized states is that in which charge carriers spend most of the time in localized states referred to as *traps* and are then thermally promoted to higher-energy extended states in which they are able to move via drift or diffusion, but are trapped again by another localized state: this is known as the *multiple trapping and release model*. The energy level that separates the localized and extended states is referred to as the *mobility edge*, since it demarks the boundary between the states in which charges are mobile and those in which they are trapped. In disordered materials, localized states play a dominant role in the charge transport mechanism.

3.3.2.1 Dispersive and Nondispersive Transport

Before entering a more specific discussion about the different methods used for modeling and analyzing charge transport in disordered organic semiconductors, it is important to understand the distinction between dispersive and nondispersive transport. The difference between the two regimes can be understood by picturing an ensemble of charge carriers generated under nonequilibrium conditions. If the mean hopping rate of the carriers is constant, the original packet will only broaden due to diffusion and the transport is nondispersive. The broadening due to diffusion follows a Gaussian shape in which the width w of the carrier packet is given by $w = (D\tau_{tr})^{1/2}$, where D is the diffusion coefficient and τ_{tr} is the typical carrier transit time through the sample. However, if the mean hopping rate of the carriers decreases with time, the packet will broaden due to dispersion caused by the change in hopping rates. The hopping rate may decrease as charge gets progressively trapped into deeper states, which are more difficult to exit. As a result, transport becomes dispersive. There is a possibility that a packet starting out in the dispersive transport regime will equilibrate after some time t_{eq}, and thus transition to a nondispersive regime. This regime transition is a characteristic feature of hopping in a lattice with a Gaussian density of states (DOS) as observed in temperature-dependent ToF mobility measurements and Monte Carlo simulations [46, 47].

Hence, the observed transport behavior is affected by the shape of the DOS as well as by experimental conditions, such as temperature and electric field.

3.3.2.2 Transport Models

The shape of the DOS is central to many transport models. The models created for the analysis of charge transport in disordered materials were first developed for inorganic amorphous semiconductors [46, 48, 49], in which the DOS is thought to have a band edge with an exponential tail of localized states:

$$g(\varepsilon) = \frac{N_0}{\varepsilon_0} e^{-\frac{(\varepsilon_{edge} - \varepsilon)}{\varepsilon_0}}, \tag{3.5}$$

where ε is the energy of the electronic states and it decreases as it moves deeper into the tail, ε_{edge} is the energy of the band edge, N_0 is the total concentration of states in the tail, and ε_0 is a positive constant that determines the width of the tail. This DOS was originally explained by the exponential energy dependence of the light absorption coefficient of such materials.

One of the most compelling arguments for the validity of an exponential DOS is the transient photocurrent experiments of Monroe and collaborators [50, 51], in which the current resulting form a short light pulse at $t = 0$ was monitored. In the proposed model, the observed decay of the current reflects the trapping of carriers in progressively deeper states through a process of multiple trapping. Since the current at time t is determined by the traps that have a release rate ν to mobile states given by $\nu = t^{-1}$, and assuming an Arrhenius form for the release rate from traps at energy ε to transport states at ε_t (with $\varepsilon_t > \varepsilon$),

$$\nu = \nu_0 e^{-\frac{\varepsilon_t - \varepsilon}{kT}} \tag{3.6}$$

one can relate the current decay to the states in the DOS that release their charge at time t as follows

$$i(t) = \frac{C}{t g(\varepsilon_d)} \tag{3.7}$$

where k is Boltzmann's constant, T is the temperature, and C is a prefactor that is independent of the general shape of the DOS. The demarcation energy ε_d is that of the states that have a release rate $\nu = t^{-1}$ and is given by

$$\varepsilon_d(t) = \varepsilon_t - kT \ln(\nu_0 t) \tag{3.8}$$

Hence, a power-law decay of the current is observed only if the DOS follows the exponential shape described above, and such power-law decay was experimentally observed over more than five orders of magnitude in current density and eight orders of magnitude in time.

For organic materials, however, it is observed that the excitonic absorption band follows a Gaussian profile, which suggests a Gaussian shape for the DOS [46].

$$g(\varepsilon) = \frac{N}{\sigma\sqrt{2\pi}} e^{-\frac{(\varepsilon - \varepsilon_c)^2}{2\sigma^2}} \tag{3.9}$$

where N is the total density of states, ε_c is the energy for the center of the distribution (which can be arbitrarily made zero), and σ is the variance of the DOS distribution. The fact that the polarization energy of organic materials is determined by a large number of internal coordinates, each varying randomly by small amounts, supports the Gaussian DOS assumption [52]. The Gaussian shape of the DOS in disordered organic semiconductors has been validated by its ability to reproduce experimental results.

As mentioned previously, charge transport in disordered organic semiconductors occurs through carrier transfer between localized states distributed according to a Gaussian DOS. It is useful to visualize the energy landscape that a charge sees when traveling through such a material, as shown in Figure 3.4. Different energy levels available for transport are represented by their energy and separation from the initial state.

Using this visualization, it is easy to see that the likelihood of a given hop is determined by an interplay of two factors: the energy difference and the spatial separation of the initial and final states. When the initial and final states have a small energy difference compared to the thermal energy kT, the transfer between those states will be favored energetically. However, for the transfer to happen, there should be a nonzero overlap integral between the wavefunctions of these states.

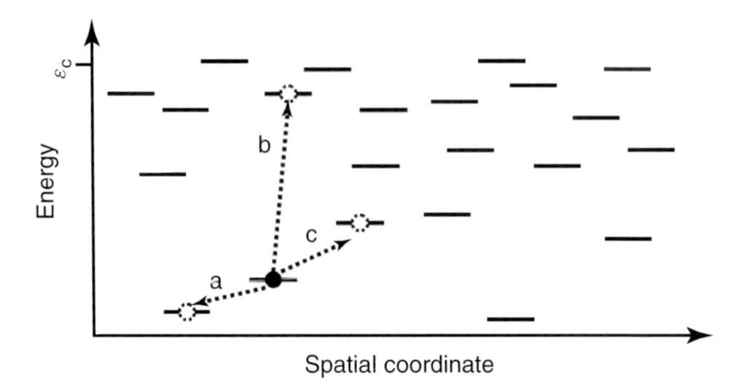

Figure 3.4 The energy landscape a charge experiences when traveling through a Gaussian band of localized states. The solid circle represents the charge, and the three arrows (a–c) pointing toward open circles represent possible hops the charge could make.

By defining a localization radius r_{loc} for the states, a characteristic distance over which the transfer probability decays can be determined. For the sake of simplicity, this localization radius is usually taken to be the same for all states, but it can be generalized to vary between states. The specific way these variables influence the transfer rate between each pair of states is discussed in a later section. For now, it is sufficient to understand that the state to which a charge transfers is the one that minimizes the penalties associated with both the energetic and spatial separations of the initial and final states. Different transport regimes can be envisioned in order of increasing temperature:

- At very low temperatures, the thermal energy is small and thus all possible transfers satisfy the relation $\Delta\varepsilon \gg kT$. This means that all upward (in energy) jumps are equally unlikely energetically and the dominant factor is the spatial dependence, favoring the charge transfer to the nearest neighbor. Not surprisingly, this regime is called "nearest-neighbor hopping" (NNH).
- At low temperatures, there is enough thermal energy to open the possibility to hop to states within a narrow energy band around the Fermi level. The constraint on the spatial coordinate is therefore relaxed, meaning it is possible for a charge to transfer to a state that is further away from the initial position but still within the accessible energy band. This regime is called "variable-range hopping" (VRH).
- At intermediate temperatures, the transport mechanism is similar to VRH, but with a wider available energy range around the Fermi level, allowing for a richer, more complex process. In this regime, concepts such as the transport energy model (TE) and percolation theory are most useful.
- At high temperatures, charges can be thermally excited above the mobility edge, from a localized state into the region of extended states. After this excitation to a mobile state, the charge moves just as it would in a band transport model, until it is trapped again by another localized state. This regime is labeled "multiple trapping and release" (MTR). MTR models, however, require the existence of extended states and are not suitable for materials in which all states are localized, such as amorphous organic semiconductors. Nevertheless, they have

been successfully used in single crystalline, polycrystalline, and semicrystalline organic materials [53–60].

A sketch illustrating each of these regimes is shown in Figure 3.5. One aspect not conveyed by the figure is the effect of temperature on transfer rates: the same hop will have a faster rate at higher temperatures. In the case of NNH, there are many hops the carrier must make, some of them comprising a large energy penalty, which means that transport in this regime will be limited. As the temperature is increased and transport operates according to the VRH regime, the carrier has a small amount of thermal energy and moves within a narrow energy range around the Fermi level. On further temperature increase, the carriers see a wider energy range and the hops occur at a faster rate. At even higher temperatures, thermal promotion to extended states and subsequent retrapping occur.

In all transport regimes, the transfer rates between each pair of states must be calculated in order to determine the transport path of charges through the sample. There are two main models used to calculate transfer rates, the Miller–Abrahams model and the polaron model. The Miller–Abrahams model calculates the transfer

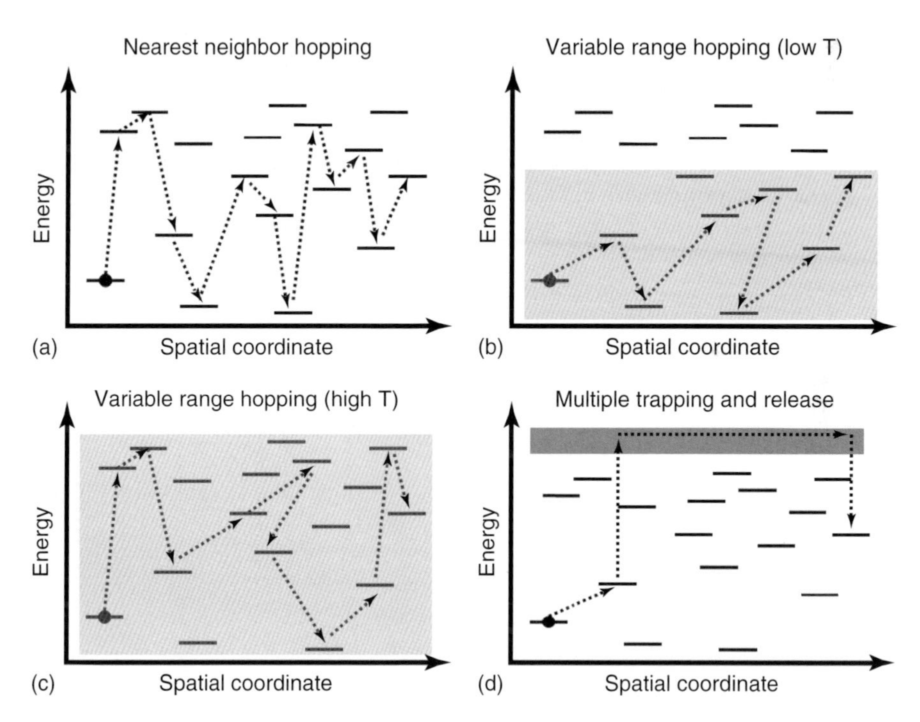

Figure 3.5 Sketch of a possible transport path for a charge carrier in different regimes: nearest neighbor hopping (a), variable range hopping (b, c), and multiple trapping and release (d). The rate of the transfers increases with temperature, but is not depicted in the figure. Backward hops are also possible, sometimes even necessary, and are illustrated in (b) and (c). The light gray bands in (b) and (c) represent the available range of energies in each case, and the dark gray band in (d) represents the extended states.

rate between two sites i and j, ν_{ij}, as follows:

$$\nu_{ij} = \nu_0 e^{-2\alpha r_{ij}} e^{-\frac{\varepsilon_j - \varepsilon_i + |\varepsilon_j - \varepsilon_i|}{2kT}} \tag{3.10}$$

where ν_0 is the attempt-to-hop frequency and is usually taken to be the phonon vibrational frequency, $\alpha = (r_{\text{loc}})^{-1}$ is the inverse localization radius, r_{ij} is the separation between i and j, and ε_i (ε_j) is the energy of the initial (final) state. Physical separation and upward differences in energy cause an exponential decrease in the transfer rate, but hops that are downward in energy are not accelerated. This framework implies that each electronic hop is accompanied by the absorption or emission of a phonon to compensate for the energy difference $\Delta\varepsilon_{ij} = \varepsilon_j - \varepsilon_i$. The need for a phonon to provide energy balance to the electronic transfer means that upward hops require a particular phonon state to be populated and are thus temperature activated, while downward hops are temperature independent. A second consequence of the role of phonons is that a phonon state with the energy $|\Delta\varepsilon_{ij}|$ must exist in order for that particular transfer to occur, limiting the validity of this approximation to values of $|\Delta\varepsilon_{ij}|$ that do not exceed the energy of the phonon modes that can effectively couple to the transfer process [61].

The polaron model incorporates the fact that charges cause a polarization of the medium around them, which induces a deformation, and thus a charge transfer event incurs some reorganization component. The charge transfer reaction between sites A and B can be expressed as $A^- + B \rightarrow A + B^-$ for electron transfer, or $A^+ + B \rightarrow A + B^+$ for hole transfer. This way, the potential energy surfaces of the process are as shown in Figure 3.6 [62].

The transfer rate for this process is usually calculated with the semiclassical Marcus theory [61–63]:

$$\nu_{ij} = \frac{|J_0|^2}{\hbar}\sqrt{\frac{\pi}{\lambda kT}} e^{-2\alpha r_{ij}} e^{-\frac{(\Delta\varepsilon_{ij} + \lambda)^2}{4\lambda kT}} \tag{3.11}$$

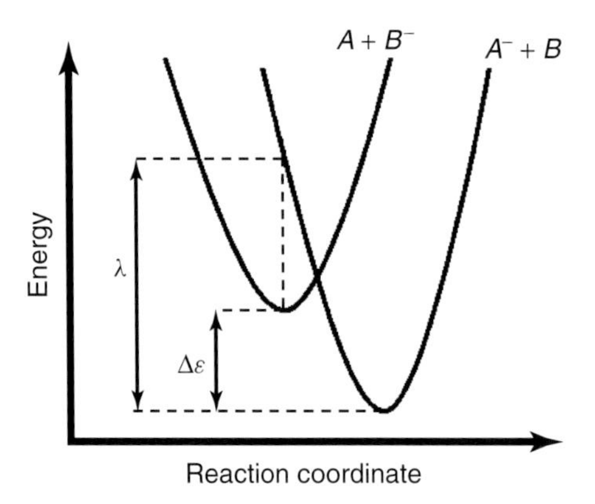

Figure 3.6 Potential energy surfaces as a function of a reaction coordinate for a charge transfer event. The energy difference $\Delta\varepsilon$ between the states and the reorganization energy λ are identified.

where J_0 is the electron transfer integral and λ is the reorganization energy as shown in Figure 3.6; the rest of the variables are as defined previously. The Marcus rate represents the probability of reaching the intersection point of the two potential energy curves (Figure 3.6) and the rate of reaction once there [62].

A key result of this model is that the transfer rate does not continue to increase for increasingly favorable energy differences $\Delta\varepsilon_{ij} < 0$. In fact, for a given reorganization energy λ, and considering a negative energetic driving force $-\lambda < \Delta\varepsilon_{ij} < 0$, the transfer rate first increases with the magnitude of $\Delta\varepsilon_{ij}$ in what is called the *normal region*, reaches a maximum for $\Delta\varepsilon_{ij} = -\lambda$, and then decreases as $\Delta\varepsilon_{ij}$ turns even more negative in what is referred to as the *inverted region* [61].

In order to include the effect of an electric field in the transfer rates, whether it is using the Miller–Abrahams or Marcus formalism, a term of the form $-q\mathbf{r}_{ij} \cdot \mathbf{F}$ must be included in the energy difference $\Delta\varepsilon_{ij} = \varepsilon_j - \varepsilon_i - q\mathbf{r}_{ij} \cdot \mathbf{F}$, where q is the charge of the carrier species and F is the electric field magnitude (vector quantities are represented in bold script). The effect of electric fields on charge transport is discussed in more detail later.

It should also be noted that both formalisms are the limiting cases of the more general case described by the time-dependent perturbation theory in the limit of weak electronic coupling [61]. The Miller–Abrahams approximation is valid in the regime of weak electron–phonon coupling and low temperatures, while the Marcus approximation describes the regime of strong electron–phonon coupling and high temperatures. As a final consideration, it is worth pointing out that the Miller–Abrahams approximation was initially developed for inorganic disordered materials, and its success in that area has made it a staple for modeling of organic materials. The molecular character and high polarizability of organic materials, however, make the Marcus formalism seem more adequate.

Calculation of the hopping attempt frequency has recently received increased attention. Historically, the hopping attempt frequency has been considered a constant only weakly dependent on experimental variables and rationalized to be at least proportional to the phonon vibration frequency. Some algebraic approaches with varying degrees of sophistication have also been proposed in calculations involved in the semiclassical Marcus theory [63]. With the advent of high-power computation tools, full quantum mechanical calculations are within reach, and hopping attempt frequencies have been shown [6, 7] to depend quite strongly on the conformation and orientation of the molecular sites involved in the transfer.

Once a fundamental form of the DOS function and the appropriate expression for the transfer rate are chosen, a methodology to connect the molecular level to a macroscopic description of charge transport is needed. A first attempt would be to include the contribution from all possible rates and obtain an average transfer rate $\langle\nu\rangle$.

$$\langle\nu\rangle = \int_0^{\infty} \mathrm{d}r \int_{-\infty}^{\infty} \mathrm{d}\varepsilon_i \int_{-\infty}^{\infty} \mathrm{d}\varepsilon_j \nu(r, \varepsilon_i, \varepsilon_j)\rho(r, \varepsilon_i, \varepsilon_j) f(\varepsilon_i)\left[1 - f(\varepsilon_j)\right], \tag{3.12}$$

where $\nu(r, \varepsilon_i, \varepsilon_j)$ is the transfer rate from a state with energy ε_i to a state with energy ε_j when they are separated by a distance r, $\rho(r, \varepsilon_i, \varepsilon_j)$ is the probability of existence of such an arrangement of states, $f(\varepsilon_i)$ is the occupancy probability of the initial state, and $1 - f(\varepsilon_j)$ is the probability of the final state being vacant. The probability $\rho(r, \varepsilon_i, \varepsilon_j)$ depends on the DOS of the system, and the occupancy probability $f(\varepsilon)$ takes into account the position of the Fermi level, temperature, and the DOS. It is then assumed that the mobility is proportional to the average hopping rate times the square of the typical displacement over a hopping event: $\mu \propto \langle \nu \rangle r_t^2$. This is a simple approach, and its results are not necessarily accurate; however, it is useful in the sense that analyzing its shortcomings leads to the understanding of more complex models.

We can try to use the average hopping rate model to study the motion of a charge in a material composed only of highly localized states at a relatively high temperature, using the Miller–Abrahams expression for the transfer rates. This means that the conditions $N(r_{\mathrm{loc}})^3 \ll 1$ and $kT \gg \sigma$ must be satisfied. Furthermore, the following approximation, that neglects the thermal activation between states, is valid:

$$\nu_{ij} \approx \nu_0 \mathrm{e}^{-2\alpha r_{ij}} \tag{3.13}$$

Since the energy of the sites does not play a role, and there are no extended states to which the charge could be excited, the rates are dominated by the spatial decay of the wavefunctions and the transport takes place via nearest-neighbor hopping. The hopping-configuration probability is then given by

$$\rho(r, \varepsilon_i, \varepsilon_j) \approx \rho(r)\delta(\varepsilon_i)\delta(\varepsilon_j) = 4\pi r^2 N \mathrm{e}^{-\frac{4}{3}\pi r^3 N} \delta(\varepsilon_i)\delta(\varepsilon_j) \tag{3.14}$$

For a single charge, the occupancy probability terms can be neglected. The average transfer rate in this case would be [46]:

$$\langle \nu \rangle \approx \int_0^\infty \mathrm{d}r \nu_0 \mathrm{e}^{-2\alpha r} \left[4\pi r^2 N \mathrm{e}^{-\frac{4}{3}\pi r^3 N} \right] \approx \frac{\pi \nu_0 N}{\alpha^3} \tag{3.15}$$

The shape of the integrand is shown in Figure 3.7: the rates are dominated by those sites separated by a distance $r \approx r_{\mathrm{loc}}$, and thus the typical distance for a hop will be $r_t \approx r_{\mathrm{loc}}$.

However, by enforcing the condition $N(r_{\mathrm{loc}})^3 \ll 1$, $r_{\mathrm{loc}} \ll N^{-1/3}$ is implied, which means that the intersite separation is much larger than the localization radius. Thus, a typical hop across a distance r_{loc} is not an accurate description of transport in this regime since the charge should be able to jump over distances much larger than that in order to hop between sites and move across the sample.

The notion of a required hopping distance in order for transport to occur in these materials is therefore important. A description of transport using percolation concepts results directly from this notion. In percolation theory, one connects the sites that have transfer rates faster than some limiting value ν_{min}, and all other sites with slower transfer rates are considered as disconnected. The value of the limiting transfer rate varies, including slower and slower hops in each iteration

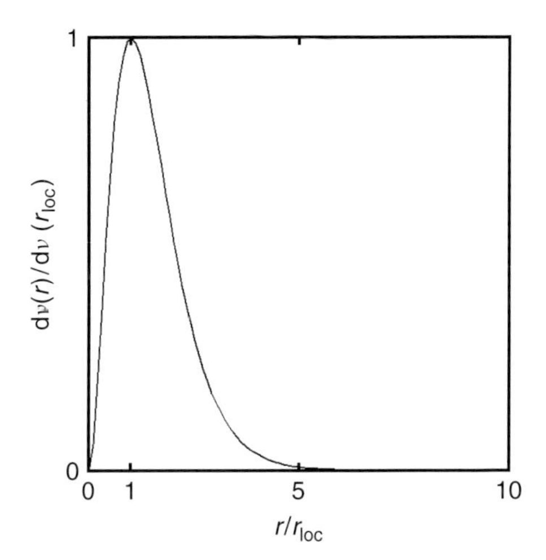

Figure 3.7 Integrand in the expression for the calculation of the average transfer rate in a system with strongly localized states at high temperatures (Eq. (3.15)).

until a continuous transport path is formed and percolation has been reached, naming this critical rate ν_{perc}. This situation is shown schematically in Figure 3.8. Careful analysis determines that at the percolation threshold, the average number of bonds per site on the transport path is $B_c = 2.7 \pm 0.1$ [46].

The exponential dependence of the transfer rates allows the argument that the electrical properties of a system are mostly determined by the limiting step in charge transport (the most difficult hop to complete), and it is assumed that the other exponentially faster rates hardly play any role as limiting steps for transport. In the case of percolation, this limiting step is clearly the percolation threshold. The methods for the analysis of percolation problems, however, are not very transparent,

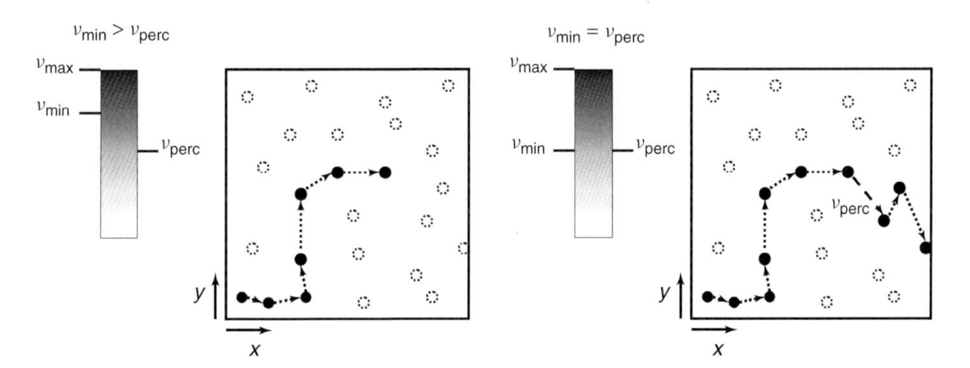

Figure 3.8 Schematic representation of the percolation of charges through a given sample. (a) There is no continuous path across the sample since the percolation threshold is not reached. At the percolation threshold (b), there exists a continuous path for transport.

and thus other approaches have been proposed to provide an explanation for the charge transport mechanism in disordered materials.

One such approach introduces the concept of the transport energy ε_t. It has been recognized that the vicinity of a certain energy level dominates the hopping of charge carriers in the band tails [46, 50, 64, 65], and the explanation is that a charge occupying a state with any energy below ε_t has a maximum transfer rate to a state at ε_t, and states above ε_t relax to a level at or below ε_t through a series of downward hops.

The position of the transport energy is determined as the solution to the expression

$$\left.\frac{\mathrm{d}\nu(\varepsilon_i, \varepsilon_j)}{\mathrm{d}\varepsilon_j}\right|_{\varepsilon_j=\varepsilon_t} = 0, \tag{3.16}$$

while satisfying the condition $\varepsilon_i < \varepsilon_j$ (the initial state ε_i is deeper in the DOS tail) since the transfer rate for upward hops in the tail must be maximized.

Following the discussion of Baranovskii *et al.* [46, 66], the Miller–Abrahams expression for the transfer rate in a Gaussian DOS can be used to obtain the equation

$$\mathrm{e}^{\frac{x^2}{2}}\left[\int_{-\infty}^{\frac{x}{\sqrt{2}}} \mathrm{d}z\mathrm{e}^{-z^2}\right]^{\frac{4}{3}} = \left[9\sqrt{2\pi}N\,(r_{\mathrm{loc}})^3\right]^{-\frac{1}{3}}\frac{kT}{\sigma}, \tag{3.17}$$

which needs to be solved for x and $\varepsilon_t = x\sigma$. The rest of the variables have their usual definitions: N is the total DOS, r_{loc} is the localization radius, kT is the thermal energy, and σ is the energy width of the DOS. The solution to this equation is only meaningful as long as the maximum determined by ε_t is sharp, since a broad maximum would mean that the transport energy is poorly defined and a transport band would need to be discussed instead. Also, a value for ε_t that is unphysical, in that it is too deep in the band tail, would point to the inapplicability of this model for that given system.

The calculation of transport energy in a Gaussian DOS cannot be done analytically as is the case for an exponential DOS [50]. It has been shown that this model provides a good description for hopping in a Gaussian band as long as $kT < \sigma$, which is valid for most systems at relevant temperatures since the typical bandwidths for Gaussian systems are of the order of 100 meV [61, 67].

The process described by this transport energy model has many similarities to the MTR model: in the MTR model, charges are thermally excited from states deep in the localized band into delocalized states above the mobility edge, where they move almost freely and are retrapped after traveling some characteristic mean free path. In the TE case, states deep in the band are excited to a narrow band around the transport level, and then relax back down to states deeper in the band so that, on average, every second hop is to the transport level. In this sense, the transport energy plays the role of the mobility edge, with the difference being that in the MTR model, the hopping between localized states is neglected, while in the TE case, only localized states are involved in the transport process (Figure 3.9).

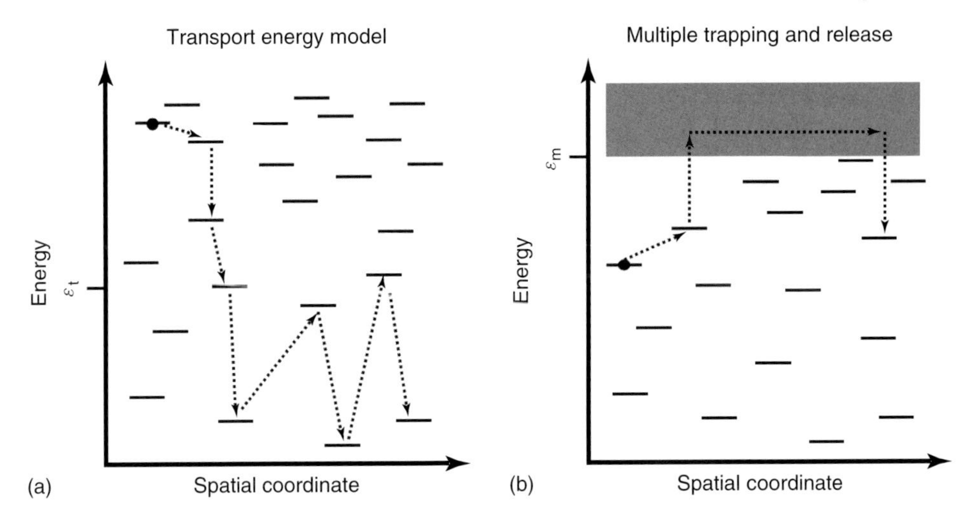

Figure 3.9 Diagram of the process described by the transport energy model (a) starting with the relaxation of a charge to deep states and the subsequent transitions between the transport level and lower-lying states. A comparison with the multiple trapping and release mechanism (b) is provided to show the conceptual similarities between them.

A further refinement to the TE model was proposed by Arkhipov *et al.* [68], introducing a distinction between the energy that defines the most probable upward hop, ε_t as defined above, and the energy of a level defining an upward hop that will most likely result in the carrier leaving the initial site, calling this second case the effective transport energy. In the hopping regime described by the TE model, the target state at the transport level is still a localized state with a limited number of neighbors, with one of the neighbors inevitably being the site where the carrier started. Therefore, it is possible for the carrier to hop to the transport level and subsequently fall back into the initial state, in a process that leads to neither transport nor energy relaxation.

For the calculation of this effective transport energy, Arkhipov *et al.* consider only the target states with a low probability of a round-trip hop, and the results obtained compare well to computational simulations and experimental data for a significant range of temperatures and densities of localized sites [68].

After discussing a series of models that describe the charge transport process in a disordered material, it is appropriate to describe how one usually goes from these descriptions to the calculation of the electrical properties of the studied system. Starting from the Einstein relation

$$\mu = \frac{q}{kT} D \tag{3.18}$$

where q is the charge of the transport species and D is its diffusion coefficient, the next step is to define the diffusion coefficient in terms of the processes involved in the charge transport mechanism.

The details of the assumed model determine the particular expression for D, most importantly, defining the states that are taken into account during the calculation of the mean jump distance and average transfer rate. As discussed previously, all possible transitions could be included, but lead to an unsatisfactory answer. Alternatively, the rate-limiting step could be considered, as is the case for a percolation-based analysis. Lastly, a population of states that determine the mechanism for charge transport can be identified and cause the analysis to be restricted to that subset of states when performing calculations of averages.

For example, in the MTR case, the diffusion coefficient is estimated as the square of the mean free path of charge carriers in the mobile states L_f^2 times the average frequency with which they are excited into those states $\langle \nu_\uparrow \rangle$

$$D \cong L_f^2 \langle \nu_\uparrow (\varepsilon_m) \rangle \tag{3.19}$$

In the TE model, the approach is the same but with a temperature-dependent transport energy instead of a fixed mobility edge. The fixed mean free path of delocalized carriers used in the MTR case is replaced by the mean jump distance of carriers, which depends on the specific characteristics of the DOS, as well as temperature.

No review of charge transport though a disordered band of localized states is complete without discussing one of the most historically important models: Mott's variable-range hopping [48, 69]. At sufficiently low temperatures, it is reasonable to assume that the relevant states for transport are within a small energy band of width 2δ around the Fermi level, so that the DOS can be approximated as constant within this interval. The typical hopping distance R of a carrier at the Fermi energy ε_F to a state with an energy $\varepsilon_F + \delta$ can then be estimated from the relation

$$g(\varepsilon_F) \cdot \delta \cdot \tfrac{4}{3}\pi R^3 \approx 1 \tag{3.20}$$

so that

$$R \approx \left[\frac{3}{4\pi\delta \cdot g(\varepsilon_F)} \right]^{\frac{1}{3}} \tag{3.21}$$

Inserting this expression into the Miller–Abrahams transfer rate (Eq. (3.10)) leads to the term in the exponential being:

$$-\left(2\alpha \left[\frac{3}{4\pi\delta \cdot g(\varepsilon_F)} \right]^{\frac{1}{3}} + \frac{\delta}{kT} \right), \tag{3.22}$$

and the optimal energy range for hopping will be that which minimizes the terms in the parentheses. Mathematically,

$$\frac{d}{d\delta}\left(2\alpha \left[\frac{3}{4\pi\delta \cdot g(\varepsilon_F)} \right]^{\frac{1}{3}} + \frac{\delta}{kT} \right) = -\frac{2\alpha}{3}\left[\frac{3}{4\pi g(\varepsilon_F)} \right]^{\frac{1}{3}} \delta^{-\frac{4}{3}} + \frac{1}{kT} = 0 \tag{3.23}$$

Solving for δ

$$\delta = \left[\frac{2\alpha^3 (kT)^3}{9\pi g(\varepsilon_F)} \right]^{\frac{1}{4}} \tag{3.24}$$

and inserting this back into the exponential term (Eq. (3.22)), the well-known temperature dependence of the hopping rate is obtained:

$$\nu \approx \nu_0 e^{-\left(\frac{T_0}{T}\right)^{\frac{1}{4}}}, \tag{3.25}$$

where T_0 is a characteristic temperature given by

$$T_0 = \frac{512}{9\pi} \cdot \frac{\alpha^3}{kg(\varepsilon_F)} \tag{3.26}$$

The prefactor varies depending on the approximations made during the derivation, but it is accepted to be a constant of order 10 [46]. One may initially agree with the approximation of the DOS as energy independent around the Fermi level, but the long-range electron–electron interactions in a population of localized states introduce a Coulomb gap centered at the Fermi energy [70], and the inclusion of this concept was one of the first corrections to the model [71].

When electrons reside in localized states, neglecting the possibility of doubly occupied states, the system is defined in terms of the available states that are filled. At sufficiently low temperatures, however, the Coulomb interactions between electrons modify the energy of these states and cause the DOS to vanish at the Fermi energy. Efros and Shklovskii [71] provide an illustrative discussion of the shape of the DOS around this gap, starting from the ground state of the system and analyzing a transition between two states $\varepsilon_i < \varepsilon_F < \varepsilon_j$. The activation energy for this transition, including the Coulomb interaction in a medium of dielectric constant κ, can be expressed as

$$\Delta H = \varepsilon_j - \varepsilon_i - \frac{q^2}{\kappa R} > 0 \tag{3.27}$$

One can then express the DOS around the Fermi level as a power-law dependence of the form

$$g(\varepsilon) = \beta\,(\varepsilon - \varepsilon_F)^{\gamma} \tag{3.28}$$

where β is an unknown positive coefficient and γ is the power-law exponent. The value of this exponent is obtained by requiring the Coulomb interaction energy for states within the energy range considered for transport to be on the order of the distance from the Fermi level (i.e., the transport bandwidth). For $\gamma < 2$, the Coulomb interaction would be larger than the energy bandwidth, and for larger coefficients, the separation between states would increase rapidly and render the Coulombic interaction negligible [71]. Thus, the shape of the DOS around the Fermi level is assumed to be parabolic. Following a similar derivation as that outlined for Mott's variable-range hopping, the following formula is obtained:

$$\nu \approx \nu_0 e^{-\left(\frac{T_c}{T}\right)^{\frac{1}{2}}}, \tag{3.29}$$

where T_c is a characteristic temperature of the form

$$kT_c = B\frac{\alpha q^2}{\kappa} \tag{3.30}$$

and B is a numerical coefficient.

The models developed by Mott [48, 69] and Efros and Shklovskii [71] are illustrative and hold historic relevance, but their validity is restricted by the initial assumptions of the shape of the DOS around the Fermi level. The required slowly varying DOS is realized in the impurity bands of lightly doped crystalline materials, where the disorder arises from the random impurities; however, for amorphous materials and particularly disordered organic materials, the dependence of the DOS on energy is much stronger, as described earlier, having an exponential form for amorphous inorganic materials and a Gaussian form for organic materials [46].

The previous discussion on the diverse analytical models should highlight the various degrees of complexity that accompany them and suggest that the use of computational tools would be of great help in understanding the processes involved in charge transport through disordered media.

3.3.2.3 **Computational Methods**

The basis of computational methods is the repeated random sampling of a distribution of states to compute their results. The process starts with the creation of a grid of sites, with each site randomly assigned an energy taken from the desired DOS. The effects of positional disorder can be approximately included by assigning a variable intersite coupling parameter $\Gamma_{ij} = 2\alpha r_{ij}$ (Eqs. (3.10) and (3.11)) to each pair of sites in the grid. This is done by splitting Γ_{ij} into two site-specific contributions, $\Gamma_{ij} = \Gamma_i + \Gamma_j$, taken from some distribution function (usually Gaussian). This is a debatable consideration, however, because it introduces a certain degree of correlation into the experiment due to the fact that all jumps from a given site i are affected by the choice of Γ_i [52]. Subsequently, a certain number of charges are placed within the grid in a manner that reflects the initial conditions of the system. They are then allowed to evolve according to the particular algorithm of the problem. Periodic boundary conditions are usually employed, and the simulation is allowed to run until a certain set of criteria is satisfied. The termination criteria are varied and can involve a limit in the simulation time, the displacement of the charges in a given direction, or a maximum number of hopping events. This calculation constitutes one "experiment," which must be repeated enough times to achieve convergence of the calculated results.

One of the most widely used algorithms to perform these computational experiments is the Monte Carlo technique, which in the context of charge transport in disordered media, was pioneered by Bässler and coworkers [47, 52, 72–76]. In this approach, after the grid is set up and the charges are placed, the transfer rates between each pair of sites is calculated. These transfer rates may be fixed throughout the experiment or, alternatively, allowed to vary during a single simulation, in which case they should be recalculated after each hopping event.[1] Once the transfer rates ν_{ij} for each pair $<ij>$ are known, the probability that a charge hops

1) An example of a case where the transfer rates evolve as the simulation runs is when they include charge–charge interactions, thus modifying the activation energies for the hopping of charges because the interaction energies depend on the instantaneous charge distribution.

from site i to site j is given by

$$P_{ij} = \frac{\nu_{ij}}{\sum\limits_{j,\text{ with } j \neq i} \nu_{ij}} \tag{3.31}$$

Since these simulations are computationally intensive, the reduction of the number of calculations necessary at each step is important, which is why the hopping range of a charge at a given site is usually reduced to a smaller number of local neighbors, and not the entire grid. This approximation is valid because the transfer probability decreases exponentially with distance.

The site to which the charge finally hops is determined by assigning a length in random number space to each neighbor, according to their transfer probabilities. Then, a random number x_R is generated, and the neighbor whose region in random number space includes x_R is selected as the destination state. After determining the site to which the charge will hop, the simulation clock is advanced a time Δt_{ij} given by

$$\Delta t_{ij} = \frac{x_t}{\sum\limits_{i \neq j} \nu_{ij}}, \tag{3.32}$$

where x_t is a randomly generated number, usually following an exponential distribution [52]. The outlined process describes the hopping of one charge, and if several charges are included in the same simulation, then it must be modified accordingly to allow for multiple hops. The Monte Carlo algorithm requires keeping track of every individual charge involved in the experiment, which means that for a large number of charges, the number of calculations can become prohibitively large.

A different algorithm that avoids the tracking of individual charges throughout the experiment is the Master Equation approach. In this procedure, the probability of occupation of each site p_i is monitored, instead of the position of each individual charge, which means that the probability evolution equation, also known as the *Master Equation*, must be solved:

$$\frac{\mathrm{d}}{\mathrm{d}t} p_i(t) = \sum_{i \neq j} \nu_{ji} p_j(t) \left[1 - p_i(t)\right] - \sum_{i \neq j} \nu_{ij} p_i(t) \left[1 - p_j(t)\right] - \zeta_i p_i(t), \tag{3.33}$$

where ν_{ij} is the transfer rate from site i to site j as usual. The first term in the right-hand side of the equation is the incoming flux, while the second term is the outgoing flux. The last term includes the recombination of charges through a decay rate ζ_i [77]. This equation is commonly solved in the steady-state regime, obtaining an occupation probability for all sites in the grid. The expected value for the velocity of charges is derived as:

$$\langle \vec{v} \rangle = \sum_{i, j \neq i} \vec{r}_{ij} \nu_{ij} p_i \left(1 - p_j\right) \tag{3.34}$$

resulting in the mobility [78]

$$\mu = \frac{\langle \vec{v} \rangle \cdot \vec{F}}{\left|\vec{F}\right|^2} \tag{3.35}$$

It is important to note that the assumption of a steady-state behavior prevents the use of this particular solution in nonequilibrium problems. Interestingly, under certain conditions and in 1-D systems, the Master Equation can be solved analytically as described by Derrida [79] and subsequently used by others [46, 80], which allows the explicit observation of effects due to correlation, electric field, and disorder, as is discussed in more detail in the following sections.

A variation of the Master Equation approach was proposed by Preezant and Tessler [81], in which the unknown is not the occupation probability of a grid of sites in position-space. Instead, the algorithm tracks the occupation of states in energy-space in a method called the Energy-Space Master Equation (ESME) approach. Here, the population density $N(\varepsilon)$ of states is found by solving the equations describing charge flow between submanifolds of different energies, and the occupational probability $p(\varepsilon)$ is calculated by using the relation between the population density and the DOS, $N(\varepsilon) = p(\varepsilon)g(\varepsilon)$. The formalism is constructed in a manner similar to the conventional Master Equation approach, and the solution process uses a small 3-D grid for performing the calculation [81].

It quickly becomes clear that the application of computational methods to the simulation of charge transport requires the optimization of the particular techniques for a given set of conditions. Hence, a wide range of variations have been proposed, such as the original Metropolis algorithm [82], or the Kinetic Monte Carlo (KMC) [83, 84], as well as variations in the implementations of a single algorithm, such as the Variable Step Size Method (VSSM), the Random Sampling Method, and the First Response Method (FRM) [85]. The main differences reside in the order of the steps to be taken for each transfer, the way to treat the time progression in the experiment, and the possibility to track several events in parallel.

It is important to highlight that one common feature among the previously described models is the ambiguity in the definition of the attempt-to-hop rate ν_0 for the Miller–Abrahams formalism (Eq. (3.10)), or the electron transfer integral J_0 in the Marcus theory approach (Eq. (3.11)). The only way to avoid this is to use a full quantum mechanical model to calculate the wavefunction overlap between every pair of hopping sites. Owing to the large number of positional, rotational, and conformational degrees of freedom, as well as the number of sites present in a typical sample, this approach is very resource intensive for complex systems. Moreover, the influence of nonlocal electron–phonon couplings and the failure of conventional first-principles methods to adequately describe weak intermolecular interactions complicates the accurate description of intermolecular electron transfer variables [61].

Yet another modeling approach stems from the realization that the phenomena involved in charge transport through disordered organic semiconductors span a large range of time and length scales, including molecular vibrations (10^{-15}–10^{-12} s), intermolecular charge transfer (10^{-12}–10^{-9} s), molecular conformational changes (10^{-12}–10^{-9} s or longer), and charge transport across a device (10^{-9}–10^{-3} s), as well as the disparity between intermolecular distances (10^{-10} m), typical grain sizes and correlation lengths (10^{-8}–10^{-6} m), and device sizes (10^{-7}–10^{-4} m) [62]. A rational consideration of all these typical scales would provide the basis for including different aspects of the charge transport process, from the determination of molecular packing and its decisive role in the determination of coupling constants and thus transfer rates between molecules [6, 7, 61], to the micro- and macroscale variations in morphology, allowing the modeling, and prediction of the transport behavior of a variety of systems from their chemical and physical structure.

Such a multiscale method has been implemented by Nelson and coworkers [62]. Starting from the chemical structure of a material and using atomistic molecular dynamics as well as coarse-grained models for molecular packing to simulate the position, orientation, and conformation of the transport units, the intermolecular charge transfer rates are calculated from quantum mechanical methods using the semiclassical Marcus theory expression for the transfer rate (Eq. (3.11)). Finally, the motion of charges in the system using a KMC algorithm is simulated. This scheme has been applied to systems consisting of conjugated small molecules, discotic liquid crystals, variable-molecular-weight polymers, and fullerenes, achieving a remarkable agreement with experimental measurements. It must be noted, however, that these systems have either a high degree of symmetry or a well-defined structure.

3.3.2.4 Comparison with Experiments

After reviewing the techniques used for modeling and simulating charge transport in disordered organic semiconductors, it is useful to review some of the ways in which the predicted properties can be measured. A very useful review is given by Coropceanu *et al.* [61], where the leading techniques are summarized. These are based on ToF measurements, FET and space-charge limited current diode (SCLC) characterization, and pulse-radiolysis time-resolved microwave photoconductivity (PR-TRMC). Each of these methods works in a different regime than the others, which illustrates the different variables that can be explored when studying charge transport in disordered organic materials. For example, ToF devices work in the low charge concentration regime, while FET and SCLC devices have charge densities several orders of magnitude higher. The differences between these measurements are not only related to the range and the ability to control variables such as temperature and charge density, but also to the length scales that are probed, with PR-TRMC measuring local transport properties of the materials, while the others probing micro- or macroscopic properties. Finally, it is important to understand whether the measurements reflect the steady-state properties of the sample as in the case of FET and SCLC, or if the charge carriers are generated in a nonequilibrium regime, as would be the case for ToF measurements. Studies of the dependence of transport on temperature or time are the most common way to test transport theories.

One of the differences between organic disordered media with a Gaussian DOS and their inorganic counterparts that have an exponential DOS lies in the relaxation process of excess charges. This means that a single charge, or a packet of noninteracting charges, which has been created out of equilibrium by photoexcitation or injection at an arbitrary energy will relax into lower-lying energy states in the band tail as time evolves. For an exponential tail, the charge will continue to fall into deeper states in the tail indefinitely. The process by which this energy relaxation happens depends on the temperature and observation time. For a given temperature, the hopping-down process will become slower than thermal activation into the transport energy after a certain time called the segregation time.

$$\tau_s = \nu_0^{-1} e^{\frac{3\varepsilon_0}{kT}} \tag{3.36}$$

It is called the segregation time because only for times longer than τ_s do the shallower states act as transport states and the deeper states as traps. As a result, depending on the typical observation time and temperature, transport in exponential tails proceeds by downward hopping in the tails or by thermal excitation into the transport energy. In fact, at all temperatures, thermal excitation into the transport energy will be the dominant factor, provided one waits long enough (Figure 3.10). It must be kept in mind that the concept of transport energy is only relevant when the initial state of the carrier is at an energy deeper than ε_t.

For a Gaussian DOS, on the other hand, the energy of photoinjected charges will eventually reach a steady-state value. This equilibration energy $\langle\varepsilon_\infty\rangle$ can be

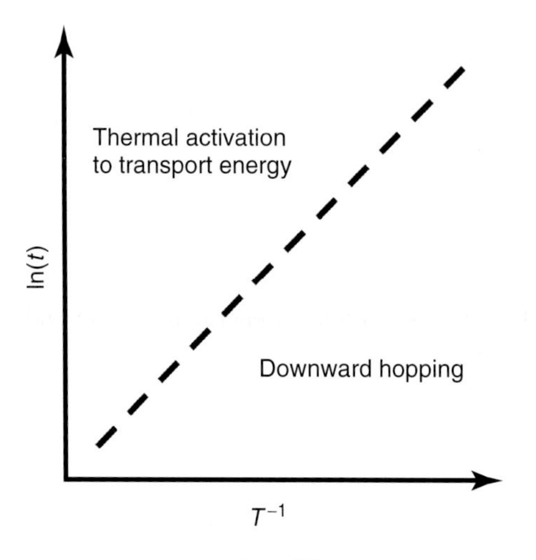

Figure 3.10 Sketch of the different transport regimes of hot carriers as they relax in an exponential distribution of localized states as a function of the observation time t and temperature T [50].

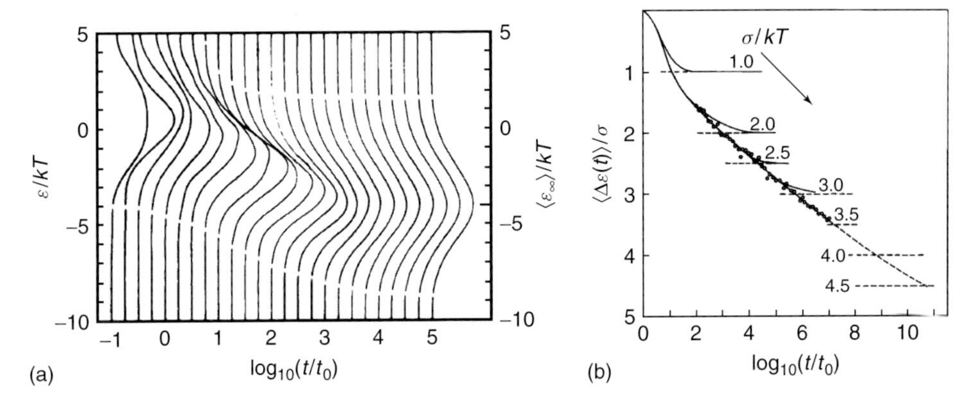

Figure 3.11 Relaxation of a packet of non-interacting charges in a Gaussian DOS. (a) The temporal distribution of energies for the charges, in the case of a DOS with a width $\sigma = 2kT$. (b) Temporal evolution of the average energy of a packet of carriers for a set of Gaussian DOS with varying σ. The time scale is defined by the dwell time t_0 of a carrier on a site in a lattice without disorder and intersite separation a, $t_0^{-1} = 6\nu_0 \mathrm{e}^{-2\alpha a}$. (Reprinted with permission from Ref. [86]. Copyright (1993) by John Wiley and Sons.)

calculated for the case of zero electric field as follows:

$$\langle \varepsilon_\infty \rangle = \lim_{t \to \infty} \langle \varepsilon(t) \rangle = \frac{\int_{-\infty}^{+\infty} \mathrm{d}\varepsilon \varepsilon g(\varepsilon) \mathrm{e}^{-\frac{\varepsilon}{kT}}}{\int_{-\infty}^{+\infty} \mathrm{d}\varepsilon g(\varepsilon) \mathrm{e}^{-\frac{\varepsilon}{kT}}} = -\frac{\sigma^2}{kT} \tag{3.37}$$

This relaxation process has been verified using Monte Carlo simulations by Bässler and coworkers as shown in Figure 3.11 [47, 52], and the average value for the energy $\langle \varepsilon(t) \rangle$ follows an approximate logarithmic behavior, as expected from analytical results [86]. The comparison of these results is not straightforward because there is no direct way to probe the relaxation of the charges in a material. However, since triplet excitations are transferred according to the same dynamics as charge carriers, they can be used as a model system to observe energy relaxation. It should be kept in mind that triplet states have a finite lifetime, which must be taken into account when comparing with charge relaxation simulations. Nevertheless, the experimental measurements agree with simulations and analytical results [52]. So, in a Gaussian DOS, the excited charges will undergo energy relaxation and eventually reach equilibrium at a given time t_{rel}, after which transport will follow by thermal excitation from the equilibration energy $\langle \varepsilon_\infty \rangle$ to the transport energy ε_t.

Another parameter that plays an important role in the charge transport mechanism is temperature. As discussed in the previous section, charge transport in single crystals is bandlike and is impeded by scattering processes, mostly by lattice phonons. Thus, the mobility decreases with a well-known power-law behavior $\mu \sim T^{-b}$. In disordered materials, conduction is not due to band transport, but due to hopping between localized states, which relies on thermal activation to overcome the energy barriers between localized states. Hence, the mobility increases strongly as temperature is increased [61, 75, 87, 88]. This effect is typically fit to an Arrhenius dependence of the form

$$\mu_{\text{Arrhenius}} = \mu_{\infty} e^{-\frac{E_a}{kT}} \tag{3.38}$$

where $E_a > 0$ is the activation energy and μ_{∞} is a temperature independent prefactor. Another parameterization of the temperature dependence is obtained by the results of theoretical simulations for transport in a Gaussian disorder model (GDM) and is of the form

$$\mu_{\text{GDM}} = \mu_{\infty} e^{-\left(\frac{T_0}{T}\right)^2} \tag{3.39}$$

where T_0 is indicative of the amount of energy disorder and is related to the DOS width. As shown previously, many theories predict a temperature dependence of mobility in the form of an exponential of a power of temperature. Unfortunately, the temperature range experimentally accessible is limited by the low melting or glass transition temperatures of these soft organic materials, and measurements of mobility in small temperature ranges yield values that can be fitted reasonably well with many functional forms [61]. In fact, ToF measurements on conjugated polymers have used both functional forms for the same experimental dataset and obtained good agreement for both cases, as shown in Figure 3.12 [75, 87].

An explanation of the Arrhenius-type dependence has been provided by Yu *et al.* [78], proposing that the DOS has a temperature-dependent width because of the variations in conformation of the molecules as the temperature is changed, and even though the energy disorder is assumed to have a Gaussian form, the temperature dependence of the mobility is $\ln(\mu) \sim T^{-1}$ instead of $\ln(\mu) \sim T^{-2}$.

In some cases, the model parameters required to fit one of these two possible temperature dependences to experimental data are physically unreasonable. One of such cases is the extremely large values of intermolecular coupling integrals needed to fit an Arrhenius-type dependence with a small-polaron model as pointed out by Kreouzis and coworkers [87]. A study by Parris and coworkers in which a small-polaron transfer rate is used on a DOS with correlated energetic disorder poses a solution to this issue. Using physically reasonable intermolecular coupling values, the predicted temperature dependence tends toward Arrhenius-type behavior when the polaron-binding energy exceeds the energetic disorder, and toward non-Arrhenius behavior in the limit where energetic disorder dominates [87, 89]. The effects of a correlated energetic disorder are discussed in more detail later.

Whenever electrical charges move within a material and their transport properties are analyzed, it is important to pay attention to the effect that an external electric field has on the behavior of observable variables such as the mobility, particularly when measurements and device-operating conditions are in a nonzero field.

Within the hopping transport mechanism, it has already been discussed that an electric field $\boldsymbol{F}$ lowers the activation barrier for a charge transfer between two localized states with different energies $(\varepsilon_i, \varepsilon_j)$ separated by a distance $\boldsymbol{r}_{ij}$ as follows $\Delta\varepsilon_{ij} = \varepsilon_j - \varepsilon_i - q\boldsymbol{r}_{ij} \cdot \boldsymbol{F}$. This presents the charge carriers easier access to shallower states as shown in Figure 3.13 and modifies their energy distribution, in a manner

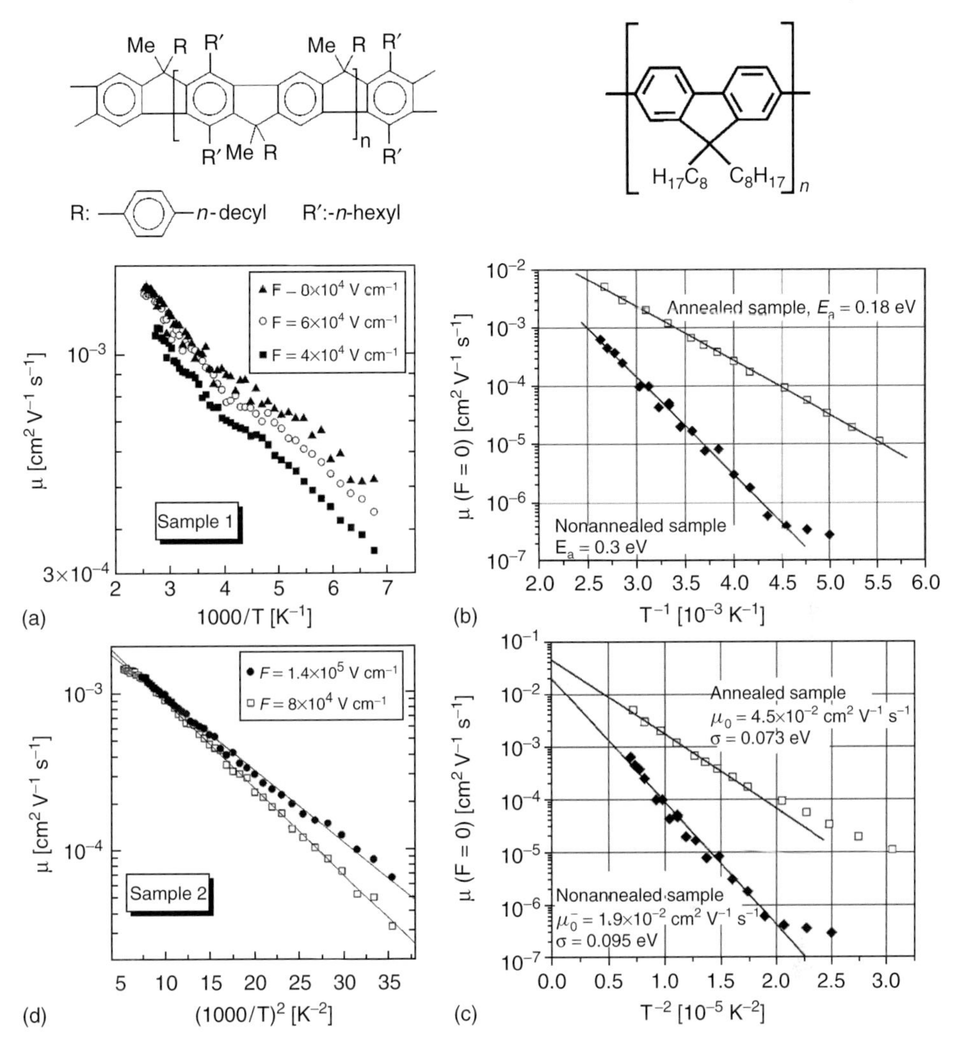

Figure 3.12 Temperature dependence of the hole mobility for conjugated polymers of the (a-b) phenylene [75] and (c-d) fluorene [87] families. For the phenylene-type polymer, the temperature dependece varies when it is measured while heating up (a) or cooling down (b). For the fluorene-type polymer, the data is the same in (c) and (d), but displayed vs. different temperature dependences. The data in (c) was modeled with an Arrhenius dependence while that of (d) is fitted to a Gaussian disorder model. (Reprinted with permission from Refs. [75] and [87]. Copyright (1999, 2006) by the American Institute of Physics and the American Physical Society, respectively.)

that has been compared to raising the temperature of the charge carriers above the lattice temperature [90, 91].

The manner in which an electric field modifies charge mobility has been modeled using a Poole–Frenkel mechanism. The Poole–Frenkel effect describes the lowering of the thermal excitation barrier for a charge in a localized state with

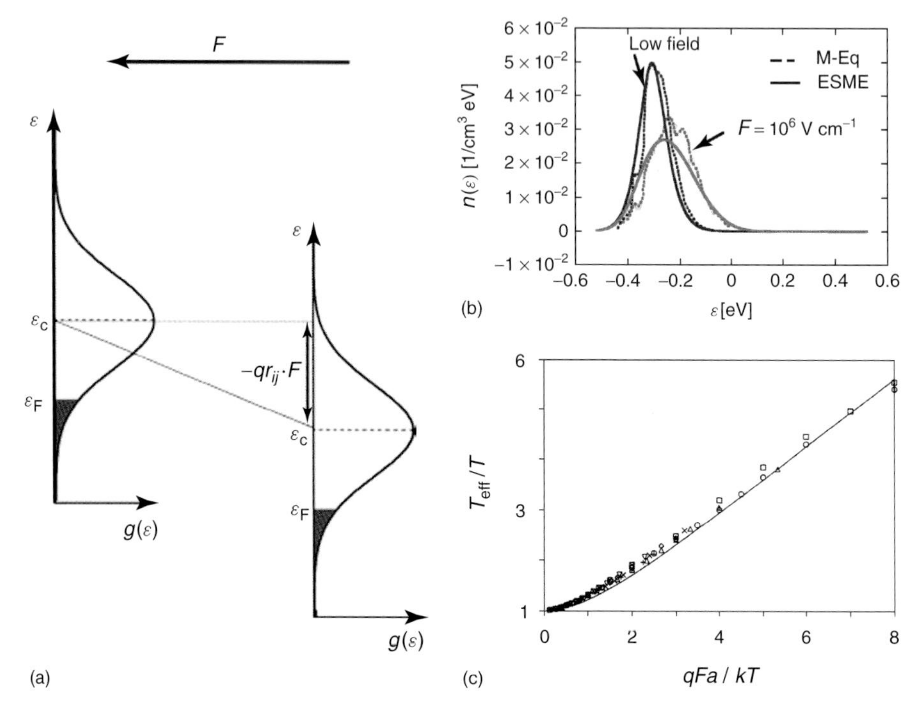

Figure 3.13 Hopping of a charge in the presence of an electric field. (a) There are more states available to the charge due to the electric field [90], which modifies the energy distribution of the carriers. The shaded region in (a) represents the occupied states. This effect is illustrated in (b) comparing the case for low and high fields using the Master Equation and the Energy-Space Master Equation approaches. (Reprinted with permission from Ref. [81]. Copyright (2006) by the American Physical Society). This is analogous to the charge carriers having an effective temperature higher than the lattice temperature (c). (Reprinted with permission from Ref. [90]. Copyright (1992) by the American Physical Society.)

a Coulombic potential, as shown in Figure 3.14. This model leads to a behavior of the form

$$\mu_{\mathrm{PF}} \propto \mathrm{e}^{\gamma_{\mathrm{PF}}\sqrt{F}}, \tag{3.40}$$

where γ_{PF} is the Poole–Frenkel factor, which has a temperature dependence

$$\gamma_{\mathrm{PF}} = C\left(\frac{1}{kT} - \frac{1}{kT_{\mathrm{PF}}}\right), \tag{3.41}$$

and C and T_{PF} are constants [80, 92]. This Poole–Frenkel behavior has been well documented in many organic systems [46, 75, 87, 93].

Such $\ln(\mu) \sim \sqrt{F}$ behavior is expected from a conduction mechanism in which charge carriers must escape from charge trapping centers, but it was realized that organic systems such as molecularly doped polymers do not contain a sufficiently large number of this type of traps, so an alternate explanation was proposed. Gartstein and Conwell [94] showed that for a wide range of electric fields, a

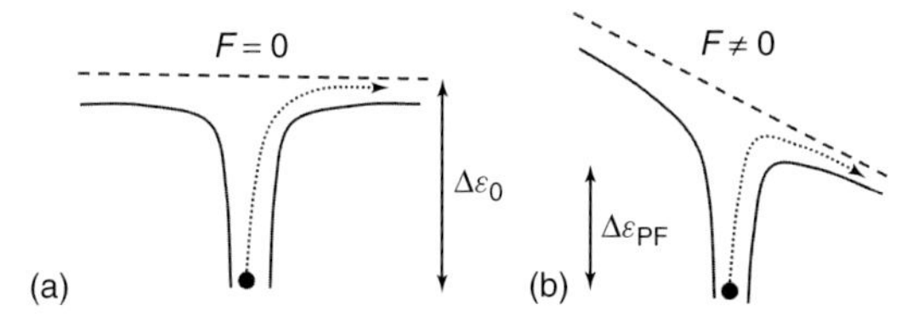

Figure 3.14 The Poole–Frenkel barrier-lowering mechanism. The case for zero electric field is shown in (a), and (b) illustrates that it is easier for a charge to escape from a trap when an external electric field is applied.

dependence close to that expected from a Poole–Frenkel behavior can also be caused by transport through a system in which the energetic disorder is spatially correlated. Such correlations can arise from fluctuations in the site energies caused by long-range interactions such as charge–dipole or dipole–dipole interactions with neighboring molecules. Also, amorphous organic materials are not entirely random systems, but they exhibit some degree of short-range order, and the resulting small correlations in position and conformation can enhance the amount of energetic correlation.

Dunlap and coworkers [80, 95] have used the closed-form expression for the steady-state drift velocity of a charge hopping in a 1-D grid developed by Derrida [79] and included the effects of correlation to obtain an analytical solution to the problem, which agreed well with numerical simulations. A 3-D generalization of this result was then used to explain the field dependence of the ToF mobility in amorphous films of a small organic molecule, phenylenediamine [95, 96]. The applicability of this correlated disorder model (CDM) to different types of disordered organic systems is still being debated, but it has been used to successfully explain the field dependence of the mobility in conjugated polymers, using both the Miller–Abrahams and the symmetric polaron transfer rates [87].

Another consequence of the presence of an external electric field, which may not be obvious at first, is the decrease in the mobility at high fields. This effect is due to the presence of dead ends in the transport path and is much more common in blends than in pure materials due to the added effect of blend morphology, with dead ends arising near the boundary between the blend components [97]. At low or moderate field strengths, charges can still escape from dead ends by hopping against the field direction, or by diffusing perpendicular to it, but at higher fields, it becomes increasingly difficult to hop against the field direction and diffusion alone may not provide an escape path for the charge. Increasing the charge density can populate the dead-end regions, and the electrostatic potential from the trapped charges will prevent further trapping at the dead ends. These processes are sketched in Figure 3.15.

Even neglecting the effect of dead ends, the transport properties still depend on the charge concentration, as has been extensively studied to understand the properties of devices that operate under high charge densities (e.g., FETs) [61, 88, 98]. It has been typically observed that for the same material, mobilities measured in FETs are at least one order of magnitude larger than those measured in a diode configuration. This discrepancy observed in amorphous materials

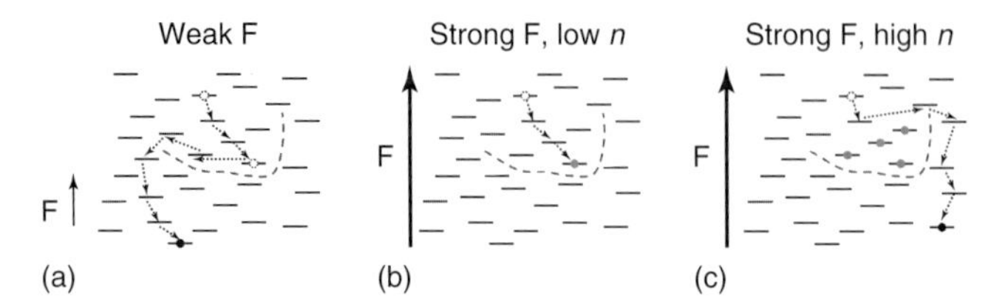

Figure 3.15 Hopping of an electron near a dead end (delimited by the curved dotted line) in the presence of an electric field ***F***. In (a), the field strength is low enough that the charge can escape from the dead end by diffusion and hopping against the field, but increasing the field intensity traps the charge by making the escape path very unfavorable energetically (b). At large carrier concentrations, it is possible to populate the sites in a dead-end region and avoid further trapping of charges (c).

(i.e., where no anisotropy in transport is to be expected) has been explained by the large difference in charge density between a FET channel and a diode [98–100].

As mentioned earlier, when charges are at equilibrium in a disordered system of localized sites they occupy states deep in the band tail of the DOS. Using the TE model, it follows that the hopping rate is directly related to the energy difference between the occupied deep states and the narrow band around the transport energy ε_t. Increasing the charge density will raise the energy distribution of charge states but will not modify the transport energy since its position is not dependent on the initial energy level of charges hopping to it. Thus, raising the energy of charges in the tail will decrease the energy difference between them and the transport energy, resulting in a strong increase in the charge mobility (Figure 3.16) [73]. The filling of states is a direct consequence of the use of Fermi–Dirac statistics, and if it is neglected by using Boltzmann statistics, the dependence of mobility with charge density is much weaker as a result, as noted by Vissenberg and Matters [88].

Apart from the level-filling effect caused by increasing the concentration of charges, their presence can modify the energy levels of the hopping sites, for

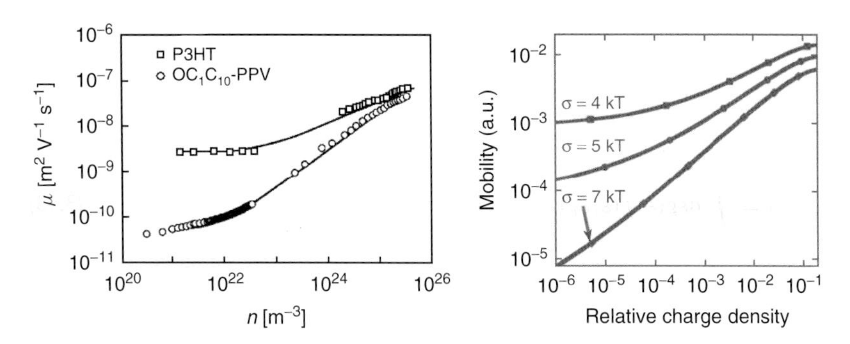

Figure 3.16 (a) Experimental measurement of the hole mobility for two different polymers in the regimes of low (diode measurements) and high (FET measurements) charge densities. (Reprinted with permission from Ref. [98]. Copyright (2005) by the American Physical Society.) (b) The mean medium approximation model of Roichman *et al.* agrees with the experimental trends. (Reprinted with permission from Ref. [77]. Copyright (2009) by John Wiley and Sons.)

example, modify the width of the DOS [73], and the Coulombic repulsion between them can affect the charge transport path. A method to account for the charge density and its effect in transport parameters was developed by Roichman *et al.* with a device-oriented motivation instead of a basic principles approach [77, 101, 102]. Such formalism is called the mean medium approximation (MMA) and it assumes a uniform energy distribution of states across the sample, giving each site a finite probability to be at any energy as defined by the system's DOS. It then calculates the probability of finding a target site at a given distance by the use of a radial correlation function, which is another input to the model along the DOS. The current is obtained by evaluating the integral over all possible distances and initial–final energy configurations, taking into account the transfer rate, as well as the Fermi–Dirac occupation probability. The merit of this model lies in the ability to reproduce experimental trends with a small number of assumptions, but it consistently overestimates the mobility values, particularly at low temperatures [103].

The dependence of mobility on charge concentration is different when polaronic effects are included, as shown by Bässler and coworkers [103]. If one uses the polaron jump rate model, a much weaker dependence of mobility on charge concentration is obtained, when compared to that observed when using the Miller–Abrahams jump rate, which neglects polaronic effects. These results show that the large change in mobility observed between ToF or diode measurements and FET devices is incompatible with a polaron energy that is large compared to the width of the DOS. Also, this observation points to the fact that a large polaron-binding energy could be responsible for the similarity in ToF and FET mobilities for some organic disordered materials such as polyfluorene copolymers [103].

Another consequence of the energy distribution of charge in the DOS is the departure from the Einstein relationship between the diffusion coefficient and mobility, even for equilibrium conditions [46]. A study of this effect for a disordered system with a Gaussian DOS in thermal equilibrium has been provided by Roichman and Tessler [104], in which a generalized relation between D and μ was derived from

$$\frac{D}{\mu} = \frac{n}{q\frac{\partial n}{\partial \varepsilon_F}}, \tag{3.42}$$

with the charge concentration determined by

$$n = \int_{-\infty}^{\infty} d\varepsilon g(\varepsilon) f(\varepsilon, \varepsilon_F) \tag{3.43}$$

Further, inserting the expressions for the DOS $g(\varepsilon)$ and Fermi–Dirac distribution function $f(\varepsilon, \varepsilon_F)$ results in the following:

$$\frac{D}{\mu} = \frac{kT}{q} \frac{\int_{-\infty}^{\infty} d\varepsilon e^{-\frac{(\varepsilon-\varepsilon_C)^2}{2\sigma^2}} \cdot \frac{1}{1+e^{\frac{\varepsilon-\varepsilon_F}{kT}}}}{\int_{-\infty}^{\infty} d\varepsilon e^{-\frac{(\varepsilon-\varepsilon_C)^2}{2\sigma^2}} \cdot \frac{e^{\frac{\varepsilon-\varepsilon_F}{kT}}}{\left[1+e^{\frac{\varepsilon-\varepsilon_F}{kT}}\right]^2}} \tag{3.44}$$

In the case when most of the charge distribution is far from the Fermi level, the usual Einstein relation is obtained. In the case of disordered organic materials, with typical DOS widths of $\sigma \sim 100$ meV and at room temperature, the peak of the charge distribution is close to the Fermi level and the material is in the degenerate semiconductor regime. If the DOS distribution was narrower, or the total charge concentration was very low, most of the charge distribution would be far from ε_F and the material would be in the nondegenerate semiconductor regime [104]. The fact that organic materials are most often found in a regime where the standard Einstein relation must be modified should be taken into account when calculating the electrical properties of such systems.

In a similar way, attention must be paid to the agreement of computational results with the standard $Dq = \mu kT$ when the simulation parameters define a situation in which it should, for example, in Monte Carlo simulations in which charges are considered as noninteracting independent entities. Some reports have claimed a deviation from the Einstein relation, but closer inspection reveals that the system under study was not in thermal equilibrium because of the choice of simulation parameters. When the equilibrium conditions for the system were satisfied, the well-known $Dq = \mu kT$ equation was obtained [46].

Nonequilibrium conditions can also be useful for studying some material properties. In the case of ToF experiments, an ensemble of optically excited charge carriers is generated out of equilibrium and its evolution is monitored over time. A characteristic feature of these experiments is the transition from dispersive to nondispersive transport that is observed as a function of temperature as described by Bässler and coworkers [47].

Also along these lines is charge recombination, an effect that is important in some organic electronic devices such as solar cells. Of the available charge transport models, the ones that can readily incorporate a pathway for charge recombination are the computational methods. In the description of the Master Equation approach, a decay rate for charge recombination is already included. For Monte Carlo experiments, one can add a recombination rate at each site in a manner equivalent to adding an extra neighbor to each point in the grid.

After reviewing the most widely used tools to describe charge transport in disordered organic materials, it is important to emphasize that whenever using these models, one should pay attention to the underlying assumptions and check that they agree with the dominant effect in the system to be studied. For example, the historical development of these concepts started with inorganic amorphous materials. Molecularly doped insulating polymer matrices were the first organic systems in which these tools were used. In covalently bound inorganic semiconductors, charges interact very weakly with phonons in the lattice, and thus transport is well described by transfer rates of the Miller–Abrahams type. For the case of molecularly doped insulating polymers, each transport site is isolated from the rest and the amount of disorder is very large, thus decreasing the importance of polaronic effects. In novel organic semiconductors such as conjugated polymers, electron–phonon interactions are no longer a perturbation but are comparable to electronic interactions, and the effect of polarons must be taken into account [61]. A great example for contrasting the assumptions of different models and comparing their results to experiment is the study by Bässler and coworkers [103], using the effective TE model to explain the strength of the dependence of mobility with charge density as a function of the importance of polaronic effects.

Another important note is the assumption of a static sample configuration for the duration of the charge transport process, which could not always be satisfied. If the time scale of intermolecular charge transport is comparable to that of the lattice vibrations, the transfer integrals will have a time-dependent component that must be taken into account. Troisi *et al.* recently extended the dynamic disorder theory to semicrystalline systems [105]. The extent of these dynamic disorder effects is yet to be determined.

3.4 Conclusions

Starting from perfect crystals and analytical models that apply to amorphous solids and continuing to advanced computer-simulation methods that model the motion of single molecules, our understanding of how charge moves through organic semiconductors has made great strides. With the rise of more sophisticated analytical tools, such as synchrotron-based X-ray diffraction or absorption methods as well as scanning probe methods, a more complete picture of the microstructure of these materials is being gained [106]. Incorporating the details of the microstructure at several length scales, from molecular packing to crystallite organization, is the next fundamental challenge facing modeling of charge transport in organic semiconductors. Only when we understand how microstructures form and how they affect carrier mobility will we be able to truly design materials to exhibit optimized transport properties.

Acknowledgments

The authors gratefully acknowledge helpful comments from L. Jimison and J. Rivnay, as well as financial support from the National Science Foundation in the form of a Career Award. This publication was partially based on work supported by the Center for Advanced Molecular Photovoltaics (Award No KUS-C1-015-21), made by King Abdullah University of Science and Technology (KAUST).

References

1. Kelley, T.W., Baude, P.F., Gerlach, C., Ender, D.E., Muyres, D., Haase, M.A., Vogel, D.E., and Theiss, S.D. (2004) *Chem. Mater.*, **16**, 4413.
2. McCulloch, I., Heeney, M., Bailey, C., Genevicius, K., Macdonald, I., Shkunov, M., Sparrowe, D., Tierney, S., Wagner, R.,

Zhang, W.M., Chabinyc, M.L., Kline, R.J., McGehee, M.D., and Toney, M.F. (2006) *Nat. Mater.*, **5**, 328.
3. Tsao, H.N., Cho, D., Andreasen, J.W., Rouhanipour, A., Breiby, D.W., Pisula, W., and Müllen, K. (2009) *Adv. Mater.*, **21**, 209.
4. Umeda, T., Kumaki, D., and Tokito, S. (2009) *J. Appl. Phys.*, **105**, 024516.
5. Pope, M. and Swenberg, C.E. (1982) *Electronic Processes in Organic Crystals*, Clarendon Press, Oxford.
6. Bredas, J.L., Beljonne, D., Coropceanu, V., and Cornil, J. (2004) *Chem. Rev.*, **104**, 4971.
7. Bredas, J.L., Calbert, J.P., da Silva, D.A., and Cornil, J. (2002) *Proc. Natl. Acad. Sci. U.S.A.*, **99**, 5804.
8. Da Silva Filho, D.A., Kim, E.G., and Brédas, J.L. (2005) *Adv. Mater.*, **17**, 1072.
9. Karl, N., (2001) Charge-carrier mobility in organic crystals, in *Organic electronic materials: conjugated polymers and low molecular weight organic solids* Chapter 8, (eds R. Farchioni and G. Grosso), Springer-Verlag.
10. Tiago, M.L., Northrup, J.E., and Louie, S.G. (2003) *Phys. Rev. B*, **67**, 115212.
11. Sato, N. (1994) *Synth. Met.*, **64**, 133.
12. Sato, N., Seki, K., and Inokuchi, H. (1981) *J. Chem. Soc. Faraday Trans.*, **2**, 1621.
13. Kittel, C. (1996) *Introduction to Solid State Physics*, John Wiley & Sons, Inc., New York.
14. Probst, K.H. and Karl, N. (1975) *Phys. Status Solidi A: Appl. Res.*, **27**, 499.
15. de Boer, R.W.I., Gershenson, M.E., Morpurgo, A.F., and Podzorov, V. (2004) *Phys. Status Solidi A: Appl. Res.*, **201**, 1302.
16. Gershenson, M.E., Podzorov, V., and Morpurgo, A.F. (2006) *Rev. Mod. Phys.*, **78**, 973.
17. Podzorov, V., Menard, E., Borissov, A., Kiryukhin, V., Rogers, J.A., and Gershenson, M.E. (2004) *Phys. Rev. Lett.*, **93**, 086602-1–086602-4.
18. Podzorov, V., Menard, E., Rogers, J.A., and Gershenson, M.E. (2005) *Phys. Rev. Lett.*, **95**, 226601-1–226601-4.
19. Sundar, V.C., Zaumseil, J., Podzorov, V., Menard, E., Willett, R.L., Someya, T., Gershenson, M.E., and Rogers, J.A. (2004) *Science*, **303**, 1644.
20. de Boer, R.W.I., Klapwijk, T.M., and Morpurgo, A.F. (2003) *Appl. Phys. Lett.*, **83**, 4345.
21. Hulea, I.N., Fratini, S., Xie, H., Mulder, C.L., Iossad, N.N., Rastelli, G., Ciuchi, S., and Morpurgo, A.F. (2006) *Nat. Mater.*, **5**, 982.
22. Molinari, A.S., Alves, H., Chen, Z., Facchetti, A., and Morpurgo, A.F. (2009) *J. Am. Chem. Soc.*, **131**, 2462.
23. Stassen, A.F., de Boer, R.W.I., Iosad, N.N., and Morpurgo, A.F. (2004) *Appl. Phys. Lett.*, **85**, 3899.
24. Menard, E., Podzorov, V., Hur, S.H., Gaur, A., Gershenson, M.E., and Rogers, J.A. (2004) *Adv. Mater.*, **16**, 2097.
25. Panzer, M.J. and Frisbie, C.D. (2008) *Adv. Mater.*, **20**, 3177.
26. Pham, P.T.T., Xia, Y., Frisbie, C.D., and Bader, M.M. (2008) *J. Phys. Chem. C*, **112**, 7968.
27. Xia, Y., Kalihari, V., Frisbie, C.D., Oh, N.K., and Rogers, J.A. (2007) *Appl. Phys. Lett.*, **90**, 162106-1–162106-3.
28. Hasegawa, T. and Takeya, J. (2009) *Sci. Technol. Adv. Mater.*, **10**, 024314-1–024314-16.
29. Takahashi, T., Takenobu, T., Takeya, J., and Iwasa, Y. (2006) *Appl. Phys. Lett.*, **88**, 033505-1–033505-3.
30. Takeya, J., Goldmann, C., Haas, S., Pernstich, K.P., Ketterer, B., and Batlogg, B. (2003) *J. Appl. Phys.*, **94**, 5800.
31. Takeya, J., Kato, J., Hara, K., Yamagishi, M., Hirahara, R., Yamada, K., Nakazawa, Y., Ikehata, S., Tsukagoshi, K., Aoyagi, Y., Takenobu, T., and Iwasa, Y. (2007) *Phys. Rev. Lett.*, **98**, 196804-1–196804-4.
32. Takeya, J., Yamagishi, M., Tominari, Y., Hirahara, R., Nakazawa, Y., Nishikawa, T., Kawase, T., Shimoda, T., and Ogawa, S. (2007) *Appl. Phys. Lett.*, **90**, 102120-1–102120-3.
33. Yamagishi, M., Soeda, J., Uemura, T., Okada, Y., Takatsuki, Y., Nishikawa, T., Nakazawa, Y., Doi, I., Takimiya, K., and Takeya, J. (2010) *Phys. Rev. B*, **81**, 161306-1–161306-4.

34. Yamagishi, M., Tominari, Y., Uemura, T., Yamada, K., and Takeya, J. (2010) *Jpn. J. Appl. Phys.*, **49**, 01AB05-1–01AB05-3.
35. Jurchescu, O.D., Subramanian, S., Kline, R.J., Hudson, S.D., Anthony, J.E., Jackson, T.N., and Gundlach, D.J. (2008) *Chem. Mater.*, **20**, 6733.
36. Troisi, A. and Orlandi, G. (2006) *Phys. Rev. Lett.*, **96**, 086601-1–086601-4.
37. Troisi, A. and Orlandi, G. (2006) *J. Phys. Chem. A*, **110**, 4065.
38. Troisi, A. (2007) *Adv. Mater.*, **19**, 2000.
39. Laarhoven, H.A.V., Flipse, C.F.J., Koeberg, M., Bonn, M., Hendry, E., Orlandi, G., Jurchescu, O.D., Palstra, T.T.M., and Troisi, A. (2008) *J. Chem. Phys.*, **129**, 044704-1–044704-5.
40. Pendry, J.B. (1982) *J. Phys. C: Solid State Phys.*, **15**, 5773.
41. McMahon, D.P. and Troisi, A. (2010) *ChemPhysChem*, **11**, 2067.
42. Anderson, P.W. (1958) *Phys. Rev.*, **109**, 1492.
43. Mott, N.F. (1988) *Philos. Mag. B: Phys. Condens. Matter Stat. Mech. Electron. Opt. Magn. Prop.*, **58**, 369.
44. Ball, M.A. (1971) *J. Phys. Part C: Solid State Phys.*, **4**, 1747.
45. Ball, M.A. (1972) *J. Phys. Part C: Solid State Phys.*, **5**, L13.
46. Baranovskii, S. (2006) *Charge Transport in Disordered Solids With Applications In Electronics*, vol. 8, John Wiley & Sons, Inc.
47. Pautmeier, L., Richert, R., and Bassler, H. (1989) *Philo. Mag. Lett.*, **59**, 325.
48. Mott, S.N.F. and Davis, E.A. (1979) *Electronic Processes in Non-Crystalline Materials*, Clarendon Press.
49. Street, R.A. (1991) *Hydrogenated Amorphous Silicon*, Cambridge University Press, New York.
50. Monroe, D. (1985) *Phys. Rev. Lett.*, **54**, 146.
51. Monroe, D. and Kastner, M.A. (1986) *Phys. Rev. B*, **33**, 8881.
52. Bassler, H. (1993) *Phys. Status Solidi B: Basic Res.*, **175**, 15.
53. Chang, J.F., Sirringhaus, H., Giles, M., Heeney, M., and McCulloch, I. (2007) *Phys. Rev. B*, **76**, 205204.
54. Horowitz, G., Hajlaoui, M., and Hajlaoui, R. (2000) *J. Appl. Phys.*, **87**, 4456.
55. Horowitz, G., Hajlaoui, R., Fichou, D., and El Kassmi, A. (1999) *J. Appl. Phys.*, **85**, 3202.
56. Salleo, A. (2007) *Mater. Today*, **10**, 38.
57. Salleo, A., Chen, T.W., Volkel, A.R., Wu, Y., Liu, P., Ong, B.S., and Street, R.A. (2004) *Phys. Rev. B*, **70**, 115311-1–115311-10.
58. Street, R.A., Northrup, J.E., and Salleo, A. (2005) *Phys. Rev. B*, **71**, 165202-1–165202-13.
59. Street, R.A., Salleo, A., Chabinyc, M., and Paul, K. (2004) *J. Non-Cryst. Solids*, **338–340**, 607.
60. Xie, H., Alves, H., and Morpurgo, A.F. (2009) *Phys. Rev. B*, **80**, 245305-1–245305-7.
61. Coropceanu, V., Cornil, J., da Silva, D.A., Olivier, Y., Silbey, R., and Bredas, J.L. (2007) *Chem. Rev.*, **107**, 2165.
62. Nelson, J., Kwiatkowski, J.J., Kirkpatrick, J., and Frost, J.M. (2009) *Acc. Chem. Res.*, **42**, 1768.
63. Barbara, P.F., Meyer, T.J., and Ratner, M.A. (1996) *J. Phys. Chem.*, **100**, 13148.
64. Grunewald, M. and Thomas, P. (1979) *Phys. Status Solidi B: Basic Res.*, **94**, 125.
65. Shapiro, F.R. and Adler, D. (1985) *J. Non-Cryst. Solids*, **74**, 189.
66. Baranovskii, S.D., Faber, T., Hensel, F., and Thomas, P. (1997) *J. Phys. Condens. Matter*, **9**, 2699.
67. Martens, H.C.F., Blom, P.W.M., and Schoo, H.F.M. (2000) *Phys. Rev. B*, **61**, 7489.
68. Arkhipov, V.I., Emelianova, E.V., and Adriaenssens, G.J. (2001) *Phys. Rev. B*, **64**, 125125-1–125125-6.
69. Mott, N.F. (1987) *Conduction in Non-Crystalline Materials*, Oxford University Press, Oxford.
70. Pollak, M. (1971) *Discuss. Faraday Soc.*, **1970**, 13.
71. Efros, A.L. and Shklovskii, B.I. (1975) *J. Phys. C: Solid State Phys.*, **8**, L49.
72. Arkhipov, V.I., Heremans, P., Emelianova, E.V., Adriaenssens, G.J., and Bassler, H. (2003) *Chem. Phys.*, **288**, 51.

73. Arkhipov, V.I., Heremans, P., Emelianova, E.V., Adriaenssens, G.J., and Bassler, H. (2003) *Appl. Phys. Lett.*, **82**, 3245.
74. Bassler, H. (1994) *Int. J. Mod. Phys. B*, **8**, 847.
75. Hertel, D., Bassler, H., Scherf, U., and Horhold, H.H. (1999) *J. Chem. Phys.*, **110**, 9214.
76. Schonherr, G., Bassler, H., and Silver, M. (1981) *Philos. Mag. B: Phys. Condens. Matter Stat. Mech. Electron. Opt. Magn. Prop.*, **44**, 47.
77. Tessler, N., Preezant, Y., Rappaport, N., and Roichman, Y. (2009) *Adv. Mater.*, **21**, 2741.
78. Yu, Z.G., Smith, D.L., Saxena, A., Martin, R.L., and Bishop, A.R. (2000) *Phys. Rev. Lett.*, **84**, 721.
79. Derrida, B. (1983) *J. Stat. Phys.*, **31**, 433.
80. Dunlap, D.H., Parris, P.E., and Kenkre, V.M. (1996) *Phys. Rev. Lett.*, **77**, 542.
81. Preezant, Y. and Tessler, N. (2006) *Phys. Rev. B*, **74**, 235202-1–235202-5.
82. Metropolis, N. and Ulam, S. (1949) *J. Am. Stat. Assoc.*, **44**, 335.
83. Battaile, C.C. (2008) *Comput. Methods Appl. Mech. Eng.*, **197**, 3386.
84. Gillespie, D.T. (1977) *J. Phys. Chem.*, **81**, 2340.
85. Kwiatkowski, J.J. (2008) From molecules to mobilities: modelling charge transport in organic semiconductors, Department of Physics, Imperial College London, London, p. 132.
86. Ries, B., Bassler, H., Grunewald, M., and Movaghar, B. (1988) *Phys. Rev. B*, **37**, 5508.
87. Kreouzis, T., Poplavskyy, D., Tuladhar, S.M., Campoy-Quiles, M., Nelson, J., Campbell, A.J., and Bradley, D.D.C. (2006) *Phys. Rev. B*, **73**, 235201-1–235201-15.
88. Vissenberg, M. and Matters, M. (1998) *Phys. Rev. B*, **57**, 12964.
89. Parris, P.E., Kenkre, V.M., and Dunlap, D.H. (2001) *Phys. Rev. Lett.*, **87**, 126601/1.
90. Marianer, S. and Shklovskii, B.I. (1992) *Phys. Rev. B*, **46**, 13100.
91. Shklovskii, B.I., Levin, E.I., Fritzsche, H., and Baranovskii, S.D. (1990) in *Transport, Correlation and Structural Defects* (ed. H. Fritzsche), World Scientific Publishing Company, Chicago, p. 161.
92. Stallinga, P. (2009) *Electrical Characterization of Organic Electronic Materials and Devices*, John Wiley & Sons, Inc.
93. Pasveer, W.F., Cottaar, J., Tanase, C., Coehoorn, R., Bobbert, P.A., Blom, P.W.M., de Leeuw, D.M., and Michels, M.A.J. (2005) *Phys. Rev. Lett.*, **94**, 206601 / 1.
94. Gartstein, Y.N. and Conwell, E.M. (1995) *Chem. Phys. Lett.*, **245**, 351.
95. Novikov, S.V., Dunlap, D.H., Kenkre, V.M., Parris, P.E., and Vannikov, A.V. (1998) *Phys. Rev. Lett.*, **81**, 4472.
96. Borsenberger, P.M. and Shi, J. (1995) *Phys. Status Solidi B: Basic Res.*, **191**, 461.
97. Koster, L.J.A. (2010) *Phys. Rev. B*, **81**, 205318-1–205318-7.
98. Coehoorn, R., Pasveer, W.F., Bobbert, P.A., and Michels, M.A.J. (2005) *Phys. Rev. B*, **72**, 155206-1–155206-20.
99. Tanase, C., Blom, P.W.M., and de Leeuw, D.M. (2004) *Phys. Rev. B*, **70**, 193202-1–193202-4.
100. Tanase, C., Meijer, E.J., Blom, P.W.M., and de Leeuw, D.M. (2003) *Phys. Rev. Lett.*, **91**, 216601 / 1.
101. Roichman, Y., Preezant, Y., and Tessler, N. (2004) *Phys. Status Solidi A: Appl. Res.*, **201**, 1246.
102. Shaked, S., Tal, S., Roichman, Y., Razin, A., Xiao, S., Eichen, Y., and Tessler, N. (2003) *Adv. Mater.*, **15**, 913.
103. Fishchuk, I.I., Arkhipov, V.I., Kadashchuk, A., Heremans, P., and Bassler, H. (2007) *Phys. Rev. B*, **76**, 045210-1–045210-12.
104. Roichman, Y. and Tessler, N. (2002) *Appl. Phys. Lett.*, **80**, 1948.
105. Cheung, D.L., McMahon, D.P., and Troisi, A. (2009) *J. Am. Chem. Soc.*, **131**, 11179.
106. Salleo, A., Kline, R.J., DeLongchamp, D.M., and Chabinyc, M.L. (2010) Microstructural Characterization and Charge Transport in Thin Films of Conjugated Polymers, *Adv. Mater.*, **22**, 3812.

4
Silylethyne-Substituted Acenes and Heteroacenes

John E. Anthony and Adolphus G. Jones

4.1
Introduction

Acenes are one of the most intensely studied classes of organic semiconductors, and materials such as pentacene [1] and rubrene [2] are considered benchmark materials in the field of organic thin-film transistors (OTFTs). The main drawback to these materials is poor solubility, which significantly complicates the application of solution-based deposition process for device fabrication. Efforts to modify the electronic or solubility properties of acenes led to the development of a variety of heteroacenes – aromatic systems where carbocyclic rings of the acene are replaced by heteroatom-containing rings. In general, these derivatives did show improved stability, but still suffered from the same solubility issues that plague pentacene. Two other approaches that have been explored to improve the solution processability of linear acenes are the precursor approach and the functionalization approach. The precursor approach is most commonly applied to pentacene (although other systems have benefitted from this method) [3] and involves the addition of expendable substituents to the pentacene π-system, yielding "bent" molecules with significantly improved solubility. After film formation, the pentacene precursors are treated either thermally [4] or photochemically [5] to regain the pentacene chromophore. Thermal annealing of the resulting film yields devices with reasonable performance, although the high temperatures required to yield best performance often preclude the use of certain processes or substrates.

Functionalization of acenes involves modification to the periphery of the aromatic chromophore and does not disrupt the electronic conjugation in the acene ring system. A myriad of functionalization approaches to solubilize acenes and heteroacenes have been investigated [6]. The simplest approach involves the attachment of straight alkyl chains to the aromatic backbone. While the approach does not typically yield stable materials when applied to pentacene, numerous heteroacene systems have yielded high-performance solution-processable materials from the addition of two alkyl chains to the heteroaromatic backbone. A more versatile approach involves the substitution of the acene with trialkylsilylethynyl groups. In this approach, the solubilizing substituent is held away from the active electronic

Organic Electronics II: More Materials and Applications, First Edition. Edited by Hagen Klauk.

component by a small, flexible spacer (the C–C triple bond), allowing the aromatic faces of adjacent molecules to achieve very close contacts. Further, adjustments to the alkyl groups on the silane allow subtle tuning of the solubility and crystal packing of the material, allowing precise tuning for optimum performance in a variety of applications.

4.2
Silylethyne-Substituted Pentacenes

Pentacene (**1**) is one of the most common small-molecule materials used for thin-film electronic device studies. Chemically, there are a number of issues related to the larger acene structures that may lead to defect or trap formation in pentacene thin films. First, pentacene is prone to oxidation of the central aromatic ring, yielding 6,13 pentacenequinone (**2**, Figure 4.1). This process can take place either by a concerted endoperoxidation reaction or by an electron-transfer/oxidation in the presence of oxygen and light [7]. This type of reaction has been postulated as a significant source of charge traps in pentacene transistors [8]. Even in the absence of oxygen, pentacene still reacts to form the so-called butterfly dimer (**3**) [9]. Studies of this material have shown that the dimerization is at least partially reversible, either thermally or photochemically, and this partial reversibility may account for the significant production of bipentacene during purification of crude pentacene by sublimation [10]. For silylethyne-substituted pentacenes, "butterfly" dimerization is also the most common degradation pathway (to give, e.g., **4**) (Figure 4.2) [11]. The chemical reactivity of silylethyne-substituted pentacenes has been used to good effect in the synthesis of triptycene-based polymers [12] and in the separation of metallic single-walled carbon nanotubes [13].

Another potential issue with pentacene is the "herringbone" packing of the aromatic backbone (Figure 4.3). Theory predicts improved charge transport from face-to-face interactions (e.g., rubrene (**5**)) rather than the edge-to-face interactions seen in pentacene [14]. Further, the herringbone arrangement allows individual molecules to "slip" along the long axis of pentacene in the solid state – the defects arising from this sort of slipping have been postulated to account for shallow traps observed in the measurements of pentacene devices [15].

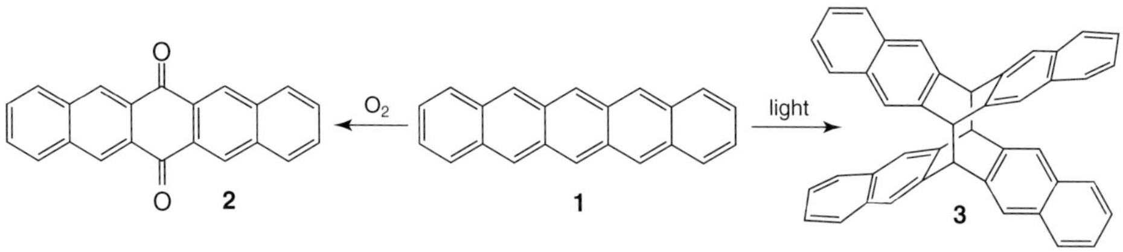

Figure 4.1 Pentacene and its decomposition pathways.

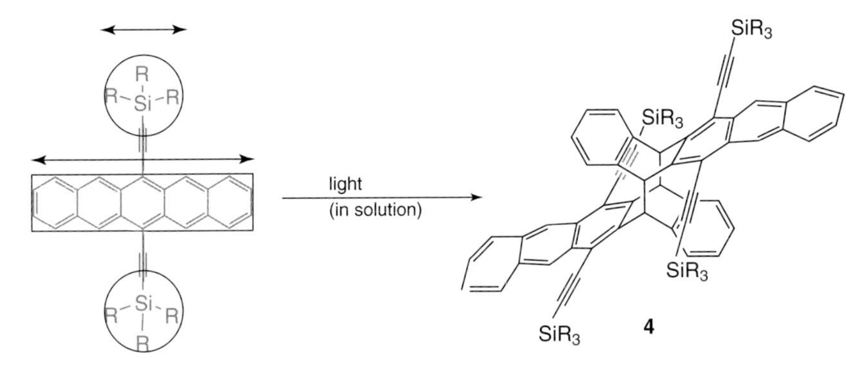

Figure 4.2 Silylethyne pentacenes and their decomposition pathway.

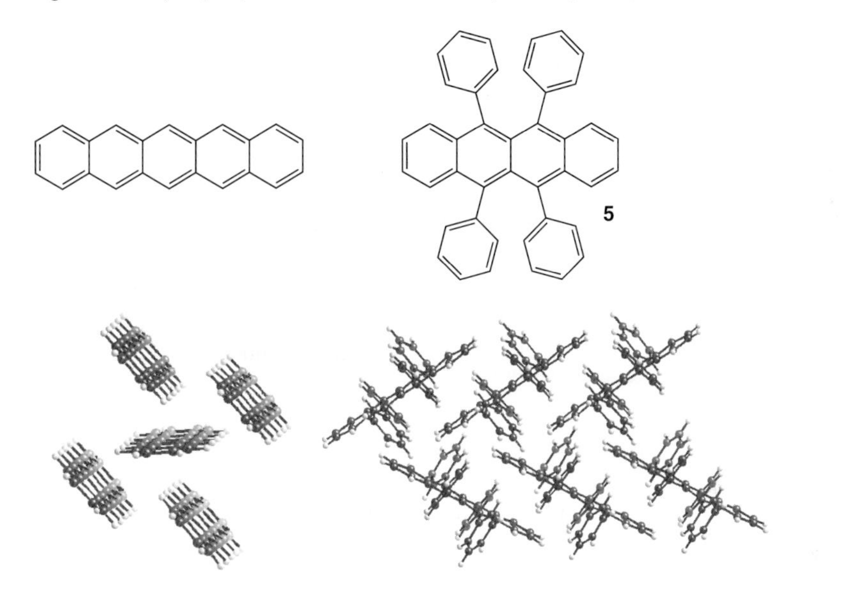

Figure 4.3 Crystal packing for pentacene and rubrene.

Substitution of pentacene with trialkylsilyl acetylenes is a simple approach to yield soluble pentacene derivatives. The trimethylsilylethynyl derivative of pentacene was first reported in passing in the year 2000 as one of a series of pentacene derivatives that reacted rapidly with C_{60} to yield isolable adducts [16]. The synthesis and study of this class of compounds for electronics applications began in 2001, with the observation that changes in crystal packing had a profound impact on the electronic properties of these pentacene materials [17]. The key to obtaining reliable performance from these materials lies in their synthesis. The straightforward preparation involves the addition of an alkyne anion to pentacenequinone, followed by deoxygenative workup with stannous chloride [18] or hydrogen iodide (Scheme 4.1) [19]. When conditions are optimized, the desired pentacene simply precipitates from solution and can be separated by filtration. Chromatography and

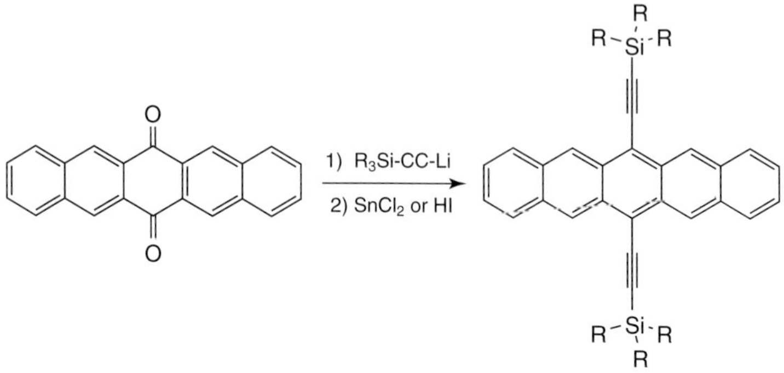

Scheme 4.1 Silylethynyl acene synthesis.

recrystallization are the most common purification methods used with this class of materials – in general, these compounds are not amenable to purification by sublimation.

The available purification method for this class of materials makes care in synthesis a top priority. Any impurities present in the starting acetylene, in particular, will be very difficult to separate from the resulting pentacene. Trace amounts of these impurities severely impact film morphology, leading to films with greatly diminished stability and devices with poor performance [20]. It is thus critically important to carefully assess starting material purity and synthetic procedures before preparing silylethyne-substituted pentacenes for electronics use.

4.3
Crystal Packing

The silylethyne approach to functionalization yields almost exclusively π-stacked (face-to-face interacting) materials. Small changes in the substitution of the silicon lead to significant changes in the crystal packing (Figure 4.4). These changes can be used to advantage in tuning materials for performance in photovoltaics (stacked electrode geometry) or in transistors (planar electrode geometries). In general, molecules with one-dimensional stacking interactions (**6**) are preferred for the former and those with two-dimensional stacking interactions (**7**, TIPS pentacene) perform better in the latter geometry. The type of stacking is closely related to the ratio of the diameter of the trialkylsilyl group to the long-axis dimension of the acene. For trialkylsilyl groups with diameter significantly greater, or significantly less, than half the length of the acene, one-dimensional stacking motifs predominate. In systems where the diameter of the silyl group is close to half the length of the acene, two-dimensional stacking motifs are common [21].

The crystal structure is very sensitive to the symmetry of the trialkylsilyl group (Figure 4.5). For example, the *tert*-butyldimethylsilyl derivative (**8**) adopts an unusual columnar stacking motif. Similarly, *n*-alkyl diisopropylsilyl derivatives (e.g., **9**), even though they are sterically very similar to the triisopropylsilyl group, adopt

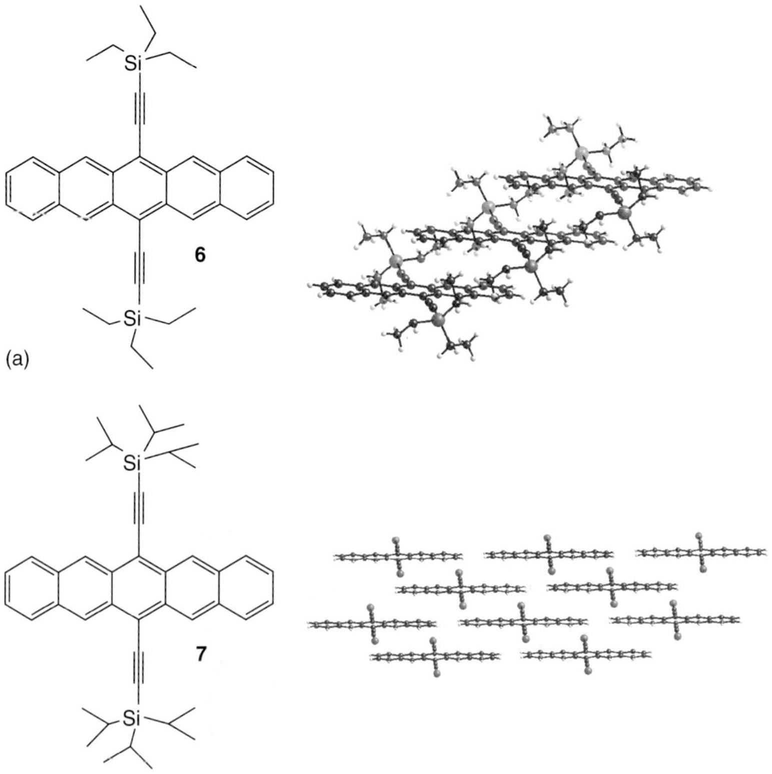

Figure 4.4 Crystal packing for triethylsilyl (a) and triisopropylsilyl (b) ethynylpentacenes, showing 1D and 2D interactions. Some alkyl substituents omitted for clarity.

one-dimensional, rather than two-dimensional, pi-stacked motifs [22]. This result points out the critical nature of alkyne purity in the synthesis of high-performance materials, since *n*-alkyl derivatives are frequent by-products in the synthesis of triisopropylsilane compounds. Even small amounts of *n*-alkyl derivatives, with their drastically different packing motifs, impede the formation of crystalline films of triisopropylsilyl derivatives.

4.3.1
Properties of Silylethyne-Substituted Pentacenes

Beyond the obvious changes in solubility, there is a clear improvement in the stability of these systems compared to the parent pentacene. The bulk of this stabilization arises from the substitution of the central, most reactive aromatic ring of the pentacene backbone, which appears to inhibit endoperoxide formation and block dimerization at this position. Dimerization can still take place at the rings adjacent to the central ring, and this is the typical photodecomposition pathway for

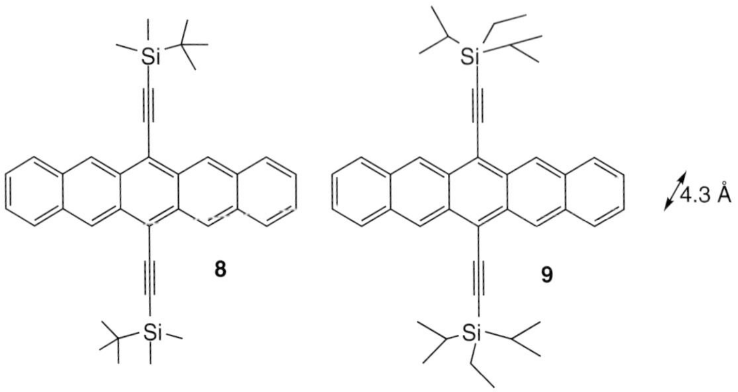

Figure 4.5 *tert*-Butyldimethylsilyl derivative (**7**) yields columnar stacks, while ethyldiisopropylsilyl derivative (**9**) forms 1D-slipped stacks.

solutions and amorphous films of this material (Figure 4.2) [11]. The trialkylsilyl groups also provide electronic stabilization of the material by selective stabilization of the LUMO of these materials, inhibiting reaction with oxygen [20]. Early studies of the stability of these materials noted a decrease in thin-film stability of these materials versus unsubstituted pentacene, but these studies compared crystalline films of pentacene with amorphous films of the silylethyne-substituted pentacene. More recent studies of crystalline films of TIPS pentacene (7) showed high thin-film stability for the crystalline material [11].

4.3.2
Electronic Structure Studies

Early calculations performed on TIPS pentacene showed a roughly 50% increase in reorganization energy compared to the unsubstituted system (although this does not appear to translate into reduction in charge-carrier mobility) [23]. More recent combined computational/experimental studies support this larger reorganization energy, which arises from strong coupling of the trialkylsilyl group through the alkyne into the acene HOMO [24]. These studies also showed that this substituent shifts the first ionization band to lower energy in the gas phase.

Band structure calculations on several silylethyne derivatives demonstrate how the width of the conduction and valence bands, as well as the dimensionality of conductivity in the crystal, depends heavily on the nature of the crystal-packing motif adopted by the molecules [25]. Surprisingly, significant dispersion was found both in the valence band and in the conduction band, implying that these materials could exhibit reasonable electron mobility. These predictions are supported by recent time-resolved electric force microscopy studies, which showed the electron mobility values of TIPS pentacene to be within an order of magnitude of the hole mobility value [26]. More recent theoretical efforts incorporated thermal motions into the band structure calculations, showing that even in the confined crystal lattice, the motions of the aromatic groups were sufficient to severely perturb intermolecular electronic coupling [27].

The lack of single crystals of ethynylpentacenes of sufficient quality for measurement of intrinsic mobility by FET methods has led to the investigation of single crystals and thin films of pentacene, TIPS pentacene, and TES-substituted pentacene (**6**) by time-resolved terahertz pulse spectroscopy, which is a contactless method for comparing carrier mobility between organic solids. These experiments confirmed the strong correlation between thin-film morphology and hole mobility for TIPS pentacene, and single-crystal experiments suggested that hole mobilities for TIPS pentacene and unsubstituted pentacene were nearly identical [28]. In contrast to the thin-film device measurements, which showed a difference in hole mobility for 2D pi-stacked TIPS pentacene and 1D pi-stacked **6** of more than a factor of 10^3 (see below), this all-optical technique yielded a hole mobility for **6** that was only a factor of 3 smaller than that measured for TIPS pentacene, likely because of the poor solution-cast film quality typically observed in materials with one-dimensional stacking interactions [29]. Using this method to perform a more detailed study of the carrier dynamics across a range of temperatures for high-purity crystals of tetracene, functionalized tetracene (rubrene), pentacene, and TIPS pentacene showed all these materials exhibited bandlike transport in the temperature range of 297–20 K, with mobility increasing with decreasing temperature [30]. At room temperature, the photoconductivity decay dynamics across this entire series of compounds were similar, but at temperatures below 70 K, the temperature dependence of the decay in transient photoconductivity is very different for TIPS pentacene (**7**) compared to pentacene, tetracene, and rubrene, indicating that either a common trap type present in these acenes is not present in **7** or **7** simply has different charge-trapping properties than unfunctionalized acenes and rubrene. Evidence for unusually long-lived trapped states in photoexcited TIPS pentacene has been reported [31], along with unusual persistent photoconductivity that may arise from trapped charge carriers [32].

Morphological studies of TIPS pentacene films showed the material forms large grains when cast from good solvents (particularly with slightly elevated substrate temperatures). The observed [0 0 1] projection for the films suggests that the materials stand upright on the surface of the substrate, with the trialkylsilyl groups serving as the "tripods" on which the acene stands [33]. This arrangement places the π-stacking axis of the self-assembled acene along the plane of the substrate, which is the ideal arrangement for planar electrode devices such as FETs. Examination of these films by hot-stage microscopy (and further studies by differential scanning calorimetry) revealed a solid-state phase transition in TIPS Pentacene **4**, occurring at $\sim$120 °C [34]. In films and crystals, the volume change associated with this phase transition led to the development of cracks in the pentacene material, creating small voids in the film. Transistor studies performed before and after heating TIPS pentacene beyond the phase transition showed that mobility dropped from 0.4–1.0 to 0.2 $cm^2\ V^{-1}\ s^{-1}$ after cracking.

4.3.3
Device Studies

The first device fabricated from a silylethyne-substituted pentacene used vapor-deposited TIPS pentacene in a bottom-contact FET structure (Figure 4.6) [35]. It was discovered that elevated substrate temperatures were critical to the formation of crystalline films of these materials, with films deposited at substrate temperatures below 45 °C showing no field-effect-controlled conductivity. Optimum substrate temperature for TIPS pentacene was found to be 90 °C, yielding devices with mobility as high as $0.4\,cm^2\,V^{-1}\,s^{-1}$. Devices based on 1D-stacked materials such as TES derivative **6**, even under optimized conditions, seldom yielded devices with mobility higher than $10^{-5}\,cm^2\,V^{-1}\,s^{-1}$. The mobility of vapor-deposited TIPS pentacene was high enough for the fabrication of some simple circuits, leading to inverters with a gain of 5.5 at a driving voltage of −10 V [36], as well as functioning *NAND* and *NOR* logic gates.

The solubility of TIPS pentacene was soon exploited for device fabrication, leading to a host of studies of the film-forming properties and device performance of this material. Mobilities as high as $1.8\,cm^2\,V^{-1}\,s^{-1}$ have been reported from films formed by drop casting [37]. Deposition methods that allow slow evaporation of solvent are required to achieve high-mobility values in TIPS pentacene solution-cast films. Spin-coating solutions of TIPS pentacene, for example, yield typical hole mobilities on the order of $10^{-3}\,cm^2\,V^{-1}\,s^{-1}$, unless a high-boiling point solvent (e.g., chlorobenzene) is used – in which case the typical mobility increases slightly to $4 \times 10^{-2}\,cm^2\,V^{-1}\,s^{-1}$ [38]. The selection of solvent has also proved important in drop-cast films, impacting parameters such as threshold voltage and also influencing device stability [39]. Drop-cast films of TIPS pentacene have proven uniform enough for the fabrication of more complicated circuitry. Working with FET with mobilities in the range of $0.2–0.6\,cm^2\,V^{-1}\,s^{-1}$, the Jackson group has produced inverters with gain of 3.5 [40], and seven-stage ring oscillators yielding oscillation frequencies >10 kHz, operating at voltages as low as −5 V. In an alternate approach, Park and coworkers used a patterning approach based on selective etching of a fluoropolymer-coated dielectric, yielding transistors with typical mobility of $0.18\,cm^2\,V^{-1}\,s^{-1}$ and inverters with gain of 5.6 [41].

As expected for typical solution-casting conditions, the films formed of TIPS pentacene tend to consist of numerous small crystalline grains, and the boundaries between grains often fall within the channel of FET devices. A combined experimental and theoretical approach examined the effect of grain boundaries on FET performance, finding that hysteresis in the electrical characteristics was related to

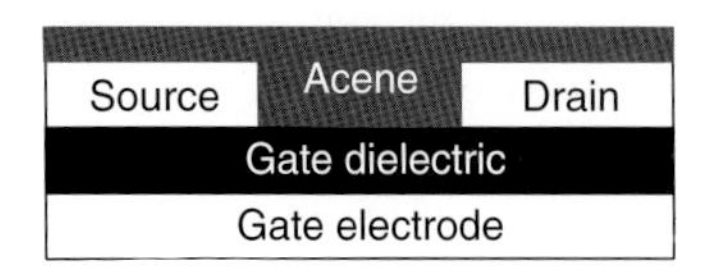

Figure 4.6 A typical bottom-contact OFET configuration.

the width of the lath-shaped grains [42]. Grains with width $>6\,\mu m$ showed higher mobility and minimal hysteresis, while those with width $<4\,\mu m$ exhibited low ($<0.01\,cm^2\,V^{-1}\,s^{-1}$) mobility and pronounced hysteresis. Following their model for boundary-limited transport, the authors estimated the mobility across grain boundaries to be $5 \times 10^{-7}\,cm^2\,V^{-1}\,s^{-1}$.

Straightforward modeling using simulation studies has also been performed on TIPS pentacene-based transistors, reproducing the device characteristics typically observed with this material, and giving insight into the compound's charge injection and transport properties [43]. In many cases, it does appear that charge injection is the major factor limiting the performance of devices. Recent studies on the modification of Ag electrodes with arenethiol-based monolayers have shown that these monolayers can significantly improve injection between TIPS pentacene and Ag, ostensibly by altering the work function of the electrode [44]. The best electrode treatment was determined to be pentafluorobenzenethiol (PFBT).

A number of different approaches to the fabrication of TIPS pentacene transistors have been reported in recent years. For example, using a polyimide gate dielectric, transistors with mobility of $0.15\,cm^2\,V^{-1}\,s^{-1}$ were fabricated. Impressively, the performance of these transistors did not degrade over the course of 60 days when stored under mild vacuum (550 T) [45]. Ink-jet printing is becoming a popular method for the deposition of organic semiconductors. The deposition of simple solutions of TIPS pentacene by this method requires careful consideration of solvent, substrate preparation, and device geometry. Using a device with electrodes in a ring geometry, and carefully optimizing solvent blends and surface polarity, devices with mobility of $0.15\,cm^2\,V^{-1}\,s^{-1}$ could be prepared by ink-jet methods [46]. Alternatively, high-boiling solvents such as *o*-dichlorobenzene were used to dissolve TIPS pentacene for ink-jet printing. In this case, it was found that elevated substrate temperatures (60 °C) were critical to the formation of adequate crystals, and mobilities as high as $0.24\,cm^2\,V^{-1}\,s^{-1}$ could be achieved [47]. An unusual approach involved the use of source and drain electrodes in a "suspended" configuration – where the electrodes hang in air over the dielectric substrate [48]. In this case, where the TIPS pentacene grows up from underneath the electrodes, ink-jet-printed TIPS pentacene provided devices with mobility as high as $0.29\,cm^2\,V^{-1}\,s^{-1}$.

The high level of crystallinity of TIPS pentacene allows detailed study on the properties of single crystals, including measurements of mobility anisotropy within single-crystalline grains. By growing crystals on a tilted substrate or using careful combinations of solvents, research groups have been able to grow large lath- or ribbon-shaped crystals from solution. Such high-quality crystals were used to fabricate FET devices, from which mobilities as high as $1.4\,cm^2\,V^{-1}\,s^{-1}$ were extracted [49]. The ability to grow oriented ribbons of TIPS pentacene by a variety of methods, including under the directed flow of an inert carrier gas [50], allowed the study of mobility anisotropy. Growing or placing the oriented ribbons across source–drain electrodes arranged in a variety of orientations shows that mobility can vary by an order of magnitude depending on the direction of crystal growth across the electrodes. Studies using four-electrode transistors and a hollow-pen approach to deposit ordered films observed similar anisotropy values [51].

4.3.4 Blends of Silylethynyl Pentacenes and Polymers

A recent innovation involves the blending of soluble pentacene derivatives with an insulating or semiconducting polymer to improve solution rheology and enhance film-forming properties (Figure 4.7) [52]. The small molecule typically segregates to one of the interfaces – either the air interface or the substrate interface – and generally forms a highly crystalline film at that interface. Recent neutron diffraction studies of blended films of poly(α-methylstyrene) and TIPS pentacene (1 : 1 weight ratio) showed that the thickest TIPS pentacene film formed at the polymer–air interface, but that there was also appreciable TIPS pentacene film thickness at the polymer–substrate surface. Bottom-gate, bottom-contact devices made from these films showed hole mobility as high as 0.54 cm^2 V^{-1} s^{-1} from spin-cast films [53]. More recently, it was found that polymers with a higher degree of crystallinity (such as isotactic poly(α-vinyl naphthalene)) yield improved segregation of the semiconductor to both interfaces from the polymer films, allowing use of as little as 10% by weight of the small-molecule semiconductor [54]. The type of polymer used to blend with the soluble acene can span a wide range of insulators or semiconductors [55]. Perhaps, the most successful polymer used has been a

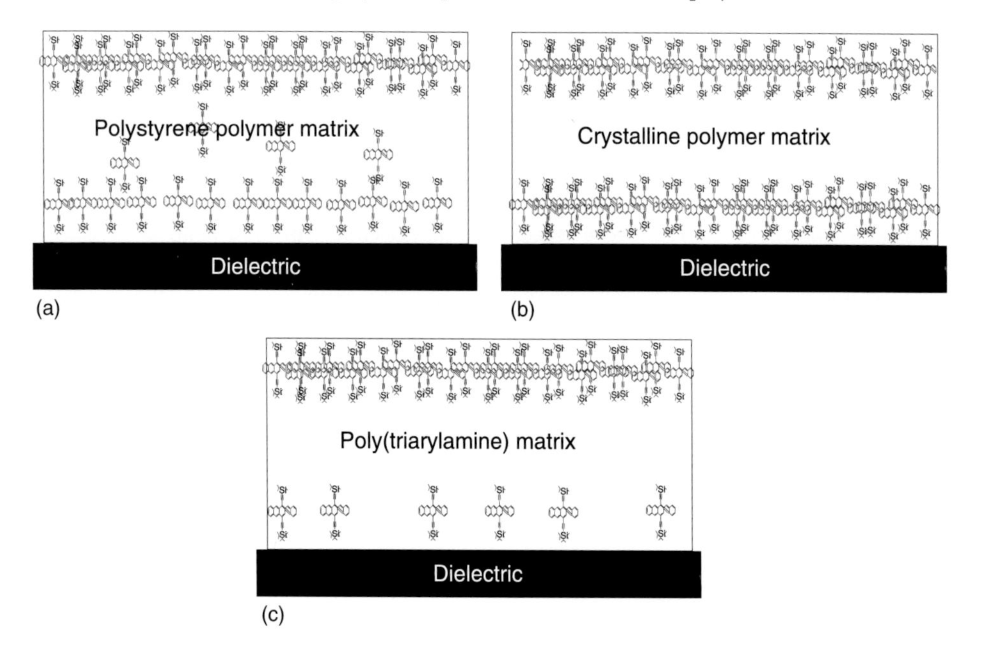

Figure 4.7 Representation of TIPS pentacene segregation in blends of a poly(α-methylstyrene) matrix (a), a crystalline polymer matrix (b), and a poly(triarylamine) matrix (c).

poly(triarylamine) (Figure 4.6c). In this case, the semiconductor segregates almost exclusively to the top (air) interface of the polymer films. Top-gate devices showed saturation mobility as high as 1.1 $cm^2\ V^{-1}\ s^{-1}$, with good device uniformity and stability [56].

4.3.5
Silylethyne Pentacene-Based Polymers

An alternative approach to modifying the morphological, rheological, and electronic properties of silylethyne-substituted acenes is to actually incorporate them into polymer structures. This can be accomplished by polymerizing the materials through conjugated linkers, to yield a more traditional semiconducting polymer, or through insulating linkers, to yield polymers that aggregate the semiconductors' performance. The nonconjugated materials (for example, **10** in Figure 4.8) are easily synthesized, and average molecular weights over 17 000 Da have been obtained. These polymers are all still highly soluble and possess many of the same optical and fluorescence characteristic of the parent silylethyne pentacene [57]. The polymeric versions appear to aggregate strongly in concentrated solution and form high-quality solution-cast films. Unfortunately, the minuscule redshift observed in the films versus solutions of the polymer (<10 nm) indicates minimal pi overlap in the solid state. For reference, the redshift between solution and thin film observed for TIPS pentacene is >100 nm [58].

Two approaches to conjugated polymers based on silylethyne-substituted pentacene have been reported to date (Figure 4.9). The first involves linkage through the 6,13-positions of the pentacene. In this case, the silylethynes act strictly as "end caps" that terminate the oligomeric compound. In thin films, reasonable redshifts are observed relative to solution absorption spectra. Further, these derivatives were subjected to study of their photoconductive yield, which gives some insight into their potential semiconductor properties. In this case, the derivative of **11** where R = *iso*-propyl and R′ = trimethylsilyl showed a photoconductive yield >10, which may indicate that the materials could possess exploitable semiconductor properties [59]. An alternative approach to larger polymeric systems keeps the basic TIPS

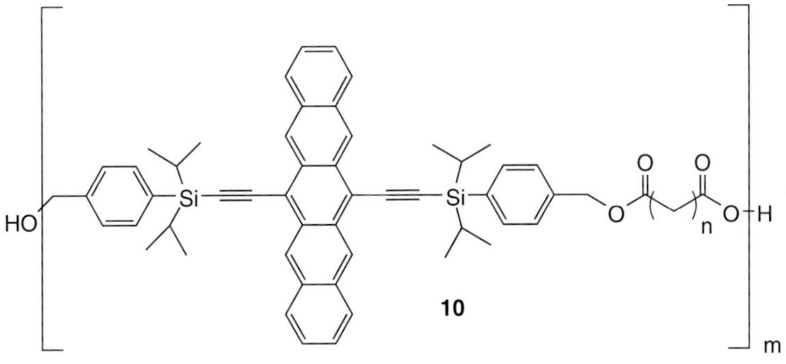

Figure 4.8 A nonconjugated polymer (**10**) based on silylethynyl pentacene.

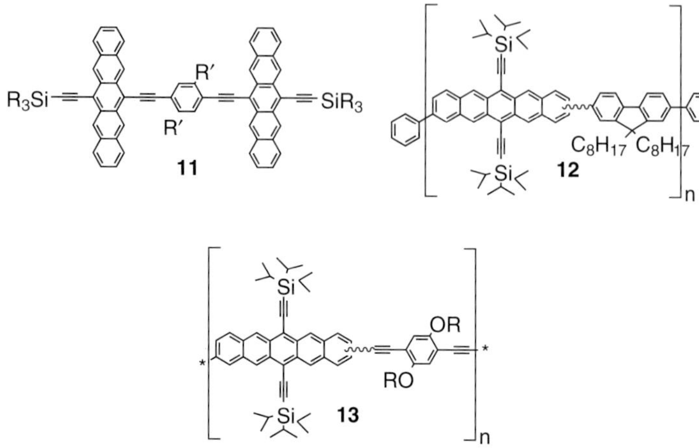

Figure 4.9 Conjugated oligomers and polymers based on silylethynyl pentacene.

pentacene chromophore intact and polymerizes from the pro-cata positions of the acene (**12**, Figure 4.9). In this case, a copolymer with a fluorene derivative was reported. The resulting polymer had number-average molecular weight of 36 400 and a polydispersity index of 3.21. The resulting highly soluble polymer has an oxidation potential similar to that of TIPS pentacene but is more easily reduced than that compound. The polymer also exhibited high thermal stability. A phenylethyne-based copolymer (**13**) has also been reported, which also exhibited high solubility and good solution stability [60].

4.3.6
Organic Light Emitting Diodes and Photovoltaics Using Silylethynyl Pentacenes

Silylethyne derivatives of pentacene have also shown promise in the fabrication of intense red organic light emitting diodes (OLEDs). In particular, they have been utilized in OLED configurations where the pentacene serves as a red-emitting dopant in a host of either Alq_3 or NPD (Figure 4.10). In a series of studies on excitation energy transfer between these hosts and a series of pentacene derivatives,

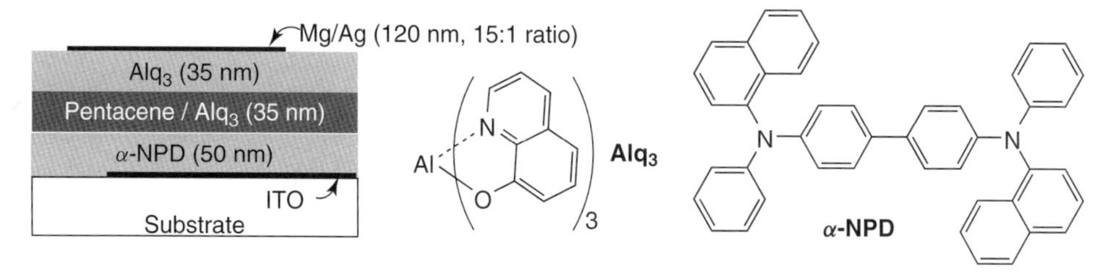

Figure 4.10 Device structure for silylethynyl pentacene light emitting diode.

it was noted that when TIPS pentacene was added to the host matrix, the emission for the pentacene redshifted, leading to significant emission beyond 700 nm.

In order to address this issue, dioxolane-substituted pentacene (**14**) was prepared (Figure 4.11) [61]. Derivative **14** had absorption and emission spectra that were significantly blue-shifted relative to TIPS pentacene, as well as significantly increased fluorescence quantum yields. At low concentrations, this derivative did prove a more efficient emitter than TIPS pentacene, but as concentrations approached 0.5 mol%, the acenes began to aggregate and emission efficiency suffered. The tetraethyl derivative (**15**) was then prepared. Crystallographic analysis of this compound showed minimal pi–face interaction in the solid state, and NPD [62] or Alq_3 [63] host films could be loaded with up to 2% of this dopant without decreasing the emission efficiency of the pentacene. In device studies, an Alq_3 host/**15** guest OLED device yielded bright red emission with an external electroluminescence quantum yield of 3.3%, which is very close to the theoretical maximum [64].

The high charge-carrier mobility and broad absorption spectrum of TIPS pentacene led to its study as donor in single-heterojunction solar cells (Figure 4.12). The earliest work used TIPS pentacene as a solution-deposited film, followed by

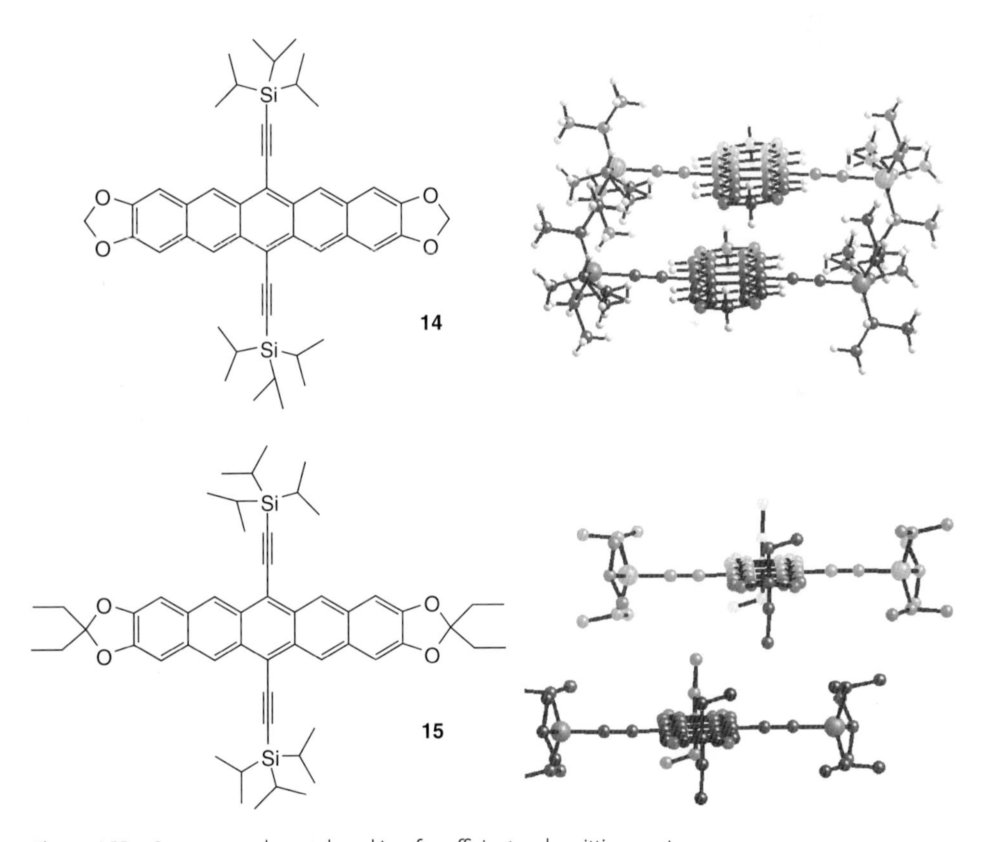

Figure 4.11 Structure and crystal packing for efficient red-emitting pentacenes.

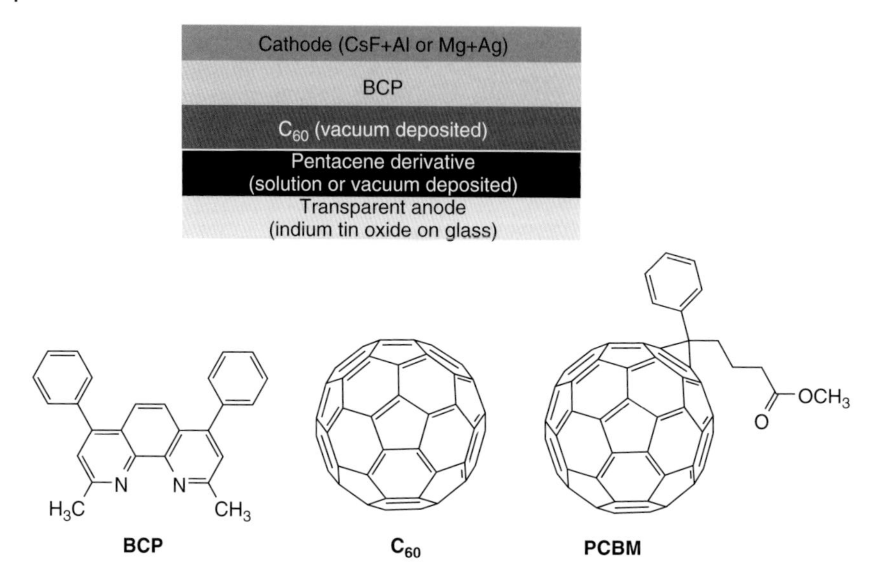

Figure 4.12 Schematic of silylethynyl pentacene solar cells.

vapor-deposited C_{60} as acceptor and CsF/Al as cathode. After extensive optimization, photovoltaic devices with power conversion efficiency of 0.52% were obtained [65]. A subsequent study of a series of silylethynyl pentacene derivatives (TIPS pentacene, **14** and **15**) used vacuum deposition for all layers of the photovoltaic device and Mg/Ag cathode. It was found that the open-circuit voltage of the cell was highly sensitive to the oxygen substituents on the pentacene ethers. After extensive optimization, dioxolane derivative (**15**) yielded the highest performing device, with open-circuit voltage of 0.72 V, short-circuit current density of 2.40 mA cm^2, and an overall power conversion efficiency of 0.74% [66]. All efforts to create bulk-heterojunction solar cells from TIPS pentacene and any fullerene derivative failed due to the rapid chemical reaction between TIPS pentacene and C_{60} or its derivatives [16].

The large dispersion in the conduction band predicted by band structure calculations on silylethyne-substituted pentacene [25] led to the investigation of these derivatives as potential *acceptors* in organic solar cells. In order to operate as effective acceptors for polymers such as poly(3-hexylthiophene) (P3HT), the HOMO and LUMO levels of these pentacenes would need to be tuned to optimize charge transfer from the polymer to the acene. Following an intensive computational study on cyanopentacenes that predicted that nitrile-substituted TIPS pentacene would be an efficient n-type material [67], a wide variety of cyano-substituted pentacenes were prepared. As shown in Figure 4.13, derivatives containing 1, 2, or 4 nitrile substituents were prepared, and the substituents on the silyl group were modified to yield derivatives with different film-forming properties. The number of nitrile substituents had the expected effect on the open-circuit voltage of the organic solar cell – more nitrile groups led to lower voltage. Thus, cells based on mononitrile

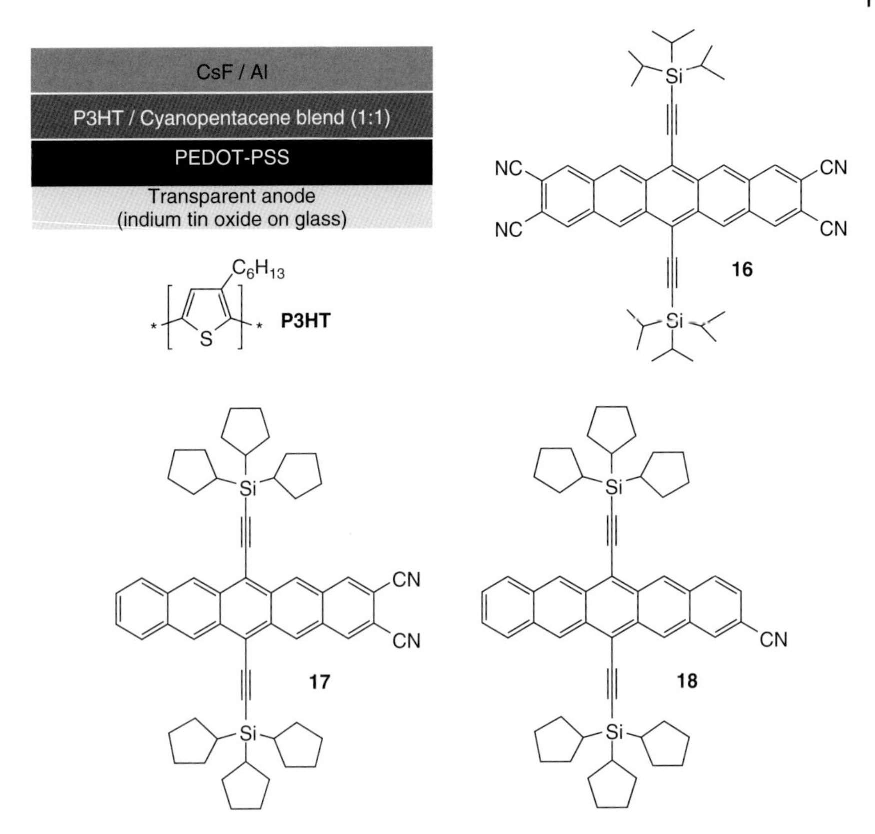

Figure 4.13 Polythiophene-based solar cell using cyanopentacene derivatives as acceptors.

derivative (**18**) exhibited relatively high voltage (0.82 V). Phase separation was optimized by tuning of the silyl substituents to improve film morphology and transport properties. The tri(cyclopentyl)silyl derivative exhibited promising performance, with a short-circuit current density of 1.11 mA cm^2 and an overall power conversion efficiency of 0.31% [68]. These early results offer promise for the development of nonfullerene acceptors for organic photovoltaic cells.

4.3.7 Silylethynyl Pentacene n-Type Semiconductors

The ability to functionalize the pentacene chromophore with electron-withdrawing substituents implies that the correct combination of substituents may yield an n-type pentacene semiconductor for use in OFETs. The earliest approach to this type of system involved partial fluorination of the pentacene backbone (**19** and **20**, Figure 4.14). Devices fabricated on unheated substrates and tested in air did not exhibit n-type behavior [69]. Subsequent studies, using heated substrates during vapor deposition of the acene and performed under inert atmosphere on carefully pretreated device substrates, showed that films of these fluorinated derivatives

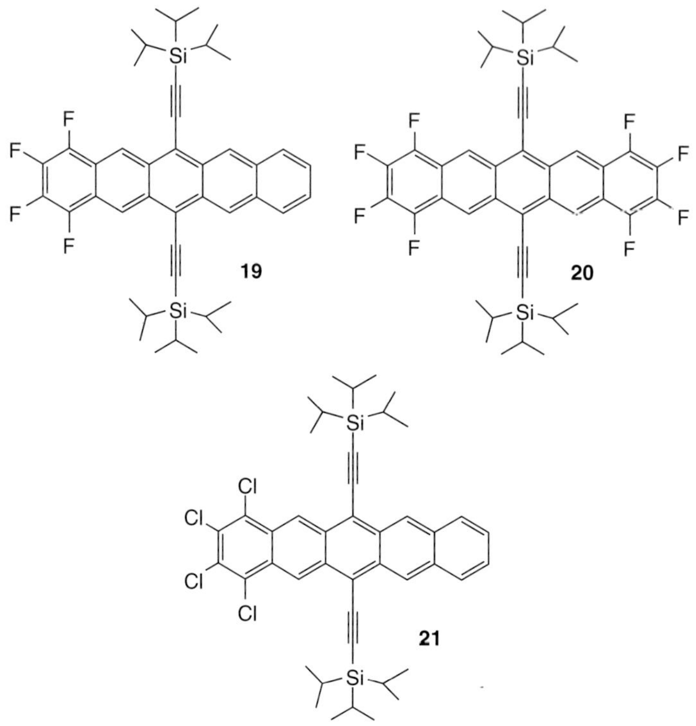

Figure 4.14 Halogenated silylethynyl pentacenes for n-type or ambipolar devices.

yielded transistors with appreciable hole *and* electron mobilities. For tetrafluoro derivative (**19**), μ_e was 0.1 $cm^2\ V^{-1}\ s^{-1}$, and μ_h was 0.07 $cm^2\ V^{-1}\ s^{-1}$. For octafluoro derivative **20**, μ_e increased to 0.4 $cm^2\ V^{-1}\ s^{-1}$, and μ_h was 0.33 $cm^2\ V^{-1}\ s^{-1}$ [70]. While fluorine substitution has always been considered the best approach to n-type semiconductors, the Bao group introduced the surprising result that *chlorine* substitution could yield stable n-type semiconductors as well. Their exploration of the tetrachloro derivative of TIPS pentacene (**21**) showed electron mobility as high as 0.054 $cm^2\ V^{-1}\ s^{-1}$ and hole mobility of 0.11 $cm^2\ V^{-1}\ s^{-1}$. The effect of chlorination on n-type behavior was applied to a wide variety of aromatic chromophores and seems to be a general approach for inducing electron transport in p-type organic semiconductors [71]. Considering that chlorination is chemically a *much* simpler process than fluorination, application of this discovery could yield numerous high-performance materials.

4.3.8
Other Silylethyne-Substituted Acenes in Organic Electronics

Bridging from the success of silylethyne-substituted pentacenes in a variety of organic electronic applications, a number of research group have begun to investigate approaches to chromophores that are approximately the same size as

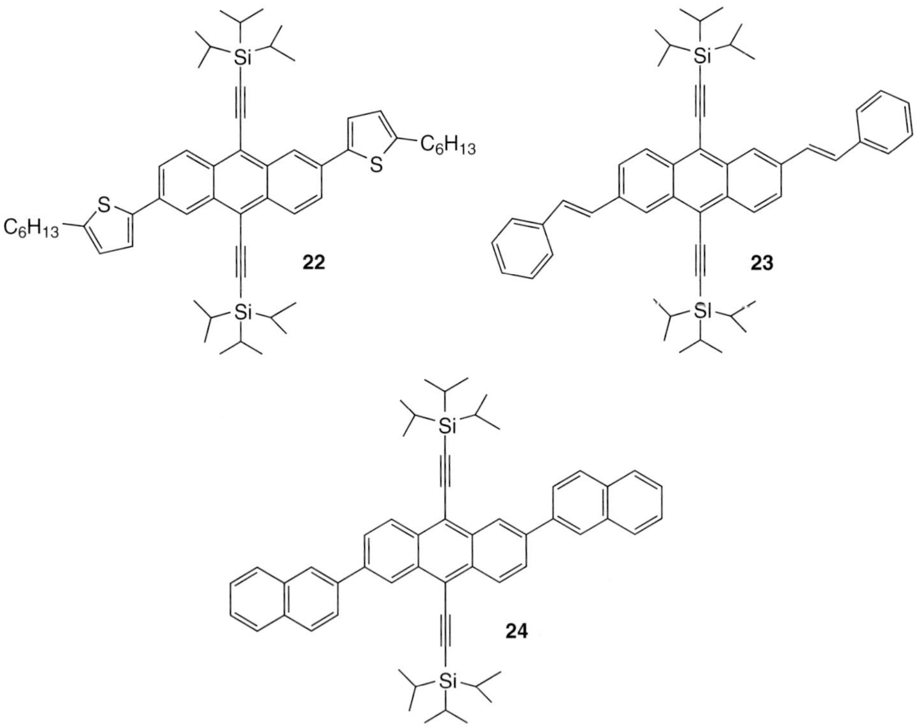

Figure 4.15 Silylethynyl anthracene based semiconductors.

pentacene but are composed of different aromatic units. The most common of these approaches begins with silylethyne-substituted anthracenes, linking on additional aromatic units to yield a chromophore of appropriate size (Figure 4.15). The thienyl derivative (**22**) has shown respectable field-effect mobilities from solution-cast films, with $\mu = 4 \times 10^{-3}$ $cm^2\ V^{-1}\ s^{-1}$ and $I_{on/off}$ of 10^6. The styryl substituent (**23**) did not prove as useful, with mobility of 3×10^{-4} $cm^2\ V^{-1}\ s^{-1}$ and $I_{on/off}$ of 10^5 [72]. Dinaphthyl derivative (**24**) was recently used in an unusual electronic device configuration, using light rather than gate voltage to modulate current. In this case, a small single crystal of **24** spanned two electrodes made from poly(ethylenedioxythiophene)/polystyrene sulfonate. The crystals showed a strong photoresponse to conductivity, yielding an "on/off" ratio of 5×10^2. The authors speculate the increase in current arises from photo-detrapping of carriers in the single crystals [73].

4.4 Heteroacenes

The replacement of carbocyclic rings of the linear acenes with heterocyclic counterparts is a compelling strategy for improving stability and altering the

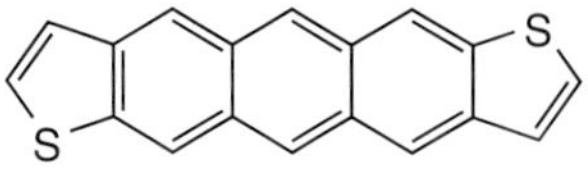

Figure 4.16 High-performance heteroacene chromophores.

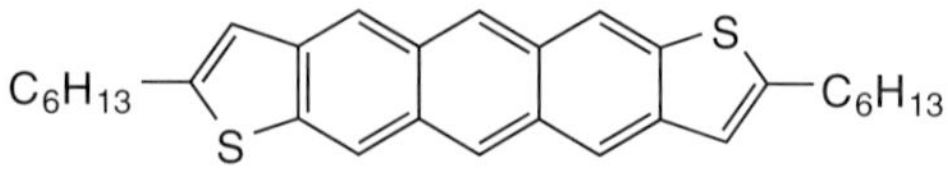

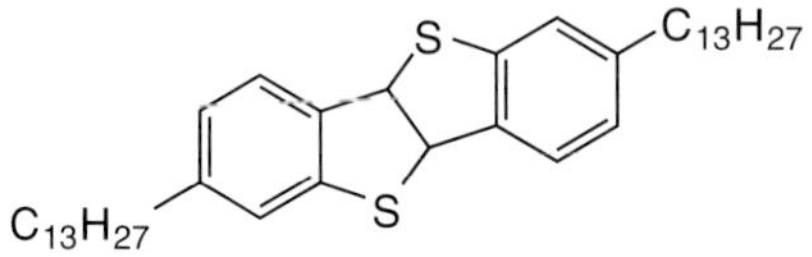

functionalization opportunities in organic electronic materials (Figure 4.16). Numerous linearly fused heteroacenes have been reported, perhaps the earliest being anthradithiophene (ADT) (**25**). Alkyl substitution of this material led to the first high-performance organic semiconductor (**26**) that could be deposited both by vapor methods (yielding transistors with hole mobility of $0.15\ cm^2\ V^{-1}\ s^{-1}$) [74] or by solution (yielding mobility of $0.02\ cm^2\ V^{-1}\ s^{-1}$) [75]. Alkyl substitution has been a very successful strategy for the formation of high-performance solution-processable organic semiconductors, yielding molecules such as (**27**) that exhibits hole mobility $>1\ cm^2\ V^{-1}\ s^{-1}$ from solution-cast films [76].

4.4.1
Silylethyne-Substituted Heteroacenes

Silylethyne substitution studies on larger heteroacenes for organic electronics applications began with the ADTs, since they were a well-known high-performance chromophore. These easily prepared derivatives are synthesized using methods analogous to their carbocyclic counterparts – addition of an alkynide to the heteroacenequinone, followed by deoxygenative workup [77]. Monothieno acenes are also demonstrating excellent device performance and are particularly appealing due to the ability to functionalize them to yield ambipolar semiconducting materials (see below). Functionalization of heteroaromatic systems is generally easier than the functionalization of the carbocyclic materials, making heteroacenes prime candidates for determining how small changes in substitution impact the electronic performance of the semiconductor. The syntheses of a number of new silylethyne-substituted heteroacenes, such as diaza- and tetraaza systems, have been reported in recent years [78]. Owing to the lack of device studies of these materials, they will not be covered here.

4.4.2
Crystal Packing

As with the carbocyclic acenes, the heteroacenes adopt a variety of strongly pi-stacked orientations in the solid state, controlled by the length and substitution pattern on the acene rings and the size of the appended trialkylsilyl groups. Representative structures are shown in Figure 4.17. TES-ADT (**28**) exhibits the two-dimensional pi-stacking common among high-performance transistor materials. One-dimensional stacks can be achieved either by increasing the size of the alkyl groups on silicon (e.g., tri(*iso*-propyl)silyl derivative **29**) or by adding alkyl groups to the terminal thiophene carbons (such as ethyl derivative **30**). A severe mismatch between substituent size and acene length leads to edge-to-face interaction with no close contact between chromophores – this is exemplified by TES thienoanthracene (**31**) [79]. As is clear from the crystal structures in Figure 4.17, there is significant scrambling of the heterocyclic ring in the crystal. In the case of ADTs, this is in part because they are synthesized as an inseparable mixture of *syn* and *anti* isomers [80]. However, the fact that the monothieno acenes show similar scrambling, even though they are synthesized as isomerically pure materials, indicates that even an isomerically pure ADT would still show scrambling of the heteroatom in the solid state.

4.4.3
Device Studies

TES-ADT (**28**) was the first silylethyne-substituted heteroacene tested in device applications. In a bottom-contact device configuration, with TES-ADT applied from a toluene solution using the doctor-blade technique, hole mobilities as high as $1.0\ cm^2\ V^{-1}\ s^{-1}$ were observed [81]. However, this solution deposition technique could not be scaled to produce large-area devices. As with TIPS pentacene, spin coating of this material yielded amorphous films with very poor mobility. However,

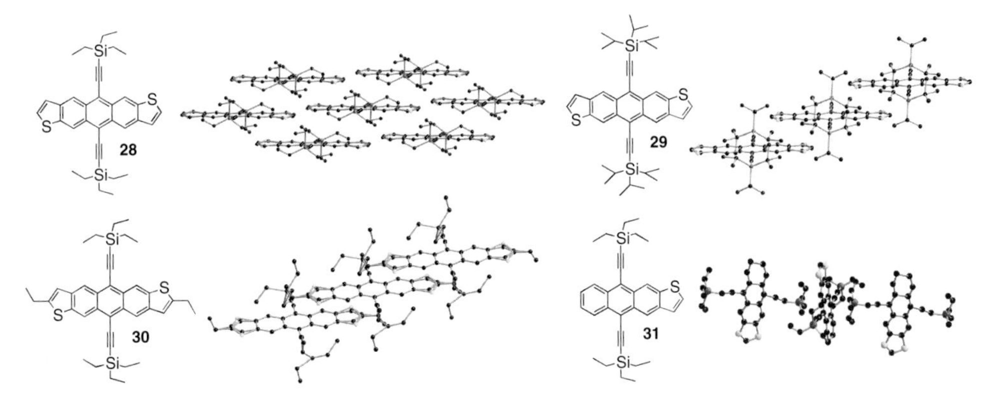

Figure 4.17 Crystal packing of representative heteroacenes.

unlike TIPS pentacene, the weaker interactions between molecules of TES-ADT make this compound highly amenable to annealing techniques. In particular, exposure of the amorphous film to solvent vapors led to rapid crystallization of the film. In bottom-contact devices fabricated via spin coating from toluene, followed by exposure to several solvent vapors for 2 min, the amorphous films were observed for transition to large crystalline domains [82]. Nonpolar solvents had the least effect with mobilities of 0.002 $cm^2\ V^{-1}\ s^{-1}$, similar to the nonannealed device. The best performance was obtained from 1,2-dichloroethane, which gave a mobility of 0.11 $cm^2\ V^{-1}\ s^{-1}$. The annealing solvent also had significant impact on threshold voltage, with nonpolar annealing solvents (hexanes) yielding near-zero V_{th} and polar annealing solvents (acetone) leading to $V_{th} > 30$ V. The mechanism of this annealing has been considered [83]. A series of devices were annealed in chloroform vapor for 3, 10, and 30 min. Shorter annealing times gave small aggregates, while spherulitic Maltese cross patterns ranging in size from a few hundred micrometers to a few millimeters were observed after 30 min of annealing time. Subsequent XRD studies showed that that solvent vapor permeation allows the low surface energy trialkylsilyl groups to adhere to the substrate. Extended annealing was observed to give a maximum mobility of 0.43 $cm^2\ V^{-1}\ s^{-1}$, with an on/off current ratio of 10^6. Solid-state crystallization of TES-ADT field-effect transistors at room temperature in the absence of solvent vapor has also been observed [84]. Devices were stored at room temperature for zero, one, three, five, and seven days under vacuum. Changes in crystallinity were observed after three days, with the formation of small aggregates. After five and seven days, the films show interconnected polycrystalline domains. Bottom-contact devices were produced under the same conditions, and the field-effect mobilities were determined. The as-produced devices gave extremely low values, that is, 0.0007 $cm^2\ V^{-1}\ s^{-1}$, while the aging after five days gave a mobility of 0.034 and 0.06 $cm^2\ V^{-1}\ s^{-1}$ after seven days. Both studies show the advantages of film morphology manipulation by appropriate annealing conditions.

The weak crystals of TES-ADT permit postdeposition manipulation of the films, allowing the patterning of devices crucial to circuit performance and the minimization of off and leakage currents. The plasticization of the films of TES-ADT during solvent vapor annealing allowed films to be patterned by UV illumination [85]. On exposure of the amorphous, spin-cast film to 1,2-dichloroethane vapor, the nonchannel regions of the substrate were selectively irradiated with 365 nm light at 540 mW cm^2 for 1–2 min. Semiconductor under the irradiated area was observed to dewet from the substrate, while the semiconductor in nonirradiated areas crystallized. Thus, selective patterning of the channel region gave large crystalline TES-ADT domains and mobilities $> 0.1\ cm^2\ V^{-1}\ s^{-1}$ with an order of magnitude lower off current than observed in nonpatterned films.

The formation of top-contact devices from these weakly crystalline materials by standard vapor deposition of metal electrodes typically results in significant damage to the organic film, leading to significant decrease in measured field-effect mobility. To ameliorate this condition, a PDMS-laminating technique was used to apply top-contact electrodes to TES-ADT films [86]. This involved depositing TES-ADT between electrode pads on the substrate and then applying the gold-coated PDMS

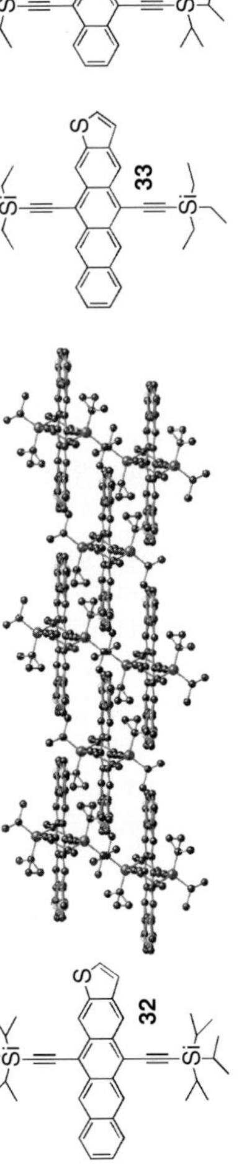

Figure 4.18 Representative monothieno acenes and crystal packing of derivative (**32**).

stamp to establish electrical connection. The resulting devices gave an average mobility of 0.19 $cm^2\ V^{-1}\ s^{-1}$.

In order to avoid potential issues with the isomerically impure ADTs, a number of research efforts have been directed toward the synthesis of acene-monothiophene derivatives. Researchers at Kodak prepared triisopropylsilylethynyl thienotetracene (**32**) and found hole mobilities as high as 0.5 $cm^2\ V^{-1}\ s^{-1}$ from spin-cast films [87]. Bao and coworkers have recently synthesized silylethyne-functionalized thienoacene derivatives with a variety of acene lengths and provided detailed reports on their crystal structures and electronic performance [79]. Of the derivatives they studied, only the TIPS thienotetracene (**32**) derivative assumes the typical 2D brickwork motif in the solid state and yielded devices with maximum mobility of 1.25 $cm^2\ V^{-1}\ s^{-1}$ from vapor-deposited films. Thienotetracene derivatives with other silyl groups (e.g., **33**, Figure 4.18), or other acene lengths, typically yielded 1D-slipped stacked structures, with corresponding thin-film mobilities of 10^{-3} $cm^2\ V^{-1}\ s^{-1}$. Materials with purely edge-to-face interactions, such as triethylsilylethynyl-substituted thienoanthracene (**31**), fared even worse in device studies – no evidence of field-modulated conductivity was observed in these materials.

In order to strengthen the interactions between molecules in ADTs, partially halogenated derivatives were synthesized and studied. The fluorine-substituted derivatives (**35**) and (**36**) (F-TES-ADT and F-TIPS ADT) proved particularly interesting, and typically underwent rapid crystallization during film formation due to the improved noncovalent interactions imparted by the fluorine atoms [88]. Both these materials adopt 2D π-stacked arrangements; F-TES-ADT (**35**) crystallizes in a motif very similar to the nonfluorinated material, while F-TIPS ADT (**36**) crystallizes in a motif similar to that of TIPS thienotetracene (**32**) (Figure 4.19). Subsequent thin-film device studies of these compounds showed dramatic differences in film morphology. TES derivative (**35**) crystallized across source-drain electrodes, even when spin-cast, and yielded devices with an average hole mobility of 0.7 $cm^2\ V^{-1}\ s^{-1}$. In contrast, film formation from the TIPS derivative (**36**) yielded large crystals and very poor surface coverage – and, consequently, no field-effect mobility could be determined. However, **36** did yield large free-standing crystals from a variety of solvents, and it proved possible to fabricate a field-effect transistor directly on the surface of the crystal. Hole mobility for this material was 0.1 $cm^2\ V^{-1}\ s^{-1}$, and this value was independent of gate voltage [89].

Optimized deposition and surface treatment conditions for **35** yielded spin-cast devices with hole mobility as high as 1.5 $cm^2\ V^{-1}\ s^{-1}$ [90]. The quality of these

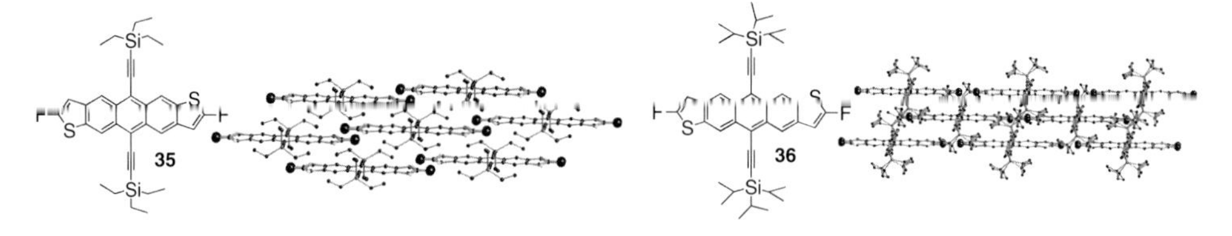

Figure 4.19 Representative fluorinated anthradithiophenes and their crystal packing.

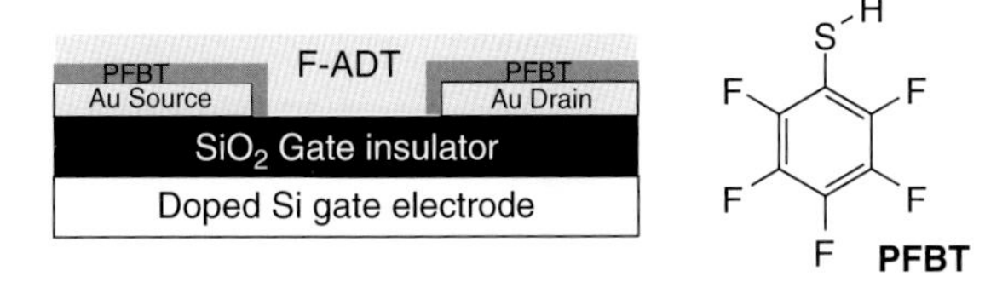

Figure 4.20 PFBT-treated electrode schematic.

films was sufficient to allow detailed studies of noise in OFETs, showing that the number of ordered domains in the channel region was closely related to the resulting low-frequency noise in the device [91]. The observation of crystal growth from gold surfaces treated with PFBT (Figure 4.20) led to more detailed studies of the interaction between **35** and these surfaces [92]. It was found that **35** nucleated on PFBT-treated gold and grew crystals that extended roughly 20 μm from the edge of the treated gold surface. Although semiconductor deposited on the surface outside this range, the material was shown to be electronically inactive. Thus, in devices with channel length $\lesssim$ 50 μm (the range at which the growing crystalline fringes can still "meld" with each other), field-effect mobilities were generally high. In devices with channel length $>$ 50 μm, the crystalline fringes growing from source and drain electrodes were separated by regions of inactive material and the observed mobility plummeted by several orders of magnitude. The channel length-dependent mobility was explored by Scanning Kelvin probe microscopy (SKPM) studies to correlate the potential and morphology with the overall device mobility values via a potential surface map [93]. Devices with channel lengths of 5, 25, and 80 μm were prepared. In the short-channel devices, it was found that source electrode injection was the greatest barrier to charge transport, since the channel was spanned by a single-crystalline grain of the semiconductor. In intermediate channel devices, grain boundaries became the significant barrier to charge transport, and in large channel devices, the grains growing off the source are effectively pinned at the source potential, due to the lack of a viable charge-transport channel from source to drain.

The surface-induced crystallization phenomenon was exploited as a form of low-cost patterning, yielding device arrays on flexible substrates exhibiting minimal cross talk. Seven-stage ring oscillators fabricated on polyimide substrates using this surface-induced crystallization approach to patterning yielded 10 kHz operation at −40 V and nearly 3 kHz oscillation frequency at −5 V [92].

Another approach to patterning films of **35** involves the selective removal of a monolayer in the channel to allow wetting only in the active device region (Figure 4.21) [94]. In this approach, the device substrate is coated with a hydrophobic monolayer (in this case, octadecyltrichlorosilane). Deep UV irradiation through a shadow mask of only the active channel area led to destruction of the monolayer in that area. Treatment of the electrodes with PFBT and spin coating of the organic semiconductor confined semiconductor growth almost exclusively to the prior irradiated areas, and nearly doubled the oscillation frequency versus similar unpatterned devices.

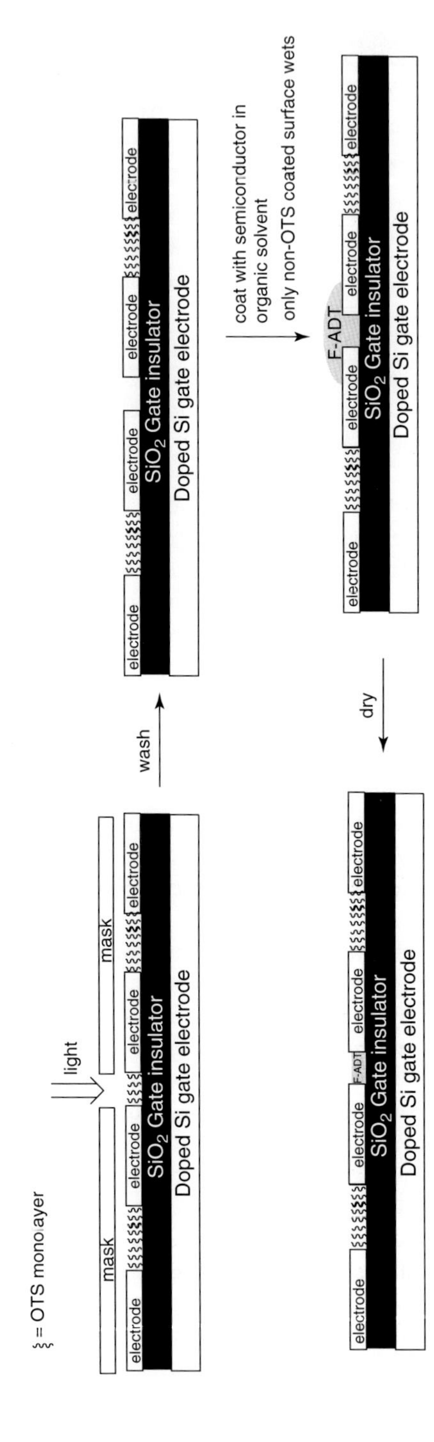

Figure 4.21 Nonrelief pattern lithography.

⸾ = OTS monolayer

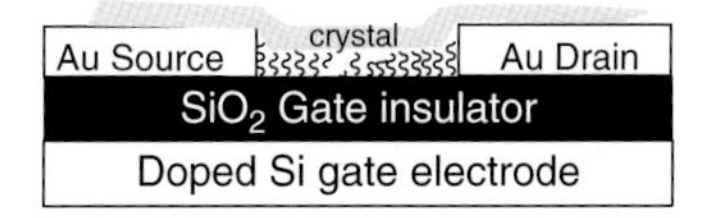

Figure 4.22 Single-crystal transistors.

Unlike the majority of soluble organic semiconductors, the fluorinated ADTs are both sufficiently stable and sufficiently volatile for the growth of high-quality crystals by vapor-transport methods. The extremely thin-single crystals of **35** could easily be laminated onto transistor devices, where they conformed to the source and drain electrodes and were electrostatically attracted to the treated dielectric, to determine mobility values approaching the defect-free limit (Figure 4.22) [95]. Transistors fabricated from **35** on OTS-treated SiO_2 dielectric yielded mobility as high as 6 $cm^2\ V^{-1}\ s^{-1}$, on/off current ratios $>10^8$, subthreshold slopes <1 V/decade, and minimal hysteresis in the current-voltage characteristics. The availability of single-crystal data for an organic semiconductor serves as an important benchmark for thin-film studies and will also allow more detailed exploration of the ways that crystal packing influence transport properties.

The facility with which the fluorinated ADTs formed crystalline films led to studies of blends of fluorinated (**35**) and nonfluorinated (**28**) ADT derivatives [96]. Using the fluorinated derivative to seed nucleation of the nonfluorinated compound, it was found that at high (>2 mol%) concentrations of the fluoro compound, numerous nucleation sites were created leading to numerous small (50–100 μm) crystals forming during spin casting of the film. At lower concentrations of the fluoro derivative (< 0.7 mol%), solvent vapor annealing was required to assist crystallization after spin coating, and much larger (>2000 μm) grains were formed. Mobility was also found to scale with typical grain size, with smaller grain films yielding mobility $< 0.1\ cm^2\ V^{-1}\ s^{-1}$, and large-grain films yielding mobilities more than three times higher.

4.4.4
Silylethynyl Heteroacenes for n-Type Applications

As with pentacene derivatives, halogenation of heteroacenes can be used to induce both ambipolar and n-type behavior. In general, there are too few sites for substitution to turn ADTs into n-type materials, but the monothienoacenes are easily modified to improve their electron transport (Figure 4.23). Tetrafluoro TIPS thienotetracene (**37**) was the first derivative explored for this purpose and yielded impressive ambipolar behavior from vapor-deposited films on octadecyltrichlorosilane-treated substrates [97]. Electron mobility for this compound was 0.13 $cm^2\ V^{-1}\ s^{-1}$ ($I_{on/off} = 10^5$) and hole mobility was 0.097 $cm^2\ V^{-1}\ s^{-1}$, for gold top-contact devices measured under nitrogen. The substitution of chlorine for the fluorine atoms (**38**) had an impressive effect on the performance of the

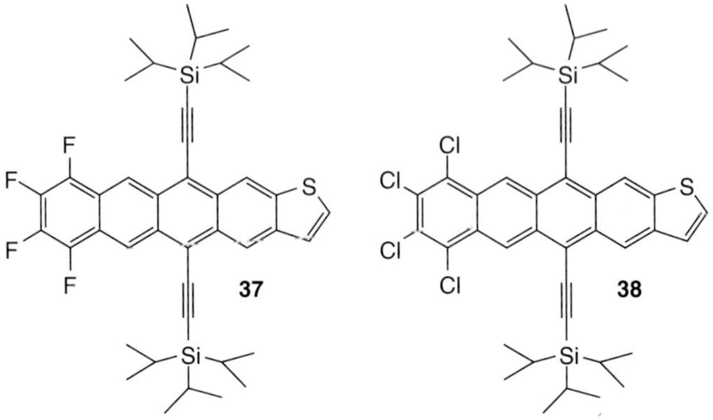

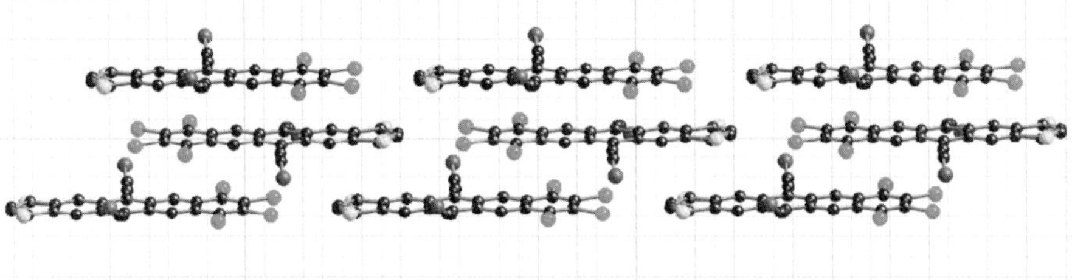

Figure 4.23 Fluoro- and chloro-thienoacenes and the crystal packing for **23**. Alkyl groups on Si omitted for clarity.

material. As shown by the structure in Figure 4.23, this derivative adopts a strongly pi-stacked arrangement, with a structure that closely approaches columnar π-stacking (although there are enough intercolumn close contacts to consider this structure a 2D π-stack). Vapor-deposited films of **38** with gold top contacts on treated SiO_2 substrates yielded devices with electron mobility as high as $0.56\,cm^2\,V^{-1}\,s^{-1}$ and $I_{on/off}$ of 10^5 and hole mobility of $0.22\,cm^2\,V^{-1}\,s^{-1}$ ($I_{on/off}$ 4×10^2). Elevated substrate temperature (70 °C) was critical to the formation of large crystalline grains and thus best device performance. The authors postulate that delocalization of the aromatic π-system into unoccupied 3d orbitals on chlorine serves to lower the LUMO, potentially facilitating electron injection.

4.4.5
Blends of Silylethyne-Substituted Heteroacenes and Polymers

Silylethyne-functionalized heteroacenes also figure prominently in devices fabricated with blends of polymers and small molecules. As with TIPS pentacene, fluorinated ADT (**35**) can also be blended with insulating or semiconducting polymers to yield materials with excellent properties for solution casting, yielding devices with impressive performance. Compound **35** was studied in a double-gate configuration (Figure 4.24) consisting of a benzocyclobutene dielectric for the bottom-gate device and a Cytop dielectric for the top gate, using a blend of

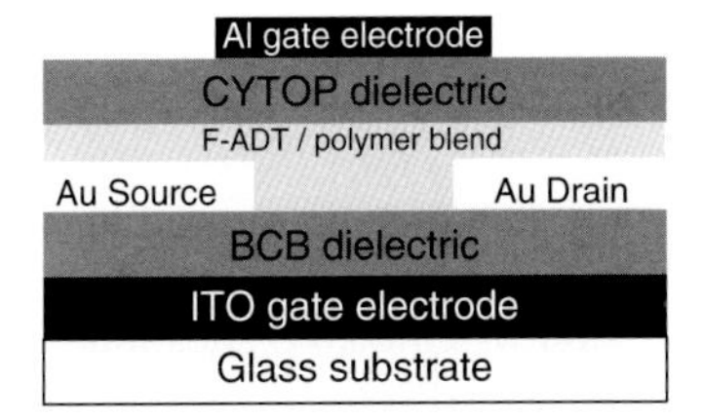

Figure 4.24 Double-gated device for testing segregation of **35** from polymers.

ADT (**35**) and poly(α-methyl styrene) as semiconductor [56]. Owing to the nature of the segregation of the fluorinated ADT, the top-channel configuration ($\mu = 0.5\ \text{cm}^2\ \text{V}^{-1}\ \text{s}^{-1}$) performed significantly better than the bottom-channel configuration ($\mu = 0.1\ \text{cm}^2\ \text{V}^{-1}\ \text{s}^{-1}$). With this information in hand, top-gate devices based on blends of fluorinated ADT and poly(triarylamine) were prepared, and yielded transistors with mobility $> 2.4\ \text{cm}^2\ \text{V}^{-1}\ \text{s}^{-1}$. These devices also exhibited good stability, reasonable threshold voltage, and minimal hysteresis. This same semiconductor blend, using bottom-contact (gold electrodes), bottom-gate devices, and a poly(vinylphenol) dielectric, was used to prepare seven-stage ring oscillators operating at over 100 kHz at drive voltages of −120 V and ∼25 kHz at −40 V [98].

4.5
Silylethynyl Heteroacene-Based Polymers

Anthradithiophenes have also been incorporated into the backbone of conjugated polymers (Figure 4.25). Copolymerizing a brominated derivative of TIPS ADT (**39**) with a fluorene derivative yielded polymer **40**, with a number-average molecular weight of 58 000 and average degree of polymerization of 56 [60]. Polymerization yielded a significant redshift in absorption, which proved beneficial in the fabrication of photovoltaic devices from this material (see below).

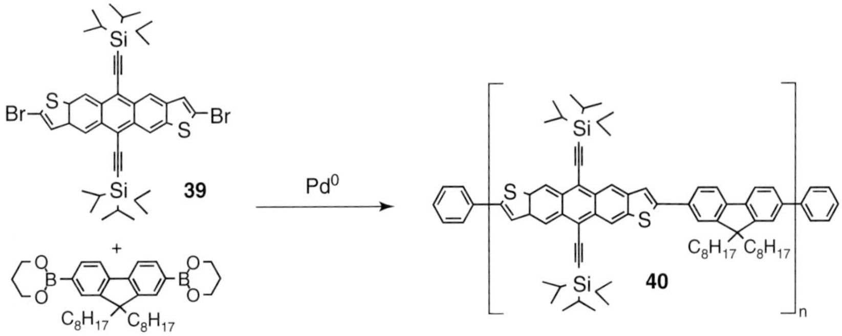

Figure 4.25 Silylethynyl anthradithiophene-based polymer.

4.6
Silylethynyl Heteroacene-Based Photovoltaics

Because many heteroacenes do not react with fullerene derivatives, they can serve as high-mobility donors in bulk-heterojunction organic photovoltaic devices. One of the first derivatives reported for this application was the alkyl-substituted ADT (**30**) (Figure 4.17). This derivative crystallizes with strong one-dimensional π-stacking interactions, leading to film morphology favorable for stacked device configurations. Blends of ethyl derivative (**30**) and PCBM (Figure 4.12) spin cast onto indium tin oxide substrates from mixtures of toluene and 1,2-dichlorobenzene were solvent vapor annealed to yield highly crystalline films. The addition of aluminum cathodes by vacuum deposition led to photovoltaic devices (for general schematic, see Figure 4.17) with short-circuit current of 2.96 mA cm^2, open-circuit voltage of 0.84 V, and fill factor of 0.40, providing an overall power conversion efficiency of 1% [99].

Polymers based on ADTs also yield promising photovoltaic devices. In a 1 : 3 blend of polymer **40** and PCBM and a device configuration similar to that shown in Figure 4.17, an open-circuit voltage of 0.75 V and short-circuit current of 2.35 mA cm^2 were achieved, leading to an overall power conversion efficiency of 0.68% [100].

4.7
Conclusion

The advantages of silylethyne functionalization to improve the stability, solubility, and device performance of organic chromophores have led to its application in a variety of systems designed for a number of electronic applications. These materials comprise a number of the best performing small-molecule semiconductors and serve as an attractive platform for the study of charge transport in organic systems. Current efforts in the derivatization of these materials are directed toward improvements in stability, enhancing film uniformity, and improving charge injection from common electrode materials. The ease with which functionalized derivatives of these materials can be made makes them versatile platforms for these studies.

References

1. Kitamura, M. and Arakawa, Y. (2008) *J. Phys. Condens. Matter*, **20**, 184011.
2. Facchetti, A. (2007) *Mater. Today*, **10**, 28.
3. Murphy, A.R., Frechet, J.M.J., Chang, P., Lee, J., and Subramanian, V. (2004) *J. Am. Chem. Soc.*, **126**, 11750.
4. See for example (a) Brown, A.R., Pomp, A., de Leeuw, D.M., Klaassen, D.B.M., Havinaga, E.E., Herwig, P., and Müllen, K. (1996) *J. Appl. Phys.*, **79**, 2136; (b) Herwig, P.T. and Müllen, K. (1999) *Adv. Mater.*, **11**, 480; (c) Afzali, A., Dimitrakopoulos, C.D., and Breen, T.L. (2002) *J. Am. Chem. Soc.*, **124**, 8812; (d) Afzali, A., Dimitrakopoulos, C.D., and Graham, T.O. (2003) *Adv. Mater.*, **15**, 2066; (e) Weidkamp, K.P., Afzali, A., Tromp,

R.M., and Hamers, R.J. (2004) *J. Am. Chem. Soc.*, **126**, 12740.
5. Yamada, H., Yamashita, Y., Kikuchi, M., Watanabe, H., Okujima, T., Uno, H., Ogawa, T., Ohara, K., and Ono, N. (2005) *Chem. Eur. J*, **11**, 6212.
6. For recent reviews, see (a) Anthony, J. (2006) *Chem. Rev.*, **106**, 5028; (b) Anthony, J. (2008) *Angew. Chem. Int. Ed.*, **47**, 452.
7. Reddy, A.R. and Bendikov, M. (2006) *Chem. Commun.*, 1179.
8. (a) Jaquith, M., Muller, E.M., and Marohn, J.A. (2007) *J. Phys. Chem. B*, **111**, 7711; (b) Jurchescu, O.D., Popinciuc, M., van Wees, B.J., and Palstra, T.T.M. (2007) *Adv. Mater.*, **19**, 688.
9. Berg, O., Chronister, E.L., Yamashita, T., Scott, G.W., Sweet, R.M., and Calabrese, J. (1999) *J. Phys. Chem. A*, **103**, 2451.
10. Roberson, L.B., Kowalik, J., Tolbert, L.M., Kloc, C., Zeis, R., Chi, X., Fleming, R., and Wilkins, C. (2005) *J. Am. Chem. Soc.*, **127**, 3069.
11. Coppo, P. and Yeates, S.G. (2005) *Adv. Mater.*, **17**, 3001.
12. (a) Satrijo, A., Kooi, S.E., and Swager, T.M. (2007) *Macromolecules*, **40**, 8833; (b) Zhao, D. and Swager, T.M. (2005) *Org. Lett.*, **7**, 4357.
13. Liu, C.-H., Liu, Y.-Y., Zhang, Y.-H., Wei, R.-R., Li, B.-R., Zhang, H.-L., and Chen, Y. (2009) *Chem. Phys. Lett.*, **471**, 97.
14. da Silva Filho, D.A., Kim, E.-G., and Brédas, J.-L. (2005) *Adv. Mater.*, **17**, 1072.
15. Kang, J.H., da Silva Filho, D., Brédas, J.-L., and Zhu, X.-Y. (2005) *Appl. Phys. Lett.*, **86**, 152115.
16. Miller, G.P., Mack, J., and Briggs, J. (2001) *Proc. Electrochem. Soc.*, **11**, 2001.
17. Anthony, J.E., Brooks, J.S., Eaton, D.L., and Parkin, S.R. (2001) *J. Am. Chem. Soc.*, **123**, 9482.
18. Ried, W., Donner, W., and Schlegelmilch, W. (1961) *Ber.*, **94**, 1051.
19. Rio, G. (1954) *Ann. Chim.*, **9**, 187.
20. Maliakal, A., Raghavachari, K., Katz, H.E., Chandross, E., and Siegrist, T. (2004) *Chem. Mater.*, **16**, 4980.
21. Anthony, J.E., Eaton, D.L., and Parkin, S.R. (2002) *Org. Lett.*, **4**, 15.
22. Chen, J., Subramanian, S., Parkin, S.R., Siegler, M., Gellup, K., Haughn, C., Martin, D.C., and Anthony, J.E. (2008) *J. Mater. Chem.*, **18**, 1961.
23. Gruhn, N.E., da Silva Filho, D.A., Bill, T.G., Malagoli, M., Coropcieanu, V., Kahn, A., and Brédas, J.-L. (2002) *J. Am. Chem. Soc.*, **124**, 7918.
24. Lobanova, O.R., Gruhn, N.E., Purushothaman, B., Anthony, J.E., and Lichtenberger, D.L. (2008) *J. Phys. Chem. C*, **112**, 20518.
25. Haddon, R.C., Chi, X., Itkis, M.E., Anthony, J.E., Eaton, D.L., Siegrist, T., Mattheus, C.C., and Palstra, T.T.M. (2002) *J. Phys. Chem. B*, **106**, 8288.
26. Jaquith, M.J., Anthony, J.E., and Marohn, J.A. (2009) *J. Mater. Chem.*, **19**, 6116.
27. Troisi, A., Orlandi, G., and Anthony, J.E. (2005) *Chem. Mater.*, **17**, 5024.
28. Ostraverkhova, O., Shcherbyna, S., Cooke, D.G., Egerton, R.F., Hegmann, F.A., Tykwinski, R.R., Parkin, S.R., and Anthony, J.E. (2005) *J. Appl. Phys.*, **98**, 033701.
29. Ostroverkhova, O., Cooke, D.G., Hegmann, F.A., Tykwinski, R.R., Parkin, S.R., and Anthony, J.E. (2006) *Appl. Phys. Lett.*, **89**, 192113.
30. Ostroverkhova, O., Cooke, D.G., Hegmann, F.A., Anthony, J.E., Podzorov, V., Gershenson, M.E., Jurchescu, O.D., and Palstra, T.T.M. (2006) *Appl. Phys. Lett.*, **88**, 162101.
31. Tokumoto, T., Brooks, J.S., Clinite, R., Wei, X., Anthony, J.E., Eaton, D.L., and Parkin, S.R. (2002) *J. Appl. Phys.*, **92**, 5208.
32. Brooks, J.S., Tokumoto, T., Choi, E.-S., Graf, D., Biskup, N., Eaton, D.L., Anthony, J.E., and Odom, S.A. (2004) *J. Appl. Phys.*, **96**, 3312.
33. Chen, J., Martin, D.C., and Anthony, J.E. (2007) *J. Mater. Res.*, **22**, 1701.
34. Chen, J., Tee, C.K., Yang, J., Shaw, C., Shtein, M., Anthony, J., and Martin, D.C. (2006) *J. Polym. Sci. B; Polym. Phys.*, **44**, 3631.
35. Sheraw, C.D., Jackson, T.N., Eaton, D.L., and Anthony, J.E. (2003) *Adv. Mater.*, **15**, 2009.

36. Park, J.G., Vasic, R., Brooks, J.S., and Anthony, J.E. (2006) *J. Appl. Phys.*, **100**, 044511.
37. Park, S.K., Jackson, T.N., Anthony, J.E., and Mourey, D.A. (2007) *Appl. Phys. Lett.*, **91**, 063514.
38. Kim, Y.-H., Lee, Y.U., Han, J.-I., Han, S.-M., and Han, M.-K. (2007) *J. Electrochem. Soc.*, **154**, H995.
39. Kim, C.S., Lee, S., Gomez, E.D., Anthony, J.E., and Loo, Y.-L. (2008) *Appl. Phys. Lett.*, **93**, 103302.
40. Park, S.K., Anthony, J.E., and Jackson, T.N. (2007) *IEEE Electron. Dev. Lett.*, **28**, 877.
41. Kim, S.H., Choi, D., Chung, D.S., Yang, C., Jang, J., Park, C.E., and Park, S.-H.K. (2008) *Appl. Phys. Lett.*, **93**, 113306.
42. Chen, J., Tee, C.K., Shtein, M., Anthony, J.E., and Martin, D.C. (2008) *J. Appl. Phys.*, **103**, 114513.
43. Gupta, D., Jeon, N., and Yoo, S. (2008) *Org. Electron.*, **9**, 1026.
44. Hong, J.-P., Park, A.-Y., Lee, S., Kang, J., Shin, N., and Yoon, D.Y. (2008) *Appl. Phys. Lett.*, **92**, 143311.
45. Sim, K., Choi, Y., Kim, H., Cho, S., Yoon, S.C., and Pyo, S. (2009) *Org. Electron.*, **10**, 506.
46. Lim, J.A., Lee, W.H., Lee, H.S., Lee, J.H., Park, Y.D., and Cho, K. (2008) *Adv. Funct. Mater.*, **18**, 229.
47. Lee, S.H., Choi, M.H., Han, S.H., Choo, D.J., Jang, J., and Kwon, S.K. (2008) *Org. Electron.*, **9**, 721.
48. Kim, Y.-H., Han, S.-M., Lee, W., Han, M.-K., Lee, Y.U., and Han, J.-I. (2007) *Appl. Phys. Lett.*, **91**, 042113.
49. (a) Lee, W.H., Kim, D.H., Jang, Y., Cho, J.H., Hwang, M., Park, Y.D., Kim, Y.H., Han, J.I., and Cho, K. (2007) *Appl. Phys. Lett.*, **90**, 132106; (b) Kim, D.H., Lee, D.Y., Lee, H.S., Lee, W.H., Kim, Y.H., Han, J.I., and Cho, K. (2007) *Adv. Mater.*, **19**, 678.
50. Chen, J., Tee, C.K., Shtein, M., Martin, D.C., and Anthony, J. (2009) *Org. Electron.*, **10**, 696.
51. Headrick, R.L., Wo, S., Sansoz, F., and Anthony, J.E. (2008) *Appl. Phys. Lett.*, **92**, 063302.
52. Stingelin-Stutzmann, N., Smits, E., Wondergem, H., Tanase, C., Plom, P., Smith, P., and de Leeuw, D. (2005) *Nat. Mat.*, **4**, 601.
53. Kang, J., Shin, N., Jang, D.Y., Prabhu, V.M., and Yoon, D.Y. (2008) *J. Am. Chem. Soc.*, **130**, 12273.
54. Madec, M.-B., Crouch, D., Llorente, G.R., Whittle, T.J., Geoghegan, M., and Yeates, S.G. (2008) *J. Mater. Chem.*, **18**, 3230.
55. (a) Kwon, J.-H., Shin, S.-I., Kim, K.-H., Cho, M.J., Kim, K.N., Choi, D.H., and Ju, B.-K. (2009) *Appl. Phys. Lett.*, **94**, 013506; (b) Ohe, T., Kuribayashi, M., Yasuda, R., Tsuboi, A., Nomoto, K., Satori, K., Itabashi, M., and Kasahara, J. (2008) *Jiro. Appl. Phys. Lett.*, **93**, 053303.
56. Hamilton, R., Smith, J., Ogier, S., Heeney, M., Anthony, J.E., McCulloch, I., Veres, J., Bradley, D.D.C., and Anthopoulos, T.D. (2009) *Adv. Mater.*, **21**, 1166.
57. (a) Lehnherr, D. and Tykwinski, R.R. (2007) *Org. Lett.*, **9**, 4583; (b) Lehnherr, D., McDonald, R., Ferguson, R.J., and Tykwinski, R.R. (2008) *Tetrahedron*, **64**, 11449.
58. Ostroverkhova, O., Shcherbyna, S., Cooke, D.G., Egerton, R.F., Hegmann, F.A., Tykwinski, R., Parkin, S.R., and Anthony, J.E. (2005) *J. Appl. Phys.*, **98**, 033701.
59. Lehnherr, D., Gao, J., Hegmann, F.A., and Tykwinski, R.R. (2008) *Org. Lett.*, **10**, 4779.
60. (a) Okamoto, T. and Bao, Z. (2007) *J. Am. Chem. Soc.*, **129**, 10308; (b) Okamoto, T., Jiang, Y., Qu, F., Mayer, A.C., Parmer, J.E., McGehee, M.D., and Bao, Z. (2008) *Macromolecules*, **41**, 6977.
61. Payne, M.M., Delcamp, J.H., Parkin, S.R., and Anthony, J.E. (2004) *Org. Lett.*, **6**, 1609.
62. Wolak, M.A., Melinger, J.S., Lane, P.A., Palilis, L.C., Landis, C.A., Anthony, J.E., and Kafafi, Z.H. (2006) *J. Phys. Chem. B*, **110**, 10606.
63. Wolak, M.A., Melinger, J.S., Lane, P.A., Palilis, L.C., Landis, C.A., Delcamp, J., Anthony, J.E., and Kafafi, Z.H. (2006) *J. Phys. Chem. B*, **110**, 7928.
64. Wolak, M.A., Delcamp, J., Landis, C.A., Lane, P.A., Anthony, J.E., and Kafafi,

Z.H. (2006) *Adv. Funct. Mater.*, **16**, 1943.

65. Lloyd, M.T., Mayer, A.C., Tayi, A.S., Bowen, A.M., Kasen, T.G., Herman, D.J., Mourey, D.A., Anthony, J.E., and Malliaras, G.G. (2006) *Org. Electron.*, **7**, 243.
66. Palilis, L.C., Lane, P.A., Kushto, G.P., Purushothaman, B., Anthony, J.E., and Kafafi, Z.H. (2008) *Org. Electron.*, **9**, 747.
67. Kuo, M.-Y., Chen, H.-Y., and Chao, I. (2007) *Chem. Eur. J.*, **13**, 4750.
68. Lim, Y.-F., Shu, Y., Parkin, S.R., Anthony, J.E., and Malliaras, G.G. (2009) *J. Mater. Chem.*, **19**, 3049.
69. Swartz, C.R., Parkin, S.R., Bullock, J.E., Anthony, J.E., Mayer, A.C., and Malliaras, G.G. (2005) *Org. Lett.*, **7**, 3163.
70. Tang, M.L., Reichardt, A.D., Wei, P., and Bao, Z. (2009) *J. Am. Chem. Soc.*, **131**, 5264.
71. Tang, M.L., Oh, J.H., Reichardt, A.D., and Bao, Z. (2009) *J. Am. Chem. Soc.*, **131**, 3733.
72. Park, J.-H., Chung, D.S., Park, J.-W., Ahn, T., Kong, H., Jung, Y.K., Lee, J., Yi, M.H., Park, C.E., Kwon, S.-K., and Shim, H.-K. (2007) *Org. Lett.*, **9**, 2573.
73. Chung, D.S., Yun, W.M., Nam, S., Kim, S.H., Park, C.E., Park, J.W., Kwon, S.-K., and Kim, Y.-H. (2009) *Appl. Phys. Lett.*, **94**, 043303.
74. Laquindanum, J.G., Katz, H.E., and Lovinger, A.J. (1998) *J. Am. Chem. Soc.*, **120**, 664.
75. Katz, H.E., Li, W., Lovinger, A., and Laquindanum, J. (1999) *Synth. Met.*, **102**, 897.
76. Ebata, H., Izawa, T., Miyazaki, E., Takimiya, K., Ikeda, M., Kuwbara, H., and Yui, T. (2007) *J. Am. Chem. Soc.*, **129**, 15732.
77. Payne, M.M., Odom, S.A., Parkin, S.R., and Anthony, J.E. (2004) *Org. Lett.*, **6**, 3325.
78. For a recent review of these systems, see Bunz, U. (2009) *Chem. Eur. J.*, **15**, 6780.
79. Tang, M.L., Reichardt, A.D., Siegrist, T., Mannsfeld, S.C.B., and Bao, Z. (2008) *Chem. Mater.*, **20**, 4669.
80. De la Cruz, P., Martin, N., Miguel, F., Seoane, C., Albert, A., Cano, H., Gonzalez, A., and Pingarron, J.M. (1992) *J. Org. Chem.*, **57**, 6192.
81. Payne, M.M., Parkin, S.R., Anthony, J.E., Kuo, C.C., and Jackson, T.N. (2005) *J. Am. Chem. Soc.*, **127**, 4986.
82. Dickey, K.C., Anthony, J.E., and Loo, Y.-L. (2006) *Adv. Mater.*, **18**, 1721.
83. Lee, W.H., Kim, D.H., Cho, J.H., Jang, Y., Lim, J.A., Kwak, D., and Cho, K. (2007) *Appl. Phys. Lett.*, **91**, 092105.
84. Lee, W.H., Lim, J.A., Kim, D.H., Cho, J.H., Jang, Y., Kim, Y.H., Han, J.I., and Cho, K. (2008) *Adv. Funct. Mater.*, **18**, 560.
85. Dickey, K.C., Subramanian, S., Anthony, J.E., Han, L.-H., Chen, S., and Loo, Y.-L. (2007) *Appl. Phys. Lett.*, **90**, 244103.
86. Dickey, K.C., Smith, T.J., Stevenson, K.J., Subramanian, S., Anthony, J.E., and Loo, Y.-L. (2007) *Chem. Mater.*, **19**, 5210.
87. Bailey, D.B., Mai, X., Scuderi, A.C., and Levy, D.H. (2006) US Patent Appl. Publ. 2006, US 2006/0,220,007.
88. Reichenbächer, K., Süss, H.I., and Hulliger, J. (2005) *Chem. Soc. Rev.*, **34**, 22.
89. Subramanian, S., Park, S.K., Parkin, S.R., Podzorov, V., Jackson, T.N., and Anthony, J.E. (2008) *J. Am. Chem. Soc.*, **130**, 2706.
90. Park, S.K., Mourey, D.A., Subramanian, S., Anthony, J.E., and Jackson, T.N. (2008) *Appl. Phys. Lett.*, **93**, 043301.
91. Jurchescu, O.D., Hamadani, B., Xiong, H.D., Park, S.K., Subramanian, S., Zimmerman, N.M., Anthony, J.E., Jackson, T.N., and Gundlach, D.J. (2008) *Appl. Phys. Lett.*, **92**, 132103.
92. Gundlach, D.J., Royer, J.E., Hamadani, B.H., Teague, L.C., Moad, A.J., Jurchescu, O.D., Kirillov, O., Richter, C.A., Kushmeric, J.G., Richter, L.J., Park, S.K., Jackson, T.N., Subramanian, S., and Anthony, J.E. (2008) *Nat. Mater.*, **7**, 216.
93. Teague, L.C., Hamadani, B.H., Jurchescu, O.D., Subramanian, S., Anthony, J.E., Jackson, T.N.,

Richter, C.A., Gundlach, D.J., and Kushmerick, J.G. (2008) *Adv. Mater.*, **20**, 4513.

94. Park, S.K., Mourey, D.A., Subramanian, S., Anthony, J.E., and Jackson, T.N. (2008) *Adv. Mater.*, **20**, 4145.
95. Jurchescu, O.D., Subramanian, S., Kline, R.J., Hudson, S.D., Anthony, J.E., Jackson, T.N., and Gundlach, D.J. (2008) *Chem. Mater.*, **20**, 6733.
96. Lee, S.S., Kim, C.S., Gomez, E.D., Purushothaman, B., Toney, M.F., Wang, C., Hexemer, A., Anthony, J.E., and Loo, Y.-L. (2009) *Adv. Mater.*, **21**, 3605.
97. Tang, M.L., Reichardt, A.D., Miyaki, N., Stoltenberg, R.M., and Bao, Z. (2008) *J. Am. Chem. Soc.*, **130**, 6064.
98. Smith, J., Hamilton, R., Heeney, M., de Leeuw, D.M., Cantatore, E., Anthony, J.E., McCulloch, I., Bradley, D.C., and Anthopoulos, T.D. (2008) *Appl. Phys. Lett.*, **93**, 253301.
99. Lloyd, M.T., Mayer, A.C., Subramanian, S., Mourey, D.A., Herman, D.J., Bapat, A.V., Anthony, J.E., and Malliaras, G.G. (2007) *J. Am. Chem. Soc.*, **129**, 9144.
100. Okamoto, T., Jiang, Y., Qu, F., Mayer, A.C., Parmer, J.E., McGehee, M.D., and Bao, Z. (2008) *Macromolecules*, **41**, 6977.

5
Conjugated Semiconductors for Organic n-Channel Transistors and Complementary Circuits

Antonio Facchetti

5.1
Introduction

Research groups exploring the field of organic/printed electronics have expanded considerably because of the potential use of organic semiconductors (and other essential organic materials such as conductors and dielectrics) as a low-cost alternative to silicon. Applications for organic semiconductors include conducting elements (after proper doping), organic photoconductors, field-effect transistors [1], light-emitting diodes [2], photovoltaic cells [3], sensors [4], and organic circuits [5] for integration into low-cost, large-area electronics [6] to cite just a few. Organic materials may be advantageous because of inexpensive materials processing and high throughput device assembly via printing methodologies. Solution-processable organic materials can eliminate the need for expensive lithography and vacuum deposition steps necessary in silicon-based microelectronics. Low-temperature solution processing also enables the use of inexpensive substrates such as flexible plastics in combination with processing options such as spin coating [7] and printing [8]. Note that realization of all-printed electronic devices requires a new set of materials, including conducting polymers/printable metals functioning as contact/electrodes, organic semiconductors, dielectric materials functioning as the gate insulator, and passive components as well as plastic flexible substrates [9].

n-channel semiconductors, a materials class in which the majority charge carriers are electrons, are essential in a variety of optoelectronic devices, including p–n junctions, bipolar transistors, and complementary circuits (CMOS (complementary metal-oxide semiconductor)). In organic CMOS, p- and n-channel transistors (vide infra) operate in concert. The great advantages of CMOS circuits compared to the corresponding single-polarity (exclusively from n-channel or p-channel TFTs (thin-film transistors)) circuits are the significantly lower static power dissipation, reliability, and stability. As a rule of thumb, the speed of complementary circuits is $\sim 10\times$ the speed of single-polarity devices, given equal charge carrier mobility. Therefore, enabling both high-performance p- and n-type organic transistors and developing an efficient methodology to integrate them in a printing organic CMOS process is the keystone for building

Organic Electronics II: More Materials and Applications, First Edition. Edited by Hagen Klauk.

faster organic circuits and electronic products. Moreover, in order to manufacture large-area and inexpensive electronic circuits on flexible, plastic substrates, solution-processable and printable semiconductors are essential [10]. During the past few years, great progress has been made in developing p-channel materials, including functionalized acenes, [1]benzothieno[3,2-b]benzothiophene derivatives (BTBT), and dithieno [2,3-d; 2′, 3′-d′]benzo[1,2-b; 4,5-b′]dithiophene (DTBDT) being among the best performing of molecular materials, and several high-performance polymeric materials, poly(3-hexylthiophene) (P3HT), poly(2,5-bis (3-alkylthiophen-2-yl)thieno[3,2-b]thiophene (PBTTT), and poly(4,8-dialkyl-2,6-bis (3-alkylthiophen-2-yl)benzo[1,2-b : 4,5-b′]dithiophene) [11]. However, the development of high-performance ambient-stable n-channel semiconductors has been far more difficult than p-channel materials. During the past five years, several high-performance n-channel semiconductors have been developed, and TFTs fabricated with vapor-deposited films of these materials exhibit excellent electron mobilities in ambient conditions. However, solution-processable n-channel semiconductors remain challenging, and only in the last two years have impressive progress been made.

Although there is tremendous research on all of the materials needed to fabricate organic complementary circuits, this chapter focuses on n-channel semiconductors for organic field-effect transistors (OFETs). In particular, I will discuss fundamental n-channel materials and focus on the results achieved during the last five years as well as on n-channel semiconductors exhibiting good charge transport in ambient conditions. After an introduction to field-effect transistor (FET) structure and operation as well as CMOS circuits, I will discuss the rationale to designing n-channel semiconductors followed by a description of the major n-channel materials families divided into small molecules and polymers. Fundamental studies reporting on organic CMOS are discussed in more detail. Finally, the challenges and opportunities that this field faces are summarized.

5.2 Basics of Field-Effect Transistors and Complementary Circuits

5.2.1 Field-Effect Transistors

The most common OFET device configuration is that of a TFT, in which a thin film of the organic semiconductor is deposited on top of a dielectric with an underlying gate (G) electrode (Figure 5.1). The acronyms FET and TFT are used in this chapter. Charge-injecting source–drain (S–D) electrodes providing the contacts are defined either on top of the organic film (top-contact configuration) or on the surface of the FET substrate prior to semiconductor deposition (bottom-contact configuration). Other device structures are shown in Figure 5.2. Minimal S–D current is measured when no voltage is applied between the G and S electrodes (device “off” state). When a voltage is applied to the gate, mobile electrons or holes are created at the

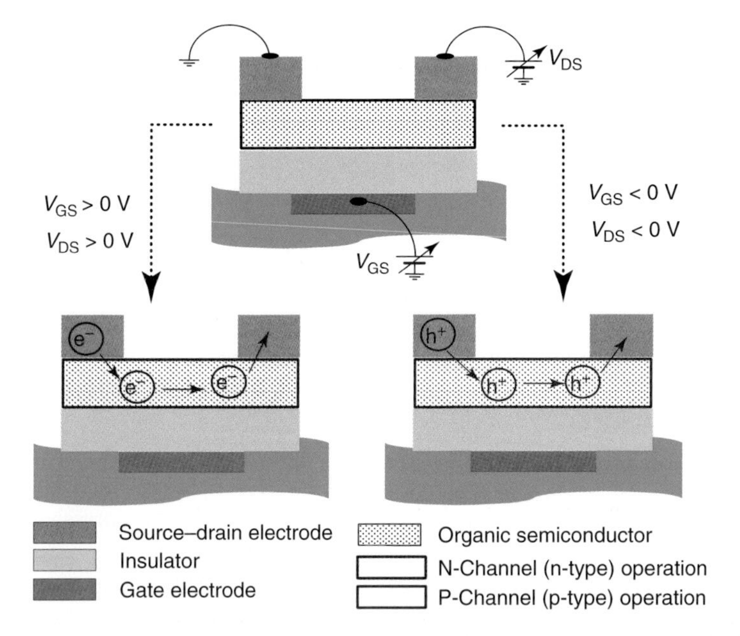

Figure 5.1 Schematic representation of a bottom-gate top-contact organic field-effect (thin-film) transistor structure with highlighted accumulation of electrons and holes for n-channel and p-channel semiconductors, respectively.

semiconducor-dielectic interface and the S–D current increases (device "on" state, Figure 5.1). The relationships describing the OFET drain current are given in Eqs. (5.1) and (5.2)

$$(I_{SD})lin = (W/L)\mu C_i\ (V_{SG} - V_T - V_{SD}/2)\ V_{SD}\ \text{(linear regime)} \tag{5.1}$$

$$(I_{SD})_{sat} = (W/2\ L)\mu C_i\ (V_{SG} - V_T)^2\ \text{(saturation regime)} \tag{5.2}$$

where μ is the field-effect mobility of the semiconductor, W the channel width, L the channel length, C_i the capacitance per unit area of the insulator layer, V_T the threshold voltage, V_{SD} the S–D voltage, and V_{SG} the source-gate voltage. On increasing the V_{SD} and V_{SG} magnitudes, a linear current regime (Eq. (5.1)) is initially observed at low drain voltages ($V_{SD} < V_{SG}$), followed by a saturation regime (Eq. (5.2)) when the drain voltage exceeds the gate voltage. In contrast to conventional Si transistors, note that OFETs normally operate in the accumulation mode, where the increase in magnitude of V_{SG} enhances channel conductivity. When an OFET is active upon the application of negative V_{SG} and V_{SD}, the organic material is said to be *p-channel* (or p-type) since holes are the majority carriers (Figure 5.2). On the other hand, when a (positive) S–D current is observed upon the application of positive V_{SG} and V_{SD}, the semiconductor is *n-channel* (or n-type) since electrons are mobile (Figure 5.1). In few cases, organic thin-film transistors (OTFTs) operate for both V_{SG} and V_{SD} polarities, and the semiconductor is said to be *ambipolar*.

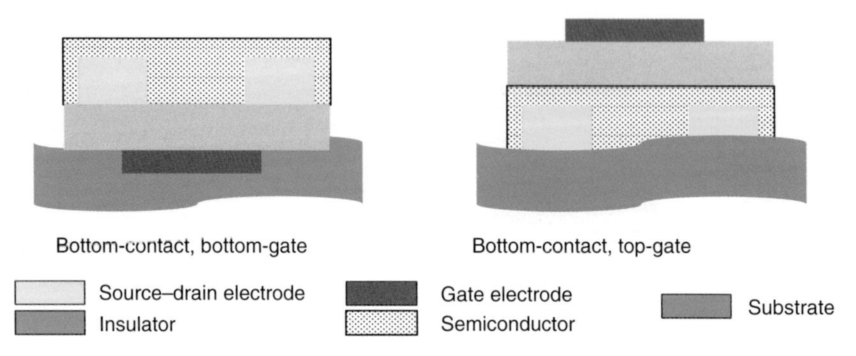

Figure 5.2 Schematic representations of two other important transistor architectures. (a) Bottom-contact bottom-gate and (b) bottom-contact top-gate configurations.

To achieve high-performance FETs, organic semiconductors must be designed to maximize both charge injection and charge transport processes via the following: (i) the electronic structure must be appropriate for holes/electron injection/extraction at accessible applied electric fields; this is usually simplified by requiring that that compound's highest occupied molecular orbital/lowest unoccupied molecular orbital (HOMO/LUMO) energies of the individual molecules have appropriate energies, (ii) the crystal structure of the material must provide sufficient overlap of frontier orbitals to allow efficient charge migration between neighboring molecules, (iii) the solid should be extremely pure since impurities act as charge carrier traps, (iv) the molecules or polymer chains must have an edge-on orientation since the most efficient charge transport occurs along the intermolecular $\pi-\pi$ stacking direction, (v) the semiconductor film morphology should be uniform and must cover the area between S and D contacts uniformly. Furthermore, for top-gate FETs a smooth film surface is desirable.

OFETs can be fabricated in different device configurations (Figure 5.2), each with their own fabrication issues, advantages, and disadvantages. In the top-contact/bottom-gate geometry, the organic semiconductor is deposited on top of the gate dielectric layer, and then the source and drain electrodes are deposited on top by either metal evaporation through a shadow mask or by printing. In this device structure, the contact resistance is usually low because of intimate contact between the semiconductor and the S–D electrodes, and the charge carrier mobilities are usually the highest among different FET structures. However, in this type of structure, small channel lengths are difficult to achieve by both metal thermal evaporation and printing (usually L $\gg$ 5 μm channels), and solution-processed metal ink/polymeric contacts may damage the semiconductor. In the bottom-contact/bottom-gate devices, the source and drain electrodes are first defined on the insulator with photolithography (allowing very small channel lengths) or other deposition processes, and only at the end the organic layer is deposited. Bottom-contact devices typically exhibit greater contact resistance than top-contact devices and high carrier mobility and typical I–V plots are usually obtained only on S–D contact treatment with thiol-based molecules to form a self-assembled monolayer [12]. However, bottom-contact devices may be more easily integrated into low-cost manufacturing processes. Furthermore, smaller device features can be obtained for the first-generation products through photolithographic techniques enhancing circuit speed. Top-gate OFETs are other technologically important OFET architectures where the dielectric and then the gate contact are deposited on top of the organic semiconductor film.

The most common device structure used by organic electronics scientists consists of a heavily doped silicon substrate as the (bottom-) gate electrode coated with thermally grown SiO_2 with a thickness of ~100–300 nm acting as the dielectric, and 30–50-nm-thick Au or Al are used as S–D (top-) contact material. While these silicon-SiO_2-metal-based devices are extremely important for comparing organic semiconductors developed by different research groups, the use of plastic substrates [13], polymeric insulators [14], and conductive inks as the source and drain electrodes [15] must be developed to enable low-cost/flexible electronics.

5.2.2
Complementary Circuits

CMOS is a technology for producing integrated circuits. Traditionally, this refers to field-effect transistors fabricated with inorganic materials and having the structure of a semiconductor coated with an oxide insulator and having a metal gate electrode placed on top. Typically, the semiconductor is silicon, and the gate insulator metal oxide is SiO_2. Aluminum was once used, but now the material is doped polysilicon. The word "complementary" refers to the fact that the typical digital design style with CMOS uses complementary and symmetrical pairs of p-type metal-oxide semiconductor (PMOS) and n-type metal-oxide semiconductor (NMOS) FETs for logic functions. The CMOS technology is used in microprocessors, microcontrollers, static random access memory (SRAM), and other digital logic circuits. CMOS technology is also used in analog applications such as image sensors, data converters, highly integrated transceivers for many types of communication, operational amplifiers as well as radiofrequency (RF) applications. Additionally, CMOS technology is used for mixed-signal (analog+digital) applications.

Conventional CMOS logic gates are fabricated with a network of PMOS transistors connecting the output to the positive supply and a network of NMOS transistors connecting the output to the negative supply (or ground); the two networks are generally designed to be complementary, in the sense that the output will also switch high or low voltage but not both at once. The PMOS transistors operate such that low channel resistance is achieved when a low gate voltage is applied and high channel resistance is achieved when a high gate voltage is applied, whereas the NMOS transistors operate in the opposite way.

For instance, Figure 5.3 shows the image of the simplest CMOS device: the inverter or NOT gate. In this device, when a low input is connected to both a PMOS transistor and an NMOS transistor, the NMOS transistor has high resistance, so it stops voltage from leaking into the ground, whereas the PMOS transistor has low resistance so it allows the voltage source to transfer voltage through the PMOS

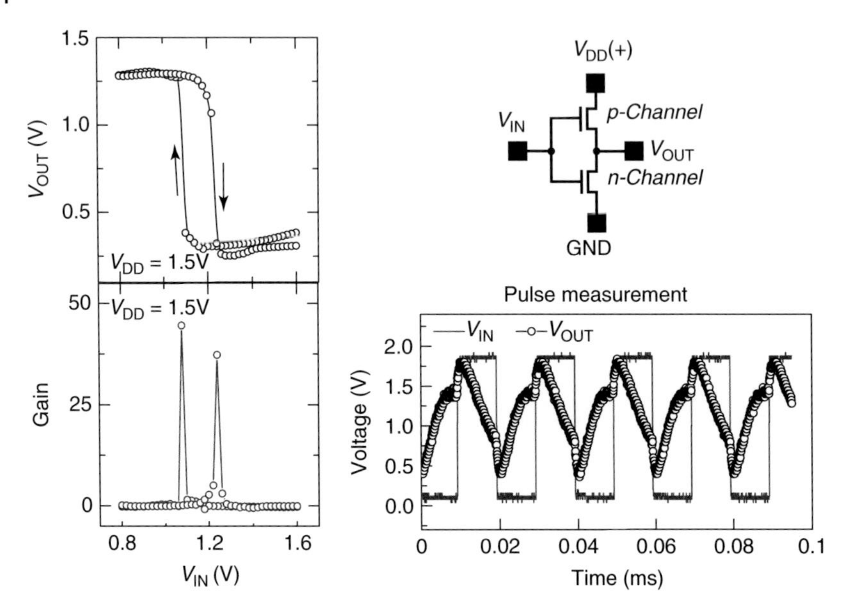

Figure 5.3 Diagram of a CMOS inverter with highlighted static and dynamic voltage modulation characteristics.

transistor to the output. The output would therefore register a high voltage. On the other hand, when the voltage of input is high, the PMOS transistor would have high resistance, so it would block the voltage source from the output, whereas the NMOS transistor would have low resistance, allowing the output to drain to the ground. This would result in the output registering a low voltage. In short, the outputs of the PMOS and NMOS transistors are complementary such that when the input is low, the output would be high, and when the input is high, the output would be low. Therefore, the output of the CMOS circuit is by default the inversion of the input. Two important characteristics of CMOS devices are high noise immunity and low static power consumption. Significant power is only drawn when the transistors in the CMOS device are switching between the on (low channel resistance) and off (high channel resistance) states. Consequently, CMOS devices do not produce as much waste heat as other forms of logic such as transistor–transistor logic (TTL) or single-polarity MOS logic, which uses only a single type of transistor. CMOS also allows a high density of logic functions on a chip. This is the primary reason why CMOS has become the most used technology to be implemented in VLSI (very large scale integration) chips.

Other logic functions such as those involving AND and OR gates require manipulating the paths between gates to represent the logic. When a path consists of two transistors in series, both transistors must have low resistance to the corresponding supply voltage, modeling an AND. When a path consists of two transistors in parallel, then either one or both of the transistors must have low resistance to connect the supply voltage to the output, modeling an OR. Another important circuit in CMOS logic is the NAND gate (Figure 5.4). If both the A and B

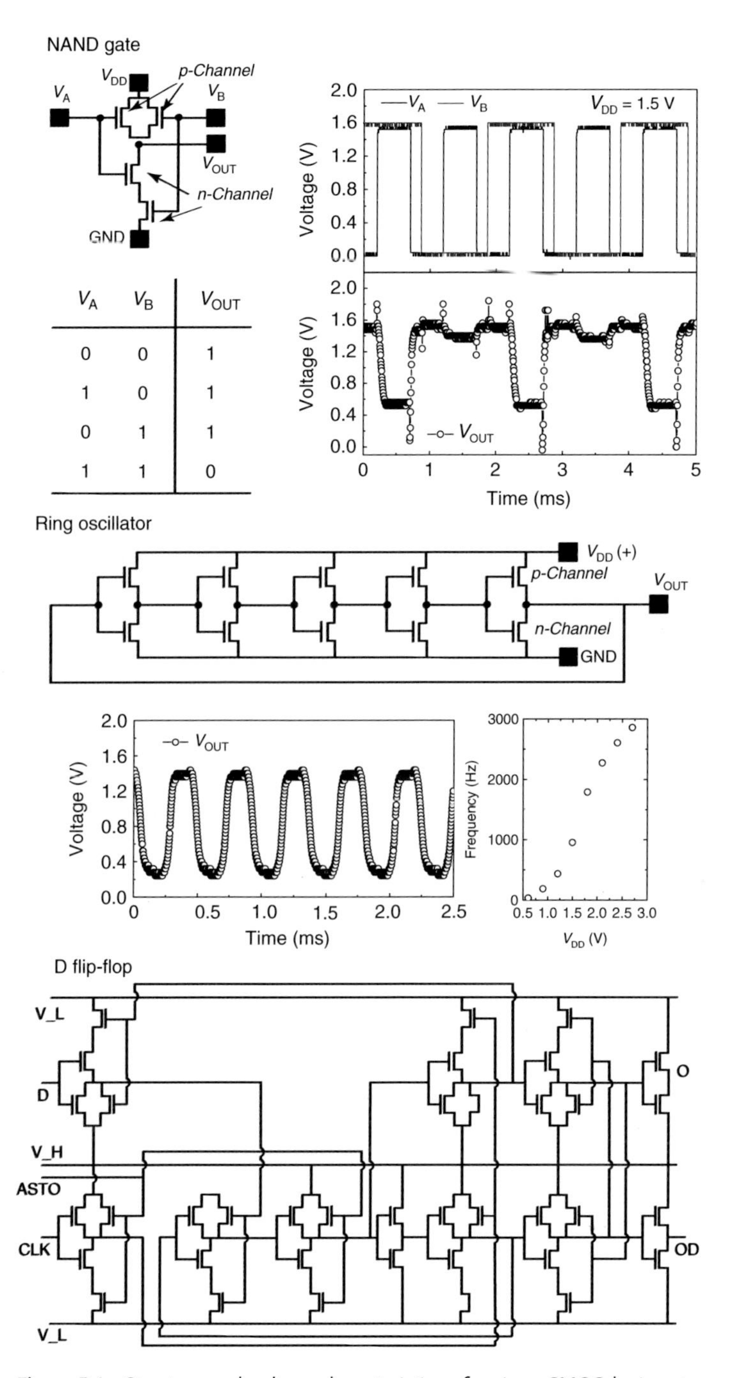

Figure 5.4 Structure and voltage characteristics of various CMOS logic gates.

inputs are high, then both the NMOS transistors (bottom half of the diagram) will conduct, neither of the PMOS transistors (top half) will conduct, and a conductive path will be established between the output and *Vss* (ground), bringing the output low. If either the A or the B input is low, one of the NMOS transistors will not conduct, one of the PMOS transistors will, and a conductive path will be established between the output and *Vdd* (voltage source), bringing the output high.

Finally, Figure 5.4 shows other important CMOS circuits such as that of a ring oscillator and D Flip-Flop. A ring oscillator is a device composed of an odd number of inverters (NOT gates), whose output oscillates between two voltage levels. The inverters are connected in a chain; the output of the last inverter is fed back into the first. The voltage-controlled oscillator in most phase-locked loops is built from a ring oscillator. A ring oscillator is often used to demonstrate a new hardware technology and to measure the effects of voltage and temperature on a chip as well as during Si wafer testing to measure the effects of manufacturing process variations. A schematic of a simple five-inverter ring oscillator is shown in Figure 5.4. Because a single inverter computes the NOT logic of its input, it can be shown that the last output of a chain of an odd number of inverters is the logical NOT of the first input. This final output is asserted a finite amount of time after the first input is asserted; the feedback of this last output to the input causes the voltage oscillation. A circular chain composed of an even number of inverters cannot be used as a ring oscillator since the last output in this case is the same as the input. However, this configuration of inverter feedback can be used as a storage element; it is the basic building block of SRAM. Ring oscillators require only power to operate, since above a certain threshold voltage, the oscillation begins spontaneously. The frequency of oscillation can be increased via increasing the applied voltage or reducing the device size. The first approach increases both the frequency of the oscillation and the power consumed, which is dissipated as heat, whereas a smaller ring oscillator results in a higher oscillation for a given power consumption.

A flip-flop is an electronic circuit having two stable states and thereby capable of serving as one bit of memory in digital circuits. A flip-flop is usually controlled by one or two control signals and/or a gate or clock signal. The output often includes the complement as well as the normal output. As flip-flops are implemented electronically, they require power and ground connections. Flip-flops can be either simple (transparent) or clocked. Simple flip-flops can be built around a pair of cross-coupled inverting elements. The more advanced clocked devices are designed for synchronous (time-discrete) systems; such devices therefore ignore their inputs except at the transition of a dedicated clock signal (known as *clocking*, *pulsing*, or *strobing*). This causes the flip-flop to either change or retain its output signal based on the values of the input signals at the transition. Some flip-flops change outputs on the rising edge of the clock, others on the falling edge. Clocked flip-flops are typically implemented as master–slave devices, where two basic flip-flops (plus some additional logic) collaborate to make them insensitive to spikes and noise between the short clock transitions. Flip-flops can be divided into types that have found common applicability in both asynchronous and clocked sequential systems:

D ("data" or "delay"), the SR ("set-reset"), T ("toggle"), and JK types are the common ones, all of which may be fabricated from (most) other types by a few logic gates. A master–slave D flip-flop is created by connecting two gated D latches in series and inverting the enable input to one of them. It is called *master–slave* because the second latch in the series changes only in response to a change in the first (master) latch. It responds on the negative edge of the enable input (usually a clock). For a positive-edge-triggered master–slave D flip-flop, when the clock signal is low (logic 0), the "enable" seen by the first or "master" D latch (the inverted clock signal) is high (logic 1). This allows the "master" latch to store the input value when the clock signal transitions from low to high. As the clock signal goes high (0 to 1), the inverted "enable" of the first latch goes low (1 to 0) and the value seen at the input to the master latch is "locked." Nearly simultaneously, the twice inverted "enable" of the second or "slave" D latch transitions from low to high (0 to 1) with the clock signal. This allows the signal captured at the rising edge of the clock by the now "locked" master latch to pass through the "slave" latch. When the clock signal returns to low (1 to 0), the output of the "slave" latch is "locked," and the value seen at the last rising edge of the clock is held while the "master" latch begins to accept new values in preparation for the next rising clock edge. A single flip-flop can be used to store one bit, or binary digit, of data. Any one of the flip-flop types can be used to build any of the others. The data contained in several flip-flops may represent the state of a sequencer, the value of a counter, an ASCII character in a computer's memory, or any other piece of information. One use is to build finite state machines from electronic logic. The flip-flops remember the machine's previous state, and digital logic uses that state to calculate the next state.

5.3 Material Design and Needs for n-Channel OTFTs

5.3.1 Electronic Structure

In an n-channel semiconductor, electron injection from the device contact and transport in the device channel occur, in the simplest picture, in the molecule's LUMO. n-Channel semiconductor design must take into account and favor these processes. Consequently, n-channel materials should have an electronic structure where the unoccupied frontier molecular orbital (FMO) energy of the semiconductor in the solid state allows electron accumulation and transport when the device is "switched on." It is intuitive to think that this should result in materials in which the LUMO energy is as low as possible. Thus, electron-depleted (or electron-poor) conjugated cores are obvious n-channel candidates. However, the design of high-performance n-channel semiconductors is a delicate balance between the stability of the gate field-induced mobile electrons from charge trapping and prevention of electron doping from donor sites, which inevitably reduces the device current on–off ratio. The former requires a very low LUMO energy, whereas the latter is prevented by higher LUMO energy. However, the first issue has been by far the most challenging, and it is the one that the organic electronics community has devoted most of the efforts.

The major sources of electron traps in a TFT are chemical functionalities on the dielectric surface and, equally relevant, air components, particularly H_2O and O_2. Rational strategies for enhancing ambient stability must prevent these trapping species from reaching the charge-transporting film (kinetic stability) area and/or involve the design of molecules or polymers in which the mobile electrons are thermodynamically resistant to trapping (thermodynamic stability) (Figure 5.5). It is generally thought that H_2O/O_2 can be excluded either by utilizing crystalline materials with sufficiently dense molecular packing to resist penetration by these species or by appropriately encapsulating the devices in an inert atmosphere, although as noted above, additional encapsulation steps should ideally be avoided. Thermodynamic stability is a complex issue. De Leeuw *et al.* have investigated the stability of electron-doped conducting polymers by evaluating standard redox potentials for reactions with air components and drawn the conclusions that the majority of radical anions cannot be thermodynamically stable in ambient conditions [16]. For unintentionally n-doped semiconductors, the LUMO energy equals the first redox potential (vs SCE) plus 4.4 eV. The possible reactions between n-type semiconductors and oxygen (and/or water) under ambient conditions include:

$$O_2 + 2H_2O + 4e \rightarrow 4OH^- \qquad E = +0.571\ \text{V (vs SCE)}$$

$$4Sec^0 + 4e^- \rightarrow 4Sec^- \qquad E_{1R}$$

$$4Sec^0 + O_2 + 2H_2O \rightarrow 4Sec^- + 4OH^- \qquad \Delta E^\circ = +0.571 - E_{1R}$$

To prevent this reaction from taking place during the charge transport process, the free energy of this redox reaction needs to be positive, that is, $\Delta G = -nF\Delta E = -nF(0.571 - E_{1R}) \geq 0$, and consequently an $E_{1R} > -0.66$ V for H_2O and $> +0.571$ V are required for air ($H_2O + O_2$) [17]. This means that the LUMO [18] of this molecule should lie at an energy below ~ -3.7 and -4.9 eV, respectively. Materials with electron affinities lower than ~5 eV are rare. However, an overpotential to the charge carrier/O_2 reaction could, in principle, prevent ambient trapping in materials in which the LUMO energies are considerably less negative. Recent experimental and theoretical studies by the Northwestern University groups empirically identified this energetic threshold at $\sim -4.0/-4.3$ eV (Figure 5.6), which scales to an overpotential versus O_2 reduction of ~0.9–0.6 V, and applies to several n-channel semiconductor families (Figure 5.6) [16,19–21].

The common design strategy to achieve sufficiently low LUMO systems is functionalizing the molecule/polymer conjugated core with strong electron-withdrawing groups [17], which include –F, –CN, –COR, –CO–NR–CO–, perfluoroalkyl, and perfluoroarenes [18]. More recently, the work undertaken by Tang *et al.* demonstrated that chlorination is also a general route to material design of n-channel organic semiconductors [19]. Another strategy is to design molecules and polymers containing electron-poor heterocycles such as azines and azoles [20].

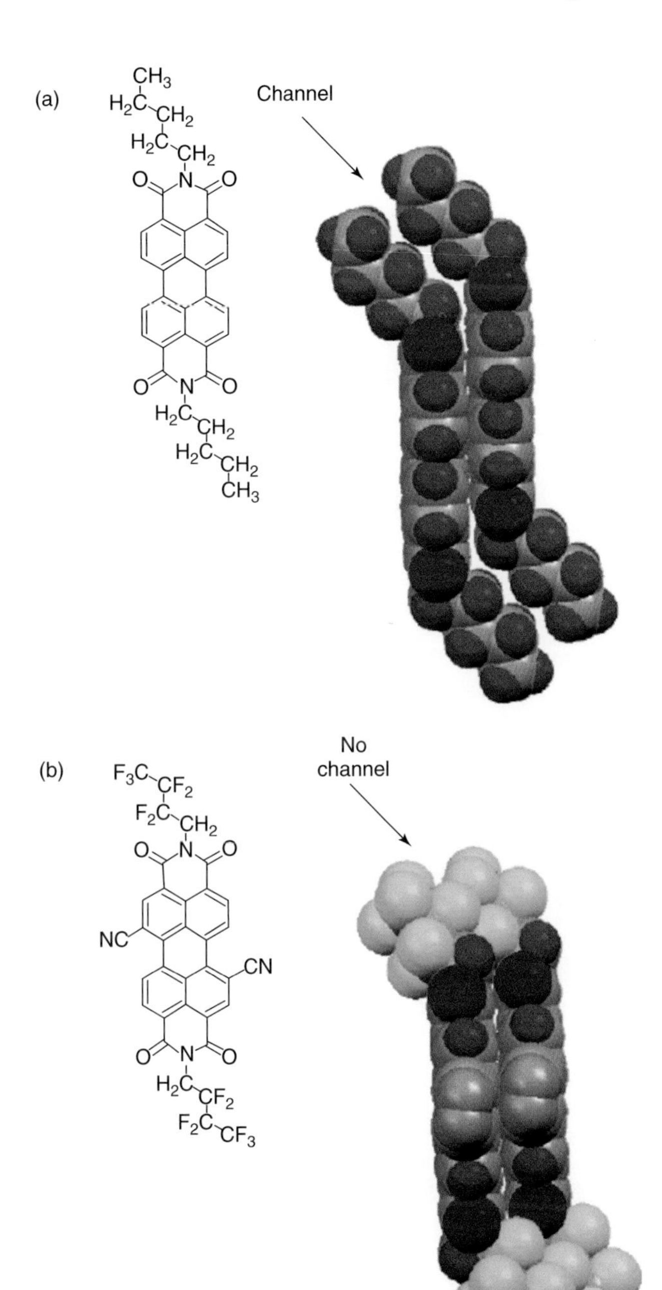

Figure 5.5 Crystal packing of two perylene diimide derivatives having hydrocarbon (a) and fluorocarbon (b) substituents on the nitrogen. (Reprinted with permission from Ref. [87b].)

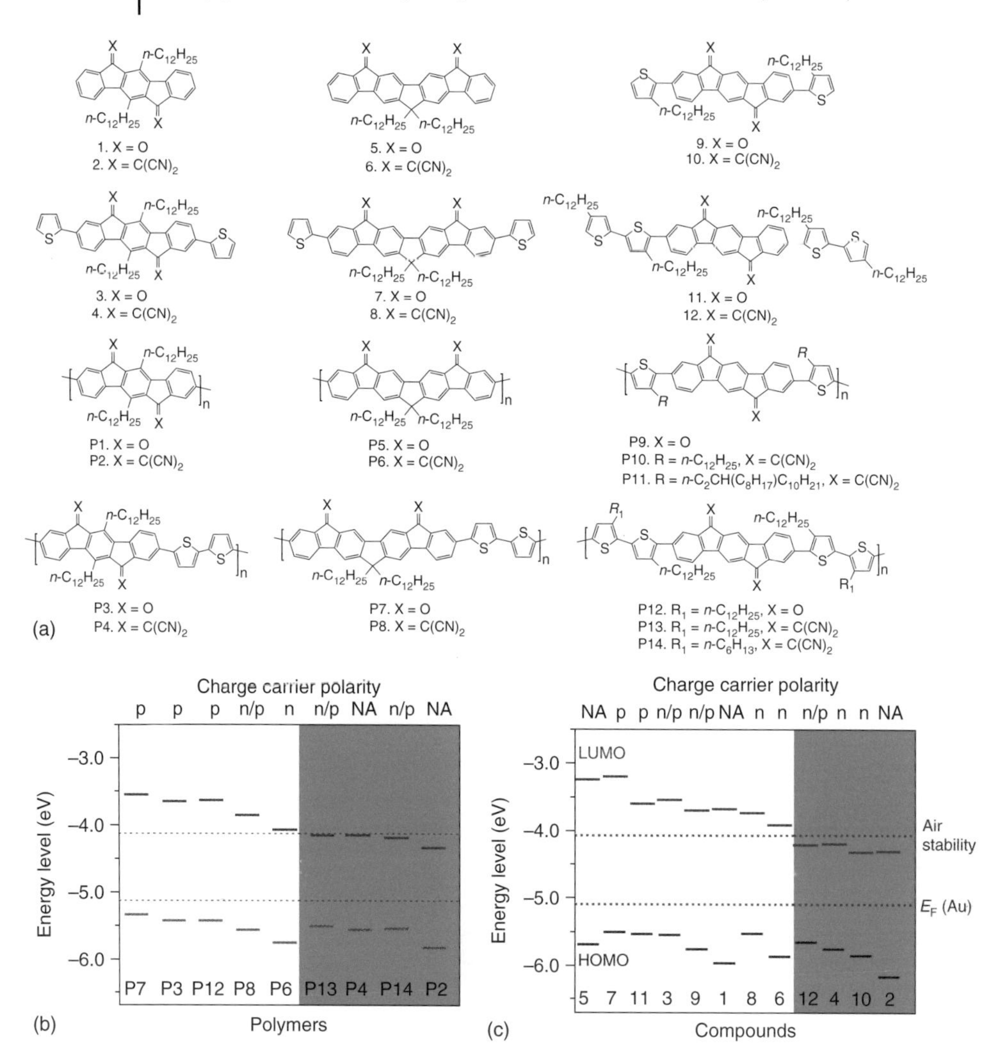

Figure 5.6 (a) Structures of ladder-type (bis)indenofluorene-based homo- and copolymers P1–P14. (b,c) Energy diagram for compounds 1–12 and polymers P2, P3, P6, P7, P8, and P12–P14, showing experimentally estimated HOMO/LUMO energy levels, and charge carrier polarity. The blue region indicates TFT devices with ambient stability. The dashed lines indicate the Fermi level of the gold electrode and the air-stability limit. (Reprinted with permission from Ref. [104].)

5.3.2
Contacts and Dielectric

Among the materials playing a key role in affecting n-channel FET performance are the electrical (S–D) contacts and the dielectric. For the contacts, the material work function is an important parameter, and several metals have been used in n-channel OTFTs. These include the high-work-function Au and Ag and the low-work-function Al, Mg, and Ca. Despite the fact that from an energetic perspective, the former metals may not be suitable for electron injection into the LUMO of organic semiconductors, Au is generally the best contact material for several n-channel semiconductors. This phenomenon is the result of the clean Au surface and possible energy level alignment at the Au/organic semiconductors interface, resulting in an interfacial electric dipole substantially reducing the injection barrier [21]. On the other hand, for the low-work-function metals, despite their formal lower injection barriers, they suffer from surface oxidation, which results in an insulating oxide layer. Furthermore, they can chemically react with the organic semiconductors, leading to charge transfer layers and decomposition products [22].

Although Au contacts have been widely used for n-channel OTFTs, [23] significant studies have been carried out to reduce the contact resistance affecting these devices. A recent study investigated the effect of Au-electrode modification on the performance of several n-channel perylene diimide-based transistors [24]. The authors used sulfur to modify Au top-contact electrodes and increase the performance of perylene diimide-based OTFTs remarkably. Alternative strategies to improve electron injection in n-channel FETs include electrode surface modifications [25], charge-transfer salt electrodes such as (TTF)(TCNQ) [26], carbon electrodes [27], graphene electrodes [28], transparent carbon films [29], and metal nanoparticles for printed transistors [30].

The other material greatly affecting electron transport in an FET device is the gate dielectric. The dielectric affects charge transport in the channel by varying the top-deposited semiconductor morphology (for bottom-gate FETs) and contributing to chemical trapping sites (in all TFT architectures) [31]. It is now well documented that acidic functionalities on the dielectric surface of an oxide dielectric such as carbinol (−C−OH), silanol (−Si−OH), and carbonyl (C=O) groups efficiently trap electrons [32]. (Figure 5.7) Thus, controlling the dielectric interface chemistry is critical to achieve high-performance n-channel OTFTs. Furthermore, this has a great impact on the device stability in ambient conditions [33]. Controlling electron trap density in the channel has been achieved by using organosilane self-assembled monolayers (SAMs) on oxide (mainly SiO_2) gate insulators [34]. An interesting study by Itaka *et al.* showed that the dielectric surface passivation can be achieved using a monolayer of an organic semiconductor such as pentacene [35]. Polymeric dielectrics have been actively pursued because of their reduced tendency to trap electrons and facile printability. Thus, several studies have addressed the use of these materials to improved n-channel semiconductor performance. Furthermore, thin polymeric films can be used in place of SAMs as passivation layers of metal-oxide dielectrics, which prevent electron trapping by the surface hydroxyl groups.

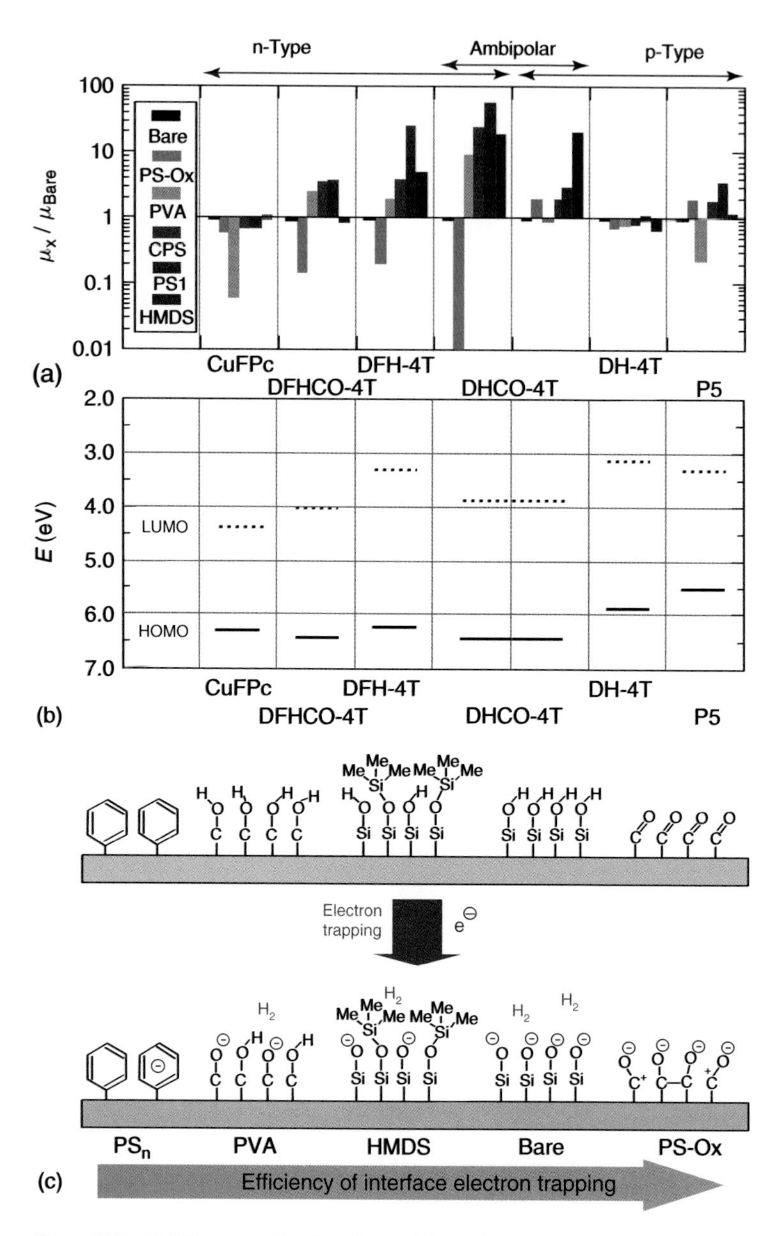

Figure 5.7 (a) Histogram showing the mobility enhancement ratio $\eta = \mu_X/\mu_{Bare}$ and (b) electrochemically derived FMO energy levels for the organic semiconductors investigated in this study. The crosshatched bars denote semiconductor–dielectric combinations for which the largest variations in semiconductor film morphology are observed. (c) Schematic diagram of functional group electron trapping efficiency on various bilayer dielectric layers. (Reprinted with permission from Ref. [32].)

5.4 n-Channel Semiconductors for OTFTs

In this section, several n-channel families are described. Furthermore, I will discuss the first and more advanced demonstration of CMOS circuits based on the most interesting derivatives.

5.4.1 Molecular Semiconductors

5.4.1.1 Phthalocyanine Derivatives

Figure 5.8 collects the chemical structure of several phthalocyanine derivatives. To the best of my knowledge, the first n-channel OFET was fabricated by Guillaud *et al.* using vapor-deposited films of lutetiumphthalocyanine ($LuPc_2$) [36]. These devices exhibit electron mobilities of 10^{-4} cm^2 V^{-1} s^{-1} in vacuum. Bao *et al.* demonstrated the first air-stable n-channel OTFTs using a semiconductor layer vacuum-deposited perfluorinated phthalocyanines ($F_{16}CuPc$) [37]. These FETs exhibit maximum electron mobilities of $\sim$0.03 cm^2 V^{-1} s^{-1} in air for sublimed

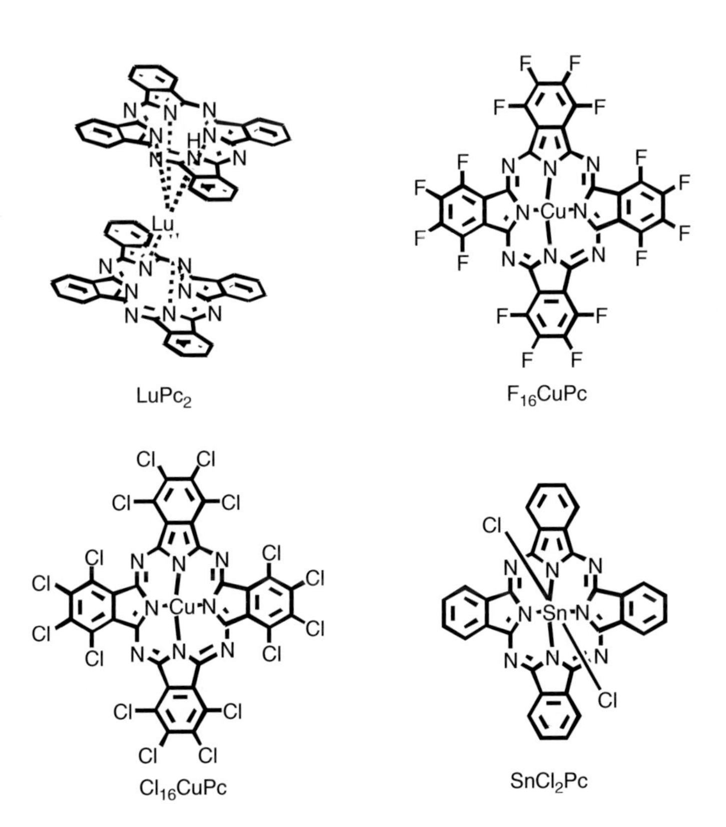

Figure 5.8 Chemical structures of various metal phthalocyanines.

films deposited at a substrate temperature of 130 °C. This is one of the first contributions where core functionalization of a p-channel material (CuPc) with strong electron-withdrawing F groups resulted in ambient-stable n-channel OTFTs. This material is also the most used for the fabrication of single-crystal devices [38]. A seed-induced growth by vapor processing was recently developed for in situ patterning of organic single-crystalline nanoribbons of $F_{16}CuPc$ on a Si/SiO_2 surface [39]. With this method, devices based on individual $F_{16}CuPc$ submicro/nanometer-sized ribbons were fabricated with Ag/Au asymmetrical source/drain electrodes, showing that the asymmetrical devices possessed much higher mobilities (0.2 $cm^2\ V^{-1}\ s^{-1}$) than those devices with Au/Au symmetrical electrodes (0.01 $cm^2\ V^{-1}\ s^{-1}$). This is due to the asymmetrical S–D electrode configuration, which built a stepwise energy level between the electrodes and the LUMO of $F_{16}CuPc$, promoting electron injection and transport. Recently, Ling *et al.* have used copper hexachlorophthalocyanine ($Cl_{16}CuPc$) as a semiconductor for n-channel OTFTs reporting electron mobilities of ~0.01 $cm^2\ V^{-1}\ s^{-1}$ [40]. Remarkably, these devices show no performance degradation after storage in air for more than 50 days. Wang *et al.* have reported n-channel FETs where the $F_{16}CuPc$ film is epitaxy grown on *para*-hexaphenyl (*p*-6P); these devices exhibit electron mobilities as high as 0.11 $cm^2\ V^{-1}\ s^{-1}$, which approach those of the single crystal-based devices [41]. Very recently, impressive electron mobilities of 0.30 $cm^2\ V^{-1}\ s^{-1}$ in air were reported by Song *et al.* using phthalocyanato tin(IV) dichloride (SnCl2Pc) as a semiconductor [42].

Complementary logic circuits based on OTFTs, such as row decoders and shift registers, were also first reported using a metalphathalocyanine. In a pioneer work [5a], Dodabalapur *et al.* used $F_{16}CuPc$ and pentacene for the fabrication of n-channel and p-channel TFTs, respectively (Figure 5.9). The largest circuit that was evaluated was a 48-stage shift register with 24 output buffers. The total number of transistors in this circuit was 864, and the channel length of each transistor is 7.5 μm. Each stage of the shift register is a D flip-flop, with the output of one stage connected to the input of the next. The clock and its complement drive all the stages. Every second stage has an output buffer, which consists of two CMOS inverters with large transistors to facilitate probing without loading the circuit. Shift registers are ubiquitous in digital systems and can perform many functions. One function is to shift a "bit" in an orderly and predictable manner from one stage to the next every clock cycle. The operation of the shift register is also illustrated in Figure 5.9A, in which the clock, data, and output of the 24 output buffers are plotted as a function of time. The data consist of a single bit, which is sequentially shifted over all stages of the register. Two-stage shift registers operated at clock rates of up to 1 kHz. The static current drawn by the 48-stage register, including the output buffers, is 70 mA, and that drawn by a smaller two-stage register (with 1 buffer per stage) is 7.6 mA. In comparison, two-stage complementary shift registers based on NOR/NAND gates required more than twice as many transistors as the two-stage shift register based on pass transistor logic and drew 20 mA. Simulations indicate that the current drawn by shift registers based on p-channel transistors alone is greater than that of complementary shift registers of comparable transistor dimensions and speed.

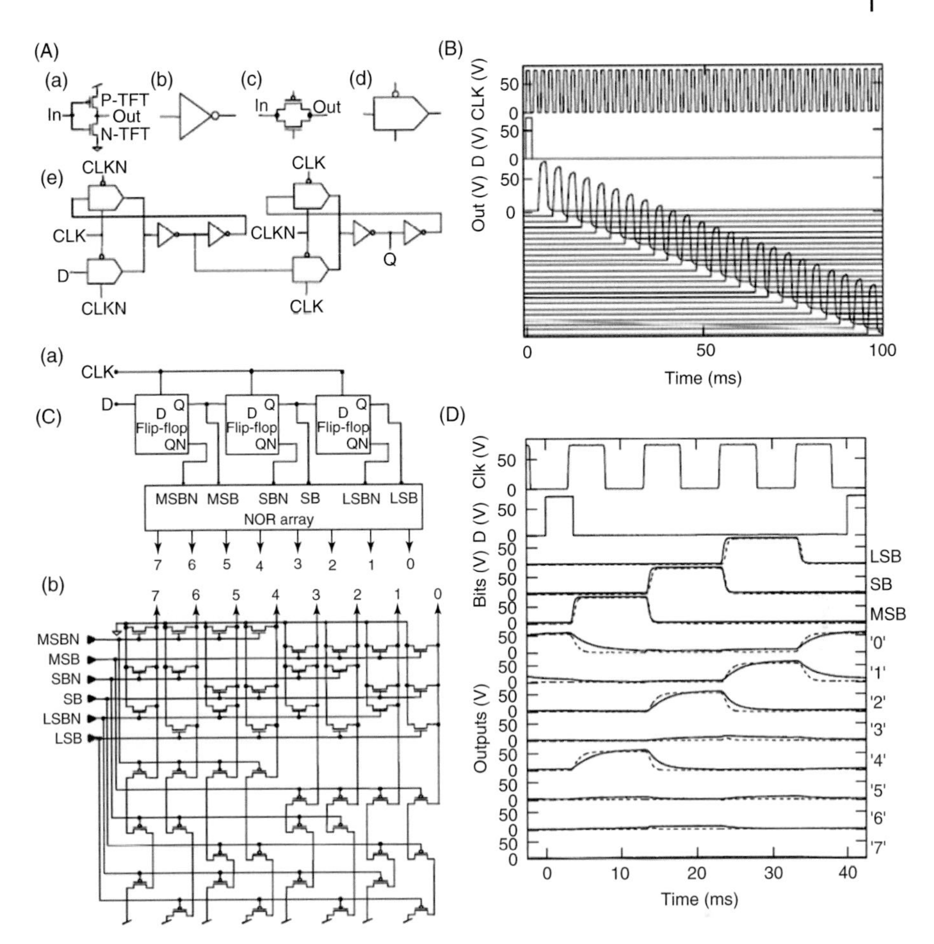

Figure 5.9 (A) Representation of circuits and their constituents. (a) A complementary inverter; (b) symbol of an inverter; (c) layout of a complementary transmission gate; (d) symbol of a transmission gate; (e) schematic of a D flip-flop showing the data input (D) and the connections of the clock (CLK) and its complement (CLKN). Also shown is the output (Q). (B) Characteristics of the 48-stage shift register. The clock (500 Hz) and data are shown along with the output voltages of the 24 output buffers as a function of time. The 24 outputs have been vertically offset for clarity. (C) Row decoder design. (a) Schematic of a three-bit row decoder with eight outputs (0 ± 7). Each square is a D flip-flop (see Figure 5.9A(e)). The bits are designated as the most significant bit (MSB), second bit (SB), and least significant bit (LSB). Complements are labeled with an N. (b) Details of the NOR array showing the manner in which individual transistors are connected. (D) Characteristics of the three-bit decoder. In descending order, the traces are as follows: clock, data, and outputs of the three D flip-flops (labeled LSB, SB, MSB on the right-hand vertical axis). The eight remaining traces, labeled 0 ± 7, show the output voltages of the eight outputs: at any time, one output at most is high. Also shown (dashed lines) are the simulated responses of the flip-flops and the outputs of the decoder. The good agreement between simulation and experiment may be noted. The simulations were performed with a set of tools developed for organic/polymer transistor circuit design. (Reprinted with permission from Ref. [5a].)

The static current drawn in p-channel TFT registers depends strongly on the nature of the load (whether enhancement or depletion), the on–off current ratio, and the ratio of the dimensions of the transistors that constitute the basic gates (NOR, NAND, and inverter). The static current drawn by complementary registers can be further reduced by reducing the off-current of the transistors. The schematic of a three-bit row decoder is shown in Figure 5.9C. This decoder is designed so that a single serial input activates one of eight outputs. It consists of three D flip-flops and a NOR array (Figure 5.9C). The output of the three flip-flops and their complements drive the NOR array, which is configured so that for any given set of inputs only one output is "high." The characteristics of the decoder are shown in Figure 5.9D. The circuit used for the NOR array is significant for another reason: four transistors are serially connected between supply and ground. Such "four-deep" connections are often employed in silicon arithmetic and logical unit (ALU) circuits. The fact that such a configuration has been shown to work for organic semiconductors suggests that complex logic gates such as those used in ALUs and simple microprocessors can be constructed out of OTFTs.

The first report on a low-voltage CMOS inverter based on $F_{16}CuPc$ (n-channel) and pentacene (p-channel) and an ultrathin polymeric dielectric (called crosslinked polymer blends, CPBs) was by our group in 2005 [43]. Inverter response is clearly observed for switching between logic "1" (2 V) and logic "0" (0 V), with the small hysteresis reflecting the transistor threshold voltage stability. The voltage gain dV_{OUT}/dV_{IN} 3.5 (>1) implies that these devices could be used to switch subsequent stages in more complex logic circuits. These inverters can be switched at frequencies up to 100 Hz, with an ~1.5 ms fall time and an ~2.3 ms rise time. Klauk *et al.* have pioneered organic complementary circuits operating at very low power using a self-assembled monolayer gate dielectric and pentacene and $F_{16}CuPc$ (Figure 5.10) [44]. The monolayer dielectric was deposited on patterned metal gates (Al) at room temperature and is optimized to provide a large gate capacitance and low gate leakage currents. By combining low-voltage p-channel and n-channel OTFTs in a complementary circuit design, the static currents are reduced to below 100 pA per logic gate. They fabricated complementary inverters, NAND gates, and ring oscillators that operate with supply voltages between 1.5 and 3 V and have a static power consumption of less than 1 nW per logic gate. The authors remarked that these organic circuits are thus well suited for battery-powered systems such as radiofrequency identification tags, portable display devices, and large-surface sensors.

More recently, Someya, Klauk, and coworkers demonstrated the feasibility of employing inkjet technology with subfemtoliter droplet volume and single-micrometer resolution for electronic device applications [45]. They manufactured p-channel and n-channel organic top-contact TFTs with S–D contacts ($L = 1\ \mu m$) prepared by subfemtoliter inkjet printing of Ag nanoparticles deposited directly on the surface of the organic semiconductor layers, without the need for any photolithographic prepatterning or any surface pretreatment (Figure 5.11). Also, in this case, the gate dielectric is a self-assembled monolayer, allowing transistors and circuits on rigid and flexible substrates to operate with very low voltages. In contrast to

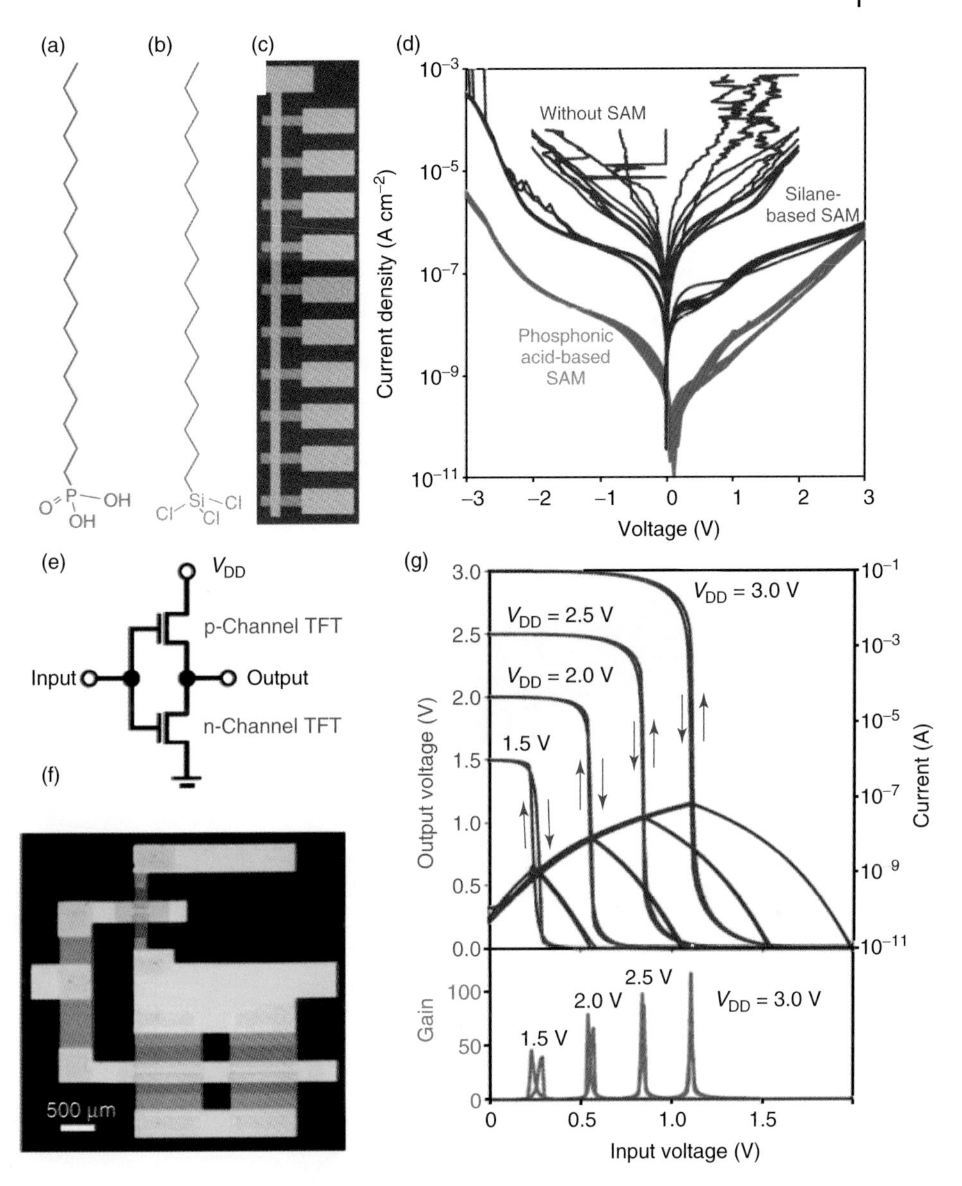

Figure 5.10 Self-assembled monolayer dielectrics on metal gates. (a) Chemical structure of n-octadecylphosphonic acid. (b) Chemical structure of n-octadecyltrichlorosilane. (c) Photograph of the leakage current test structure with shadow-mask-patterned aluminum bottom and gold top electrodes. (d) Leakage current density as a function of applied voltage. Each measurement was repeated on 10 junctions to evaluate the uniformity. (e) Complementary inverter with SAM gate dielectric. (f) Photograph of the inverter. (g) Output voltage, current, and small-signal gain as a function of input voltage for supply voltages between 1.5 and 3.0 V. The inverter shows rail-to-rail output switching, a maximum static current of 100 pA, and a small-signal gain as large as 100. (Reprinted with permission from Ref. [44].)

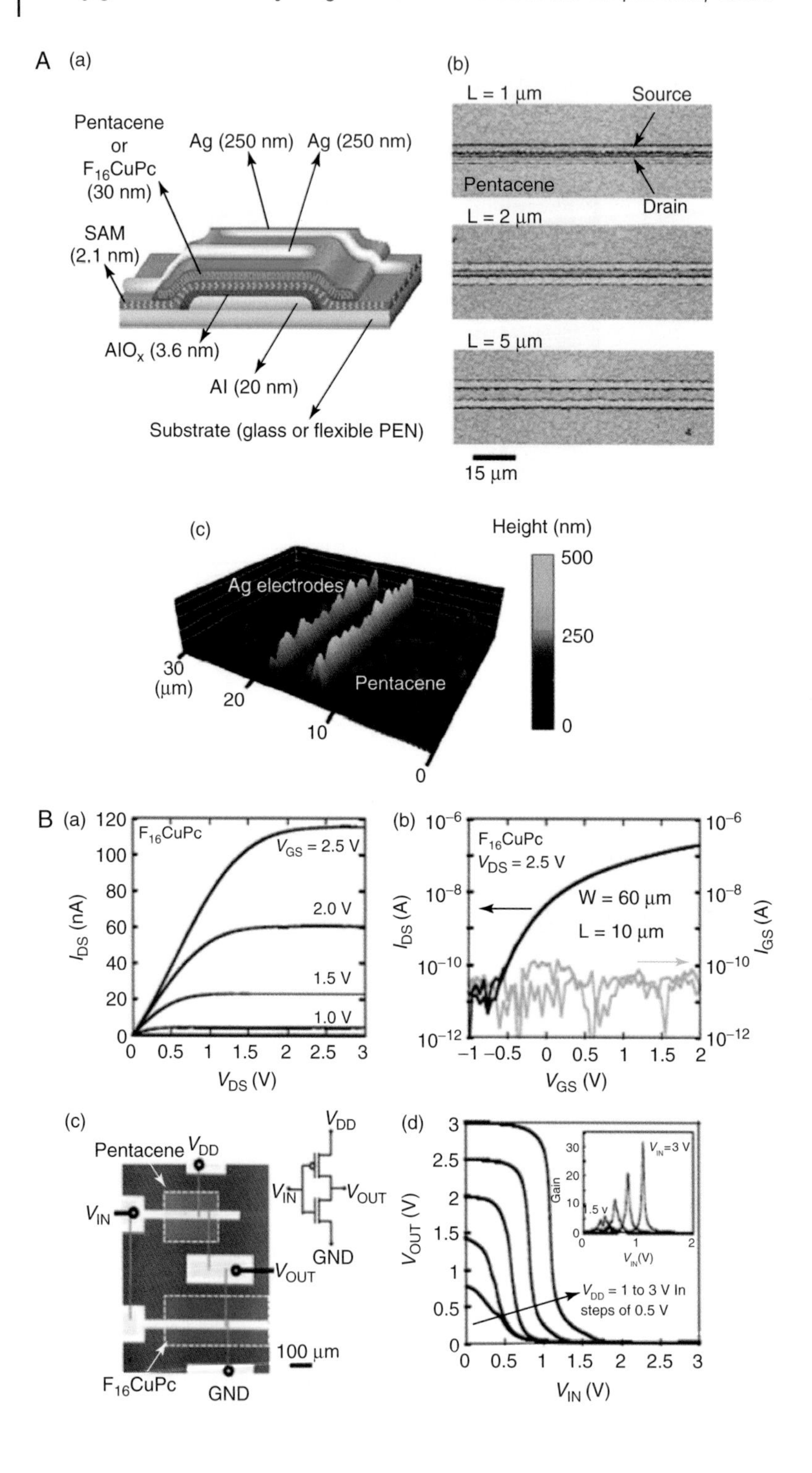
A (a)
Pentacene or $F_{16}CuPc$ (30 nm)
Ag (250 nm)
Ag (250 nm)
SAM (2.1 nm)
AlO_x (3.6 nm)
Al (20 nm)
Substrate (glass or flexible PEN)
(b)
L = 1 μm
Source
Pentacene
Drain
L = 2 μm
L = 5 μm
15 μm
(c)
Height (nm)
500
250
0
Ag electrodes
Pentacene
30 (μm)
20
10
0
B (a)
$F_{16}CuPc$
V_{GS} = 2.5 V
2.0 V
1.5 V
1.0 V
I_{DS} (nA)
V_{DS} (V)
(b)
$F_{16}CuPc$
V_{DS} = 2.5 V
W = 60 μm
L = 10 μm
I_{DS} (A)
I_{GS} (A)
V_{GS} (V)
(c)
Pentacene
V_{DD}
V_{IN}
V_{OUT}
GND
$F_{16}CuPc$
100 μm
(d)
V_{OUT} (V)
V_{IN} (V)
V_{DD} = 1 to 3 V In steps of 0.5 V
Gain
V_{IN}=3 V
1.5 V

bottom-contact TFTs (in which the contacts are defined before the deposition of the organic semiconductor and hence the channel length can be very small by the use of photolithography or electron-beam lithography), top-contact TFTs benefit from significantly lower contact resistance but require a contact patterning technique that does not harm the organic semiconductor. Because the amount of organic solvent dispensed during subfemtoliter inkjet printing is extremely small and the Ag nanoparticle calcination temperature after subfemtoliter inkjet printing is low (130 °C), the morphology of the organic semiconductors is not damaged. As a result, TFTs with short channel lengths exhibit small contact resistance (5 kΩ cm) and small parasitic capacitance. Using p-channel pentacene TFTs and n-channel $F_{16}CuPc$ TFTs with patterned Al gates, SAM-based gate dielectric, and ink-jet-printed S–D contacts, the authors fabricated organic complementary inverters. A photograph and the electrical transfer characteristics of such an inverter are shown in Figure 5.11B(c,d). The pentacene TFT has a channel length of 50 μm, the $F_{16}CuPc$ TFTs has a channel length of 5 μm, and both TFTs have a channel width of 60 μm. The difference in channel length is necessary to achieve similar drain currents for both TFTs despite the significant difference in carrier mobility (0.1 $cm^2\ V^{-1}\ s^{-1}$ for the pentacene TFT, 0.02 $cm^2\ V^{-1}\ s^{-1}$ for the $F_{16}CuPc$ TFT). The inverter operates with supply voltages between 1.5 and 3 V and with a small signal gain 10.

5.4.1.2 **Thiophene Derivatives**

Figure 5.12 shows the chemical structures of several n-channel oligothiophenes. The first n-channel oligothiophene-based transistor was discovered at Northwestern University in 2000. In a series of papers, our group has described the synthesis, comparative physicochemical properties, and solid-state structures of several oligothiophenes substituted with perfluorohexyl chains [46]. These series include the n-channel α,ω-diperfluorohexyl-nTs (DFH-nTs) and β,β'-diperfluorohexyl-nTs (isoDFH-nTs), which were compared to the corresponding p-channel hexyl-substituted and unsubstituted oligothiophenes (αnTs,

Figure 5.11 (A) Structure, micrograph, and AFM image of organic transistors. (a) Schematic cross-section of the organic thin-film transistors with patterned Al gates, ultrathin gate dielectric, vacuum-deposited organic semiconductor, and subfemtoliter ink-jet-printed Ag nanoparticle source/drain contacts. (b) Optical microscopic images of pentacene TFTs with channel lengths of 1, 2, and 5 μm after calcination (the linewidth of the ink-jet-printed contact lines is 5 μm). (c) AFM image of a pentacene TFT with a channel length of 5 μm and a contact linewidth of 2 μm. (B) Electrical characteristics of n-type F16CuPc TFTs and organic CMOS inverter. Output (a) and transfer (b) characteristics of an n-channel F16CuPc TFT with a channel length of 10 μm and a channel width of 60 μm. The measurements were carried out in air. Optical microscopic image (c) and transfer characteristics (d) of an organic complementary inverter. The pentacene TFT has a channel length of 50 μm, the F16CuPc TFT has a channel length of 5 μm, and both TFTs have a channel width of 60 μm. The inverter operates with supply voltages between 1.5 and 3 V and with a small-signal gain $>$ 10. (Reprinted with permission from Ref. [45].)

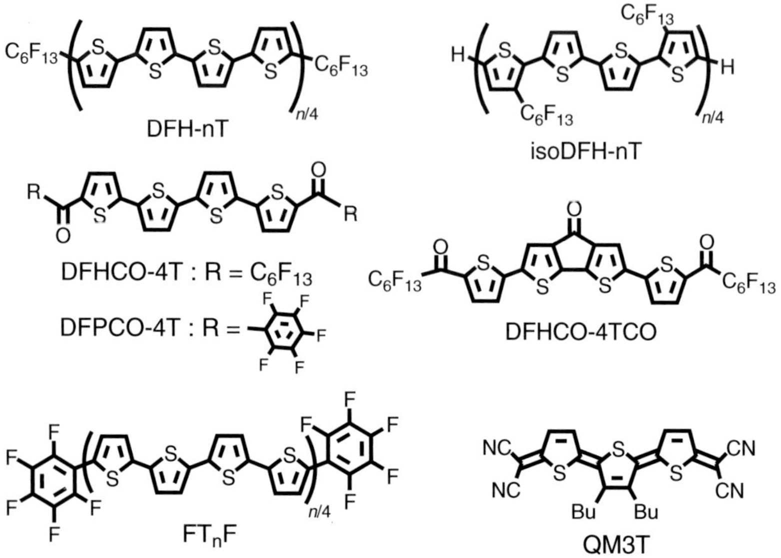

Figure 5.12 Chemical structures of various n-channel oligothiophenes.

n = 2–6). The crystal structures of key fluorocarbon-substituted oligomers (Figure 5.13) were also analyzed and evidenced close $\pi-\pi$ intermolecular interactions between the aromatic cores, whereas the fluorocarbon chains segregate into lamellar structures. X-ray structural analysis was performed for DFH-3T ($n = 3$) and DFH-4T ($n = 5$), which exhibit an all-*anti*, fully planar geometry, with dihedral angles between the mean plane of the rings $<2^\circ$. This value compares well to that reported for α4T but is much smaller than the 6–9° reported for α3T. The perfluorohexyl substituents exhibit a zigzag helical conformation characteristic of fluorocarbon chains [47] and are positioned at $\sim140^\circ$ with respect to the oligothiophene backbone axes in both cases. The molecular packing of DFH-4T shares the familiar herringbone (HB) motif with an angle of 50° between mean planes of adjacent molecules. Typical HB angles for oligothiophene αnTs ($n = 4-6, 8$) range between 55 and 70°. The minimum interplanar distances between neighboring molecules in DFH-4T is 3.52 Å, comparable to 3.5–3.9 Å in analogous oligothiophenes. The crystal structures and packing characteristics of β,β'-disubstituted systems isoDFH-5T are considerably different from those of the end-capped compounds. The effect of $\alpha,\omega \rightarrow \beta,\beta'$ regiochemical substitution on the π-core structure is dramatic with a large torsional angle (up to 64°) forced between adjacent thiophene rings compared to the DFH-nTs series. To a lesser degree, this has been observed as well for some β-alkylsubstituted oligothiophenes [48] and is due to steric repulsion between the fluorohexyl chain α-CF_2 group and either the sulfur atom or the C_β-H moiety on the adjacent thiophene ring. Finally, it was observed that, regardless of the differences in molecular structure and packing characteristics, all fluorocarbon-substituted oligothiophenes exhibit similar intermolecular separations between the oligothiophene core and fluorocarbon regions.

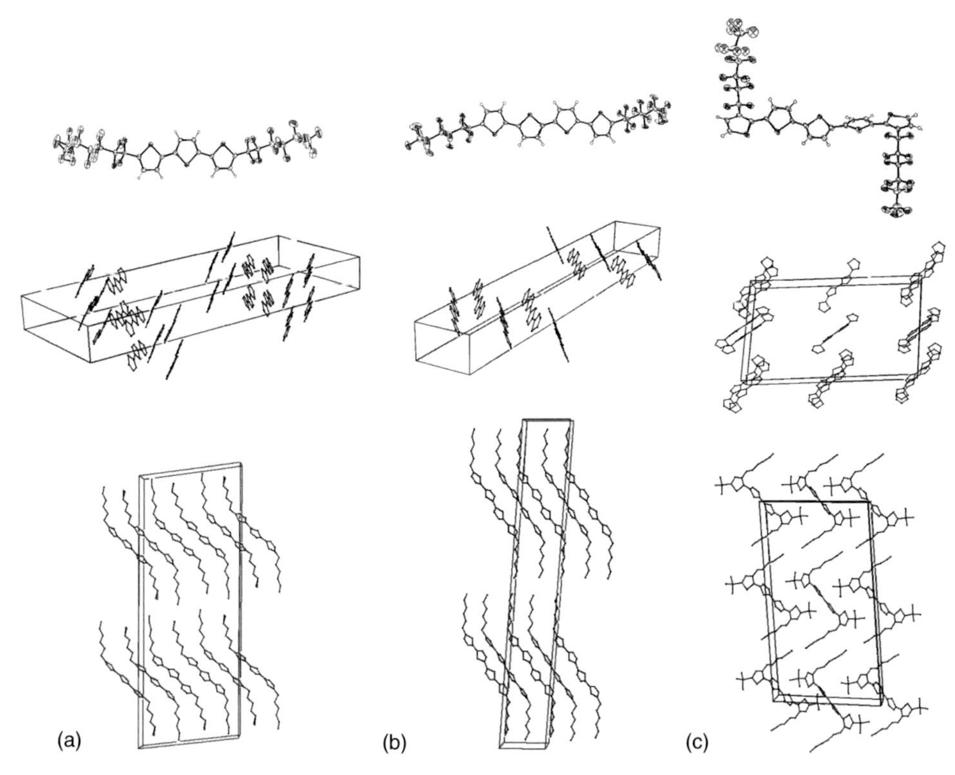

Figure 5.13 Crystal structures of (a) DFH-3T, (b) DFH-4T, and (c) isoDFH-5T. (Reprinted with permission from Ref. [46b].)

From the electrochemical and optical data, the energy positions of the HOMOs and LUMOs of these oligothiophenes were estimated and, as expected, fluorocarbon functionalization of the oligothiophene core reduces both HOMO and LUMO energies ($\sim$0.1–0.5 eV depending on the core length). The shift of HOMO and LUMO energies of DFH/isoDFH-nTs and the corresponding alkyl-substituted oligothiophenes with respect to αnTs is in quantitative agreement with the expected effects of fluorocarbon and hydrocarbon substituent-derived Hammett σ parameters. Overall, the molecular orbital energy trends within each series are determined by a balance between the substituent nature and density versus the extension of the aromatic core. The net results from this interplay are FMOs in which LUMO and HOMO energies are practically independent of conjugation length for fluorocarbon- and alkyl-substituted nTs, respectively. The experimental results were confirmed by density-functional computational modeling, the results of which closely parallel experimental trends. OFET devices were fabricated and characterized on Si–SiO_2 substrates treated with HMSO. The results clearly demonstrate the crucial importance of oligothiophene core substitution in determining the p- or n-channel activity of the semiconductor. Fluorocarbon-substituted systems are the majority carrier electron-transporting materials, whereas oligothiophenes (with or without $-C_6H_{13}$ chains) are majority carrier hole-transporting semiconductors. Trends in field-effect mobilities and I_{on}:I_{off} ratios between and within these series were explained by the interplay of molecular structure and film microstructure and morphology. The films that exhibited the greatest carrier mobilities possessed the appropriate combination of large grain size and smooth, interconnected morphology, exhibited molecular orientation directed along the substrate normal, and had sufficient core conjugation/length. The second distinctive material is DFH-4T, which was shown to exhibit one of the highest n-type carrier mobility ($0.22\ cm^2\ V^{-1}\ s^{-1}$) and I_{on}:I_{off} ratio reported for an n-type transporting material. More recently, an extensive IR/electrochemical characterization of these fluorocarbon-substituted oligothiophenes has been performed by Casado and Navarrete [46e].

The NU group also reported the comparative properties of arene–thiophene and fluoroarene–thiophene (FT_nF, $n = 2–4$) semiconductors [49] with respect to regiochemical modifications of the core and oligothiophene core shortening. These extensive studies included thermal analysis, optical spectroscopy, cyclic voltammetry, single-crystal X-ray diffraction (XRD) structural data for the majority of these compounds, film microstructure and morphology, and OTFT fabrication and characterization. For the fluoroarene–thiophene series, the majority charge transport type and mobilities ($0.00001–0.5\ cm^2\ V^{-1}\ s^{-1}$) and the I_{on}:I_{off} current ratios ($10^1–10^8$) change dramatically with molecular regiochemistry as well as with the substrate deposition temperature. The large electron mobility of FTTTTF was attributed to the favorable interaction between the electron-rich and electron-deficient moieties consenting to achieve optimal molecular $\pi-\pi$ overlap and excellent film texture. The shorter fluoroarene–thiophene oligomers ($n = 2, 3$) are also n-type semiconductors, demonstrating that fluoroarene end-substitution promotes majority charge flip independently of the thiophene core extension. DFT band structure calculations are shading light at the base of the majority charge carrier flip in this family [50].

An effective strategy to tune oligothiophene core energy levels, optical characteristics, and solid-state packing is to functionalize the oligothiophene core with strong inductive/mesomeric carbonyl (C=O) groups. In the first systems, the investigated carbonyl chemical functionalities included perfluorohexylcarbonyl groups (e.g., DFHCO-4T) installed at the oligothiophene α,ω positions as well as bridged carbonyl groups positioned in the center of the molecular core (e.g., DFHCO-4TCO) [51]. OFETs were fabricated on HMDS-treated $Si–SiO_2$ substrates, and DFHCO-4T shows monopolar n-type activity (in vacuum) with an exceptionally high mobility of $0.32\ cm^2\ V^{-1}\ s^{-1}$ for semiconducting films deposited at a substrate temperature of 25 °C. Note that similar carbonyl group effect on n-type (electron) transport was previously demonstrated in electroactive aromatic polyketones and polyesters [52]. On additional carbonyl group introduction into the quaterthiophene core, DFHCO-4TCO exhibited stable n-type activity even in air, although the ambient electron mobility ($0.01\ cm^2\ V^{-1}\ s^{-1}$) is somewhat lower than that recorded under vacuum ($0.08\ cm^2\ V^{-1}\ s^{-1}$). After proper dielectric surface modification, n-type mobilities were substantially improved to $1.7\ cm^2\ V^{-1}\ s^{-1}$ for DFHCO-4T.

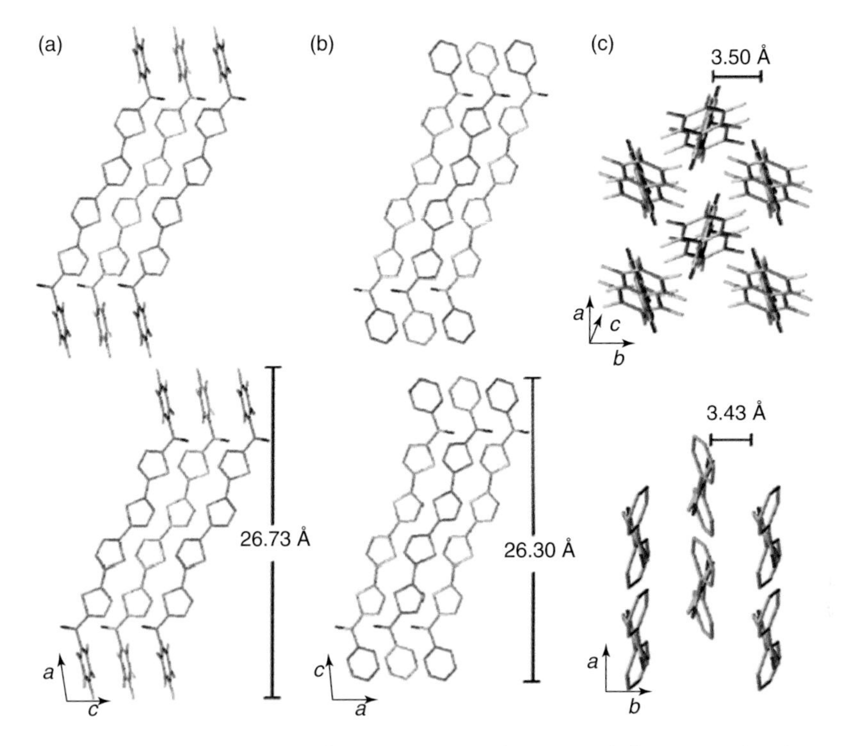

Figure 5.14 Crystal structures of DFCO-4T (a,c) and DPCO-4T (b,d). (Reprinted with permission from Ref. [49a].)

More recently, Schols *et al.* have improved electron mobilities of DFHCO-4T Au top-contact transistors to 4.6 $cm^2\ V^{-1}\ s^{-1}$, and the drastically reduced performance using Al/LiF as top contact is also explained as a consequence of an electron transfer reaction occurring at the metal/ DFHCO-4T interface [53]. Another promising carbonyl-containing oligothiophene is DFCO-4T (Figure 5.14) [54]. DFCO-4T crystallizes in a herringbone motif, with the shortest intercore distance being 3.50 Å and the average dihedral angle between the phenyl substitutent and the adjacent thiophene subunit at ~53°. Field-effect transistors of DFCO-4T were fabricated with Au top-contact electrodes. Semiconductor films (50 nm) were deposited onto temperature-controlled HMDS-treated SiO_2/p^+-Si substrates by vapor deposition and drop casting. A 50 nm layer of Au was then deposited through a shadow mask to define the source and drain electrodes. OFET characterization was performed under Argon. High electron mobilities of ~0.5 $cm^2\ V^{-1}\ s^{-1}$ were observed for vapor-deposited DFCO-4T films ($T_D = 80°C$) with a threshold voltage of ~30 V (I_{on}:$I_{off} > 10^8$). In solution-cast devices, electron mobilities were exceptionally high, with a maximum of ~0.25 $cm^2\ V^{-1}\ s^{-1}$ ($I_{on} : I_{off} = 10^5$; $V_T = 50–70$ V).

Recently, the NU group has analyzed the temperature dependence of FET mobility for a series of n-channel oligothiophenes along with those of other p-channel and ambipolar organic semiconductors [55]. The materials

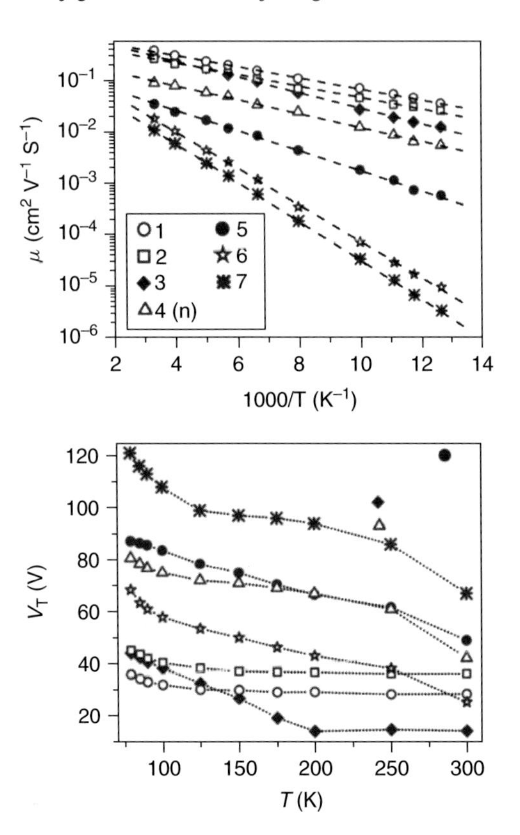

Figure 5.15 Plots of FET μ_{eff} versus inverse temperature with E_A for each of the indicated organic semiconductors (a) and V_T as a function of temperature (b, V_T is plotted for p-channel materials). The dashed lines in (a) are least-squares fits to the Arrhenius relationship, $\mu_{eff} = \mu_0 \exp\left(\frac{-E_A}{kT}\right)$, while the dotted lines in (b) are drawn as a guide to the eye. (Reprinted with permission from Ref. [55].)

(and dominant carrier type, Figure 5.15) studied are: 5,5‴-bis(perfluorophenacyl)-2,2′:5′, 2″:5″, 2‴-quaterthiophene (DFCO-4T, compound **1**); 5,5‴-bis(perfluorohexyl carbonyl)-2,2′:5′, 2″:5″, 2‴-quaterthiophene (DFHCO-4T, compound **2**); pentacene (compound **3**, p-channel); 5,5‴-bis(hexylcarbonyl)-2,2′:5′, 2″:5″, 2‴-quaterthiophene (compound **4**, ambipolar); 5,5‴-bis-(phenacyl)-2,2′:5′, 2″:5″, 2‴-quaterthiophene (compound **5**, p-channel); 2,7-bis((5-perfluorophenacyl)thiophen-2-yl)-9, 10-phenanthrenequinone (compound **6**, n-channel); and poly(*N*-(2-octyldodecyl)-2,2′-bithiophene-3,3′-dicarboximide) (compound **7**, n-channel). Fits of the effective field-effect mobility (μ_{eff}) data assuming a discrete trap energy within a multiple trapping and release (MTR) model reveal low activation energies (E_As) for high-mobility semiconductors **1**–**3** of 21, 22, and 30 meV, respectively. Higher E_A values of 40–70 meV are exhibited by **4**–**7**-derived FETs having lower mobilities (μ_{eff}). Analysis of this data reveals little correlation between the conduction state energy level and E_A, while there is an inverse relationship between E_A and μ_{eff}. The first variable-temperature study of an ambipolar OFET reveals that although n-channel behavior exhibits $E_A = 27$ meV, the p-channel regime exhibits significantly more trapping with $E_A = 250$ meV. Interestingly, calculated free carrier mobilities (μ_0) are in the range ~0.2–0.8 $cm^2\ V^{-1}\ s^{-1}$ in this materials set, largely independent of μ_{eff}. This indicates that in the absence of charge traps, the inherent magnitude of carrier mobility is comparable for each of these materials. Finally, the effect of temperature on threshold voltage (V_T) reveals two distinct trapping regimes, with the change in trapped charge exhibiting a striking correlation with room temperature μ_{eff}. The observation that E_A is independent of conduction state energy and that changes in trapped charge with temperature correlate with room temperature μ_{eff} support the applicability of trap-limited mobility models such as an MTR mechanism to this materials set.

Frisbie *et al.* reported that quinomethane terthiophene (QM3T) exhibits $\mu \cdot s \sim$ 0.002–0.5 $cm^2\ V^{-1}\ s^{-1}$ [56]. An extended series showing even greater performance and ambipolar transport has been recently reported [57, 58]. Handa *et al.* have shown a new hybrid-type dicyanomethylene-substituted terthienoquinoid compound for use as a solution-processable n-channel semiconductor; electron mobilities up to 0.16 $cm^2\ V^{-1}\ s^{-1}$ were obtained upon annealing the spin-coated films [59]. To my knowledge, CMOS devices based on n-channel oligothiophenes have never been fabricated.

5.4.1.3 **Fullerenes**

Figure 5.16 shows the chemical structures of several fullerene derivatives. The first report of C_{60}-based TFT evidenced low electron mobilities of ~10^{-4} $cm^2\ V^{-1}\ s^{-1}$ [60]. Following that report, C_{60} thin films were grown and studied in ultrahigh vacuum (UHV) (Figure 5.17) [61]. TFT were fabricated without air exposure and consisted of heavily n-type doped silicon wafers, which were oxidized to leave a 3000 Å thick layer of silicon dioxide dielectric and bottom-contact Cr/Au electrodes. The authors reported electron mobilities of 0.56 ± 0.2 $cm^2\ V^{-1}\ s^{-1}$, comparable to that obtained with time-of-flight measurements on C_{60} single crystals. n-channel field-effect transistors based on C_{70} were first reported a few years later [62] and exhibited field-effect mobilities up to 2×10^{-3} $cm^2\ V^{-1}\ s^{-1}$ in UHV and current on–off ratios as high as 10^5. Unfortunately, the performance of C_{60}- and C_{70}-based FETs degrades quickly on exposure to air. More recently, Itaka *et al.* have enhanced the crystallinity of vapor-deposited C_{60} films by coating the dielectric with a thin layer of pentacene [35]. These FETs exhibit electron mobilities of ~2.0–5.0 $cm^2\ V^{-1}\ s^{-1}$. Jang *et al.* have recently reported air-stable C_{60} thin-film transistors with electron mobilities of ~0.05 $cm^2\ V^{-1}\ s^{-1}$ by using a perfluoropolymer as the gate dielectric [63]. Very recently, Anthopoulos *et al.* have reported the greatest field-effect electron mobilities for C_{60} transistors, which approach μ~6 $cm^2\ V^{-1}\ s^{-1}$ [25]. He also demonstrated ring oscillators based on C_{60} films grown by hot wall epitaxy. Low-voltage C_{60} n-channel OTFTs with high electron mobilities of 2.3 $cm^2\ V^{-1}\ s^{-1}$ were fabricated by engineering the electrode/semiconductor and dielectric/semiconductor interfaces [64].

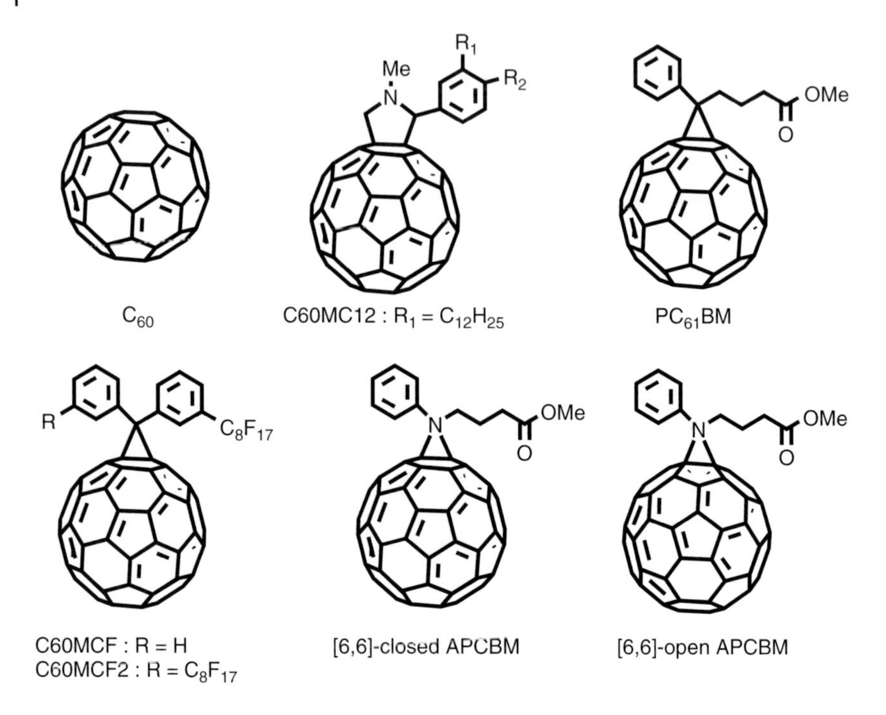

Figure 5.16 Chemical structures of several fullerene derivatives.

Functionalized fullerenes are promising candidates for solution-processed n-channel OTFTs. Chikamatsu *et al.* have synthesized soluble C_{60}-fused *N*-methylpyrrolidine-*meta*-C12 phenyl (C_{60}MC12) and fabricated FETs exhibiting high electron mobilities of $\sim$0.07 $cm^2\,V^{-1}\,s^{-1}$ [65]. More recently, Wöbkenberg *et al.* developed several fluorine-containing C_{60} derivatives. Solution-processed OTFTs based on these compounds exhibit electron mobilities up to 0.15 $cm^2\,V^{-1}\,s^{-1}$ [66]. These devices show enhanced stability in ambient conditions as compared to standard methanofullerene OTFTs. Far more ambient-stable C_{60}-based FETs were developed by Chikamatsu *et al.*, who synthesized new soluble perfluoroalkyl-substituted C_{60} derivatives (Figure 5.17) exhibiting electron mobilities as high as 0.25 $cm^2\,V^{-1}\,s^{-1}$ [67]. TFTs based on [6,6]-phenyl-C61-butyric acid ester ([60]PCBM) and [6,6]-phenyl-C71-butyric acid Me ester ([70]PCBM), materials, widely used for the fabrication of organic photovoltaic devices, have also been recently reported [68]. Despite the fact that both derivatives form glassy films when processed from solution, their electron mobilities are high at $\sim$0.21 and $\sim$0.1 $cm^2\,V^{-1}\,s^{-1}$, for [60]PCBM and [70]PCBM, respectively. Although the derived mobility of [60]PCBM is comparable to the best values reported in the literature, the electron mobility of [70]PCBM is the highest value reported to date for any C_{70}-based derivative. Finally, Wudl reported the one-pot preparation of two isomeric imino-PCBMs, that is, [5,6]-open azafulleroid

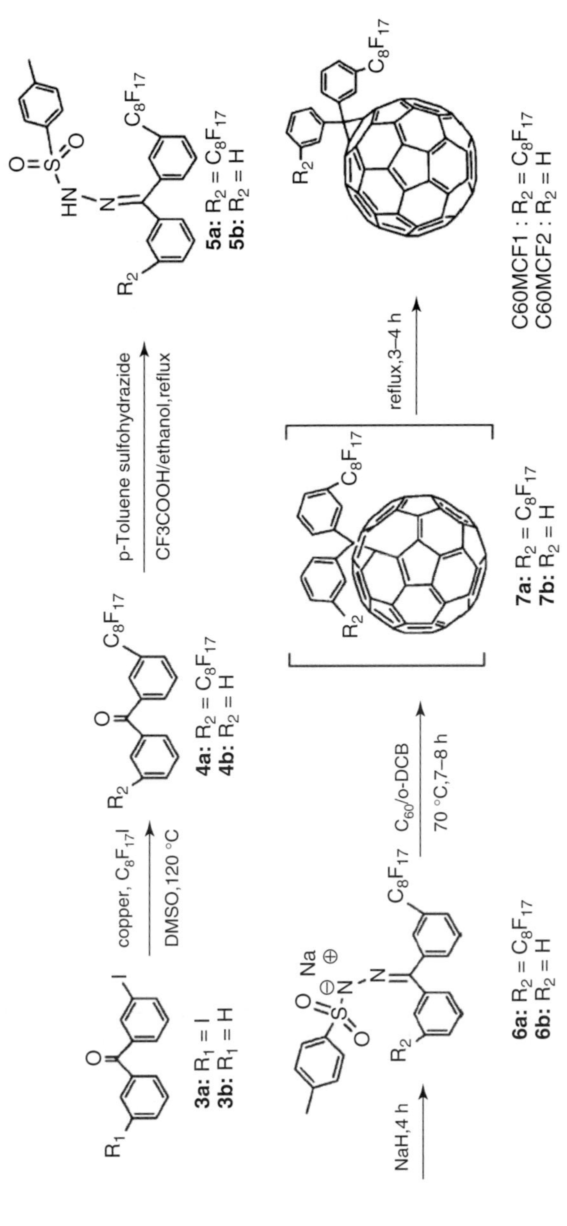

Figure 5.17 Synthesis of some fullerene derivatives. (Reprinted with permission from Ref. [67].)

([5,6]-open APCBM; PCBM = [6,6]-phenyl-C61-butyric acid methyl ester, A = aza) and [6,6]-closed aziridinofullerene ([6,6]-closed APCBM). OFETs were fabricated on heavily doped Si wafers with a 200 nm thick SiO_2 layer with top-contact geometry, yielding electron mobilities of ~0.04, ~0.02, and 0.03 $cm^2\ V^{-1}\ s^{-1}$ for [5,6]-open APCBM, [6,6]-closed APCBM, and PCBM, respectively. The higher electron mobility in the [5,6]-open APCBM OFET can be attributed to its 60 π-electron nature, which affords a stronger electron-acceptor strength than that of the [6,6] junctions [69].

CMOS NOT logic circuits composed of C_{60} and pentacene were first fabricated by Kanbara *et al.* [70]. In these devices, the source of the n-channel FET is grounded, while the source of the p-channel FET is connected to a power supply (V_{DD} = 70 V). The voltage gain demonstrated is ~4. Low-voltage CMOS inverters with C_{60} (and pentacene) were reported by Kitamura *et al.* on glass substrates [71]. The inverter operated at low voltages of 1–5 V, and the C_{60} and pentacene TFTs had high field-effect mobilities of 0.68 and 0.59 $cm^2\ V^{-1}\ s^{-1}$ and threshold voltages of 0.80 and −0.84 V, respectively. As expected from the threshold voltages of the individual transistors, the inverter operates at supply voltages of only $V_{DD} > 1.64$ V. The signal gain calculated is in the range of −50 to −150. More recently, Kippelan *et al.* demonstrated complementary inverters on polyethylene naphthalate (PEN) substrates [72]. These inverters were fabricated in a top-contact configuration, as shown in Figure 5.18. Source/drain electrodes of Au for p-channel and Ca for n-channel transistors were selected to minimize the contact resistance. A 50 nm thick layer of Au with a 10 nm thick layer of Ti was deposited using an e-beam deposition system and served as a common gate $G_{n,p}$ for both n- and p-channel OFETs. This common gate also serves as the input V_{IN} node of the inverter. A 200 nm thick Al_2O_3 gate insulator (Ci = 26.5 nF/cm^2), passivated with PS, was formed by atomic layer deposition at 100 °C. The C_{60} and pentacene layers were deposited at rates of 0.3 and 1 Å/s, respectively, followed by the deposition of the appropriate contacts. A photograph of the flexible inverters is shown in Figure 5.18c. As shown in the circuit diagram in Figure 5.18b, the drain terminals of the two OFETs were connected to form the output node, V_{OUT}, of the inverter. A supply voltage V_{DD} was applied to the source of p-channel OFETs Sp, while a low power supply V_{SS}= 0 V in this case was applied to the source of n-channel OFETs S_n. With low threshold voltages for both transistors, the inverters could be operated at a supply voltage as low as 3 V. The voltage transfer characteristics (VTCs) with a supply voltage of V_{DD}= 3, 4, and 5 V are shown in Figure 5.18d. The noise margin low and high are 2.0 and 2.3 V, respectively, for V_{DD} = 5, 1.7; 1.8 V for V_{DD} = 4 V; and 1.2 V for V_{DD} = 3 V. These results are believed to be among the highest noise margin values obtained in organic complementary inverters. The maximum dc voltage gain, defined as dV_{OUT}/dV_{IN}, is dependent on V_{DD}. A high dc gain of 180 was achieved with V_{DD} = 5 V. These inverters were subjected to bending experiments to test their stability under flexing, and it was observed that the VTC of the inverter shifted by a negligible value of 0.6% under the tensile strain and by a small fraction of 3% under the compressive strain (Figure 5.18e). Accordingly, the maximum dc gain increased by 11 and 20%, respectively.

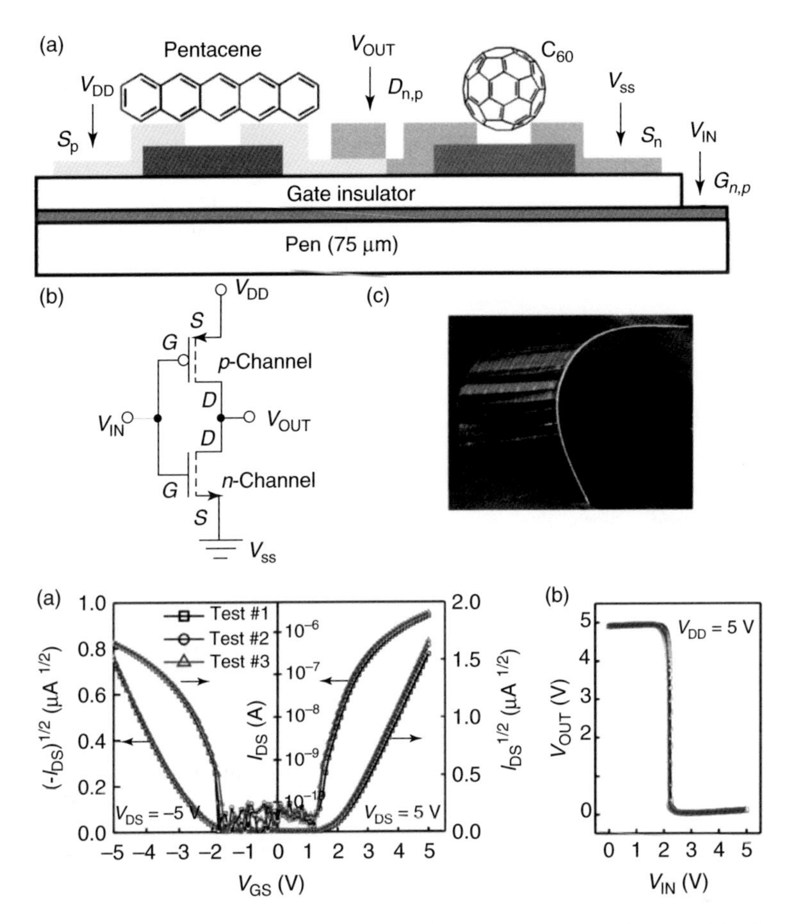

Figure 5.18 (A) (a) Schematics of the complementary inverter built on a PEN substrate with pentacene and C60 as active semiconductors. (b) Corresponding inverter circuit diagram. (c) Photograph of the inverters under tensile stress, with the transistors on the outer bent surface. (B) Superposition of (a) the transfer characteristics of p-channel and n-channel OFETs and of (b) the VTCs of the inverters measured before/after tensile and compressive stress cycles. (Reprinted with permission from Ref. [72].)

5.4.1.4 **Rylene and Other Diimide Derivatives**

Benzene, Naphthalene, and Anthracene Diimides Figure 5.19 shows the chemical structures of these derivatives. Katz *et al.* reported the only example of an OFET based on a pyromellitic diimide [73]. The synthesis of pyromellitic diimide derivatives involves one simple conventional reaction between amines and pyromellitic dianhydride in dimethylformamide (DMF) at 110 °C. Single crystals of PMDIF were obtained by slowly cooling hot, saturated DMF solutions. The unit cell of the single crystal is monoclinic with a = 10.24 Å, b = 11.53 Å, c = 9.28 Å, with the crystal packing exhibiting a close $\pi-\pi$ packing between the side chain benzene ring and

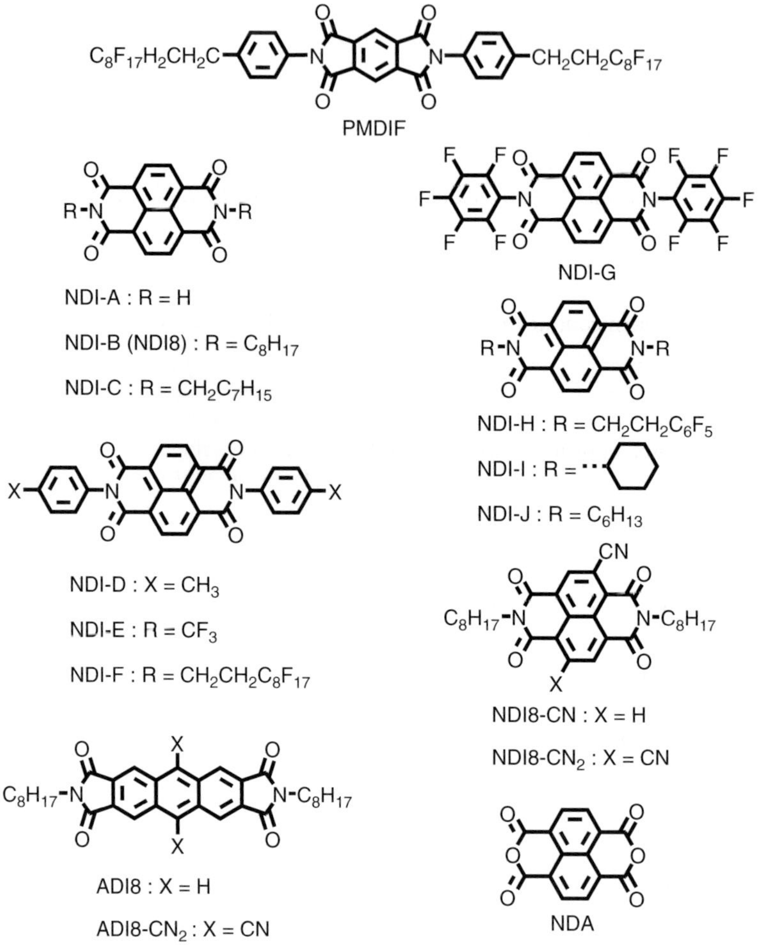

PMDIF

NDI-A : R = H

NDI-B (NDI8) : R = C_8H_{17}

NDI-C : R = $CH_2C_7H_{15}$

NDI-G

NDI-H : R = $CH_2CH_2C_6F_5$

NDI-I : R = cyclohexyl

NDI-J : R = C_6H_{13}

NDI-D : X = CH_3

NDI-E : R = CF_3

NDI-F : R = $CH_2CH_2C_8F_{17}$

NDI8-CN : X = H

NDI8-CN_2 : X = CN

ADI8 : X = H

ADI8-CN_2 : X = CN

NDA

Figure 5.19 Chemical structures of various rylene diimides.

the pyromellitic diimide core. PMDIF exhibited a mobility of 0.054 $cm^2\ V^{-1}\ s^{-1}$ in air and very high on–off ratios.

The first OFET based on a rylene diimide was fabricated using the unsubstituted naphthalenediimide, NDI-A, and yielded electron mobilities of $\sim 10^{-4}$ $cm^2\ V^{-1}\ s^{-1}$ [74]. Later, Katz *et al.* demonstrated that the OFET performance could be greatly improved by functionalizing the NDI core at the nitrogen positions with *n*-octyl groups. Compound NDI-B (or NDI-8) exhibits electron mobilities of $\sim$0.16 $cm^2\ V^{-1}\ s^{-1}$ in vacuum, although almost no FET activity was measurable in air [75]. It was found that N,N′ substitution with *n*-$CH_2C_7F_{15}$ groups in NDI-C significantly improve device air stability, with mobilities of 0.05–0.1 $cm^2\ V^{-1}\ s^{-1}$ in air [76]. Replacing the methyl substituents on the *N,N′* benzyl groups of NDI-D with CF_3 groups in NDI-E leads to $10^5\times$ enhancement of the mobility in air;

elongating CF_3 to n-$CH_2CH_2C_8F_{17}$, further enhancing the mobility from 0.12 (NDI-E, Figure 5.3) to 0.57 $cm^2\ V^{-1}\ s^{-1}$ (NDI-F) [27]. Strikingly, inserting an ethylene bridge between the nitrogen atoms and the perfluorophenyl substituents of NDI-G to give NDI-H leads to a crystalline rather than an amorphous material and increases the mobility from $<10^{-6}$ to 0.31 $cm^2\ V^{-1}\ s^{-1}$ [77]. Shukla *et al.* reported that, compared to linear *n*-hexyl chains, cyclohexyl substituents assist in directing intermolecular $\pi-\pi$ stacking, affording a dramatic increase in mobility from 0.70 (NDI-J) to 6.2 $cm^2\ V^{-1}\ s^{-1}$ (NDI-I) in vacuum [78].

The studies described above show the great potential of core-unsubstituted naphthalene diimides. Wasielewski *et al.* reported two new core-cyanated naphthalene diimide semiconductors, NDI8-CN and NDI8-CN_2, which are air-stable, high-mobility, and transparent organic n-type semiconductors [79]. The syntheses of NDI8-CN and NDI8-CN_2 were achieved via a new NDI core bromination, cyanation sequence. NDI8-CN and NDI8-CN_2 electronic structures were examined by cyclic voltammetry, optical spectroscopy, and photoluminescence. Electrochemical reduction potentials in dichloromethane versus SCE are −0.22 V for NDI8-CN and +0.08 V for NDI8-CN_2, consistent with systematic LUMO energy depression with increasing cyanation. Importantly, NDI8-CN_2 has a reduction potential similar to that of N-N-dialkylsubstituted core-cyanated perylenes (perylene diimide PDI, vide infra) (−0.07 V vs SCE); therefore, the LUMO/charge carrier energies in the NDI and PDI materials should be similar. Optical and photoluminescence spectroscopy of these NDI derivatives reveals a band gap of >3 eV, reflecting the smaller conjugated core dimensions. Thus, thin films of these NDIs are transparent in the visible region (Figure 5.20). OFET measurements performed in vacuum ($<10^{-6}$ Torr) reveal optimal average electron mobilities for NDI8-CN and NDI8-CN_2 films of $4.7 \times 10^{-3}\ cm^2\ V^{-1}\ s^{-1}$ and 0.15 $cm^2\ V^{-1}\ s^{-1}$, for film vapor deposited at 130 and 110 °C, respectively. Interestingly, OFET operation in ambient atmosphere reveals that the NDI8-CN devices undergo severe I–V curve degradation, whereas the NDI-8CN2 devices exhibit stable operation with only a slightly lower maximum average mobility of 0.11 $cm^2\ V^{-1}\ s^{-1}$. The current on–off ratios (I_{on}/I_{off}) can be as high as $\sim10^5$ for NDI-8CN and $\sim10^3$ for NDI8-CN_2 thin films. The lower I_{on}/I_{off} ratio of NDI8-CN_2 was ascribed to high I_{off}, which is likely due to dopants in the NDI8-CN_2 thin films from contacts or donor sites in the dielectric. Top-contact bottom-gate transparent channel flexible n-type OFETs were fabricated with NDI8-CN_2 to demonstrate the unique material's properties. Thin NDI8-CN_2 films (50 nm) were vapor deposited onto overhead transparency film coated with a spin-cast PEDOT:PSS polymeric gate, a P-UV-013 (Polyera Corporation) polymer dielectric, and Au contacts. This air-stable, flexible, transparent OFET exhibits a mobility of 0.03 $cm^2\ V^{-1}\ s^{-1}$ in ambient atmosphere (Figure 5.20). An analogous rigid device fabricated on an ultrasmooth ITO/glass substrate as a gate gives mobility of ~0.08 $cm^2\ V^{-1}\ s^{-1}$ in ambient atmosphere.

Marks *et al.* reported the only examples of OFETs based on anthracene-2,3:6,7-tetracarboxylic diimides [80]. Devices based on ADI8 exhibit a mobility of 0.02 $cm^2\ V^{-1}\ s^{-1}$ in vacuum, 10× lower than its NDI counterpart. However, OFETs based on ADI8 do not operate in air because of its relatively low electron affinity. Also,

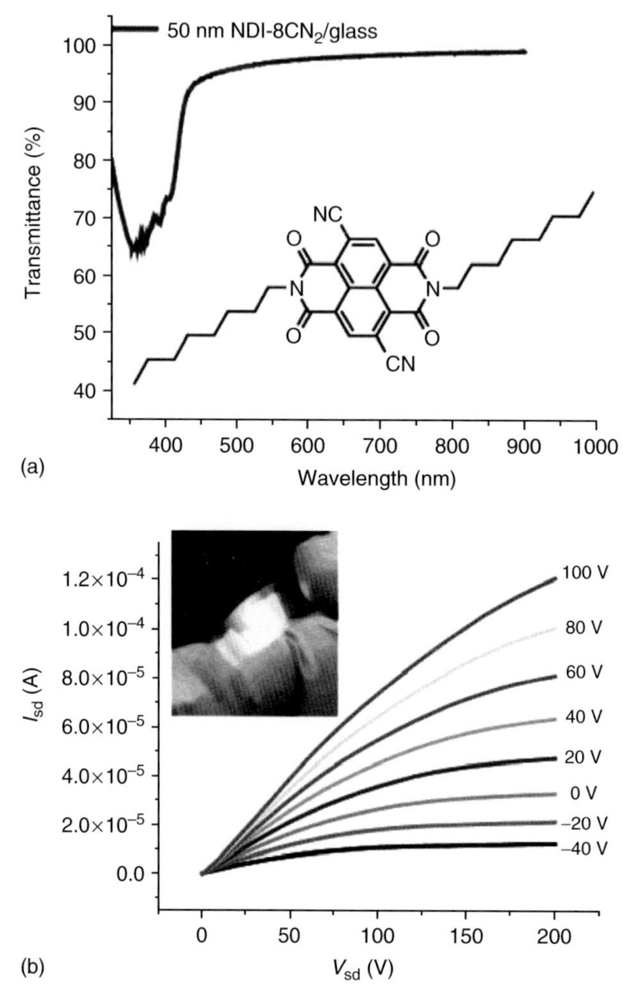

Figure 5.20 (a) Transmission optical spectrum of a 50 nm vapor-deposited thin film of NDI8-CN_2 on glass demonstrating the impressive transparency of this material between 400 and 800 nm. (b) Output plot for a transparent, flexible OFET having a PEDOT: PSS gate, polymer gate dielectric, NDI8-CN_2 semiconductor, and Au source and drain, and exhibiting an electron mobility of 0.03 $cm^2\ V^{-1}\ s^{-1}$ in air. Inset: photograph of an array of 100 devices fabricated on overhead transparency film demonstrating transparency and flexibility. (Reprinted with permission from Ref. [79].)

in this case, introduction of cyano groups in the 9,10 positions of the anthracene ring significantly increasess the electron affinity and, therefore, yields improved air stability; devices based on ADI8-CN_2 exhibit a mobility of 0.02 $cm^2\ V^{-1}\ s^{-1}$ but very high current on–off ratios ($>10^7$).

The first demonstration of a CMOS inverter with a rylene derivative used napthalenetetracarboxylicdianhydride (NDA) as the n-channel semiconductors and pentacene or copperphthalocyanine as the p-channel materials [23]. These inverters

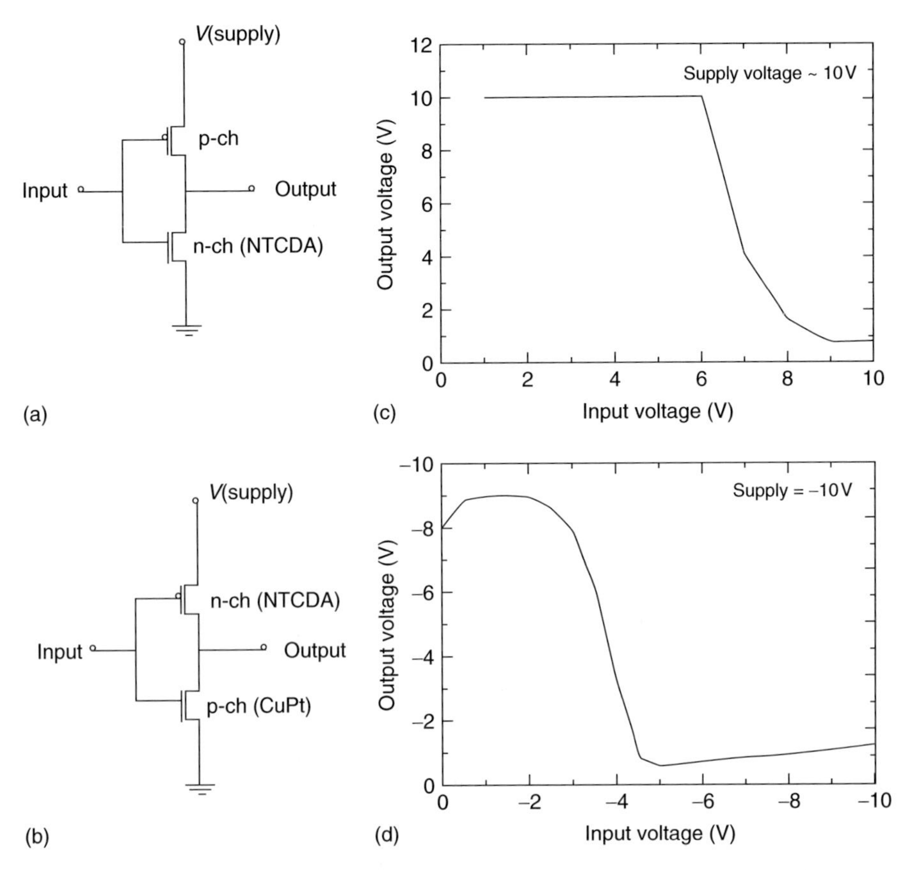

Figure 5.21 Inverter circuit configurations. In (a), the n-channel TFT is the driver and the p-channel TFT is the load, and in (b), the transistor types are reversed. (c) Transfer characteristics of the former (c) and latter (d) inverter circuit configurations. (Reprinted with permission from Ref. [23].)

were fabricated using two different configurations, where the n-channel TFTs have been used as the load in one circuit configuration and as the driver in the second configuration. The two circuit configurations used are shown in Figure 5.21 along with the inverter characteristics. Although the results are poor for current standards, it was a very important discovery for the time. No other NDI-based CMOS devices have been reported recently.

Perylene and Higher Rylene Diimides The chemical structure of several perylenes and quaterylene derivatives is shown in Figure 5.22. Horowitz *et al.* first demonstrated electron mobilities of $\sim 10^{-5}$ cm^2 V^{-1} s^{-1} with an N,N′-diphenyl-substituted perylene, PDI-A [81]. In 2001, Malenfant *et al.* reported n-channel OFETs based on N,N'-dioctyl PDI-B (or PDI8) (Figure 5.23), with electron mobility of 0.6 cm^2 V^{-1} s^{-1} under nitrogen, but with a high threshold voltage of +75 V, which was attributed to a large trap density [82]. Later, Chesterfield *et al.* demonstrated

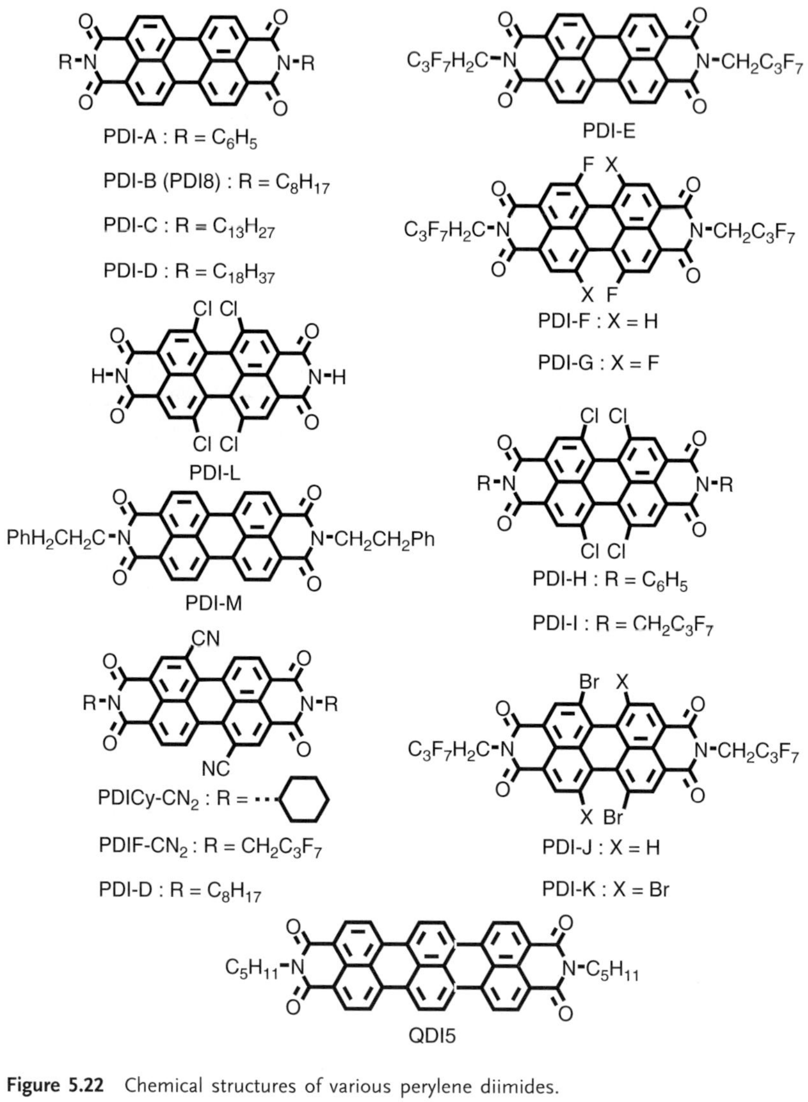

PDI-A : R = C_6H_5

PDI-B (PDI8) : R = C_8H_{17}

PDI-C : R = $C_{13}H_{27}$

PDI-D : R = $C_{18}H_{37}$

PDI-E

PDI-F : X = H

PDI-G : X = F

PDI-L

PDI-M

PDI-H : R = C_6H_5

PDI-I : R = $CH_2C_3F_7$

PDICy-CN_2 : R = cyclohexyl

PDIF-CN_2 : R = $CH_2C_3F_7$

PDI-D : R = C_8H_{17}

PDI-J : X = H

PDI-K : X = Br

QDI5

Figure 5.22 Chemical structures of various perylene diimides.

that devices with a maximum mobility of 1.7 $cm^2\ V^{-1}\ s^{-1}$, an on–off ratio of 10^7, and threshold voltages of 10–15 V can be obtained using PDI-B by coating the SiO_2 gate dielectric with poly(α-methylstyrene) [83]. Coating of the dielectric with polymers also considerably improves the air stability of device operation for PDI-B, presumably by passifying acidic silanol groups on the SiO_2 surface that can act as electron traps [84]. Ichikawa *et al.* demonstrated that mobility of OFETs based on *N, N'*-bis-tridecyl PDI-C can be increased $10^3\times$ to 2.1 $cm^2\ V^{-1}\ s^{-1}$ by thermal annealing [85]; the thermal treatments improve both the thin-film crystallinity and morphology. Bao, Würthner, and coworkers reported that OFETs based on

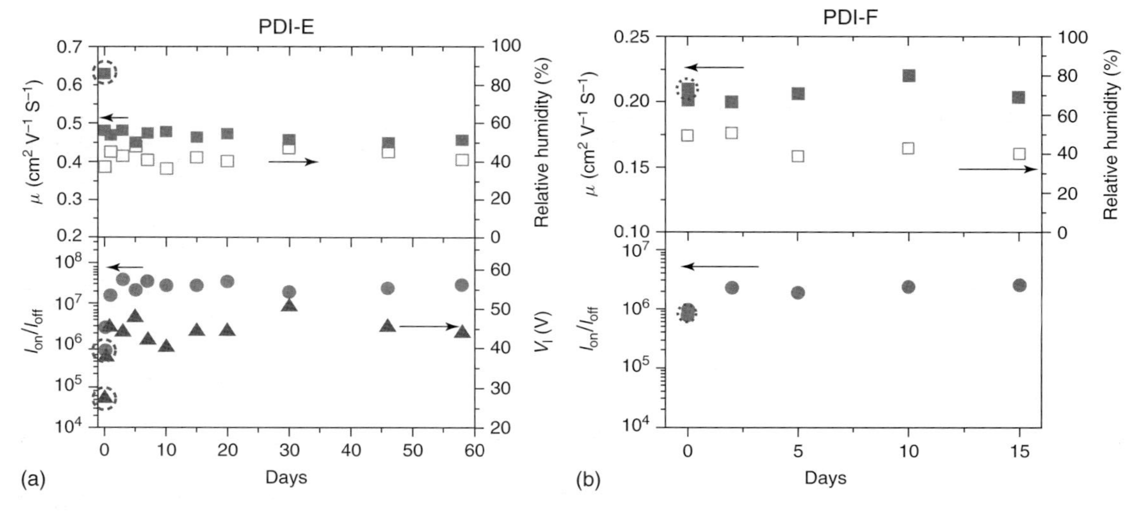

Figure 5.23 Air-stability measurements of PDI-based TFTs based on (a) PDI-E and (b) PDI-E. Top: μ (solid squares) and relative humidity (open squares). Bottom: I_{on}/I_{off} (solid circles) and Vt (solid triangles). The dotted circle values at initial time correspond to the electrical performance in a nitrogen-filled glovebox. (Reprinted with permission from Ref. [86].)

PDI-E exhibit mobilities as high as 0.72 cm^2 V^{-1} s^{-1}, which decrease only slightly after air exposure and remain stable for more than 50 days [86]. Since the partial fluorination has only a small effect on the redox potential (LUMO energy) relative to *N*, *N*′-di(alkyl) analogs, the stability was attributed to hindrance of O_2 and H_2O diffusion by dense packing of the cores and by the fluoroalkyl chains.

Core-cyanated perylene diimides were first synthesized by Wasielewski *et al.* [87] These systems are significantly more readily reduced than their unsubstituted analogs (by about 0.36 V); the associated high electron affinity is believed to be a factor contributing to the high electron mobility (0.10 cm^2 V^{-1} s^{-1}) achieved in air for OFETs based on PDICy-CN_2. Combining partial fluorination of the N, N′ substituents and 1,7-dicyano substitution in PDIF-CN_2 affords a still higher electron mobility (0.64 cm^2 V^{-1} s^{-1}). The effects of PDIF-CN_2 film-growth conditions on n-channel OFET performance have also been investigated [44]; dramatic enhancements of the on–off ratio and the mobility are obtained with increased substrate temperature (T_s) during film growth, the increased mobility being correlated with higher levels of molecular ordering and with minimization of film-surface irregularities [88]. In addition, the effects modifying the SiO_2 surface of the gate dielectric with octadecyltrichlorosilane- or hexamethyldisilazane-derived monolayers, as well as with polystyrene, were investigated for PDIF-CN_2 films deposited at $T_s = 130\,^\circ C$; the SiO_2 surface treatments substantially modulate the mobility and growth morphology of PDIF-CN_2 films. Recently, Morpurgo *et al.* fabricated OFETs based on PDIF-CN_2 single crystals (Figure 5.24), with poly(methyl methacrylate) as the gate dielectric that exhibited electron mobilities approaching 6 cm^2 V^{-1} s^{-1}, 10× greater than those of the corresponding thin-film devices, both in air and in vacuum [89]. Furthermore, these devices exhibit near-zero threshold voltage and subthreshold slopes and current on–off ratios (10^3–10^4) comparable to the very best p-channel single-crystal devices when the same gate dielectric is employed.

In related work, Weitz *et al.* reported air-stable n-channel OFETs based on five dicyano PDIs with fluorinated linear and cyclic N, N′-substituents (mobilities up to 0.1 cm^2 V^{-1} s^{-1}) and investigated the relationships between molecular structure, thin-film morphology, substrate temperature, device performance, and air stability [90]. Interestingly, the mobility degradation rate in air was found to be similar for all compounds and at all substrate temperatures, raising the question of whether air stability can always be explained on the basis of kinetic barriers to O_2/H_2O diffusion formed by densely packed fluorine substituents. In addition to core cyanation, core halogenation is an effective way to functionalize the perylene "bay" positions. Würthner *et al.* reported that a 1,6,7,12-tetrachloro *N*, *N*′-didodecyl PDI exhibited PR-TRMC mobilities as high as 0.1 cm^2 V^{-1} s^{-1} [91]. More recently, Würthner *et al.* studied a series of core halogenated *N*, *N*′-bis(heptafluorobutyl) PDIs. Although introduction of halogens in the "bay" positions facilitates reduction, with 1,6,7,12-tetrahalo derivatives being slightly more readily reduced than their 1,7-dihalo analogs, the more highly substituted examples tend to exhibit lower mobilities; this can be attributed to disruption of core planarity and, therefore, of effective π–π overlap due to steric interactions (Figure 5.25) [92]. Thus, both the difluoro compound PDI-F and the parent PDI-E exhibit densely π-stacked,

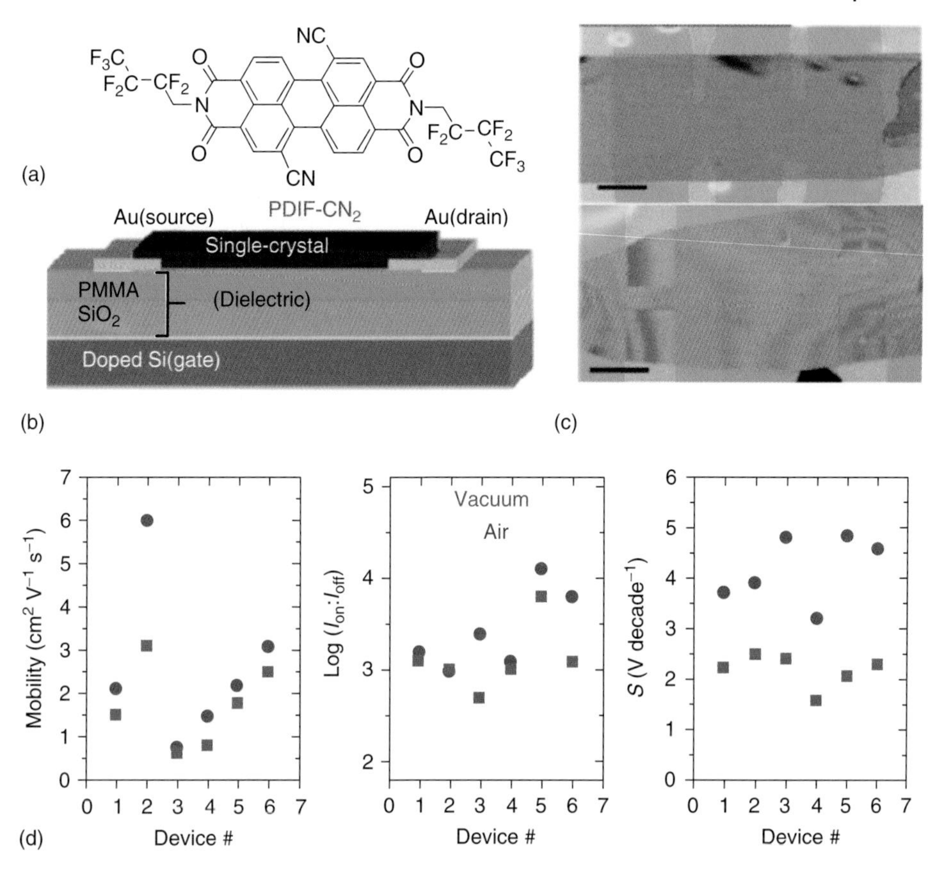

Figure 5.24 (a) Molecular structure of PDIF-CN2; (b) schematic layout of a single-crystal field-effect transistor; and (c) optical microscopic images of PDIF-CN2 devices used in our investigations. The bar is 200 μm. (d) Electron mobilities, I_{on}/I_{off}, and S for six single-crystal PDIFCN2 FETs measured in vacuum (light symbols) and in the air (dark symbols). (Reprinted with permission from Ref. [89].)

more-or-less planar PDI units (torsion angles of 3.0 and 1.5 °, respectively, between the two constituent naphthalene units according to X-ray crystal structures), with the FET electron mobility of PDI-F being around half that of PDI-E. On the other hand, the perylene core of PDI-G is distinctly nonplanar because of F–F steric interactions (torsion angles 20–25°), leading to less dense and less regular packing and to mobility that is 1 order of magnitude lower than that of PDI-F. Tetrabromo derivative PDI-J also exhibits about 10× lower mobility than its dibromo analog, PDI-K. Furthermore, 1,6,7,12-tetrachloro and bromo derivatives, PDI-H and PDI-J, exhibit mobilities $10^3\times$ lower than that of their tetrafluoro counterpart, PDI-G, presumably because of the increased bulk of the substituents leading to significantly increased torsion angles and reduced intermolecular π–π interactions. Interestingly, replacing the N, N′-fluoroalkyl substiutents of the 1,6,7,12-tetrachloro PDI-H by *N*, *N*′-pentafluorophenyl groups in PDI-I leads to an $\sim 10^4\times$ increase

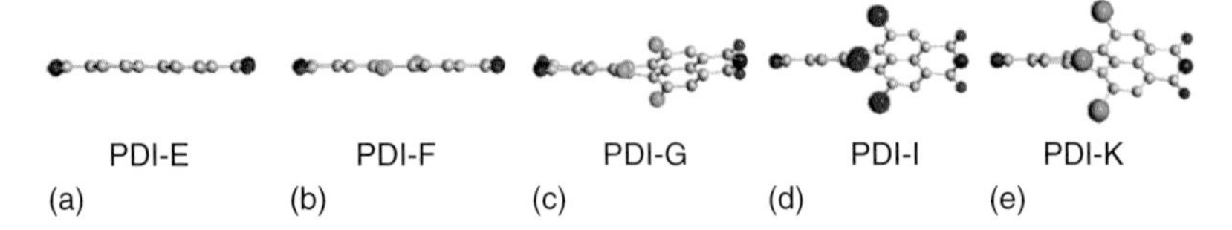

Figure 5.25 Crystal structure side view onto the aromatic systems for five perylene diimides (a–e) showing a progressive twist of the perylene core. The angles are 1.5° for PDI-F, 3.0° for PDI-F, 19°/26° for PDI-G 1c, 35°/36° for PDI-I, and 37° for PDI-K. (Reprinted with permission from Ref. [92].)

in the mobility. Müllen *et al.* have pioneered the synthesis of higher rylene diimide dyes and other species based on extended PDI cores, investigating in detail their thermotropic behavior and optical properties as well as the details of their microstructure [93]. However, exploitation of higher rylene diimides in organic electronics is limited to a report by Petit *et al.* on the FET properties of vapor-deposited films of N, N'-dipentylterrylene-3,4:11,12-tetracarboxylic diimide derivative QDI5 on Si–SiO_2 substrates [94]; a maximum electron mobility of $\sim$0.07 $cm^2\ V^{-1}\ s^{-1}$ and an on–off ratio in excess of 10^4 was obtained.

The first CMOS based on a perylene derivative was reported using the core-cyanated perylene diimide PDI8-CN_2 [95]. In that study, Dodabalapur *et al.* first described the electrical characteristics of bottom-contact PDI8-CN_2 OFETs fabricated with PDI-8CN2, which exhibited a mobility of 0.14 $cm^2\ V^{-1}\ s^{-1}$ in air. The effect of electrode/dielectric surface treatment on these devices was also examined, with a combination of 1-octadecanethiol and hexamethyldisilazane. Organic complementary five-stage ring oscillators were fabricated using pentacene and PDI8-CN_2 and operated at an oscillation frequency of 34 kHz and a propagation delay per stage of 3 s. In a related publication, the fabrication of an organic complementary D flip-flop using PDI8-CN_2 n-channel transistors and pentacene p-channel transistors was also demonstrated (Figure 5.26) [96]. The measured clock-to-output delay is 25 μs for 1 kHz and 14 μs at a clock frequency of 5 kHz. Recently, ambient-stable inverters were fabricated with PDI derivatives PDI-L and PDI-M [97]. Electron mobility greater than 0.1 $cm^2\ V^{-1}\ s^{-1}$ was achieved, and complementary inverters based on pentacene exhibited gains of about 12 and 10 for PDI-L/pentacene and PDI-M/pentacene inverters, respectively.

Most studies involving PDI-based FETs have been carried out using films deposited by vacuum evaporation; however, recent studies demonstrate the great potential of these materials for solution-processed/printed FETs (Figure 5.25). For example, Dodabalapur *et al.* demonstrated the first organic complementary circuits (CMOS) fabricated on Si–SiO_2 substrates, using solution-deposited films of PDI8-CN_2 (as n-channel FETs) and poly-3-hexylthiophene (P3HT, as p-channel FETs) [52]. They reported ring oscillators operating at a frequency of $\sim$2 kHz without passivation or packaging. More recently, Loi *et al.* compared the properties of solution-processed films of flourocarbon- versus hydrocarbon-functionalized core-cyanated perylenes [98]. Note that fluorocarbon- versus hydrocarbon-functionalization of aromatic cores usually leads to far

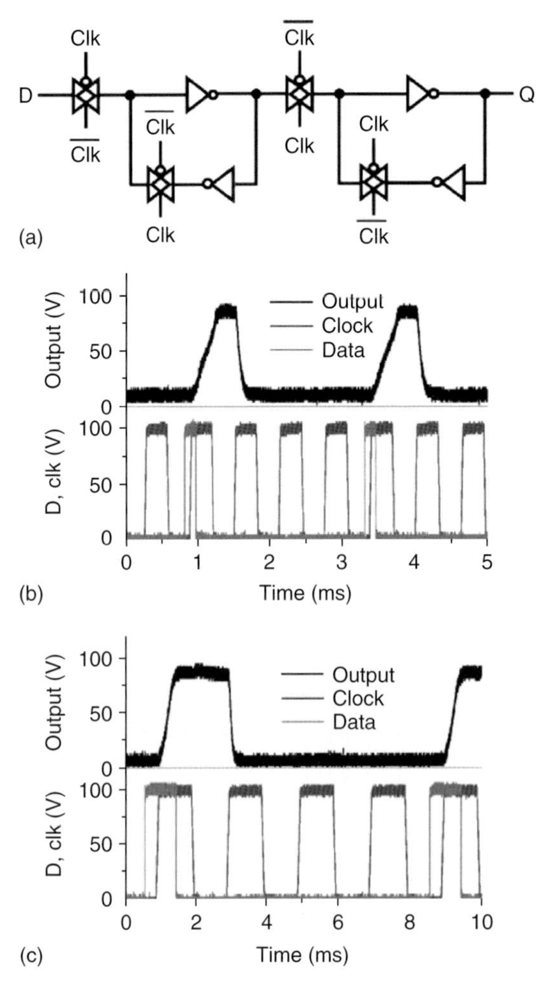

Figure 5.26 Output characteristics of complementary D flip-flops. The supply voltage is 100 V (W/L = 2000/7.5 lm). One input data pulse is applied at the positive edge of the clock cycle, and the second input data pulse is applied after four clock cycles. All measurements are carried out in air. (a) Schematic diagram of conventional D flip-flop. Operation at (b) 500 Hz and (c) 1.6 kHz. (Reprinted with permission from Ref. [52].)

less processable materials in conventional organic solvents. Bottom-contact, bottom-gate OFETs based on spin-coated films of PDI8-CN_2 and PDIF-CN_2 were fabricated on Si–SiO2 substrates and Au contacts. It was found that PDIF-CN_2 combines solution-processability characteristics and excellent semiconductor properties with electron mobilities of up to 0.15 $cm^2\ V^{-1}\ s^{-1}$ in vacuum. This high mobility value is obtained only after a mild thermal annealing of the spin-coated films, resulting in a highly enhanced crystalline organization of the molecules. Electron mobilities of ~0.08 $cm^2\ V^{-1}\ s^{-1}$ are still measured after

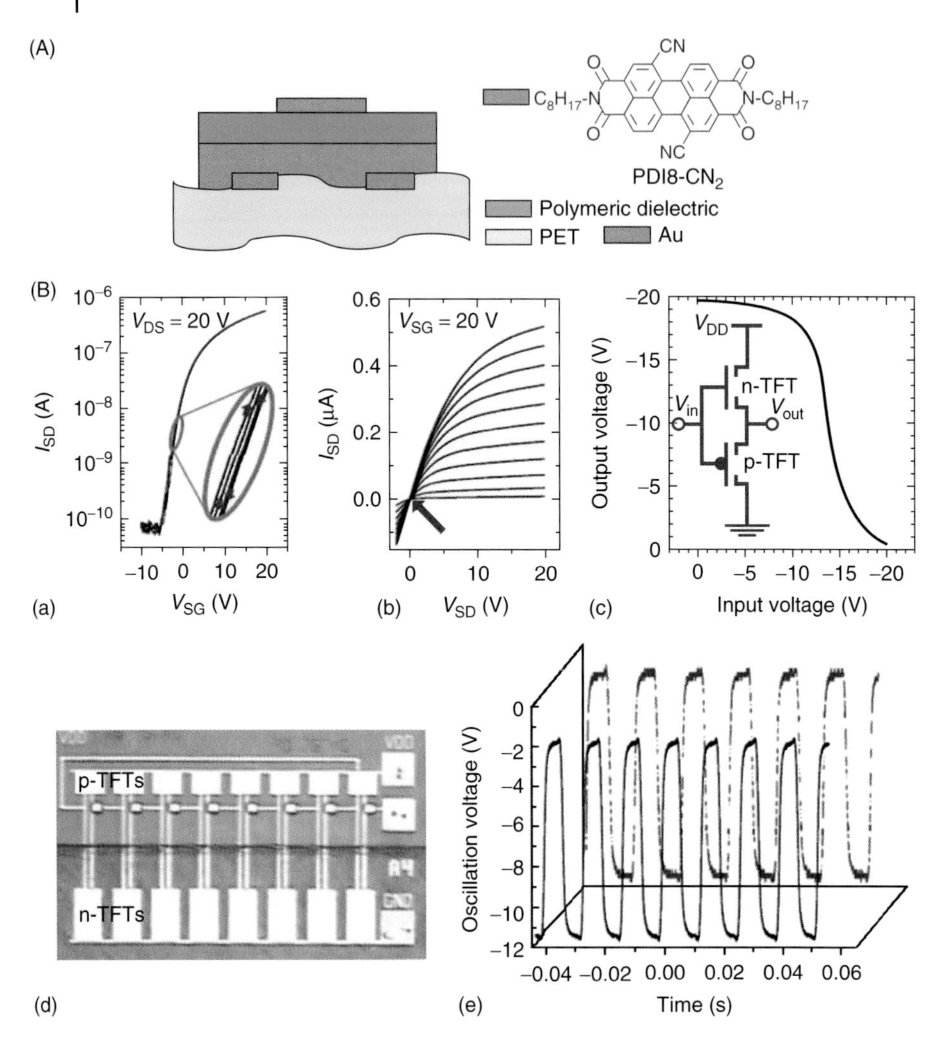

Figure 5.27 (A) Illustration of the top-gate TFT structure and chemical structure of PDI8-CN2. (B) (a) Transfer and (b) output plots for PDI8-CN2 top-gate TFTs. (c) Inverter structure and corresponding output characteristics. (d) Picture of a seven-stage ring oscillator on PET. (e) CMOS ring oscillator responses of the as-prepared device (dark line) and after nine months from fabrication (light line). (Reprinted with permission from Ref. [99].)

20 days of continuous exposure to air. It was observed that an almost complete recovery of the mobility is obtained by retesting the device under vacuum. Polyera Corporation has reported the first demonstration of solution-processed top-gate FETs with PDI8-CN_2 exhibiting electron mobilities $\sim$0.1 $cm^2\ V^{-1}\ s^{-1}$ on plastic substrates [99], as well as CMOS inverters and ring oscillators on plastic substrates operating at 50 Hz and using PDI8-CN_2 and P3HT in combination with solution-processed polymer dielectrics. Finally, our group and Arias *et al.*

demonstrated inkjet-patterned FETs and inverters based on a PDI derivative and a poly(2,5-bis(3-alkylthiophen-2-yl)thieno[3,2-*b*]thiophene) and analyzed the details of bias stress (a parameter related to FET operation in a circuit) for digital and analog electronic circuit applications (Figure 5.27) [100]. The inverter TFTs show mobilities approaching $10^{-2}\,cm^2\,V^{-1}\,s^{-1}$ for both p- and n-channel semiconductors using a bilayer dielectric and silver S–D electrodes. The corresponding inverters achieve a gain of −4.4 at $V_{DD} = +10$ V and a −3 dB cutoff at 100 kHz with 0.02 pF load. The results showed that incomplete compensation of threshold voltage shifts presents stability problems for complementary OTFT analog circuits. As for digital operations, since the threshold voltage shifts in the p- and n-TFTs partially cancel each other, the complementary inverters demonstrate improved stability compared to unipolar inverters and exhibit an estimated noise margin >1.1 V at $V_{DD} = +15$ V. This is a promising result toward adopting complementary technology over unipolar circuits for organic electronics.

5.4.1.5 **Other Small Molecular n-Channel Semiconductors**

In the last few years, several unconventional structures have been designed and synthesized as n-channel semiconductor candidates. Figure 5.28 shows the chemical structures of these derivatives. Yamashita *et al.* have developed several molecular materials [101], such as functionalized anthracenes exhibiting electron mobilities up to $\sim 5 \times 10^{-3}\,cm^2\,V^{-1}\,s^{-1}$, trifluoromethylphenyl-subtituted

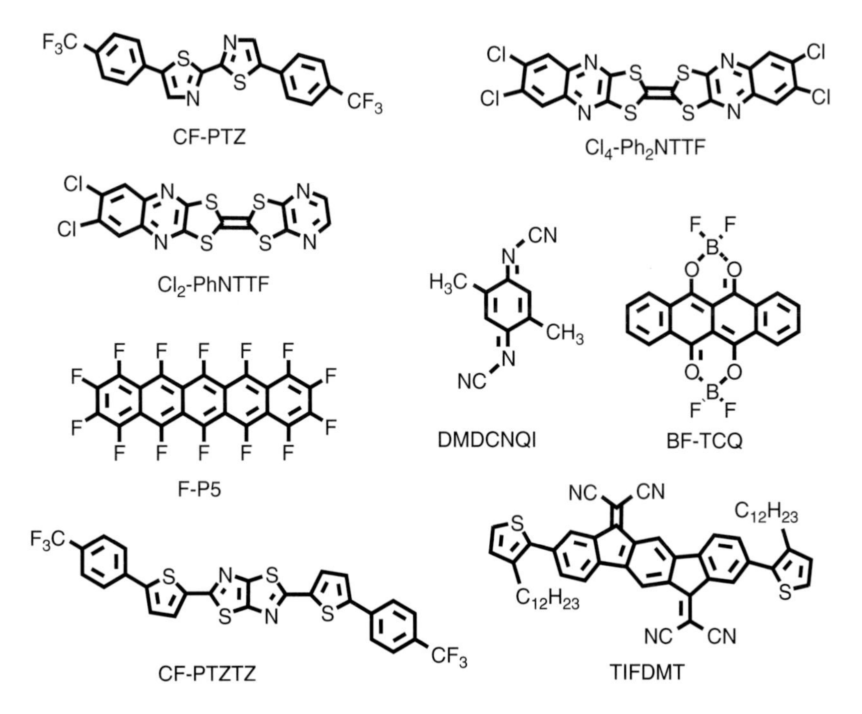

Figure 5.28 Chemical structures of several n-channel semiconductors.

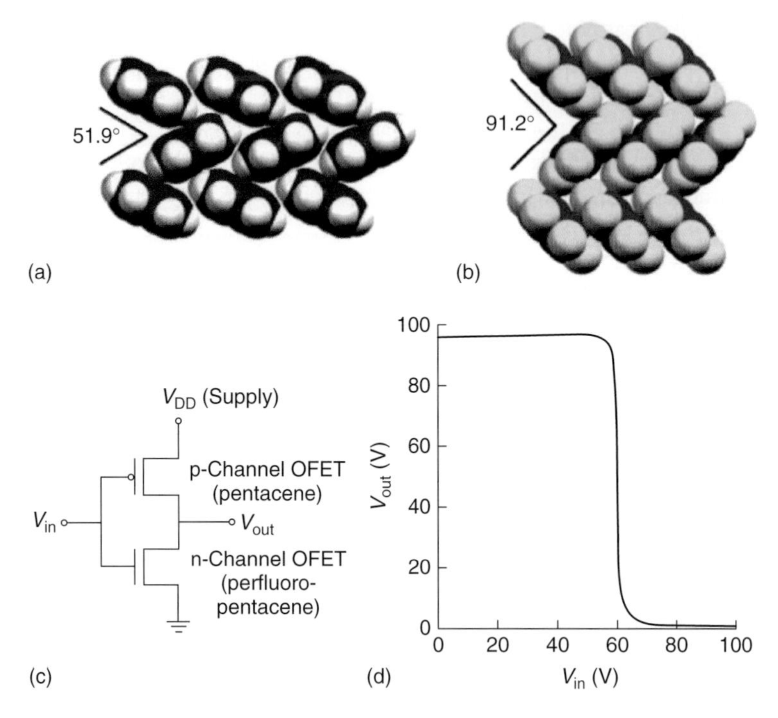

Figure 5.29 Molecular packing diagrams of (a) pentacene and (b) perfluoropentacene. (c) Inverter circuit configuration. (d) Transfer characteristics of a pentacene/perfluoropentacene complementary inverter with a 100 V supply. (Reprinted with permission from Ref. [103].)

thiazole and thiazolothiazole oligomers enabling electron mobilities up to $\sim$2 cm^2 V^{-1} s^{-1}, and tetrathiafulvalene derivatives with electron mobilities up to $\sim$0.1 cm^2 V^{-1} s^{-1}. Interesting are the performances of indenofluorenedione- and diindenopyrazinedione-based TFTs, which exhibit electron mobilities approaching 0.2 cm^2 V^{-1} s^{-1}. Tetracene and perylene BF_2 complexes are also very interesting new electron-deficient arene semiconductors. Perfluoropentacene was developed by Sakamoto *et al.* and exhibits the highest electron mobility of 0.11 cm^2 V^{-1} s^{-1} [102]. The same group also demonstrated the first CMOS circuit using pentacene as the p-channel counterpart (Figure 5.29). Particularly interesting are two new trifluoromethyltriphenodioxazine derivatives, the devices of which exhibit electron mobilities approaching 0.1 cm^2 V^{-1} s^{-1} in ambient conditions [103]. Recently, dimethyldicyanoquinonediimine (DMDCNQI) was developed as an n-channel semiconductor to achieve ambient-stable TFTs with electron mobilities of $\sim$0.01 cm^2 V^{-1} s^{-1} [25].

Our group has recently developed new molecular and polymeric (vide infra) electron-depleted cores based on indenofluorene and bisindenofluorene functionalized with dicyanovinylene units [104]. TFTs fabricated with vapor-deposited and solution-processed films of TIFDMT exhibit similar electron mobilities up to $\sim$0.2 cm^2 V^{-1} s^{-1} in air. Furthermore, this study provided an interesting

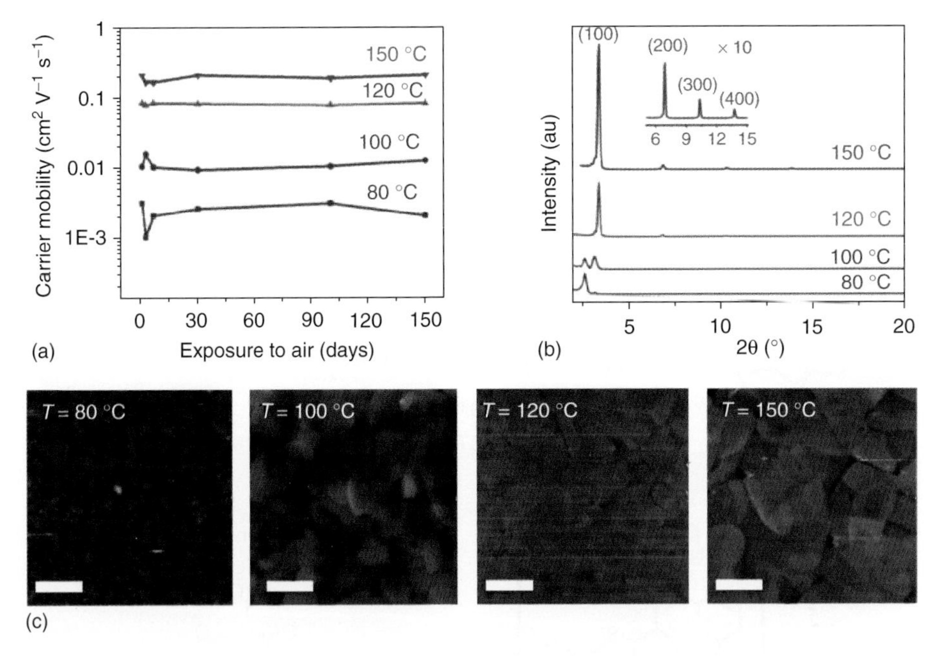

Figure 5.30 (a) Electron mobilities measured in ambient for TFTs fabricated with spin-coated 10 and annealed at the indicated temperatures (80–150 °C) versus storage time of the devices in air without excluding light or ambient humidity (25–40%). (b) $\theta-2\theta$ X-ray diffraction (XRD) scans of spin-coated 10 thin films after annealing at the indicated temperatures (Note that peak intensities are not normalized for comparative purposes). (c) Tapping mode AFM images of 10 thin films after annealing at the indicated temperatures. Scale bars denote 1 μm. (Reprinted with permission from Ref. [104].)

observation about the effect of the n-channel semiconductor film morphology on the TFT response in ambient temperature over a long period of time. The results show that when an n-channel semiconductor has a proper electronic structure (sufficiently low LUMO energy), the film morphology has no effect on the stability of the corresponding devices (Figure 5.30).

5.4.2 Polymeric Semiconductors

The chemical structures of several n-channel polymers are shown in Figure 5.31. Highly purified vacuum-deposited small molecules have traditionally shown the greatest transistor performance. However, π-conjugated polymeric semiconductors present several advantages such as solution processability (when properly functionalized) and reduced sensitivity of the device performance to film morphology. Furthermore, the viscosity of solutions based on polymeric materials can be tuned to a greater extent, enabling printability using methodologies spanning from inkjet printing to flexographic printing.

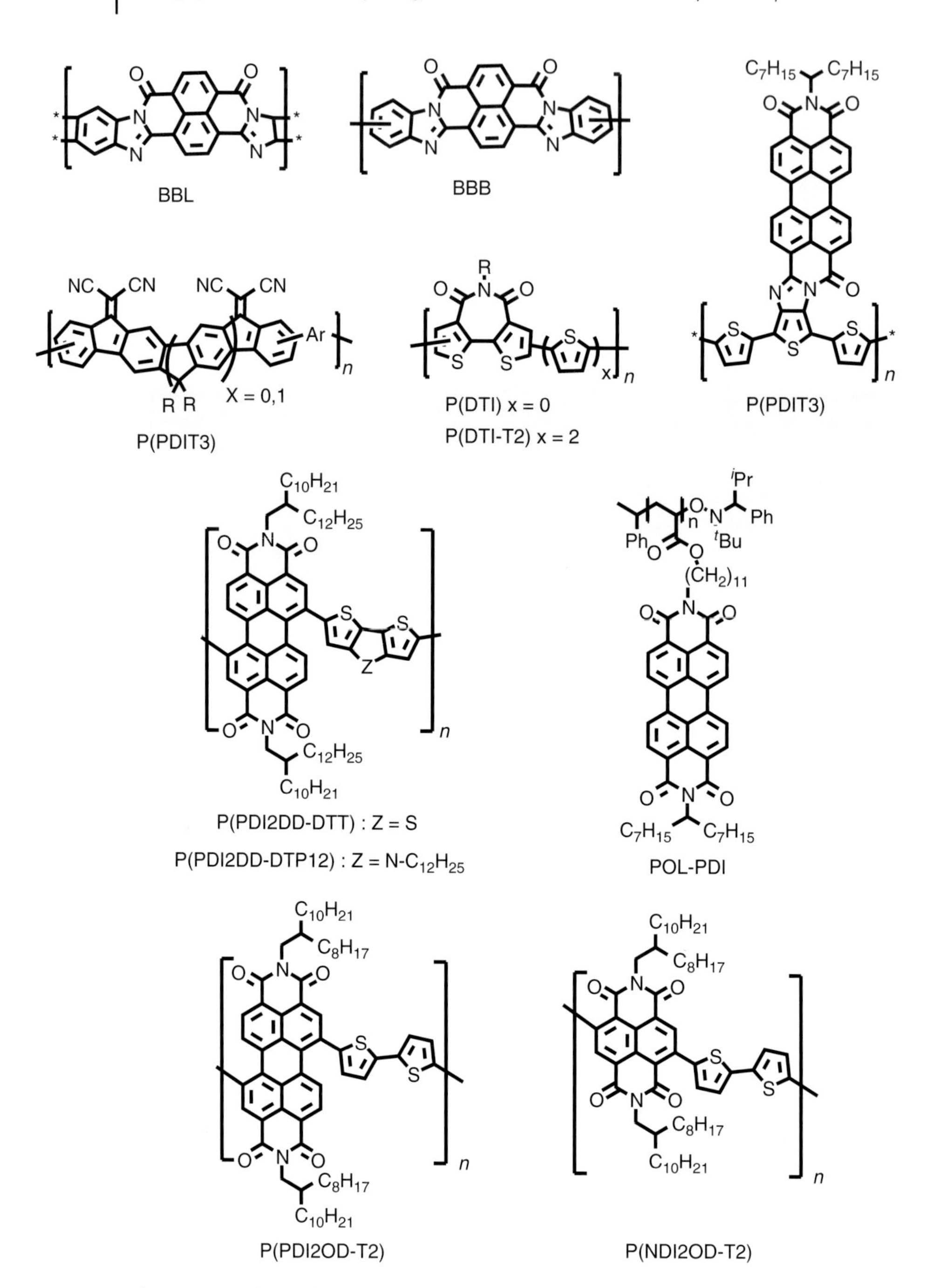

Figure 5.31 Chemical structures of several n-channel polymers.

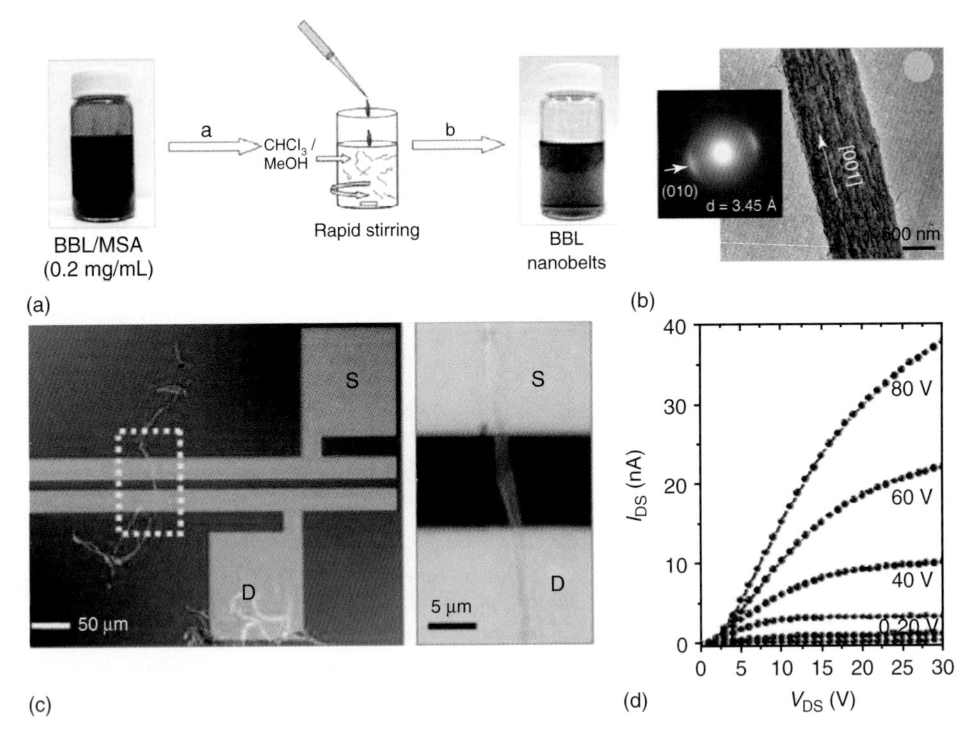

Figure 5.32 (A) Schematic preparation of BBL nanobelts. In step "a," a BBL/methanesulfonic acid (MSA) solution (~0.2 mg/ml) is added dropwise to a rapid-stirring vial containing a $CHCl_3$/MeOH (4 : 1) solvent mixture. In step "b," the nanobelts are washed and centrifuged two times with deionized water. The last frame shows a vial containing an aqueous dispersion of BBL nanobelts for use in device fabrication via solution deposition. (B) Representative TEM images of an individual BBL nanobelt. The inset is the corresponding selected area electron diffraction pattern. A notable observation is the textured morphology of the polymer chains that are aligned along the [001] direction or the long axis of the nanobelt, thus demonstrating the extent of orientation. (C) Typical single BBL nanobelt transistor and a close-up showing the nanobelt bridging the source-drain electrodes. (D) TFT output characteristics showing well-resolved current-voltage saturation. (Reprinted with permission from Ref. [106].)

The first report of an n-channel TFT-active polymer was poly(benzobisimidazobenzophenanthroline) (BBL). This ladder-type polymer exhibits high electron mobility of 0.1 $cm^2\ V^{-1}\ s^{-1}$ after annealing the solution-cast film [105]. More recently, it was found that BBL nanobelts can be prepared by a simple high-yield, solution-phase process, which enables dispersions of the nanobelts from a large number of solvents, including environmentally benign solvents such as methanol and water (Figure 5.32) [106]. Characterization of the nanobelts by transmission electron and force microscopies, electron diffraction, and X-ray diffraction showed that the BBL polymer chains are oriented parallel to the long axis of each nanobelt. This unique packing motif is unlike the reported packing of polymer chains in other nanostructures, such as poly(3-hexylthiophene) nanowires, where the polymer backbone packs face-to-face along the nanowire direction. This chain packing in BBL nanobelts is explained by the rather strong intermolecular interactions, which are a result of the rigid and planar polymer chains. The authors investigated electron transport in single nanobelts, and nanobelt networks in TFT architectures show electron mobilities up to $\sim$0.01 cm^2 V^{-1} s^{-1} and current on–off ratios of $\sim 10^4$. The n-channel nanobelt transistors showed stability and repeatability in air for more than six months, among the most stable n-channel polymer transistors. These results demonstrate that the BBL nanobelts are promising n-channel polymer candidates.

Thiophene-based polymeric structures containing electron-poor PDI cores have been synthesized (Figure 5.33). The interesting donor–acceptor polythiophene P(PDIT3) was synthesized by electropolymerization of the corresponding peryleneamidine monoimide-fused terthiophene precursor. This polymer exhibits facile p- and the n-doping processes. Although OFET data are not yet available, P(PDIT3) is the first example of a p-type conjugated polymer in direct conjugation with n-type perylenemonoimide moieties [107]. Our group has designed new homopolymers and copolymers based on the dithenodiimide core [108]. A novel design approach was employed using computational modeling to identify favorable monomer properties such as core planarity, solubilizing substituent tailorability, and appropriate electron affinity, with gratifying results. Monomeric model compounds were also synthesized to confirm these properties, and a crystal structure reveals a short 3.43 Å $\pi-\pi$ stacking distance with favorable solubilizing substituent orientations. A family of 10 homopolymers and bithiophene copolymers was synthesized via Yamamoto and Stille polymerizations, respectively. Two of these polymers are processable in common organic solvents: the homopolymer poly(*N*-(2-octyldodecyl)-2,2′-bithiophene-3,3′-dicarboximide) [P(DTI)] exhibits n-channel FET activity, and the copolymer poly(*N*-(2-octyldodecyl)-2,2′:5′, 2″:5″, 2‴-quaterthiophene-3,3′-dicarboximide) [P(DTI-T2)]) exhibits air-stable p-channel FET operation. After annealing, P1 films exhibit a very high degree of crystallinity and an electron mobility >0.01 cm^2 V^{-1} s^{-1} with a current on–off ratio of 10^7, which is remarkably independent of film-deposition conditions. Extraordinarily, P(DTI) films also exhibit terracing in AFM images, with a step height matching the X-ray diffraction d spacing, a rare phenomenon for polymeric organic semiconductors. Another fascinating property of these materials is the air-stable p-channel FET performance. In a series of papers, we also reported polymers based on the indenofluorene and bisindenofluorene core having C=O and C=C(CN)$_2$ substituents (see structures in Figure 5.6). However, although some of the polymer building blocks are excellent n-channel semiconductors, the polymers are mainly ambipolar or are poorly active.

Relevant n-channel polymers are those based on rylene diimide cores, particularly perylene and naphthalene. The first perylene-based polymer [P(PDI2DD-DTT)] was synthesized by Stille coupling of *N*, *N*′-dialkyl-1,7-dibromo-3,4,9,10-perylene diimide with a distannyl derivative of dithienothiophene [109]. The polymer was found to be soluble in chloroform, THF, and chlorobenzene and could readily be processed from solution. The molecular weight of P(PDI2DD-DTT) was not very high, $\sim$15 kD

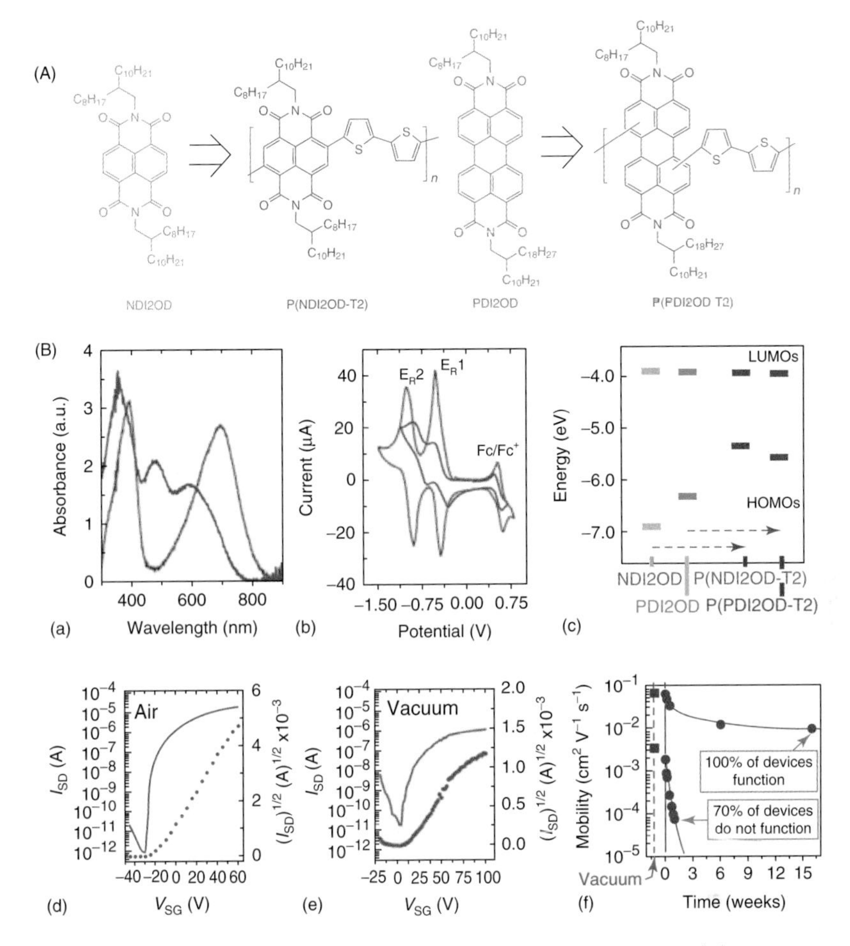

Figure 5.33 (A) Chemical structures of the formal polymer precursors and the corresponding polymers. (B) (a) Optical absorption spectra of spin-coated P(NDI2OD-T2) (red line) and P(PDI2OD-T2) (blue line) films (~30 nm thick) on glass. (b) Thin-film cyclic voltammetries [Fc (+0.54 V vs SCE) internal standard)] of P(NDI2OD-T2) (red line) and P(PDI2OD-T2) (blue line) thin films on a Pt electrode. The ER1 values of NDI2OD and PDI2OD (not shown) are −0.49 and 0.46 V versus SCE, respectively. (c) Energy diagram for the specified rylene monomers and polymers. (d) I–V transfer plots for P(NDI2OD-T2) TFT in air for 1 h, and (e) P(PDI2OD-T2) TFT in vacuum. (f) Polymer TFT electron mobility plots in vacuum and ambient (RH) 20–40%, T ≈ 25 °C) versus time. (Reprinted with permission from Ref. [112].)

using GPC (polystyrene standards). DSC showed a glass-transition temperature of 215 °C, whereas TGA suggested excellent thermal stability with an onset decomposition temperature under nitrogen of 410 °C. Polymer P(PDI2DD-DTT)-based OFETs (Al S–D electrodes, top-contact/bottom-gate geometry) were measured under nitrogen and exhibited electron mobilities as high as ~0.01 $cm^2 V^{-1} s^{-1}$ and I_{on}:I_{off} > 10^4. Very recently, a dithienopyrrole analog, P(PDI2DD-DTP12), was reported to show an electron mobility of 7.4×10^{-4} cm^2 V^{-1} s^{-1}, which increases to 1.2×10^{-3} cm^2 V^{-1} s^{-1} on annealing at 100 °C for 60 min under inert atmosphere [110]. The lower mobility observed for P(PDI2DD-DTP12) may be related to dilution of the PDI ET units by the presence of the additional N-substituents of the dithienopyrrole donors or because of the disruption of PDI–PDI interactions caused by these groups. Thelakkat *et al.* have reported OFETs based on polymers containing perylene unit as pendant groups [111]. For POL–PDI after thermal annealing at 210 °C for 60 min, the threshold voltage drops significantly to 7 V, while the current and charge carrier mobility both increase by 100×, approaching 1.2×10^{-3} cm^2 V^{-1} s^{-1}. Unfortunately, OFETs based on these polymers are unstable in ambient conditions.

In a recent communication, our group at Polyera Corporation reported the synthesis, characterization, and comparative properties of N,N'-dialkylperylenedicarboximide-dithiophene (PDIR-T2) and N,N'-dialkylnaphthalenedicarboximide-dithiophene (NDIR-T2) copolymers and the fabrication of the corresponding bottom-gate TFTs on Si–SiO_2 substrates (Figure 5.33) [112]. The results of that paper demonstrate that the choice of the NDIR versus PDIR comonomer is strategic to achieve both high-performance bottom-gate n-channel TFTs and a device functioning under ambient conditions. The rylene building block and the polymer structural design rationale were the following. (i) The electron-poor NDIR comonomer was selected because of the large electron affinity of this core, comparable to that of the far more π-extended PDIR systems. (ii) Equally important, NDIR-Br_2 can be easily isolated as pure 2,6-diastereoisomers, enabling the synthesis of a regioregular polymeric backbone. Note that isolation of PDIR-Br_2 regioisomers, although demonstrated, is tedious. Therefore, compared to PDIR-based polymers, it should lead to a more π-conjugated structure and, consequently, better charge transport efficiencies. (iii) Proper alkyl (R) functionalization at the rylene nitrogen atoms, in that study on 2-octyldodecyl (2OD), should result in highly soluble and processable, yet charge transport-efficient, polymers. (iv) Finally, the dithiophene (T2) unit is utilized because of the commercial availability, stability, and known electronic structure and geometric characteristics of this core, likely providing highly conjugated, planar, and rodlike polymers. The new NDIR- and PDIR-based polymers were synthesized in high-yields via a Pd-catalyzed Stille polymerization. Using the reported synthetic procedure, polymer M_ws are larger for P(NDI2OD-T2) (~250 K, PD ~ 5) than for P(PDI2OD-T2) (~32 K PD ~ 3). The optical and electrochemical properties of these new systems reveal important aspects of the polymer electronic structures and NDIR versus PDIR comonomer effects (Figure 5.33a–c). Bottom-gate top-contact OTFTs were fabricated on n^{++}-Si/SiO_2/OTS substrates, on which the semiconducting polymer solutions (~3–10 mg/ml in DCB-$CHCl_3$) were spin coated to afford ~100 nm thick films. The films were annealed at 110 °C for 4 h before the TFT structure was completed by Au S–D vapor deposition. Electrical measurements were performed both under high vacuum and ambient conditions. Electron mobilities of ~0.08–0.06 cm^2 V^{-1} s^{-1} for P(NDI2OD-T2) and ~0.003–0.001 cm^2 V^{-1} s^{-1} for P(PDI2OD-T2) are measured in vacuum.

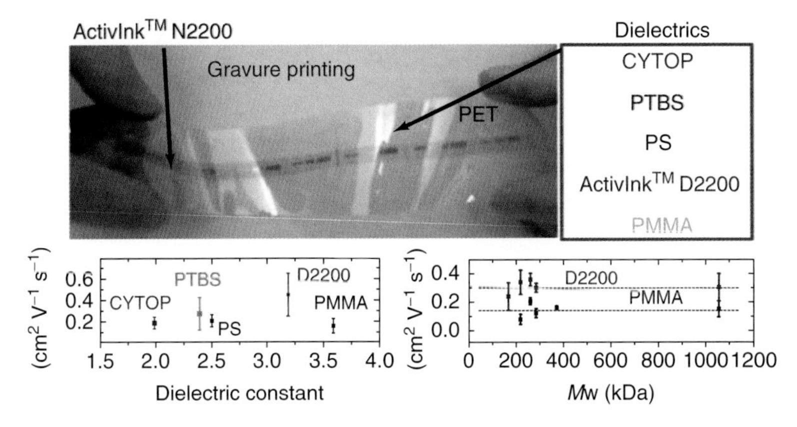

Figure 5.34 Top: Optical image of the first gravure-printed n-channel polymeric TFTs on a PET web fabricated at Polyera. Center: Mobility versus dielectric constant (k) of the polymeric gate dielectric (left) and molecular weight (Mw) (right). Bottom: Temporal (left) and humidity (right) stability of ActivInk™ N2200 TFTs with different gate dielectrics. (Reprinted with permission from Ref. [113].)

However, when the same TFT array is measured under ambient conditions, the P(NDI2OD-T2)-based devices continue to function until after 16 weeks of fabrication ($\mu = 0.01$ cm^2 V^{-1} s^{-1}), while P(PDI2OD-T2) mobility drops to $\sim 2 \times 10^{-4}$ cm^2 V^{-1} s^{-1} within one week, in agreement with previous studies on PDI-based polymers (Figure 5.33d,e).

Very recently, we also developed high-performance polymeric top-gate bottom-contact (TGBC, Figure 5.34) TFTs and the first all-polymeric CMOS circuit functioning in ambient based on P(NDI2OD-T2) [113]. These TGBC TFTs were fabricated on glass or PET and have the structure substrate/Au(S–D contacts)/P(NDI2OD-T2)/polymeric dielectric/Au(gate contact). This structure was selected because of the superior injection characteristics of typical staggered (top-gate) architectures and considering the facile channel miniaturization for bottom-contact TFTs, which could lead to high-frequency circuits. These devices were fabricated with the P(NDI2OD-T2) film deposited by spin coating as well as gravure, flexographic, and inkjet printing and with the dielectric layer deposited by spin coating. Furthermore, TFTs in which both the semiconductor and the dielectric layers were gravure printed are demonstrated. All device fabrication processes were performed in ambient conditions, with the exception of the Au contact vapor deposition and the film drying steps ($\leq 110\,^{\circ}$C). The TGBC TFTs based on this polymer exhibit excellent n-channel OTFT characteristics in ambient, with electron mobilities up to $\sim$0.45–0.85 cm^2 V^{-1} s^{-1}, I_{on}:I_{off} > 105, V_{on} $\sim$0–5 V. Importantly, the carrier mobility of P(NDI2OD-T2)-based TFTs is insensitive to the dielectric constant (k) of the gate dielectric material (Figure 5.34). This is of great importance to broaden the compatibility of this n-channel semiconductor family with several p-channel materials *using the same gate dielectric*. Furthermore, this polymer's TFT properties are independent of the polymer molecular weight

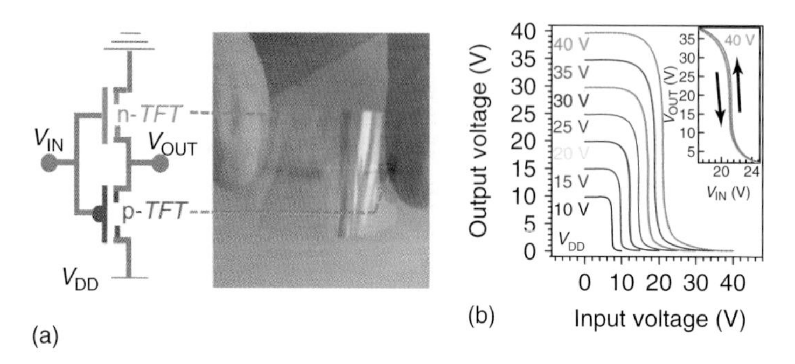

Figure 5.35 (a,b) Schematic, optical image, and static characteristics of the fist gravure-printed polymeric CMOS inverter. (Reprinted with permission from Ref. [113].)

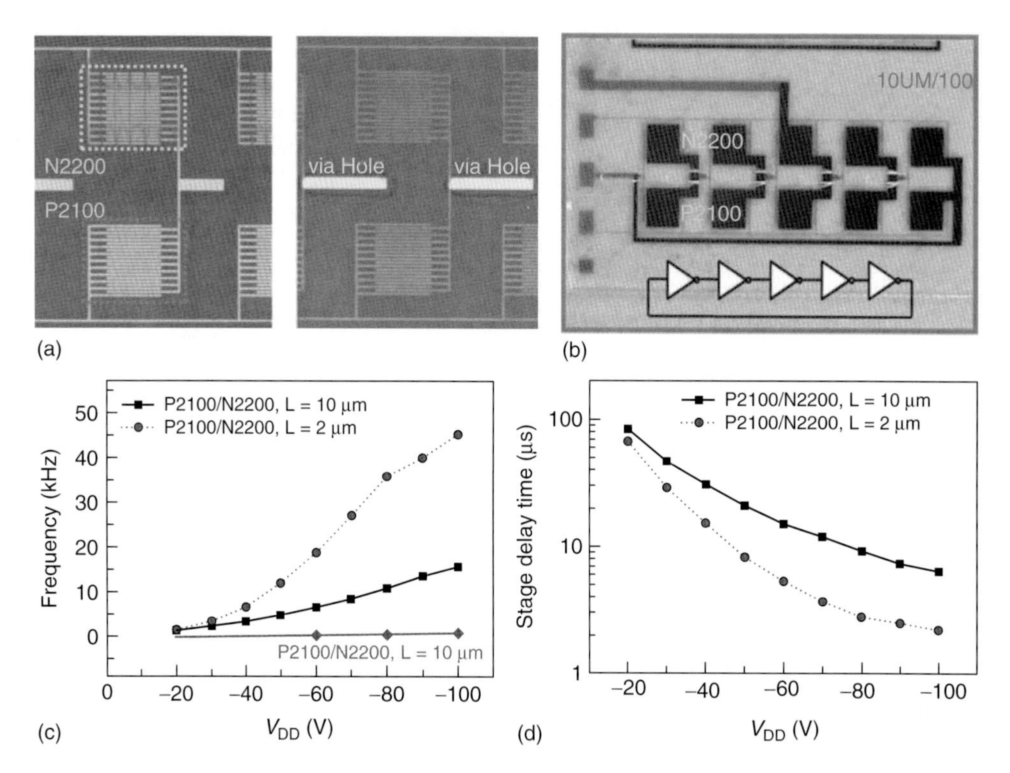

Figure 5.36 Optical microscopic image of ink-jet-printed five-stage complementary polymer ring oscillators showing (a) before and after gate dielectric layer (PMMA) coating and via-hole formation and (b) the completed ring oscillator. (c) Oscillation frequency and (d) stage delay time as functions of VDD for the indicated ring oscillators. (Reprinted with permission from Ref. [114].)

(Mw) over a large range of values (Mw ~ 200 KDa to >1 MDa, Figure 5.34). The insensitivity of the device performance on the polymer chain-length extension is of extreme importance for large-scale synthesis and batch-to-batch reproducibility of the TFT characteristics. P(NDI2OD-T2)-based TFTs are also exceptionally stable under ambient conditions up to ~70% relative humidity (RH).

Owing to the stability of this n-channel polymer family, excellent TFT performance with high-work-function metal contacts, and compatibility of our UV-curable dielectrics with both p- and n-channel semiconductors, we also enabled the first polymeric complementary logic. Figure 5.35 shows an optical image and the corresponding switching characteristics of a printed inverter fabricated at Polyera with P(NDI2OD-T2) (ActivInk™ N2200, n-channel), P3HT (p-channel), and ActivInk™ D2200 (gate dielectric). These inverters show remarkably small hysteresis reflecting the transistor threshold voltage stability. The voltage gains for the gravure-printed devices are very large (dV_{OUT}/dV_{IN}(max) > 25–60).

Finally, Noh *et al.* fabricated monolithically integrated polymeric complementary circuits using P(NDI2OD-T2 and two p-type polymers P3HT and a dithiophene-based polymer (Polyera ActivInk P2100) (Figure 5.36). Ink-jet-printed TGBC FETs exhibit very high hole and electron mobilities (μ_{FET}) of 0.2–0.5 $cm^2 V^{-1} s^{-1}$. The FET active regions were patterned and via-holes were defined by direct inkjet printing of the conjugated polymer solutions and the polymer gate dielectric solvent, respectively, enabling high-performance CMOS inverters (gain > 30) and ring oscillators (f_{osc} up to ~50 kHz) [114].

5.5 Conclusions and Outlook

In this chapter, several organic semiconductors for n-channel transistors were reviewed. These materials are essential for the fabrication of organic complementary circuits. During the past year, impressive results in developing new n-channel molecular and polymeric semiconductors were achieved and the performance difference with the p-channel counterpart has been dramatically reduced. I have shown several examples where modification of the core results in a dramatic variation and majority charge carrier flip from p- to n-channel. Importantly, from the technology perspective, some of these materials exhibit, as solution-processed films, carrier mobilities and I_{on}:I_{off} surpassing those of amorphous Si. However, several issues remain before organic semiconductor-based FETs will be implemented in commercial devices, first of all device yields and reliability. Note that also the simplest electronic circuits are composed of thousands of FETs, which must all function within a strict operational parameter range. An excellent review addresses issues related to the use of organic materials in electronic circuits [115]. Furthermore, uniform coverage of a large area of semiconductor film is generally difficult to realize for high-performance oligomeric structures, while polymers generally suffer from either low carrier mobility and/or facile doping unless special film-deposition conditions are utilized. However, these deposition conditions either are not compatible with a roll-to-roll process or are expensive to realize. Yet, impressive progresses have been made in developing methodology to control resolution and registration over large area. It will be exciting to see how they will work on a production line. Finally, a tremendous jump in several aspects concerning organic electronics science and technology has been achieved recently, which encompass materials design, synthesis, multilayer thin-film fabrication, thin-film morphology, and device physics [116]. These results are exciting and are essential to move the field forward toward market applications.

References

1. (a) Braga, D. and Horowitz, G. (2008) *Adv. Mater.*, **20**, 423; (b) de Boer, B. and Facchetti, A. (2008) *Polym. Rev.*, **48**, 423; (c) Gao, X., Qiu, W., Liu, Y., Yu, G., and Zhu, D. (2008) *Pure Appl. Chem.*, **80** (11), 2405; (d) Klauk, H. (2008) *Nanotechnology*, **4**, 125; (e) Panzer, M.J. and Frisbie, D.C. (2008) *Adv. Mat.*, **20**, 3177; (f) Ong, B.S., Wu, Y., Li, Y., Liu, P., and Pan, H. (2008) *Chemistry*, **14** (16), 4766; (g) Di, C.-A., Yu, G., Liu, Y., and Zhu, D. (2007) *J. Phys. Chem. B*, **111** (51), 14083; (h) Facchetti, A. (2007) *Mater. Today*, **10**, 28; (i) Murphy, A.R. and Frechet, J.M. (2007) *J. Chem. Rev.*, **107**, 1066; (j) Locklin, J., Roberts, M., Mannsfeld, S., and Bao, Z. (2006) *Polym. Rev.*, **46**, 79; (k) Anthopoulos, T.D., Setayesh, S., Smits, E., Colle, M., Cantatore, E., de Boer, B., Blom, P.W.M., and de Leeuw, D.M. (2006) *Adv. Mat.*, **18**, 1900; (l) Chabinyc, M. and Loo, Y.L. (2006) *J. Macromol. Sci. Polym. Rev.*, **46**, 1; (m) Tulevski, G.S., Miao, Q., Afzali, A., Graham, T.O., Kagan, C.R., and Nuckolls, C. (2006) *J. Am. Chem. Soc.*, **128**, 1788; (n) Muccini, M. (2006) *Nature Mater.*, **5**, 605; (o) Sirringhaus, H. (2005) *Adv. Mat.*, **17**, 2411; (p) Dimitrakopoulos, C.D. and Malenfant, P.R.L. (2002) *Adv. Mater.*, **14**, 99; (q) Horowitz, G. (1998) *Adv. Mater.*, **10**, 365; (r) Bao, Z., Rogers, J.A., and Katz, H.E. (1999) *J. Mater. Chem.*, **9**, 1895.
2. (a) Armstrong, N.R., Wang, W., Alloway, D.M., Placencia, D., Ratcliff, E., and Brumbach, M. (2009) *Macromol. Rapid Commun.*, **30** (9–10), 717–731; (b) Kovac, J., Peternai, L., and Lengyel, O. (2003) *Thin Solid Films*, **433**, 22; (c) Burroughes, J.H., Bradley, D.D.C., Brown, A.R., Marks, R.N., Mackay, K., Friend, R.H., Burns, P.L., and Holmes, A.B. (1990) *Nature*, **347**, 539.
3. (a) Cheng, Y.-J., Yang, S.-H., and Hsu, C.S. (2009) *Chem. Rev.*, **109** (11), 5868; (b) Heremans, P., Cheyns, D., and Rand, B.P. (2009) *Acc. Chem. Res.*, **42** (11), 1740; (c) Armstrong, N.R., Veneman, A.P., Ratcliff, E., Placencia, D., and Brumbach, M. (2009) *Acc. Chem. Res.*, **42** (11), 1748; (d) Coakley, K.M. and McGehee, M.D. (2004) *Chem. Mater.*, **16**, 4533; (e) Brabec, C.J., Sariciftci, N.S., and Hummelen, J.C. (2001) *Adv. Funct. Mater.*, **11**, 15.
4. (a) Roberts, M.E., Sokolov, A.N., and Bao, Z. (2009) *J. Mater. Chem.*, **19** (21), 3351; (b) Someya, T., Katz, H.E., Gelperin, A., Lovinger, A.J., and Dodabalapur, A. (2002) *Appl. Phys. Lett.*, **81**, 3079; (c) Crone, B., Dodabalapur, A., Gelperin, A., Torsi, L., Katz, H.E., Lovinger, A.J., and Bao, Z. (2001) *Appl. Phys. Lett.*, **78**, 2229.
5. (a) Crone, B., Dodabalapur, A., Lin, Y.Y., Filas, R.W., Bao, Z., LaDuca, A., Sarpeshkar, R., Katz, H.E., and Li, W. (2000) *Nature*, **403**, 521; (b) Drury, C.J., Mutsaers, C.M.J., Hart, C.M., Matters, M., and de Leeuw, D.M. (1998) *Appl. Phys. Lett.*, **73**, 10811; (c) Brown, A.R., Pomp, A., Hart, C.M., and Deleeuw, D.M. (1995) *Science*, **270**, 972.
6. Forrest, S.R. (2004) *Nature*, **428**, 911.
7. (a) Sirringhaus, H., Tessler, N., and Friend, R.H. (1998) *Science*, **280**, 1741;

(b) Garnier, F., Hajlaoui, R., Yassar, A., and Srivastava, P. (1994) *Science*, **265**, 1684.

8. (a) Speakman, S.P., Rozenburg, G.G., Clay, K.J., Milne, W.I., Ille, A., Gardner, I.A., Bresler, E., and Steinke, J.H.G. (2001) *Org. Electron.*, **2**, 65; (b) Sirringhaus, H., Kawase, T., Friend, R.H., Shimoda, T., Inbasekaran, M., Wu, W., and Woo, E.P. (2000) *Science*, **290**, 2123; (c) Rogers, J.A., Bao, Z., Baldwin, K., Dodabalapur, A., Crone, B., Raju, V.R., Kuck, V., Katz, H., Amundson, K., Ewing, J., and Drzaic, P. (2001) *Proc. Natl. Acad. Sci. U.S.A.*, **98**, 4835; (d) Rogers, J.A., Bao, Z., Meier, M., Dodabalapur, A., Schueller, O.J.A., and Whitesides, G.M. (2000) *Synth. Met.*, **115**, 5.
9. Gamota, D., Brazis, P., Kalyanasundaram, K., and Zhang, J. (2004) *Printed Organic And Molecular Electronics*, Kluwer Academic Publishers, Norwell.
10. Sirringhaus, H. (2009) *Nature*, **457**, 667.
11. (a) Dhoot, A.S., Yuen, J.D., Heeney, M., McCulloch, L., Moses, D., and Heeger, A.J. (2006) *Proc. Natl. Acad. Sci. U.S.A.*, **03**, 11834; (b) Pan, H., Li, Y., Wu, Y., Liu, P., Ong, B.S., Zhu, S., and Xu, G. (2007) *J. Am. Chem. Soc.*, **129**, 4112; (c) Mcculloch, I., Heeney, M., Bailey, C., Genevicius, K., Macdonald, I., Shkunov, M., Sparrowe, D., Tierney, S., Wagner, R., Zhang, W., Chabinyc, M.L., Kline, R.J., Mcgehee, M.D., and Toney, M.F. (2006) *Nat. Mater.*, **5**, 328.
12. (a) Sirringhaus, H. (2005) *Adv. Mater.*, **17**, 2411; (b) Halik, M., Klauk, H., Zschieschang, U., Schmid, G., Ponomarenko, S., Kirchmeyer, S., and Weber, W. (2003) *Adv. Mater.*, **15**, 917; (c) Sirringhaus, H., Tessler, N., and Friend, R.H. (1998) *Science*, **80**, 1741.
13. (a) Gamota, D.R., Brazis, P., Kalyanasundaram, K., and Zhang, J. (2004) *Printed Organic and Molecular Electronics*, Kluwer Academic Publishers, New York; (b) Dimitrakopoulos, C.D., Purushothaman, S., Kymissis, J., Callegari, A., and Shaw, J.M. (1999) *Science*, **283**, 822.
14. Facchetti, A., Yoon, M.H., and Marks, T.J. (2005) *Adv. Mater.*, **17**, 1705.
15. (a) Huang, D., Liao, F., Molesa, S., Redinger, D., and Subramanian, V.J. (2003) *Electrochem. Soc.*, **150**, G412; (b) Loo, Y.L., Someya, T., Baldwin, K.W., Bao, Z.N., Ho, P., Dodabalapur, A., Katz, H.E., and Rogers, J.A. (2002) *Proc. Natl. Acad. Sci. U.S.A.*, **99**, 10252.
16. de Leeuw, D.M., Simenon, M.M.J., Brown, A.R., and Einerhand, R.E.F. (1997) *Synth. Met.*, **87**, 53.
17. (a) Jones, B.A., Facchetti, A., Marks, T.J., and Wasielewski, M.R. (2007) *Chem. Mater.*, **19**, 2703; (b) See, K.C., Landis, C., Sarjeant, A., and Katz, H.E. (2009) *Chem. Mater.*, **20**, 3609; (c) Jung, B.J., Sun, J., Lee, T., Sarjeant, A., and Katz, H.E. (2009) *Chem. Mater.*, **21**, 94; (d) Jones, B.A., Ahrens, M.J., Yoon, M.H., Facchetti, A., Marks, T.J., and Wasielewski, M.R. (2004) *Angew. Chem. Int. Ed.*, **43**, 6363; (e) Piliego, C., Jarzab, D., Gigli, G., Chen, Z., Facchetti, A., and Loi, M.A. (2009) *Adv. Mater.*, **21**, 1573.
18. (a) Chikamatsu, M., Itakura, A., Yoshida, Y., Azumi, R., and Yase, K. (2009) *Chem. Mater.*, **20**, 7365; (b) Letizia, J.A., Facchetti, A., Stern, C.L., Ratner, M.A., and Marks, T.J. (2005) *J. Am. Chem. Soc.*, **127**, 13476; (c) Yoon, M.H., Facchetti, A., Stern, C.E., and Marks, T.J. (2006) *J. Am. Chem. Soc.*, **128**, 5792; (d) Handa, S., Miyazaki, E., Takimiya, K., and Kunugi, Y. (2007) *J. Am. Chem. Soc.*, **129**, 11684.
19. Tang, M.L., Oh, J.H., Reichardt, A.D., and Bao, Z. (2009) *J. Am. Chem. Soc.*, **131**, 3733.
20. Yamashita, Y. (2009) *Chem. Lett.*, **38** (9), 870.
21. Ishii, H., Sugiyama, K., Ito, E., and Seki, K. (1999) *Adv. Mater.*, **11**, 605.
22. (a) Hill, I.G. and Kahn, A. (1998) *Proc. SPIE*, **168**, 3476; (b) Hirose, Y., Kahn, A., Aristov, V., Soukiassian, P., Bulovic, V., and Forrest, S.R. (1996) *Phys. Rev. B*, **54**, 13748.
23. Dodabalapur, A., Laquindanum, J., Katz, H.E., and Bao, Z. (1996) *Appl. Phys. Lett.*, **69** (27), 4227.
24. Wen, Y., Liu, Y., Di, C.A., Wang, Y., Sun, X., Guo, Y., Zheng, J., Wu, W.,

Ye, S., and Yu, G. (2009) *Adv. Mater.*, **21**, 1631.
25. Anthopoulos, T.D., Singh, B., Marjanovic, N., Sarciftci, N.S., Montaigne, A., Sitter, H., Cölle, M., and de Leeuw, D.M. (2006) *Appl. Phys. Lett.*, **89**, 213504.
26. Wada, H., Shibata, K., Bando, Y., and Mori, T. (2008) *J. Mater. Chem.*, **18**, 4165.
27. Wada, H. and Mori, T. (2008) *Appl. Phys. Lett.*, **93**, 213303.
28. (a) Di, C.-A., Wei, D., Yu, G., Liu, Y., Guo, Y., and Zhu, D. (2008) *Adv. Mater.*, **20**, 3289; (b) Kim, K.S., Zhao, Y., Jang, H., Lee, S.Y., Kim, J.M., Kim, K.S., Ahn, J.H., Kim, P., Choi, J.-Y., and Hong, B.H. (2009) *Nature*, **457**, 706.
29. Wang, X., Zhi, L., Tsao, N., Tomović, Z., Li, J., and Müllen, K. (2008) *Angew. Chem. Int. Ed.*, **47**, 2990.
30. (a) Cho, J.H., Lee, J., Xia, Y., Kim, B., He, Y., Renn, M.J., Lodge, T.P., and Frisbie, C.D. (2008) *Nat. Mater.*, **7**, 900; (b) Ko, S.H., Pan, H., Grigoropoulos, C.P., Luscombe, C.K., Fréchet, J.M.J., and Poulikakos, D. (2007) *Appl. Phys. Lett.*, **90**, 141103; (c) Kim, D., Jeong, S., Lee, S., Park, B.K., and Moon, J. (2007) *Thin Solid Films*, **515**, 7692; (d) Wu, Y., Li, Y., and Ong, B.S. (2007) *J. Am. Chem. Soc.*, **129**, 1862.
31. (a) Shtein, M., Mapel, J., Benziger, J.B., and Forrest, S.R. (2002) *Appl. Phys. Lett.*, **81**, 268; (b) Sirringhaus, H. (2005) *Adv. Mater.*, **17**, 2411; (c) Facchetti, A., Yoon, M.-H., and Marks, T.J. (2005) *Adv. Mater.*, **17**, 1705.
32. Yoon, M.H., Kim, C., Facchetti, A., and Marks, T.J. (2006) *J. Am. Chem. Soc.*, **128**, 12851.
33. (a) Chua, L.L., Zaumseil, J., Chang, J.-F., Ou, E.C.-W., Ho, P.K.-H., Sirringhaus, H., and Friend, R.H. (2005) *Nature*, **434**, 194; (b) Chen, F.C. and Liao, C.H. (2008) *Appl. Phys. Lett.*, **93**, 103310.
34. Kobayashi, S., Nishikawa, T., Takenobu, T., Mori, S., Shimoda, T., Mitani, T., Shimotani, H., Yoshimoto, N., Ogawa, S., and Iwasa, Y. (2004) *Nat. Mater.*, **3**, 317.
35. Itaka, K., Yamashiro, M., Yamaguchi, J., Haemori, M., Yaginuma, S., Matsumoto, Y., Kondo, M., and Koinuma, H. (2006) *Adv. Mater.*, **18**, 1713.
36. Guillaud, G., Al Sadound, M., and Maitrot, M. (1990) *Chem. Phys. Lett.*, **67**, 503.
37. Bao, Z., Lovinger, A.J., and Brown, J. (1998) *J. Am. Chem. Soc.*, **120** (1), 207–208.
38. (a) Tang, Q., Jiang, L., Tong, Y., Li, H., Liu, Y., Wang, Z., Hu, W., Liu, Y., and Zhu, D. (2008) *Adv. Mat.*, **20**, 2497; (b) Reese, C. and Bao, Z. (2007) *Mater. Today*, **10** (3), 20; (c) de Boer, R.W.I., Gershenson, M.E., Morpurgo, A.F., and Podzorov, V. (2005) *Phys. Org. Semicond.*, **3**, 393.
39. (a) Tang, Q.X., Li, H.X., Song, Y.B., Xu, W., Hu, W.P., Jiang, L., Liu, Y., Wang, X., and Zhu, D.B. (2006) *Adv. Mater.*, **18**, 3010; (b) Tang, Q.X., Li, H.X., Liu, Y.L., and Hu, W.P. (2006) *J. Am. Chem. Soc.*, **128**, 4634; (c) Tang, Q., Jiang, L., Tong, Y., Li, H., Liu, Y., Wang, Z., Hu, W., Liu, Y., and Zhu, D. (2008) *Adv. Mater.*, **20**, 2947–2951.
40. Ling, M.M., Bao, Z., and Erk, P. (2006) *Appl. Phys. Lett.*, **89**, 163516.
41. Wang, H., Zhu, F., Yang, J., Geng, Y., and Yan, D. (2007) *Adv. Mater.*, **19**, 2168.
42. Song, D., Wang, H., Zhu, F., Yang, J., Tian, H., Geng, Y., and Yan, D. (2008) *Adv. Mater.*, **20** (11), 2142.
43. Facchetti, A., Yoon, M.H., Yan, H., and Marks, T.J. (2005) *J. Am. Chem. Soc.*, **127**, 10388.
44. Klauk, H., Zschieschang, U., Pflaum, J., and Halik, M. (2007) *Nature*, **445** (7129), 745.
45. Sekitani, T., Noguchi, Y., Zschieschang, U., Klauk, H., and Someya, T. (2008) *Proc. Natl. Acad. Sci. U.S.A.*, **105** (13), 4976.
46. (a) Facchetti, A., Mushrush, M., Yoon, M.-H., Hutchison, G.R., Ratner, M.A., and Marks, T.J. (2004) *J. Am. Chem. Soc.*, **126**, 13859; (b) Facchetti, A., Yoon, M.-H., Stern, C.L., Hutchison, G.R., Ratner, M.A., and Marks, T.J. (2004) *J. Am. Chem. Soc.*, **126**, 13480; (c) Facchetti, A.,

Mushrush, M., Katz, H.E., and Marks, T.J. (2003) *Adv. Mater.*, **15**, 33; (d) Facchetti, A., Deng, Y., Wang, A., Koide, Y., Sirringhaus, H., Marks, T.J., and Friend, R.H. (2000) *Angew. Chem., Int. Ed.*, **39**, 4547; (e) Casado, J., Ponce Ortíz, R., Hernández, V., López Navarrete, J.T., Facchetti, A., and Marks, T.J. (2005) *J. Am. Chem. Soc.*, **127**, 13364.

47. Bunn, C.W. and Howells, E.R. (1954) *Nature*, **174**, 549.
48. (a) Barbarella, G., Zambianchi, M., Bongini, A., and Antolini, L. (1992) *Adv. Mater.*, **4**, 282; (b) Herrema, J.K., Wildeman, J., van Bolhuis, F., and Hadziioannou, G. (1993) *Synth. Met.*, **60**, 239; (c) Dihexyl-α6T: Sato, T., Fujitsuka, M., Shiro, M., and Tanaka, K. (1998) *Synth. Met.*, **95**, 143.
49. (a) Yoon, M.H., Facchetti, A., Stern, C.F., and Marks, T.J. (2006) *J. Am. Chem. Soc.*, **128**, 5792; (b) Facchetti, A., Yoon, M.H., Stern, C.L., Katz, H.E., and Marks, T.J. (2003) *Angew. Chem., Int. Ed.*, **42**, 3900.
50. Koh, S.E., Delley, B., Medvedeva, J.E., Facchetti, A., Freeman, A.J., Marks, T.J., and Ratner, M.A. (2006) *J. Phys. Chem. B*, **110**, 24361.
51. (a) Yoon, M.H., DiBenedetto, S.A., Russell, M.T., Facchetti, A., and Marks, T.J. (2007) *Chem. Mater.*, **19**, 4864; (b) Yoon, M.-H., DiBenedetto, S., Facchetti, A., and Marks, T.J. (2005) *J. Am. Chem. Soc.*, **127**, 1348.
52. (a) Chiechi, R.C., Sonmez, G., and Wudl, F. (2005) *Adv. Funct. Mater.*, **15**, 427; (b) Donat-Bouillud, A., Mazerolle, L., Gagnon, P., Goldenberg, L., Petty, M.C., and Leclerc, M. (1997) *Chem. Mater.*, **9**, 2815.
53. Schols, S., Willigenburg, L.V., Müller, R., Bode, D., Debucquoy, M., Jonge, S.D., Genoe, J., Heremans, P., Lu, S., and Facchetti, A. (2008) *Appl. Phys. Lett.*, **93**, 263303.
54. Letizia, J.A., Facchetti, A., Stern, C.L., Ratner, M.A., and Marks, T.J. (2005) *J. Am. Chem. Soc.*, **127**, 13476.
55. Letizia, J.A., Rivnay, J., Facchetti, A., Ratner, M.A., and Marks, T.J. (2010) *Adv. Funct. Mater.*, **20**, 50–58.
56. Pappenfus, T.M. *et al.* (2002) *J. Am. Chem. Soc.*, **124**, 4184.
57. Casado, J. *et al.* (2002) *J. Am. Chem. Soc.*, **124**, 12380.
58. Chesterfield, R.J. (2003) *Adv. Mater.*, **15**, 1278.
59. Handa, S., Miyazaki, E., Takimiya, K., and Kunugi, Y. (2007) *J. Am. Chem. Soc.*, **129**, 11684.
60. Kastner, J., Paloheimo, J., and Kuzmany, H. (1993) *Solid State Sciences*, Springer, New York.
61. Haddon, R.C., Perel, A.S., Morris, R.C., Palstra, T.T.M., Hebard, A.F., and Fleming, R.M. (1995) *Appl. Phys. Lett.*, **67** (1), 121.
62. Haddon, R.C. (1996) *J. Am. Chem. Soc.*, **118** (12), 3041.
63. Jang, J., Kim, J.W., Park, N., and Kim, J.J. (2008) *Org. Electron.*, **9**, 481.
64. Zhang, X.H. and Kippelen, B. (2008) *Appl. Phys. Lett.*, **93**, 133305.
65. Chikamatsu, M., Nagamatsu, S., Yoshida, Y., Saito, K., Yase, K., and Kikuchi, K. (2005) *Appl. Phys.Lett.*, **87**, 203504.
66. Wöbkenberg, P.H., Ball, J., Bradley, D.D.C., Anthopoulos, T.D., Kooistra, F., Hummelen, J.C., and de Leeuw, D.M. (2008) *Appl. Phys. Lett.*, **92**, 143310.
67. Chikamatsu, M., Itakura, A., Yoshida, Y., Azumi, R., and Yase, K. (2008) *Chem. Mater.*, **20**, 7365.
68. Wobkenberg, P.H., Bradley, D.H., Kronholm, D., Hummelen, J.C., de Leeuw, D.M., Colle, M., and Anthopoulos, T.D. (2008) *Synth. Met.*, **158** (11), 468.
69. Yang, C., Cho, S., Heeger, A.J., and Wudl, F. (2009) *Angew. Chem. Int. Ed.*, **48** (9), 1592.
70. Kanbara, T., Shibata, K., Fujiki, S., Kubozono, Y., Kashino, S., Urisu, T., Sakai, M., Fujiwara, A., Kumashiro, R., and Tanigaki, K. (2003) *Chem. Phys. Lett.*, **379** (3–4), 223.
71. Kitamura, M. and Arakawa, Y. (2007) *Appl. Phys. Lett.*, **91** (5), 53505.
72. Zhang, X.-H., Potscavage, W.J. Jr., Choi, S., and Kippelen, B. (2009) *Appl. Phys. Lett.*, **94** (4), 43312.
73. Zheng, Q., Huang, J., Sarjeant, A., and Katz, H.E. (2008) *J. Am. Chem. Soc.*, **130**, 14410.

74. Laquindanum, J.G., Katz, H.E., Dodabalapur, A., and Lovinger, A.J. (1996) *J. Am. Chem. Soc.*, **118**, 11331.
75. Katz, H.E., Johnson, J., Lovinger, A.J., and Li, W. (2000) *J. Am. Chem. Soc.*, **122**, 7787.
76. Katz, H.E., Lovinger, A.J., Johnson, J., Kloc, C., Siegrist, T., Li, W., Lin, Y.-Y., and Dodabalapur, A. (2000) *Nature*, **404**, 478.
77. Jung, B.J., Sun, J., Lee, T., Sarjeant, A., and Katz, H.E. (2009) *Chem. Mater.*, **21**, 94.
78. Shukla, D., Nelson, S.F., Freeman, D.C., Rajeswaran, M., Ahearn, W.G., Meyer, D.M., and Carey, J.T. (2008) *Chem. Mater.*, **20**, 7486.
79. Jones, B.A., Facchetti, A., Marks, T.J., and Wasielewski, M.R. (2007) *Chem. Mater.*, **19**, 2703.
80. Wang, Z., Kim, C., Facchetti, A., and Marks, T.J. (2007) *J. Am. Chem. Soc.*, **129**, 13362.
81. Horowitz, G., Kouki, F., Spearman, P., Fichou, D., Nogues, C., Pan, X., and Garnier, F. (1996) *Adv. Mater.*, **8**, 242.
82. Malenfant, P.R.L., Dimitrakopoulos, C.D., Gelorme, J.D., Kosbar, L.L., Graham, T.O., Curioni, A., and Andreoni, W. (2002) *Appl. Phys. Lett.*, **80**, 2517.
83. Chesterfield, R.J., McKeen, J.C., Newman, C.R., Ewbank, P.C., da Silva Filho, D.A., Brédas, J.L., Miller, L.L., Mann, K.R., and Frisbie, C.D. (2004) *J. Phys. Chem. B*, **108**, 19281.
84. Chen, F.-C. and Liao, C.H. (2008) *Appl. Phys. Lett.*, **93**, 103310.
85. Tatemichi, S., Ichikawa, M., Koyama, T., and Taniguchi, Y. (2006) *Appl. Phys. Lett.*, **89**, 112108.
86. Oh, J.H., Liu, S., Bao, Z., Schmidt, R., and Würthner, F. (2007) *Appl. Phys. Lett.*, **91**, 212107.
87. (a) Jones, B.A., Ahrens, M.J., Yoon, M.H., Facchetti, A., Marks, T.J., and Wasielewski, M.R. (2004) *Angew. Chem., Int. Ed.*, **43**, 6363; (b) Jones, B. A., Facchetti, A., Wasielewski, M. R., Marks, T. J. (2007) *J. Am. Chem. Soc.*, **129**, 15259–15278.
88. Jones, B.A., Facchetti, A., Wasielewski, M.R., and Marks, T.J. (2008) *Adv. Funct. Mater.*, **18**, 1329.
89. Molinari, A.S., Alves, H., Chen, Z., Facchetti, A., and Morpurgo, A.F. (2009) *J. Am. Chem. Soc.*, **131**, 2462.
90. Weitz, R.T., Amsharov, K., Zschieschang, U., Villas, E.B., Goswami, D.K., Burghard, M., Dosch, H., Jansen, M., Kern, K., and Klauk, H. (2008) *J. Am. Chem. Soc.*, **130**, 4637.
91. Chen, Z., Debije, M.G., Debaerdemaeker, T., Osswald, P., and Würthner, F. (2004) *Chem. Phys. Chem.*, **5**, 137.
92. Schmidt, R., Oh, J.H., Sun, Y.S., Deppisch, M., Krause, A.-M., Radacki, K., Braunschweig, H., Könemann, M., Erk, P., Bao, Z., and Würthner, F. (2009) *J. Am. Chem. Soc.*, **131**, 6215.
93. (a) Nolde, F., Pisula, W., Mueller, S., Kohl, C., and Müllen, K. (2006) *Chem. Mater.*, **18**, 3715; (b) Nolde, F., Qu, J., Kohl, C., Pschirer, N.G., Reuther, E., and Müllen, K. (2005) *Chemistry*, **11**, 3959.
94. Petit, M., Hayakawa, R., Shirai, Y., Wakayama, Y., Hill, J.P., Ariga, K., and Chikyow, T. (2008) *Appl. Phys. Lett.*, **92**, 163301.
95. Ling, M.-M., Bao, Z., Erk, P., Koenemann, M., and Gomez, M. (2006) *Appl. Phys. Lett.*, **88**, 082104.
96. Yoo, B. Madgavkar, A., Jones B.A., Nadkarni, S., Facchetti, A., Dimmler, K., Wasielwski, M.R., Marks, T.J., and Dodabalapur A. (2006) *IEEE Electron Device Lett.*, **27** (9), 737.
97. Ling, M.-M., Bao, Z., Erik, P., and Koenemann, M. (2007) *Appl. Phys. Lett.*, **90**, 093508.
98. (a) Piliego, C., Cordella, F., Jarzab, D., Lu, S., Chen, Z., Facchetti, A., and Loi, M.A. (2009) *Appl. Phys. A: Mater. Sci. Proc.*, **95** (1), 303; (b) Piliego, C., Jarzab, D., Gigli, G., Chen, Z., Facchetti, A., and Loi, M.A. (2009) *Adv. Mater.*, **21**, 1573.
99. Yan, H., Zheng, Y., Blache, R., Newman, C., Lu, S., Woerle, J., and Facchetti, A. (2008) *Adv. Mater.*, **20**, 3393.
100. Ng, T.N., Sambandan, S., Lujan, R., Arias, A.C., Newman, C.R., Yan, H., and Facchetti, A. (2009) *Appl. Phys. Lett.*, **94**, 233307.

101. (a) Ando, S., Nishida, J.-I., Fujiwara, E., Tada, H., Inoue, Y., Tokito, S., and Yamashita, Y. (2005) *Chem. Mater.*, **17**, 1261; (b) Ando, S., Nishida, J.-i., Fujiwara, E., Tada, H., Inoue, Y., Tokito, S., and Yamashita, Y. (2005) *J. Am.Chem. Soc.*, **127**, 5336; (c) Ando, S., Murakami, R., Nishida, J.-i., Fujiwara, E., Tada, H., Inoue, Y., Tokito, S., and Yamashita, Y. (2005) *J. Am. Chem. Soc.*, **127**, 14996; (d) Naraso, J.-i., Nishida, D., Kumaki, S., Tokito, S., and Y. Yamashita (2006) *J. Am. Chem. Soc.*, **128**, 9598; (e) Nakagawa, T., Nishida, J.-I., Tokito, S., and Yamashita, Y. (2008) *Chem. Mater.*, **20**, 2015; (f) Ono, K., Yamaguchi, H., Taga, K., Saito, K., Nishida, J.-i., and Yamashita, Y. (2009) *Org. Lett.*, **11**, 149.

102. Sakamoto, Y., Suzuki, T., Kobayashi, M., Gao, Y., Fukai, Y., Inoue, Y., Sato, F., and Tokito, S. (2004) *J. Am. Chem. Soc.*, **126**, 8138.

103. Di, C.-A., Li, J., Yu, G., Xiao, Y., Guo, Y., Liu, Y., Qian, X., and Zhu, D. (2008) *Org. Lett.*, **10**, 3025.

104. Usta, H., Facchetti, A., and Marks, T.J. (2008) *J. Am. Chem. Soc.*, **130**, 8580.

105. Babel, A. and Jenekhe, S.A. (2003) *J. Am. Chem. Soc.*, **125**, 13656.

106. Briseno, A.L., Mannsfeld, S.C.B., Shamberger, P.J., Ohuchi, F.S., Bao, Z., Jenekhe, S.A., and Xia, Y. (2008) *Chem. Mater.*, **20** (14), 4712.

107. Blanco, R., Gomez, R., Seoane, C., Segura, J.L., Mena-Osteritz, E., and Baeuerle, P. (2007) *Org. Lett.*, **9**, 2171.

108. Letizia, J., Salata, M., Tribout, C., Facchetti, A., Ratner, M.A., and Marks, T.J. (2008) *J. Am. Chem. Soc.*, **130**, 9679.

109. Zhan, X., Tan, Z., Domercq, B., An, Z., Zhang, X., Barlow, S., Li, Y., Zhu, D., Kippelen, B., and Marder, S.R. (2007) *J. Am. Chem. Soc.*, **129**, 7246.

110. Zhan, X., Tan, Z., Zhou, E., Li, Y., Misra, R., Grant, A., Domercq, B., Zhang, X., An, Z., Zhang, X., Barlow, S., Kippelen, B., and Marder, S.R. (2009) *J. Mater. Chem.*, **19**, 5794.

111. Hüttner, S., Sommer, M., and Thelakkat, M. (2008) *Appl. Phys. Lett.*, **92**, 093302.

112. Chen, Z., Zheng, Y., Yan, H., and Facchetti, A. (2009) *J. Am. Chem. Soc.*, **131**, 8.

113. Yan, H., Chen, Z., Zheng, Y., Newman, C.E., Quin, J., Dolz, F., Kastler, M., and Facchetti, A. (2009) *Nature*, **457**, 679.

114. Baeg, K.-J., Khim, D., Kim, D.-Y., Jung, S.-W., Koo, J.B., You, I.-K., Yan, H., Facchetti, A., and Noh, Y.-Y. (2010) *J. Poly. Sci. B: Poly. Phys.*, in press.

115. Sheats, J.R. (2004) *J. Mater. Res.*, **19**, 1974.

116. (a) DiBenedetto, S.A., Facchetti, A., Ratner, M.A., and Marks, T.J. (2009) *Adv. Mater.*, **21**, 1407; (b) Someya, T., Pal, B., Huang, J., and Katz, H.E. (2008) *MRS Bull.*, **33** (7), 690; (c) Sirringhaus, H. and Ando, M. (2008) *MRS Bull.*, **33** (7), 676; (d) Loo, Y.-L. and McCulloch, I. (2008) *MRS Bull.*, **33** (7), 653.

6
Low-Voltage Electrolyte-Gated OTFTs and Their Applications

Yu Xia and C. Daniel Frisbie

6.1
Overview

For many conceivable applications of organic electronics, it is desirable to operate circuits at low voltages, as has been pointed out in previous chapters. Wearable biometric devices and smart drug delivery patches, for example, will likely be powered by thin-film batteries that deliver current at 1.5 V or so. While current-to-voltage converters can be used to step up the battery supply voltage, it will be simpler and potentially more power efficient to design organic circuits that can operate directly at low supply biases.

In light of this goal, high-capacitance gate insulators for organic thin-film transistors (OTFTs) have drawn increased attention in organic electronics in order to realize sufficient channel conductivity under low-voltage operation. The specific capacitance (F m^{-2}) of the gate dielectric in an OTFT can be expressed as: $C' = k\varepsilon_0/d$, where k is the dielectric constant, ε_0 is the vacuum permittivity, and d is the thickness of the dielectric layer [1]. Therefore, in order to increase the gate capacitance and to demonstrate low-voltage OTFTs, researchers have utilized high dielectric constant (high-k) materials [2–6] and ultrathin (small d) films including self-assembled monolayers/multilayers and cross-linked polymers [7–12] as gate insulators.

This chapter discusses another strategy, namely the use of solid electrolytes as high-capacitance gate dielectrics. Semiconductor gating with electrolytes is an old idea [13], but recent progress in printable, highly polarizable solid polymer electrolytes has enabled the demonstration of low-voltage digital circuits operating with sub-millisecond signal propagation delay times. Furthermore, electrolyte-gated transistors have well-recognized potential as chemical sensors, which can facilitate the expansion of organic electronics into biodetection and diagnostics [14, 15]. The large capacitance of electrolytes (~10 μF cm^{-2}) also enables very large charge densities to be achieved in organic semiconductors which is important for fundamental transport investigations. Electrolyte gating is thus very attractive from several perspectives and interest in this approach is growing rapidly [16].

Organic Electronics II: More Materials and Applications, First Edition. Edited by Hagen Klauk.

The chapter begins with an introduction to the structure and operation of electrolyte-gated transistors and then briefly reviews the historical development of these devices. Considerable space is devoted to electrical characterization and emerging applications.

6.2 Introduction to Electrolyte-Gated Organic Transistors

6.2.1 Structure and Operating Mechanisms

Figure 6.1 shows a scheme of an OTFT employing an electrolyte film as the gate dielectric layer. Electrolytes are ionic conductors and electronic insulators. Free ions are therefore available in the dielectric layer. When a gate voltage is applied, the ions in the electrolyte move to screen the field in the bulk of the electrolyte film; this electrolyte polarization gives rise to the accumulation of carriers in the semiconductor. However, there are two different scenarios for charge accumulation [17]. In one case, application of the gate voltage results in the formation of electrical double layers (Helmholtz layers) at the semiconductor/electrolyte and gate/electrolyte interfaces. In these double layers, induced charges in the semiconductor channel and

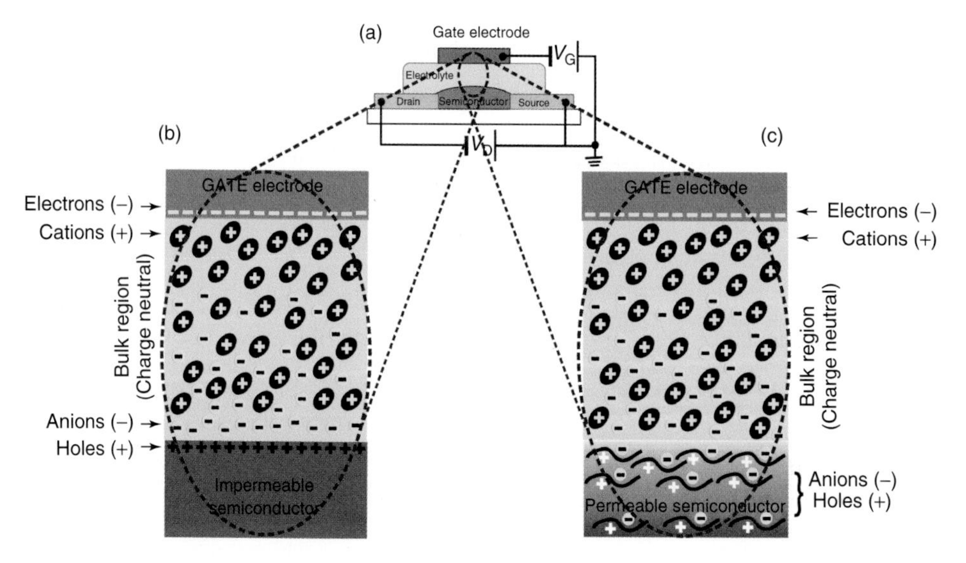

Figure 6.1 (a) Organic thin-film transistor (OTFT) employing an electrolyte film as the gate dielectric layer. The magnified scheme (b) illustrates the accumulation of ions and induced charge carriers (holes in this case) in an impermeable semiconductor when a gate bias is applied. This is the electrostatic or double-layer charging regime. The magnified scheme (c) illustrates the ion and carrier distributions in the electrochemical regime, that is, when the semiconductor is permeable to ions.

in the gate electrode, respectively, serve to balance the ionic charges. The double layers at the two interfaces can be viewed as ultrathin capacitors in series. The specific capacitance of each double layer is $C' \sim k\varepsilon_0/\lambda$, where λ is the screening length or effective thickness of the double layer, on the order of 1 nm. One can show that the expected capacitance is on the order of $1-10\,\mu\text{F cm}^{-2}$, which is easily several times larger than the capacitance achieved using ultrathin monolayers or high-k dielectrics [18, 19]. Note that the fields inside the double layers are very strong, on the order of $10^5-10^6\,\text{V cm}^{-1}$, so that one can view the double-layer charging mechanism as a variant of the usual field effect in conventional TFTs.

In the second scenario, termed *electrochemical gating*, a double layer is formed at the gate/electrolyte interface, but ions penetrate the semiconductor layer. In this case, one views the semiconductor charging process as a *reversible* electrochemical reaction in which the neutral semiconductor is electrochemically doped upon application of a gate voltage. Partially electrochemically oxidized organic semiconductors are hole conductors ("oxidation" in this case simply means hole injection); partially reduced organic semiconductors are electron conductors. It is clear that in electrochemical gating the solid-state packing of the semiconductor is disrupted, while in the double-layer gating mechanism the solid-state structure is not perturbed. As in rechargeable batteries, the electrochemical transformation is reversible in that removal of the gate bias restores the semiconductor to its undoped, OFF state. In the case of a p-type semiconductor, this occurs by diffusion of the anions out of the film or by diffusion of compensating cations into the film. Any structural disorder induced by the electrolyte penetration is not reversible, however.

For electrolyte-gated organic TFTs, the switching mechanism is often electrochemical because the ions can penetrate the organic semiconductor. However, electric double-layer gating is also possible. The dominance of one mechanism over the other depends on the permeability of the organic semiconductor, which in turn depends on the state of order (e.g., crystalline vs amorphous films), the size of the ions, the magnitude of the gate electric field, and the time scale of the gating process (fast vs slow switching). This will be discussed further in Section 6.2.3.

It is important to note that electrolyte-gated devices require the use of electrochemically stable source, drain, and gate electrodes. Typically, this means that these electrodes must be fabricated from noble metals (gold, platinum) or carbon-based conductors. Electropositive metals, such as silver or copper, are easily corroded or dissolved in electrolytes, particularly upon application of positive bias voltages. If such reactive metals are used, they must be isolated from the electrolyte, which can be done by changing the transistor geometry (e.g., a bottom contact geometry), but in practice, it is difficult to achieve complete isolation.

6.2.2
The Development of Electrolyte-Gated Transistors

The idea of using electrolytes as high-capacitance dielectrics can be traced back more than 50 years to shortly after the invention of the first transistor [13].

Electrolytes have been used to gate silicon devices (in the double-layer mode), and today, electrolyte gates are still being employed to examine charge transport phenomena in conventional semiconductors and semiconducting oxides [20–24]. The concept of electrolyte gating also facilitated the development of the so-called ion-sensitive field-effect transistors (ISFETs) for chemical sensing in the 1970s [25–27].

In the 1980s, soon after the discovery of the first conductive polymer [28], Wrighton and others developed microelectrochemical transistors (Figure 6.2) in which conductive polymers were doped and dedoped electrochemically [29–35]. As systematically studied by Wrighton *et al.*, semiconducting polymers such as polyaniline, polyacetylene, polythiophene, polypyrrole, and their derivatives could be electrochemically oxidized, or doped, in the presence of an electrolyte [29–35]. During oxidation (negative gate bias), positively charged, mobile holes were introduced into the conducting polymer backbones; this charge was compensated by anions that diffused into the polymers from the electrolyte. This process is reversible and an opposite gate voltage undoped the polymer back to its neutral, low conductance state. For poly(3-hexylthiophene) (P3HT)-based microelectrochemical transistors, this reversible electrochemical process can be expressed as: $\text{P3HT} + \text{A}^- \leftrightarrow [\text{P3HT}]^+\text{A}^- + \text{e}^-$(to electrode), where A^- is the anion from the electrolyte. The partially oxidized P3HT ($[\text{P3HT}]^+\text{A}^-$) is conductive. More recently, another electrochemical transistor demonstrated by Berggren *et al.* involves the reversible electrochemical dedoping of conducting poly(3,4-ethylenedioxythiophene):poly(styrenesulfonate) (PEDOT:PSS) [36–38]. The electrochemical process can be expressed as $\text{PEDOT:PSS} + \text{C}^+ + \text{e}^- \leftrightarrow \text{PEDOT}^0 + \text{C}^+ : \text{PSS}^-$, where C^+ is the cation from the electrolyte. A wide variety of electrochemical transistors have been developed for chemical and electrochemical sensor applications [14, 15], as will be detailed in Section 6.3.3.

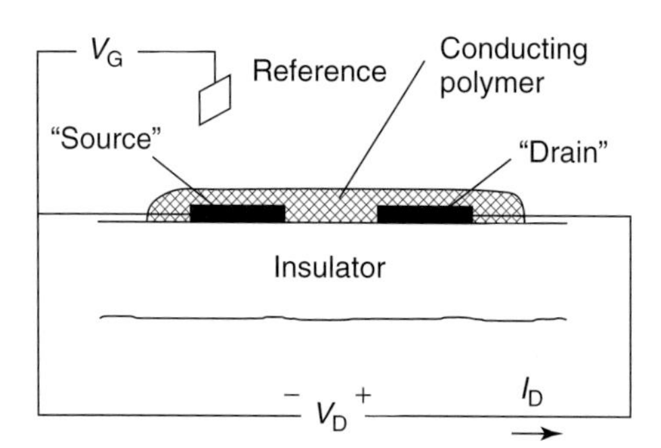

Figure 6.2 Configuration of a conducting polymer-based transistor immersed in electrolyte (not shown) that covers the conducting polymer as well as the gate (reference) electrode [35]. Adapted with permission from Ofer, D., Crooks, R.M., and Wrighton, M.S. Potential dependence of the conductivity of highly oxidized polythiophenes, polypyrroles, and polyaniline: finite windows of high conductivity. *J. Am. Chem. Soc.*, **112**, 7869–7879 (1990). Copyright 1990 American Chemical Society.

The early microelectrochemical transistors of the 1980s primarily used liquid electrolytes but it was recognized that solid-state devices employing solid polymer electrolytes were necessary for most practical applications [32, 39, 40]. In the 1980s, the best solid polymer electrolytes were based on the polymers poly(ethylene oxide) (PEO) or poly[bis(2-(2-methoxyethoxy)ethoxy)phosphazene] (MEEP) (Figure 6.3a) [32, 41]. Conventional salts such as $LiClO_4$ or $Li(CF_3SO_3)$ could be dissolved in these polymers to produce an electrolyte with modest room temperature ionic conductivity, on the order of $10^{-5}-10^{-4}$ S cm^{-1} [42, 43]. During transistor fabrication, the polymer electrolyte was cast from solution over the conducting polymer and the metal source, drain, and gate electrodes. The transistors exhibited switching and amplification, but the switching times were limited by the low conductivity (i.e., slow polarization response) of the solid electrolytes.

With the resurgence of interest in electrolyte gates in the last decade, polymer electrolytes, in particular $LiClO_4$ in PEO, have been widely adapted to gate a variety of novel high-quality organic semiconductors including organic single crystals [44, 45], small-molecule thin films [46, 47], carbon nanotubes [48], and polymers [49–54]. However, there has also been increased interest in developing new solid electrolytes that have higher ionic conductivity and shorter polarization times [55]. In particular, Berggren *et al.* introduced solid acidic polyelectrolytes based on immobile polyanions and fast, mobile protons for gating of OTFTs (Figure 6.3b) [56–59]. Österbacka *et al.* have also demonstrated the use of protonated polyvinylphenol (PVP) as a hygroscopic gate dielectric, in which water can be dissolved to facilitate proton conduction [60–63].

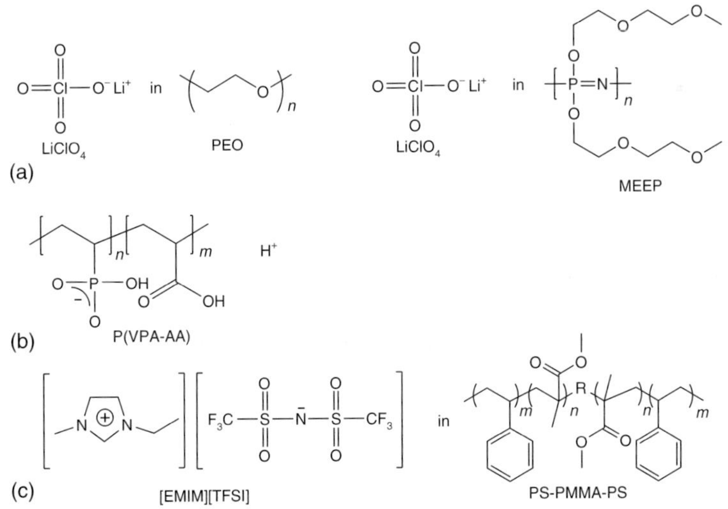

Figure 6.3 Molecular structure of the three commonly reported solid electrolytes: (a) $LiClO_4$ in PEO (polymer electrolyte) and $LiClO_4$ in MEEP, (b) poly(vinyl phosphonic acid-*co*-acrylic acid), P(VPA-AA) (polyelectrolyte), and (c) [EMIM][TFSI] in PS-PMMA-PS (ion gel).

With respect to electrolyte development, the advent of room temperature ionic liquids and especially rubbery ionic gels has also enhanced the development of electrolyte-gated OTFTs. Ionic liquids are liquid organic salts typically consisting of nitrogen-containing cations and organic or inorganic anions [64, 65]. They are generally chemically inert and they have very large ionic conductivities (up to 10^{-2} S cm^{-1}) and large electrochemical stability windows, which facilitate their use in electrochemistry and as gate dielectrics. They also have negligible vapor pressure at room temperature. Ionic liquids were used by MacFarlane and colleagues to gate conducting polymers in microelectrochemical transistors [66]. To facilitate their use in solid-state devices, ionic liquids can also be solidified by the addition of cross-linking polymer agents [67–70]. A particularly convenient solidification strategy pioneered by Lodge *et al.* is to add 5–10 wt% of a self-assembling triblock copolymer such as polystyrene-*b*-polymethylmethacrylate-*b*-polystyrene (PS-PMMA-PS) [71]. The PS blocks are insoluble in ionic liquids such as 1-ethyl-3-methylimidazolium [EMIM] [TFSI] (TFSI = bis(trifluoromethylsulfonyl)-imide) (Figure 6.3c) whereas the PMMA blocks are soluble. Addition of this triblock results in a rubbery network, a so-called ion gel consisting of glassy PS cores (the cross-links) and interconnecting polar PMMA chains that are swollen with the ionic liquid [71–74]. These ion gels have large ionic conductivities and high thermal stability, up to 200 °C. They are rubbery solids with a modulus of 10–30 kPa depending on the polymer content. Moreover, the gels can be printed or spin-coated by co-dissolving the ionic liquid and triblock polymer in a common co-solvent to make an "ink." Printing or coating of this solution, followed by solvent evaporation, results in the spontaneous formation of the gel. The ionic conductivity of the ion gels is higher ($\sim 10^{-3}$ S cm^{-1}) than PEO-based polymer electrolytes ($\sim 10^{-4}$–10^{-5} S cm^{-1}) [43, 72]. This high ionic conductivity favors both high capacitance and fast polarization, hence considerably shortening the response time of the electrolyte-gated transistors [75]. Recently, ion gels have been employed to gate various polymer transistors and circuits, with switching speeds as fast as 10 kHz, as will be discussed later [75–78]. Overall, improvements in polymer and gel electrolytes since the 1980s have enhanced the opportunities for designing fast and sensitive electrolyte-gated OTFTs.

6.2.3
More on the Gating Mechanism in Electrolyte-Gated Transistors

In electrolyte-gated transistors, the accumulation of ions at the semiconductor/electrolyte interface or within the semiconductor itself ultimately controls the density of the carriers in the channel (Figure 6.1). A key feature that separates electrochemical and electrostatic (double-layer) gating processes is the permeability of the semiconductor. For instance, on the surface of organic and inorganic single crystals, where ions cannot penetrate, the electrostatic process predominates [21–24, 44, 45, 79–83]. However, even for penetrable polymer semiconductors, both double-layer and electrochemical charging mechanisms can be operative depending on the timescale of the gating process [78, 84]. For example, Figure 6.4a shows the channel current (I_D) of an [EMIM][TFSI] ion-gel-gated P3HT transistor

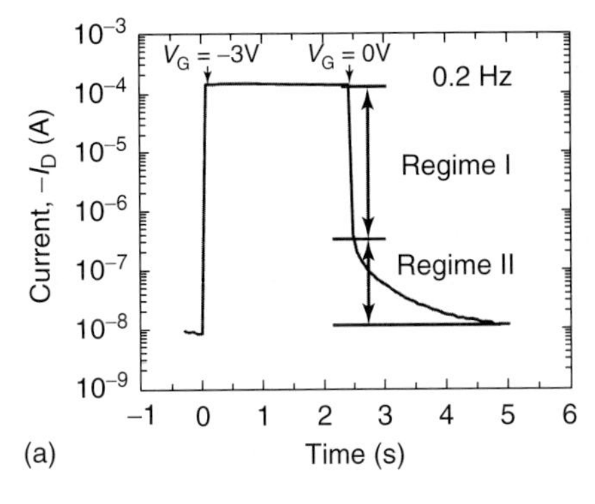

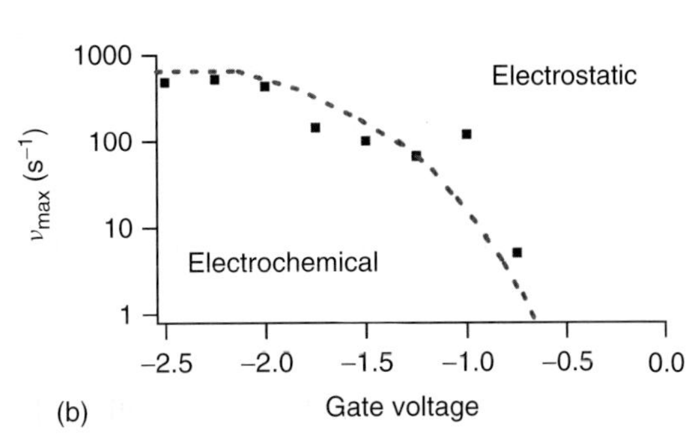

Figure 6.4 (a) The transient output current response of [EMIM][TFSI] ion-gel-gated P3HT transistors at 0.2 Hz square-wave gate voltage switching [76]. Cho, J.H., Le, J., He, Y., Kim, B., Lodge, T.P., and Frisbie, C.D. High-capacitance ion-gel-gate dielectrics with faster polarization response times for organic thin-film transistors. *Adv. Mater. (Weinheim, Ger.)*, **20**, 686–690 (2008). Copyright Wiley-VCH Verlag GmbH & Co. KGaA. Reproduced with permission.(b) Ion gel charging mechanisms as a function of transistor switching frequency (ν_{max}) and gate voltage [78]. Adapted with permission from Lee, J., Kaake, L.G., Cho, J.H., Zhu, X.Y., Lodge, T.P., and Frisbie, C.D. Ion-gel-gated polymer thin-film transistors: operating mechanism and characterization of gate dielectric capacitance, switching speed, and stability. *J. Phys. Chem. C*, **113**, 8972–8981 (2009). Copyright 2009 American Chemical Society.

that is switched ON and OFF [76]. As can be seen, as the gate voltage is switched OFF from −3 to 0 V, the decay of I_D can be divided into two regimes. First, there is a 1000-fold current drop within a few milliseconds, likely caused by the quick motion of the ions at the interface, that is, an electrostatic gating process. The current then undergoes a slow decay over several seconds, which is mainly attributed to electrochemical dedoping. A similar observation of two transient current modulation regimes was reported by Berggren *et al.* on other solid electrolyte-gated P3HT transistors [85]. It was found that timescales vary with the ionic conductivity in the electrolytes, as well as the diffusivity of different ions in various semiconductors. In general, since bulk diffusion is involved in the electrochemical gating process, it is considerably slower compared to electrostatic/double-layer gating.

In transistors based on P3HT and other polythiophene derivatives, the injection of holes can be monitored by examining the thiophene ring stretching mode using *in-situ* infrared (IR) spectroscopy such as attenuated total internal reflection (ATR) IR–near IR spectroscopy [52, 78]. The concentration of the holes can be analyzed to distinguish two different gating mechanisms, as the electrochemical bulk doping process induces a much higher number of holes compared to the interface electrostatic gating. It was found, as summarized in Figure 6.4b, that the degree of electrochemical doping depended strongly on the gate voltage (V_G) and the gate voltage switching frequency (υ) [78]. Higher V_G and lower υ tends to favor electrochemical doping, while low V_Gs and high υs favor the double-layer/electrostatic mechanism. Importantly, even under a gate voltage as high as −2.5 V, the electrostatic charging mechanism may still dominate if the switching frequency is sufficiently fast. IR spectroscopy has also been applied to study polymer electrolyte gating of *N*, *N′*-dioctyl-3,4,9,10-perylene tetracarboxylic diimide (PTCDI-C8), an n-channel organic semiconductor, leading to similar conclusions [86].

Another approach to effectively reduce the ion doping in p-channel semiconductors is to utilize electrolytes in which only one ion is mobile. An example is the polyelectrolyte proton conductor poly(vinyl phosphonic acid-*co*-acrylic acid) (P(VPA-AA)) investigated by Berggren [56–58]. In these electrolytes, the mobile protons are driven toward the gate electrode under a negative gate voltage, leaving immobile polyanions that are too large to penetrate through the semiconductor at least on short timescales. The measured switching response time for P(VPA-AA)-gated P3HT transistors was reported to be as short as 3.5 ms, which may be consistent with the formation of an electrical double layer at the dielectric/semiconductor interface [57]. Overall, increasing the transistor operation speed by decreasing the polarization time of the electrolyte remains a key challenge in the development of electrolyte-gated OTFTs.

6.2.4 Electrical Characterization of Electrolyte-Gated OTFTs

6.2.4.1 Low-Voltage Operation

Figure 6.5 shows a comparison of the transfer characteristics ($I_D - V_G$) for a polymer electrolyte ($LiClO_4$ in PEO)-gated P3HT transistor and a conventional silicon

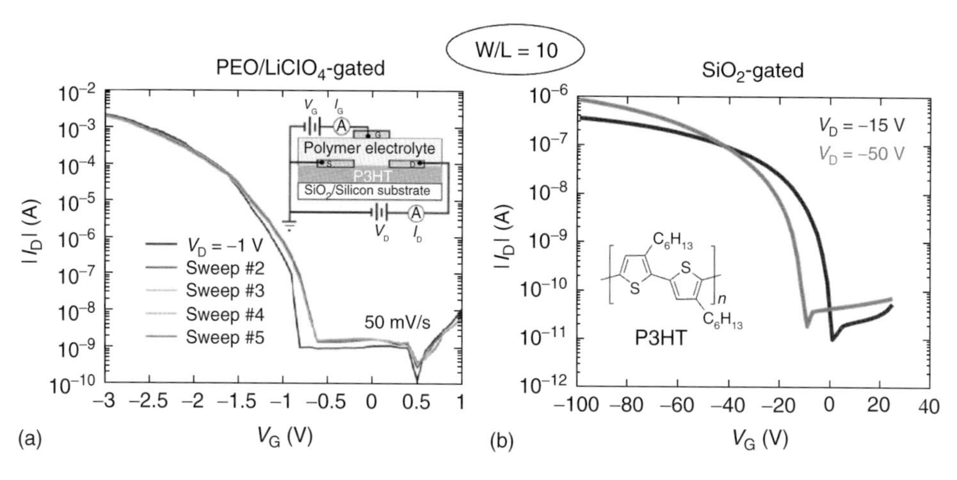

Figure 6.5 (a) Transfer characteristics of a polymer electrolyte-gated P3HT transistor. Five consecutive measurements are shown. The inset shows the cross-section of the transistor. (b) Transfer characteristics of a typical silicon dioxide-gated P3HT transistor. The SiO_2 is 300 nm thick [49]. The current and voltage scales are very different. Panzer, M.J. and Frisbie, C.D. (2006) High carrier density and metallic conductivity in poly(3-hexylthiophene) achieved by electrostatic charge injection. *Adv. Funct. Mater.*, **16**, 1051–1056. Copyright Wiley-VCH Verlag GmbH & Co. KGaA. Reproduced with permission.

dioxide-gated P3HT device [49]. The devices have the same width to length ratio ($W : L = 10$). It is clear that the electrolyte-gated transistor delivers orders of magnitude greater current at a gate bias of only -3 V and a drain bias of only -1 V, as compared to $V_G = -100$ V and $V_D = -50$ V for the SiO_2-gated transistor. The higher current and lower operating voltage both reflect the greatly increased capacitance of the electrolyte relative to SiO_2; more current flows in the electrolyte-gated device simply because there is much more charge in the channel. The comparison shown in Figure 6.5 is the principal motivation for developing electrolyte-gated OTFTs and circuits. These performance gains are universal features for various electrolyte-gated polymers [17, 35–39, 49–54, 56–62, 70, 75–78, 85, 87–94], organic small molecules [44–47, 55, 79–83], carbon nanotubes [48, 95–99], and inorganic transistors [20–22, 26, 55, 100].

6.2.4.2 Use of a Reference Electrode

The possibility of inserting a reference electrode into the gate dielectric is an additional unique aspect of the electrolyte-gated OTFTs that provide valuable information about the device operation. The reference electrode allows direct measurement of the electrochemical potential at the organic semiconductor/electrolyte interface while the device is operating [88]. Figure 6.6 shows the

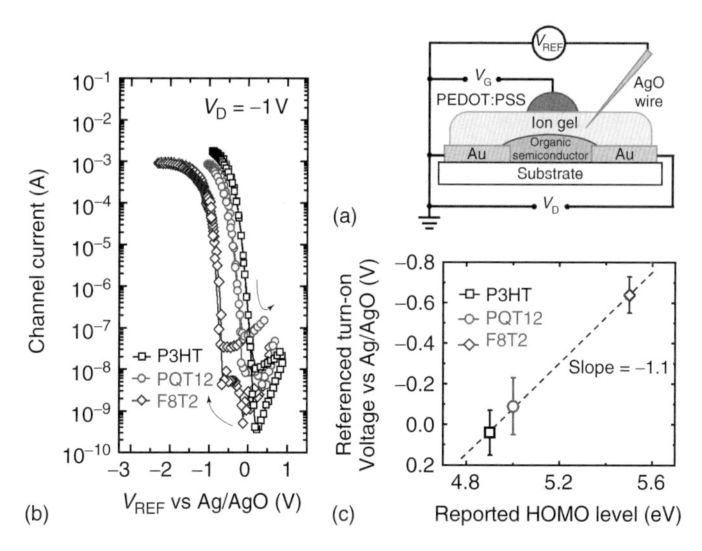

Figure 6.6 (a) Schematic cross-section of an ion-gel-gated transistor, with an oxidized silver wire inserted into the ion gel as quasi-reference electrode. (b) $I_D - V_{Ref}$ characteristics of ion-gel-gated P3HT, PQT-12, and F8T2 transistors, respectively. (c) Linear correlation between the referenced turn-on voltage of P3HT, PQT-12, and F8T2 transistors and the reported HOMO level (ionization potential) of these polymer semiconductors [88]. Reprinted with permission from Xia, Y., Cho, J., Paulsen, B., Frisbie, C.D., and Renn, M.J. Correlation of on-state conductance with referenced electrochemical potential in ion-gel-gated polymer transistors. *Appl. Phys. Lett.*, **94**, 013304/1–013304/3 (2009). Copyright 2009, American Institute of Physics.

I_D versus referenced gate voltage (V_{Ref}) characteristic for ion-gel-gated OTFTs based on P3HT, poly(3,3‴-didodecylquaterthiophene) (PQT12), and poly(9,9′-dioctylfluorene-*co*-bithiophene) (F8T2) [88]. Two important observations can be made about the $I_D - V_{Ref}$ characteristics in Figure 6.6b. First, the lack of hysteresis between forward and reverse sweeps indicates that on the timescale of the experiment (~60 s), the gating process at the semiconductor/electrolyte interface is highly reversible. Second, a clear linear correlation is observed between the referenced turn-on voltages and the reported HOMO levels of these organic semiconductors. This means that the referenced turn-on voltages reflect the ionization potential of the polymers, an observation that is not easily made with conventional OTFTs as the turn-on voltages are large because of the low gate capacitance. The correlation demonstrates that gating these polymers with the [EMIM][TFSI]-based electrolyte does not induce significant interface dipoles or trap states that perturb the original electronic structures of the semiconductors. In general, the option to insert a reference electrode opens opportunities to perform a kind of "transport spectroscopy" in the same sense that electrochemists employ cyclic voltammograms as a kind of electronic spectroscopy [101].

6.2.4.3 **Determination of Accumulated Charge**

The accurate determination of the channel charge concentration is important for understanding OTFT operation. In the case of electrolyte-gated OTFTs, it is especially important to confirm the carrier density as it is expected to be very high. The 2-D carrier density can be determined by $p' = C'(V_G - V_{TH})/e$, where C' is the 2-D sheet capacitance of the dielectric, V_{TH} is the threshold voltage, and e is the elementary charge. The key is thus to determine the value of C'. Figure 6.7 shows the

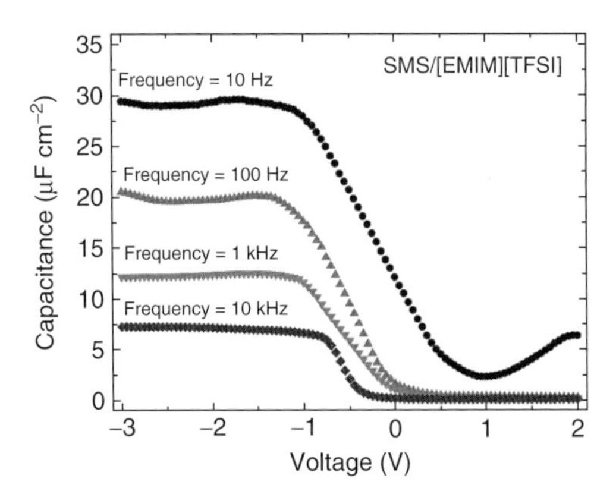

Figure 6.7 Capacitance–voltage characteristics of an ion gel ([EMIM][TFSI] in PS-PMMA-PS) dielectric at different frequencies measured using a p-Si/ion gel/metal testing structure [78]. Adapted with permission from Lee, J., Kaake, L.G., Cho, J.H., Zhu, X.Y., Lodge, T.P., and Frisbie, C.D. Ion-gel-gated polymer thin-film transistors: operating mechanism and characterization of gate dielectric capacitance, switching speed, and stability. *J. Phys. Chem. C*, **113**, 8972–8981 (2009). Copyright 2009 American Chemical Society.

$C-V$ characteristics for a p-Si/ion gel/metal sandwich structure [78]. Capacitance increases upon sweeping the bias voltage on the metal contact as expected because of the onset of hole accumulation in the p-Si substrate. Sheet capacitance as high as $30\,\mu\text{F cm}^{-2}$ is observed at -3 V and 10 Hz measurement frequency, but the capacitance values decrease with increasing measurement frequency, limited by the polarization response time of the ion motion. The strong frequency dependence of the capacitance illustrates the difficulty of estimating the carrier density in an electrolyte-gated OTFT. Ideally, one integrates the $C - V$ as $p' = \int C' \mathrm{d}V/e$ to get the hole density. However, because C is frequency (time) dependent, the accumulated charge in a transistor channel depends on the speed of the gate voltage sweep (or the gate modulation frequency).

An alternative approach is to estimate the carrier density by recording the gate displacement current (I_G) as a function of V_G, which can be measured simultaneously with the $I_D - V_G$ characteristics [102, 103]. As demonstrated in Figure 6.8, the turn-on of an [EMIM][TFSI]-based ion-gel-gated P3HT transistor correlates with a pronounced increase in I_G caused by a significant increase in displacement current according to:

$$\Delta I_G \approx \Delta I_{\text{Disp}} = \Delta \frac{\mathrm{d}Q}{\mathrm{d}t} = \Delta C' \frac{\mathrm{d}V_G}{\mathrm{d}t} \tag{6.1}$$

where $\Delta C'$ represents the change of capacitance when the transistor channel becomes conductive, and $\mathrm{d}V_G/\mathrm{d}t$ is the gate voltage sweep rate. The total injected carrier density in the channel can then be calculated from the integration of

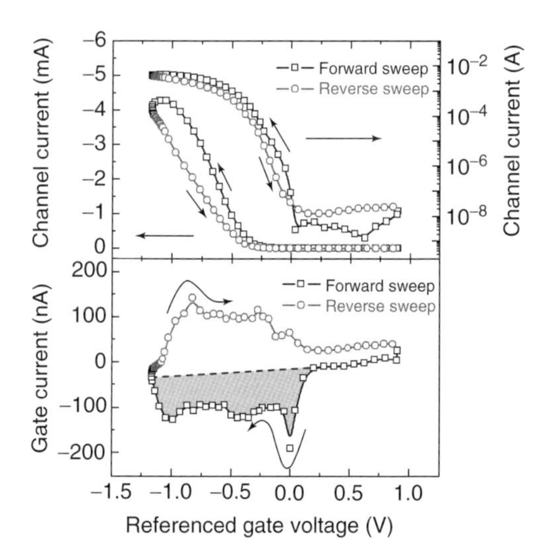

Figure 6.8 $I_D - V_{\text{Ref}}$ (a) and $I_G - V_{\text{Ref}}$ (b) characteristics measured simultaneously for an ion-gel-gated P3HT transistor, acquired under a gate voltage sweep rate of 75 mVs^{-1}. The shaded area in (b) indicates the total injected carrier density.

ΔI_{Disp} as

$$p' = \frac{Q}{eA} = \frac{\int \Delta I_{\text{Disp}} \times \mathrm{d}V_{\text{G}}}{eA \times \mathrm{d}V_{\text{G}}/\mathrm{d}t} \tag{6.2}$$

where A is the area of the transistor channel.

The displacement current approach has been adopted to analyze the hole density of various polymer transistors gated by different polymer electrolytes and ion gels [49, 50, 76, 78, 83, 88]. Values between 10^{14} and 10^{15} cm^{-2} under a few volts of gate bias were obtained. Note these "2-D" hole sheet densities also include the contribution of electrochemical doping; that is, the hole distribution in these devices is actually 3-D due to diffusion of ions into the polymer semiconductor film.

Nevertheless, the effective 2-D concentration can be used to calculate the mobility in the OTFT linear regime according to:

$$\mu = \frac{j_{\text{D}}}{ep'E} = \frac{I_{\text{D}}}{ep'} \times \frac{L}{WV_{\text{D}}} \tag{6.3}$$

Electrolyte-gated transistors yield hole mobilities for polymer semiconductors on the order of 1 cm^2 Vs^{-1} [83, 88]. In fact, using the standard field-effect transistor equations and estimating a reasonable average capacitance, similar high mobility values have been commonly reported as a typical figure of merit for electrolyte-gated polymer transistors [49–51, 61, 76, 77]. The origin of such high mobility will be discussed in detail in the next section but is related to the large gate-induced carrier densities. It is worth noting that I_{G} is actually a combination of currents originating from dielectric leakage, parasitic capacitance, as well as channel capacitance, all of which can be a function of gate voltage and operation speed. To be sure that one is sensitive to the channel capacitance, displacement currents should scale linearly with channel dimensions ($W \times L$). Also, the shape of the $I_{\text{D}} - V_{\text{G}}$ characteristic confirms the quality of the data (Figure 6.8).

Ultimately, the carrier density and mobility in electrolyte-gated transistors can be separately determined by means of Hall effect measurements. Iwasa *et al.* studied the Hall effect on $KClO_4$/PEO polymer electrolyte-gated ZnO transistors under various gate voltages [21], and they reported a 2-D electron density of $\sim 4 \times 10^{13}$ cm^{-2} under 1.2 V effective gate voltage (i.e., $V_{\text{G}} - V_{\text{TH}}$). A similar analysis on a $SrTiO_3$ transistor yielded a higher electron density of 1×10^{14} cm^{-2} under the same effective gate voltage [22]. Furthermore, by gating the ZnO single crystals with ionic liquid *N,N*-diethyl-*N*-(2-methoxyethyl)-*N*-methylammonium bis(trifluoromethylsulfonyl)-imide ([DEME][TFSI]), electron densities of 4.5×10^{14} cm^{-2} were achieved at 2.5 V effective gate voltage [24]. At temperatures below 230 K, it was found that the ionic liquid could be supercooled such that a higher gate voltage (up to 6 V) could be applied without dielectric breakdown, which facilitated the injection of ultrahigh electron density up to 8×10^{14} cm^{-2}. These are phenomenal carrier densities corresponding to ~1 charge per unit cell, and opening the door to a variety of interesting transport measurements at high carrier densities. However, Hall measurements and their interpretation are clearly challenging for OTFTs because the carrier mobilities are generally low, so that in the near term, the best way is to measure charge density with electrolyte gates is by integrating displacement current.

6.2.4.4 Switching Time

Switching response time is another key parameter for transistors, which is generally characterized by monitoring the transient channel current in response to the switching of the gate voltage (i.e., turning the transistor ON and OFF) at different frequencies. The switching time of conventional organic transistors is limited by the carrier transport through the channel, that is, $t \propto L^2/\mu V_D$, where L is the channel length and μ is the carrier mobility [1]. In electrolyte-gated transistors, however, the polarization of ions in the electrolyte and the possible ion diffusion into the semiconductor can introduce additional important timescales. In Figure 6.9, Berggren *et al.* studied the switching response of P(VPA-AA) – polyelectrolyte-gated P3HT transistors with a variety of channel lengths, which are presumably operating under the electrostatic gating mechanism [59]. It was found that when the gate voltage was switched from 0 to −1 V, the time required for the channel current to rise to 70% of its maximum current response was essentially the same for transistors with $L < 9\,\mu m$ (i.e., the switching is restricted by the speed of ion migration in the electrolyte). On the other hand, for longer channel transistors, the rise time increased with increasing channel length according $t \propto L^{1.6}$, suggesting that hole transport in the semiconductor started to limit the transistor switching speeds. From the rise time value, the authors estimated that their polyelectrolyte-gated short

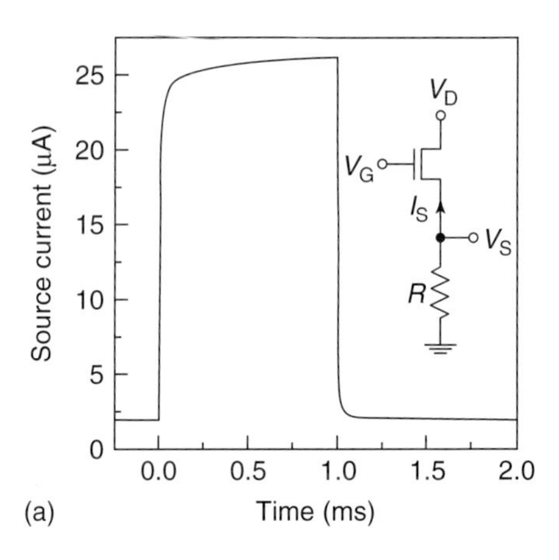

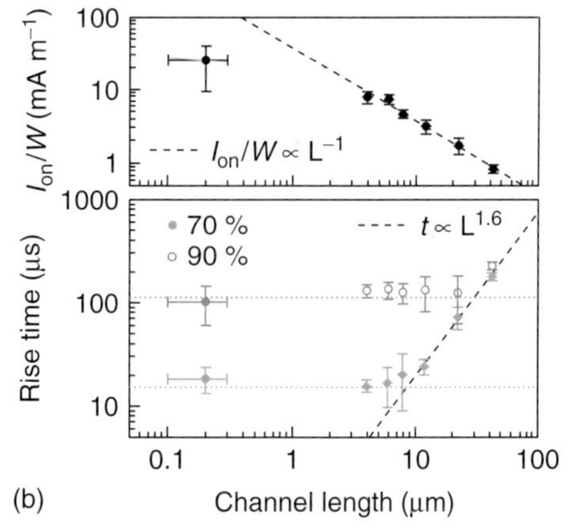

Figure 6.9 (a) Current response of a polyelectrolyte-gated P3HT transistor with gate voltage switching between 0 and −1 V to the drain electrode. The inset shows the circuit schematic of the measurement setup and (b) source current density after 1 ms (top) and rise times for 70 and 90% of the response (bottom) as a function of channel length [59]. Herlogsson, L., Noh, Y.-Y., Zhao, N., Crispin, X., Sirringhaus, H., and Berggren, M. Downscaling of organic field-effect transistors with a polyelectrolyte gate insulator. *Adv. Mater. (Weinheim, Ger.)*, **20**, 4708–4713 (2008). Copyright Wiley-VCH Verlag GmbH & Co. KGaA. Reproduced with permission.

channel transistors could operate at a switching frequency of 10 kHz. This operation speed is considerably faster than many electrolyte-gated polymer transistors where low ionic conductivity polymer electrolytes can limit the response times to seconds or tenths of seconds [36, 60, 75].

As mentioned previously, ion gel electrolytes have higher ionic conductivity that leads to faster ion polarization. The switching response for PQT-12 transistors gated by two ion gels, [EMIM][TFSI] with PS-PMMA-PS gelating agent and 1-butyl-3-methylimidazolium hexafluorophosphate ([BMIM][PF_6]) also with PS-PMMA-PS, were studied for gate voltages switching between −2.5 and +0.5 V [78]. As shown in Figure 6.10a, the channel ON/OFF current ratios (I_{ON}/I_{OFF}) for these transistors decreased with increasing frequency. [EMIM][TFSI]-gated transistors operated with current gain up to a few kilohertz, while [BMIM][PF_6]-gated transistors were a little slower. Figure 6.10b suggests that the switching speed limitation was mainly caused by the increase of OFF current, that is, under higher frequency, the transistors became more and more difficult to turn OFF. One can rationalize this observation by recognizing that any electrolyte will have a characteristic polarization time (an *RC* time constant). If the switching time is shorter than the characteristic *RC* polarization time, then the gate current will be larger compared to the drain current. Thus, improving electrolyte-gated transistor switching times depend on continuing to minimize the electrolyte polarization time. This can be accomplished by developing more conductive electrolytes or by decreasing the electrolyte layer thickness, for example [104].

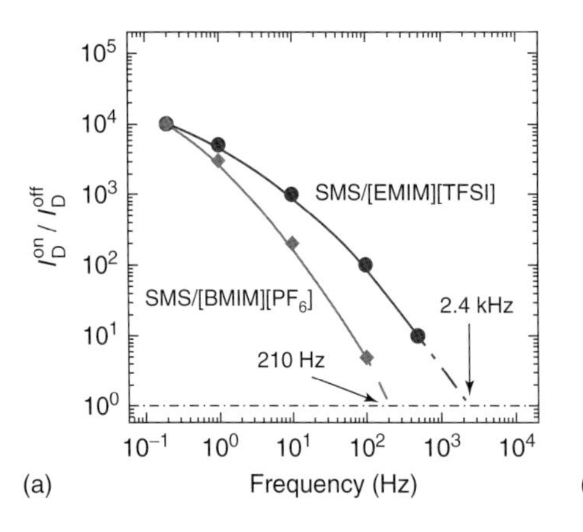

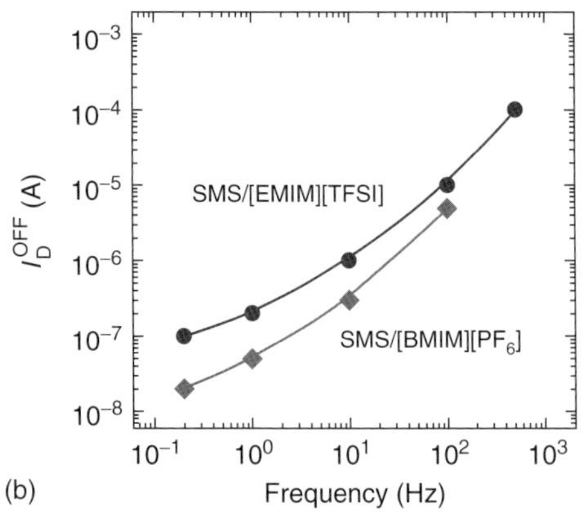

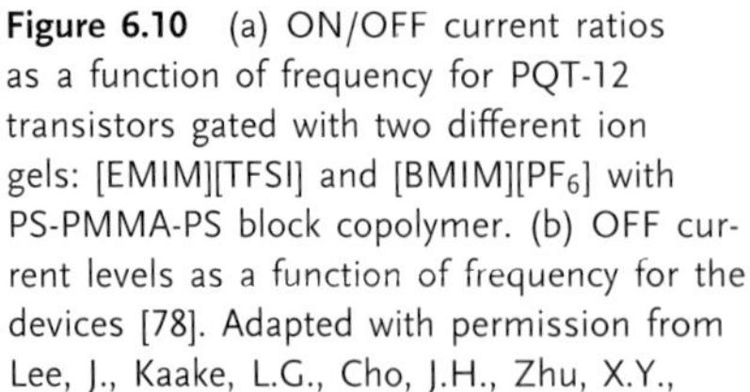

Figure 6.10 (a) ON/OFF current ratios as a function of frequency for PQT-12 transistors gated with two different ion gels: [EMIM][TFSI] and [BMIM][PF_6] with PS-PMMA-PS block copolymer. (b) OFF current levels as a function of frequency for the devices [78]. Adapted with permission from Lee, J., Kaake, L.G., Cho, J.H., Zhu, X.Y., Lodge, T.P., and Frisbie, C.D. Ion-gel-gated polymer thin-film transistors: operating mechanism and characterization of gate dielectric capacitance, switching speed, and stability. *J. Phys. Chem. C*, **113**, 8972–8981 (2009). Copyright 2009 American Chemical Society.

It is useful to try to estimate a realistic minimum RC time constant associated with electrolyte polarization, where R is the resistance due to ion motion in electrolyte film, and C is the capacitance. Using a 100 nm thick ion gel dielectric as a model, with resistivity of $10^3\ \Omega$cm and a specific capacitance of $10\,\mu F\ cm^{-2}$, RC is on the order of 100 ns. On the basis of this calculation, it seems reasonable to expect that megahertz switching frequencies can be achieved for electrolyte-gated OTFTs, which is certainly fast enough for many applications.

6.2.5
Charge Transport at Ultrahigh Carrier Densities

With exceptionally high specific capacitances, electrolyte dielectrics allow examination of transistor performance under ultrahigh carrier densities that are unattainable by any other dielectric. This opens opportunities for increased understanding of fundamental charge transport physics in organic semiconductors. Recently, Xia *et al.* have exploited ionic liquids and other liquid dielectrics to examine the mobility-carrier density relationship for polymer TFTs and organic single-crystal transistors (Figure 6.11) [83]. A large induced hole density variation of nearly 5 orders of magnitude was realized by gating these materials with vacuum, silicon oil, and ionic liquid ([EMIM][TFSI]) dielectrics, respectively.

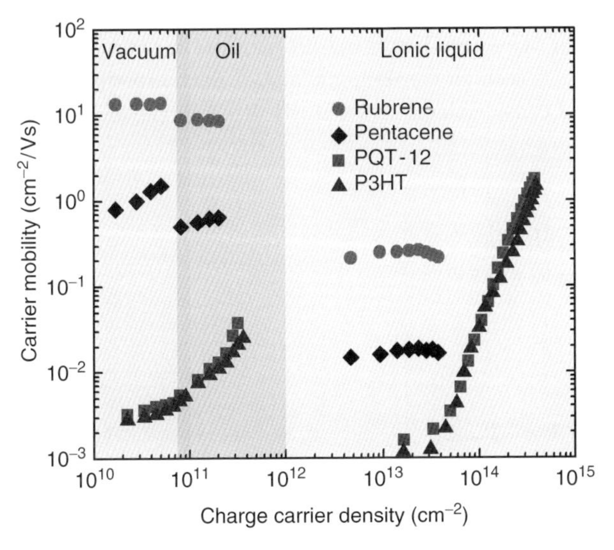

Figure 6.11 Mobility-carrier density relationship for transistors based on rubrene and pentacene single crystals as well as PQT-12 and P3HT thin films. The background colors represent different types of gate dielectric materials at various ranges of carrier density [83]. Xia, Y., Cho, J.H., Ruden, P.P., and Frisbie, C.D. Comparison of the mobility–carrier density relation in polymer and single-crystal organic transistors employing vacuum and liquid gate dielectrics. *Adv. Mater. (Weinheim, Ger.)*, **21**, 2174 (2009). Copyright Wiley-VCH Verlag GmbH & Co. KGaA. Reproduced with permission.

Striking differences were found in the transport behavior for polymer versus single-crystal semiconductors. In polymer transistors, the hole mobility increased with increasing carrier density over the whole range (despite a large initial drop in mobility upon switching from oil to ionic liquid caused by the nature of electrochemical doping) and exceeded 1 $cm^2\ Vs^{-1}$ at a carrier density of $4 \times 10^{14}\ cm^{-2}$. This phenomenon was explained by a combination of Coulomb potential screening and the gradual filling of numerous trap states in the polymer bandgap with increasing carrier density, which together enhance the carrier hopping rates [105]. Similar observations on liquid electrolyte-gated P3HT transistors have also been reported by Iwasa *et al.* [89]. On the other hand, the mobility in single-crystal transistors did not change significantly with carrier density, indicating the lack of trap states at the surface of high purity single crystals. The mobility variation in single crystals upon switching dielectrics was related to the effect of interface polarons, that is, the polarization of the dielectric medium at the semiconductor/dielectric interface [106, 107]. Similarly, Takeya *et al.* reported rubrene single-crystal transistors gated by various ionic liquids and the mobility was found to decrease with increasing dielectric capacitance in a power-law relationship, which is also in good agreement with the interface polaron model [45].

Electrolyte gating has also been employed to examine the transition to metallic behavior in organic semiconductors [49, 51]. It has been shown, for example, that for polymer semiconductors, the dependence of resistance on temperature can become much flatter as gate-induced charging is increased (Figure 6.12) [49]. The hallmark of a metallic state would be a marked *decrease* in resistance with decreasing temperature, as is seen for conventional metals. Such experiments are in early stages, but there is considerable interest in whether truly metallic or even

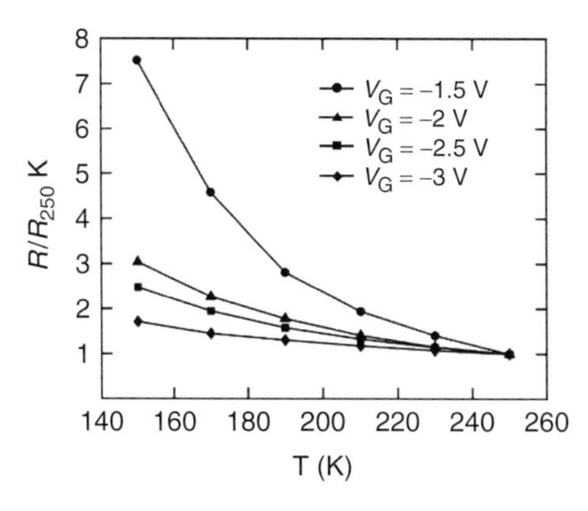

Figure 6.12 Channel resistance scaled by the resistance at 250 K versus temperature for a polymer electrolyte-gated P3HT transistor at four different gate voltages. Data acquired at $V_D = -1$ V [49]. Panzer, M.J. and Frisbie, C.D. High carrier density and metallic conductivity in poly(3-hexylthiophene) achieved by electrostatic charge injection. *Adv. Funct. Mater.*, **16**, 1051–1056. (2006). Copyright Wiley-VCH Verlag GmbH & Co. KGaA. Reproduced with permission.

superconducting behavior can be observed in organic semiconductors that are "super-charged" with electrolyte gates.

6.3 Applications of Electrolyte-Gated Organic Transistors

6.3.1 Printable Low-Voltage Polymer Transistors and Circuits

Electrolyte-gated polymer transistors feature both high carrier mobility and low operation voltage. In addition, compared to other high-capacitance dielectrics [2–12], the printability of electrolytes makes these transistors very promising candidates for incorporation into low-cost digital circuits on flexible substrates.

Recently, Cho *et al.* demonstrated all-printed transistors and circuits, where all components (i.e., metal source and drain electrodes, polymer semiconductor, ion gel dielectric, and conductive polymer gate electrode) were printed sequentially with good registration using a commercial aerosol jet printing technique [77]. Figure 6.13a illustrates a printed array of ion-gel ([EMIM][TFSI] in

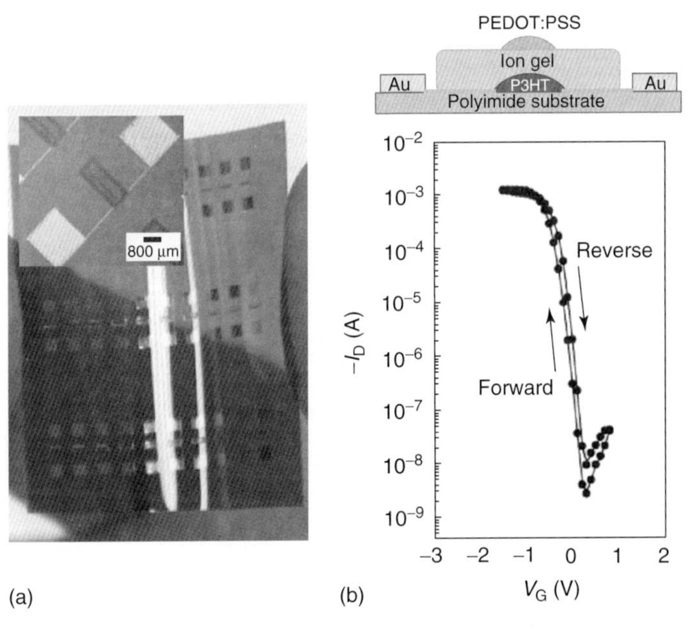

Figure 6.13 (a) Optical image of an aerosol jet printed ion-gel-gated P3HT transistor array fabricated on a flexible polyimide substrate. The source and drain electrodes are gold and the gate electrode is PEDOT:PSS. (b) Transfer characteristics of a typical ion-gel-gated P3HT transistor acquired at $V_D = -1$ V. The top of the panel shows the device schematic diagram in cross-section [77]. Adapted by permission from Macmillan Publishers Ltd: *Nature Materials*, **7**, 900–906 (2008).

PS-PMMA-PS)-gated P3HT transistors on a flexible polyimide substrate. As illustrated by the $I_D - V_G$ characteristics of a typical printed transistor (Figure 6.13b), the devices showed excellent subthreshold (turn-on) behavior, ON/OFF current ratios of 10^5, and negligible hysteresis. The output currents of the devices were higher than 1 mA (transconductances > 1 mS) at $V_G = -1$ V, with mobility values larger than 1 cm^2 Vs^{-1}.

Reproducibility and stability are two of the most important figures of merit for electronic devices. To assess these factors, Cho *et al.* examined an array of 30 printed transistors fabricated in one printing batch [77]. As shown in Figure 6.14a, 28 of the transistors demonstrated nearly identical behavior, while the other two had a slightly higher OFF current. An overall yield of ~90% was reported for printed batches. Also, the performance of a typical transistor for over six months (Figure 6.14b) showed no pronounced degradation, indicating the robustness of the printed devices. The device parameters for a 100 transistors fabricated in different batches are summarized in Figure 6.15, where 60 of them were measured in air, and the rest were measured under vacuum. As can be seen, there was no significant variance in mobility and ON/OFF current ratio between the transistors operated in different environments. Testing in air did result in a small positive shift in turn-on voltage, as expected because it is known that O_2 can hole dope polythiophenes [108].

Investigation of low-voltage circuits based on the ion gel dielectric began with fabrication of an all-printed inverter (Figure 6.16) [77]. As shown in Figure 6.16c, the output voltage (V_{OUT}) of the inverter switched from −1.5 V (same as the applied V_{DD} value) to nearly 0 V when the input voltage (V_{IN}) was swept from 0 to −1.5 V and vice versa. The switching took place with a maximum gain of over six. Additionally, no hysteresis was observed between the forward and reverse V_{IN} sweeps. Figure 6.16d shows the dynamic response of a typical inverter which tracked a 1 kHz square-wave input signal. A stability test showed that the inverter

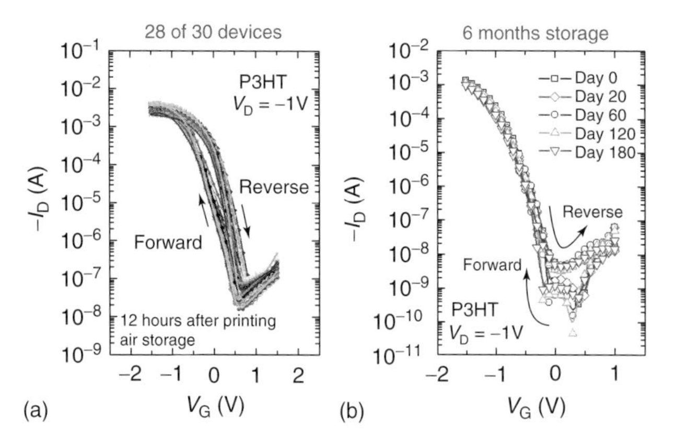

Figure 6.14 (a) Transfer characteristics of 28 ion-gel-gated P3HT transistors of a 30-device array measured in ambient conditions [77]. Adapted by permission from Macmillan Publishers Ltd: *Nature Materials*, **7**, 900–906 (2008). (b) Transfer characteristics of a typical ion-gel-gated transistor upon six months storage in nitrogen.

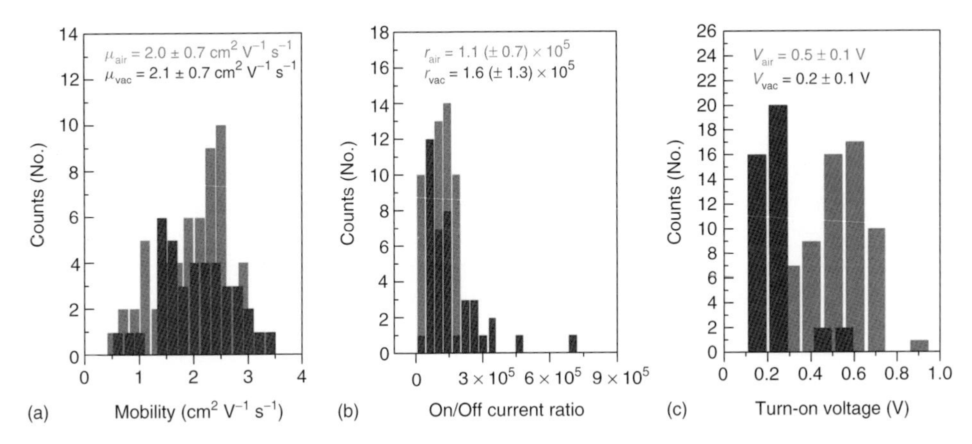

Figure 6.15 Summary of (a) mobility, (b) ON/OFF current ratio, and (c) turn-on voltage of 100 ion-gel-gated P3HT transistors printed in different batches. Sixty transistors were measured in air (gray) and 40 were measured under vacuum (black) [77]. Adapted by permission from Macmillan Publishers Ltd: *Nature Materials*, **7**, 900–906 (2008).

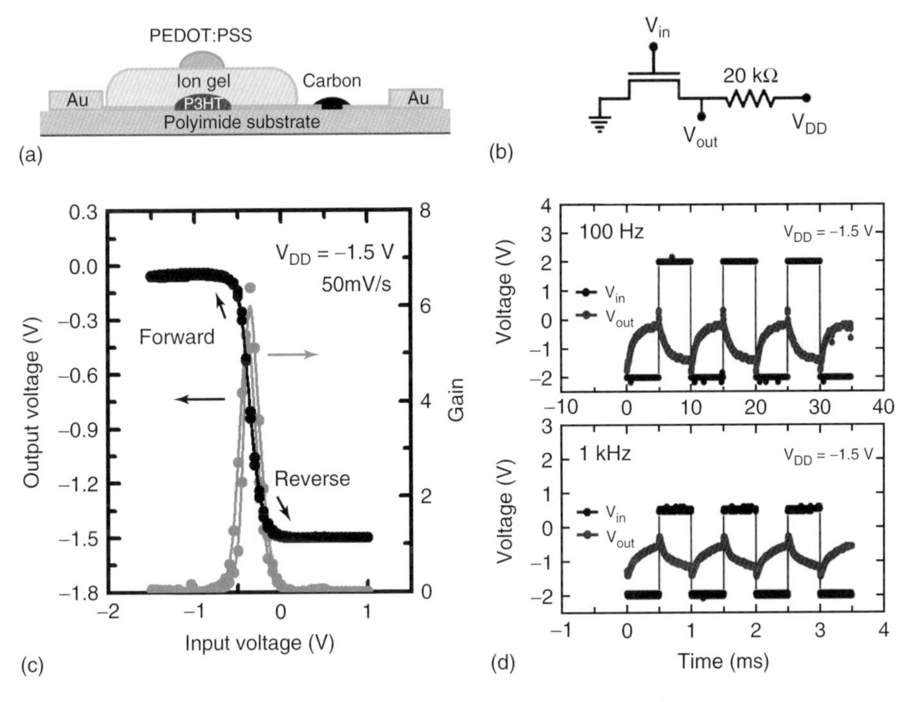

Figure 6.16 All-printed resistor loaded inverters. (a) Device schematic diagram in cross-section, (b) circuit diagram, (c) input–output voltage characteristics, and (d) output responses when the input voltage was switched at 100 Hz and 1 kHz [77]. *Adapted by* permission from Macmillan Publishers Ltd: *Nature Materials*, **7**, 900–906 (2008).

operated at 100 Hz switching frequency for more than 9 h, equivalent to over 3 million cycles, with less than 30% overall drop in output voltage. By optimizing the configurations of inverters, for example, by replacing the resistor with another printed transistor with individual bias control, switching frequencies as high as 10 kHz have also been realized [109].

NAND logic gates have been fabricated by Xia *et al.* using two identical printed ion gel transistors and a load resistor, as shown in the circuit diagram in Figure 6.17 [109]. Figure 6.17b displays the NAND logic gate response, in which the output voltage (V_{OUT}) switched to "0" only when the gate voltages of both transistors (V_A and V_B) were set at "1." In all other configurations, V_{OUT} remained in the "1"

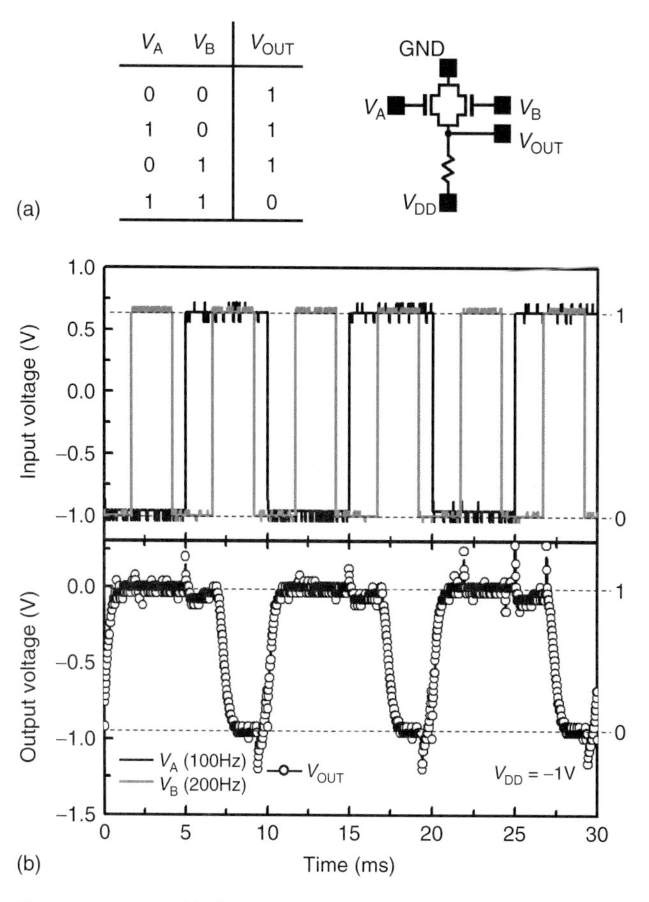

Figure 6.17 Printable low-voltage NAND logic gate. (a) Logic sequences and circuit diagrams. (b) Dynamic response of the NAND gate. The red and black lines in the top panel represent two input voltage, V_A and V_B, respectively., and the open circles in the bottom panel show the output response [109]. Xia, Y., Zhang, W., Ha, M., Cho, J.H., Renn, M.J., Kim, C.H., and Frisbie, C.D. Printed sub-2 V gel-electrolyte-gated polymer transistors and circuits. *Adv. Funct. Mater.*, **20**, 587–594 (2010). Copyright Wiley-VCH Verlag GmbH & Co. KGaA. Reproduced with permission.

state. Additionally, the value of V_{OUT} was fully switched between the applied V_{DD} and ground within 2 ms. This is a key result, given that by combining inverters and NAND gates, one can construct all sorts of logic circuits. For example, Figure 6.18 demonstrates a D-flip-flop circuit consisting of eight NAND gates and three inverters [109]. As shown in the output characteristics, the output (Q) of this flip-flop read and kept the state of data input (D) only at the moment of a falling edge of the clock signal (CLK). The subsequent changes of D did not influence Q until the next falling clock edge. Such a circuit serves as one bit of volatile memory and is widely used in modern circuits for functional electronics such as

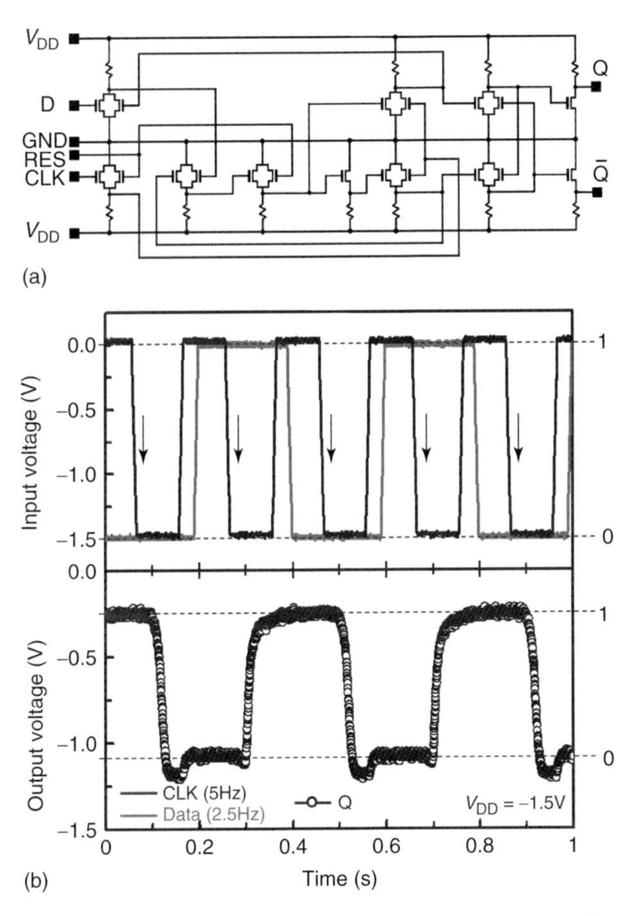

Figure 6.18 Printable low-voltage D-flip-flop circuit. (a) Circuit diagram. (b) Dynamic response of the D-flip-flop circuit. The gray and black lines in the top panel represent the Data signal (D) and the Clock signal (CLK), respectively. The open circles in the bottom panel show the output response (Q) [109]. Xia, Y., Zhang, W., Ha, M., Cho, J.H., Renn, M.J., Kim, C.H. and Frisbie, C.D. Printed sub-2 V gel-electrolyte-gated polymer transistors and circuits. *Adv. Funct. Mater.*, **20**, 587–594 (2010). Copyright Wiley-VCH Verlag GmbH & Co. KGaA. Reproduced with permission.

shift registers and counters [110]. The measured output-to-clock delay (t_{C-Q}, the response time of Q upon the falling edge of clock signal) and setup time (t_{setup}, the time for the circuit to efficiently read the data input) in these devices were both less than 50 ms.

A benchmark circuit for thin-film electronics is the ring oscillator because it can be used to determine the signal propagation delay time for each transistor stage. In addition, ring oscillators are used as clocks to synchronize different circuit units and to generate oscillation sequences [110]. Xia *et al.* printed five-stage ring oscillators based on ion-gel-gated polymer TFTs [109]. The ring oscillator started to oscillate spontaneously with a drain supply voltage (V_{DD}) less than -1 V, and reached a maximum frequency of nearly 150 Hz at $V_{DD} = -2$ V (Figure 6.19). This corresponds to a signal delay time of $\sim$1 ms. Recently, Berggren and colleagues made similar seven-stage ring oscillators based on the P(VPA-AA) electrolytes; their propagation delays were as small as 100 μs, which is a milestone for printed electronics operating at such low supply voltages [111].

Berggren *et al.* have also reported printable rectifiers [37] and transistors [36] based on the electrochemical switching of the conductivity of poly(3,4-ethylenedioxythiophene):poly(styrenesulfonate) (PEDOT:PSS) in contact with various electrolytes. As discussed previously, under different gate biases, the ions in the electrolytes reversibly dope and undope PEDOT chains, hence modulating the output current of the devices. Low operation voltage inverters, NAND and NOR logic gates, and ring oscillators based on PEDOT:PSS have been demonstrated [38]. Moreover, Berggren and Inganäs *et al.* have successfully demonstrated the incorporation of 3-D electrolyte-gated PEDOT:PSS transistors and P3HT transistors on textile fibers, which could be woven into electronically functional fabrics [85, 112]. As illustrated in Figure 6.20, each transistor is formed at the junction of a fiber and a gold wire. The semiconductor-coated fiber consists of several transistor channels (i.e., source-drain gaps), and the wire is threaded perpendicularly on top of the individual channels. Each junction is completed

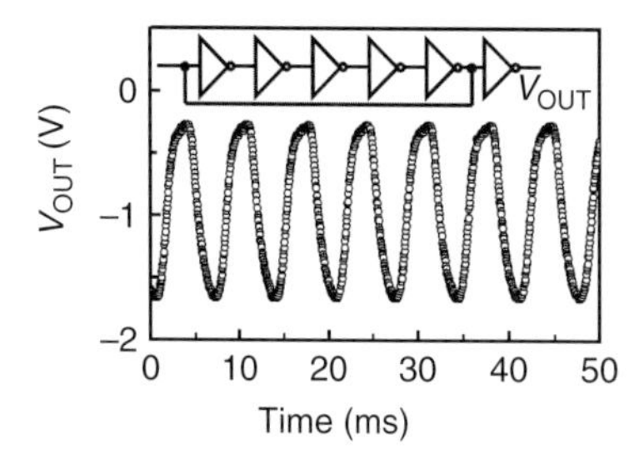

Figure 6.19 Output characteristics of a printed five-stage ring oscillator based on ion-gel-gated P3HT transistors. Oscillation frequency is 150 Hz. Data acquired at $V_{DD} = -2$ V. The inset shows the circuit diagram of the ring oscillator, where the sixth inverter serves as an output buffer [109]. Xia, Y., Zhang, W., Ha, M., Cho, J.H., Renn, M.J., Kim, C.H., and Frisbie, C.D. Printed sub-2 V gel-electrolyte-gated polymer transistors and circuits. *Adv. Funct. Mater.*, **20**, 587–594 (2010). Copyright Wiley-VCH Verlag GmbH & Co. KGaA. Reproduced with permission.

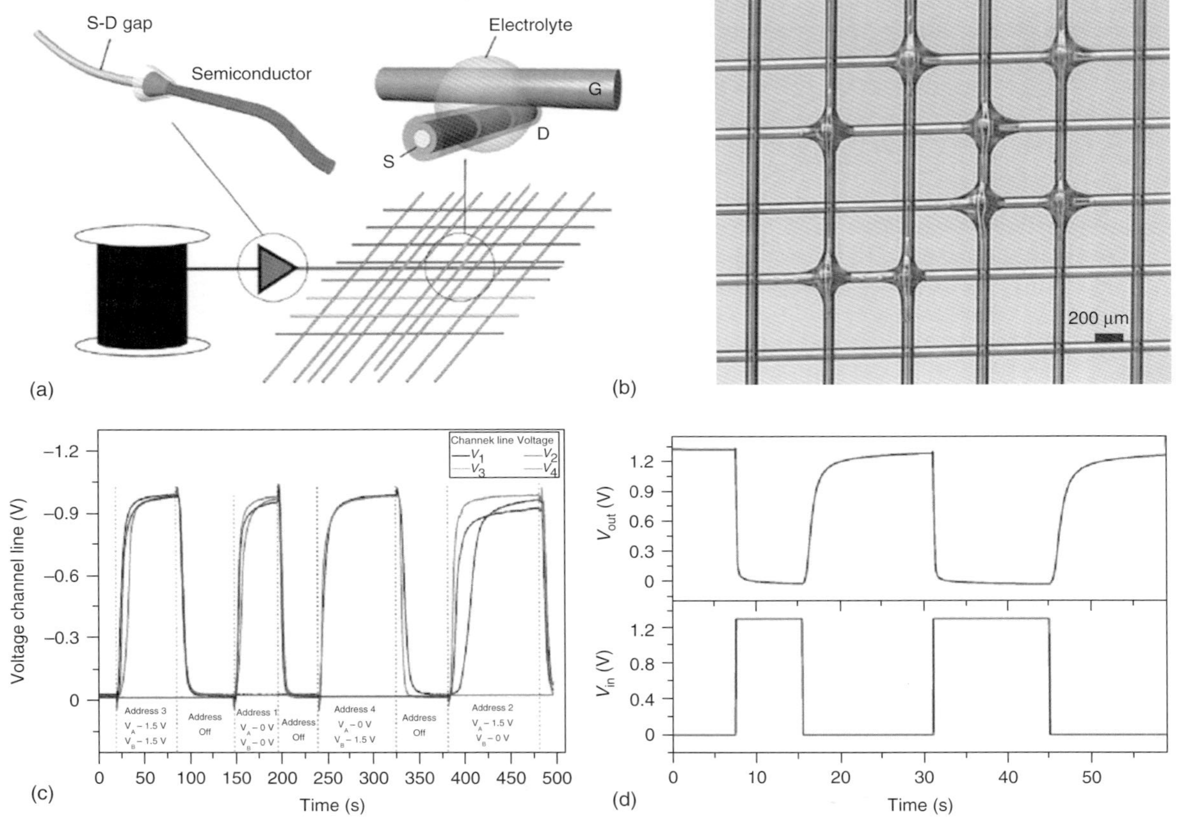

Figure 6.20 Electronically active textiles. (a) Schematic of a fiber transistor fabricated by weaving a conductive wire over the source-drain gap of a semiconductor-coated fiber. The junction is completed with an electrolyte droplet to couple the gate to the channel. (b) Optical image of a binary tree multiplexer on a fabric mesh. (c) Dynamic electrical characteristics of the multiplexer. (d) Dynamic switching characteristics of a fiber integrated inverter [112]. Adapted by permission from Macmillan Publishers Ltd: *Nature Materials*, **6**, 357–362 (2007).

by a droplet of electrolyte. Functioning digital logic circuits such as a binary tree multiplexer and inverter have been demonstrated by weaving the wires and fibers to form proper junction arrays. The response times for these devices were relatively slow (as shown in Figure 6.20 c,d, most circuits were switched in a few seconds), presumably caused by the non-optimal device geometry. However, this demonstration illustrates that electrolyte gating allows unconventional transistor designs that can facilitate new and unusual applications.

6.3.2 Active-Matrix Display Backplanes

Electrolyte-gated OTFTs are naturally very compatible with electrochromic displays, which are themselves electrochemical devices. On the basis of PEDOT:PSS, Berggren *et al.* demonstrated printable active-matrix electrochromic displays on flexible substrates [113, 114]. In order to individually address and effectively switch the displays in the matrix, as shown in Figure 6.21a, each smart pixel of the device consists of a PEDOT:PSS electrochromic display cell connected with a PEDOT:PSS electrochemical transistor. Figure 6.21b illustrates a circuit layout for a 2×2 matrix, where the row lines connect the gates of all transistors in the same row, while the column lines connect the working electrodes of all display cells in the same column. To switch the color of an individual display pixel, a V_G close to 0 V must be applied (i.e., address the row) to turn the transistor ON, so that an appropriate V_{DC} from a separate power source can bias the display cell and change the electrochromic state of the PEDOT:PSS film. When V_G is set at 3 V, the transistors are turned OFF and the pixels in the entire row are not addressed. Figure 6.21c shows an example of an operating 5×5 active-matrix electrochromic display. The optimized active-matrix displays demonstrated filling factors as high as 65%, color contrast around 30 (CIE coordinates), and switching times of a few seconds.

Besides various electrochromic displays [16], electrolytes are also widely incorporated in the so-called polymer light-emitting electrochemical cells (PLECs) [115–121]. In these devices, electrolytes are generally mixed with electroactive polymers and sandwiched by two electrodes. Under bias, the migration of mobile ions tends to oxidize polymer chains near one electrode, and reduce the chains near the other electrode. Electrons and holes then recombine and emit light. Compared to polymer light-emitting diodes (PLEDs), the introduction of mobile ions in PLECs lowers the turn-on voltage for light emission, and thereby reduces the thickness and cathode material requirements of the devices [120]. To the best extent of our knowledge, no active-matrix PLECs addressed by electrolyte-gated transistors have been reported, yet an electrolyte-gated backplane should be fully compatible with PLECs.

6.3.3 Organic Electrochemical Transistors as Chemical Sensors

The basic principle of chemical sensing with transistors is that the presence of the analyte at the gate or at the gate insulator/semiconductor interface creates

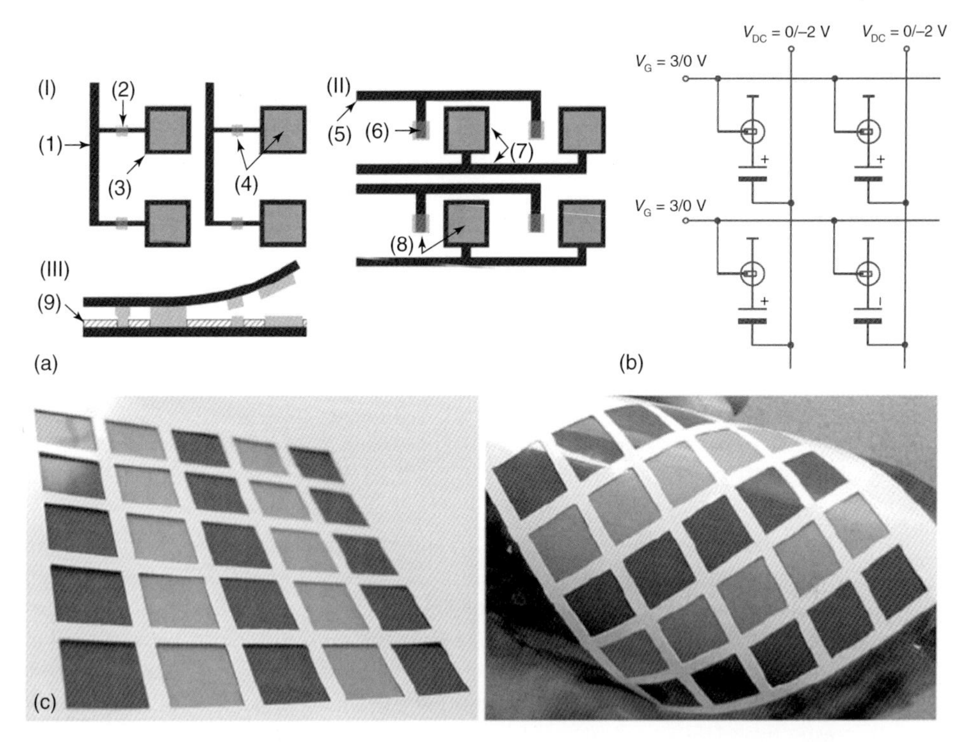

Figure 6.21 (a) Schematic illustration describing a route to manufacture a 2 × 2 printed active-matrix display. In the lower (I) and upper (II) layers, the black areas represents PEDOT:PSS and the gray areas are covered by electrolyte. Two layers are then laminated together (III) to form the display. (b) The addressing scheme of 2 × 2 electrochemical smart pixel. (c) Image of a flexible 5 × 5 active-matrix display with 65% filling factor [113]. Andersson, P., Forchheimer, R., Tehrani, P., and Berggren, M. Printable all-organic electrochromic active-matrix displays. *Adv. Funct. Mater.*, **17**, 3074–3082 (2007). Copyright Wiley-VCH Verlag GmbH & Co. KGaA. Reproduced with permission.

a large change in the source to drain current – this is the critical amplification step. In the past 40 years, electrolytes have been commonly used as an *additional* dielectric layer in inorganic transistors for chemical sensing applications [25]. In the so-called ISFETs, analyte ions diffuse into the electrolyte and affect the gate dielectric polarization, thus modifying the carrier accumulation and transport in the semiconductor; small changes in the dielectric polarization (surface potential) due to the ions result in large changes in source-drain current. Organic ISFETs based on similar dielectric configurations (Figure 6.22a) have also been demonstrated for pH and glucose sensing applications [122–124].

The biggest advantage for organic semiconductors in chemical sensing is their permeability to various ions (as compared to impermeable inorganic crystals). These ions may change the oxidation state of the molecules or conformation of the structures, thus directly influencing the conductivity of the materials. The

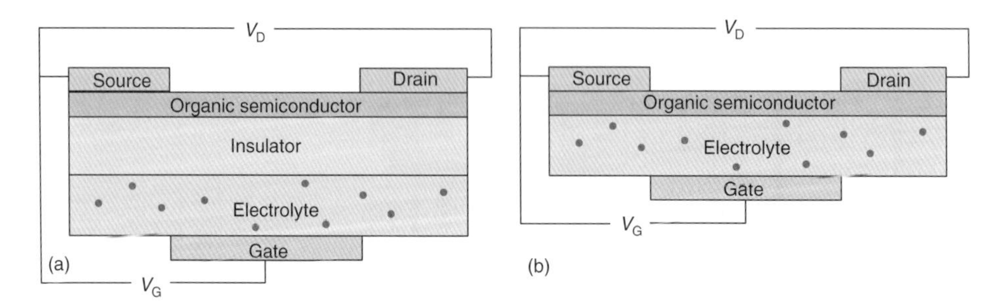

Figure 6.22 (a) Schematic of an organic ion-sensitive field-effective transistor (ISFET) in cross-section. (b) Schematic of an organic electrochemical transistor sensor in cross-section. The filled circles in both schemes represent the analyte of interest [15]. With kind permission from Springer Science+Business Media: Mabeck Jeffrey, T. and Malliaras George, G. Chemical and biological sensors based on organic thin-film transistors. *Anal. Bioanal. Chem.*, **384**, 343–353 (2006).

utilization of electrolyte dielectrics in an electrochemical transistor configuration accelerates the ion migration and amplifies the conductance modulation, which effectively increases the sensitivity of organic chemical and biochemical sensors [14, 125, 126]. Figure 6.22 compares the structural difference between an organic ISFET and an organic electrochemical transistor sensor, where the analyte present in the electrolyte affects the output current via faradaic or galvanic effects, respectively [15].

Electrochemical transistor-based sensors (Figure 6.22b) were first investigated by Wrighton *et al.* [31, 127, 128]. The output currents of these devices were shown to be sensitive to the pH value of the electrolyte, and the presence of chemical oxidants ($Ru(NH_3)_6^{3+}$, $IrCl_6^{2-}$) and reductants ($Fe(CN)_6^{4-}$), and O_2 and H_2. SO_2 in aqueous electrolyte has also been effectively detected with polyaniline-based electrochemical transistors [129]. In order to achieve selective sensing of particular ions, Contractor *et al.* incorporated 18-crown-6 ethers into polyaniline transistors [130]. The narrow distribution of the crown cavity radius (1.33–1.43 Å) allowed the selective capture of K^+ ions (radius of 1.33 Å). These bound ions induced a strong local electrostatic field that caused conformational changes of polyaniline and affected the output current of the devices. Also, Prakash *et al.* demonstrated polycarbazole electrochemical transistors that were sensitive to Cu^{2+} ions [92]. Instead of incorporating a recognition element (e.g., 18-crown-6 ethers in the previous example), the nitrogen-atom-surrounded cavities in the polymeric matrix selectively bound Cu^{2+} and changed the conformation and hence the conductivity of the polycarbazole chains. In addition, Malliaras *et al.* introduced a bilayer lipid membrane (BLM) containing gramicidin ion channels adjacent to the electrolyte of PEDOT:PSS transistors (Figure 6.23a) [91]. The gramicidin dimer channel is permeable to monovalent ions but blocks most divalent species. As shown in Figure 6.23b, under moderate gate voltage ($V_G > 0.3$ V ruptures the BLM), only monovalent ions penetrated through the BLM and doped the PEDOT:PSS, which caused a pronounced modulation in channel current, while the divalent

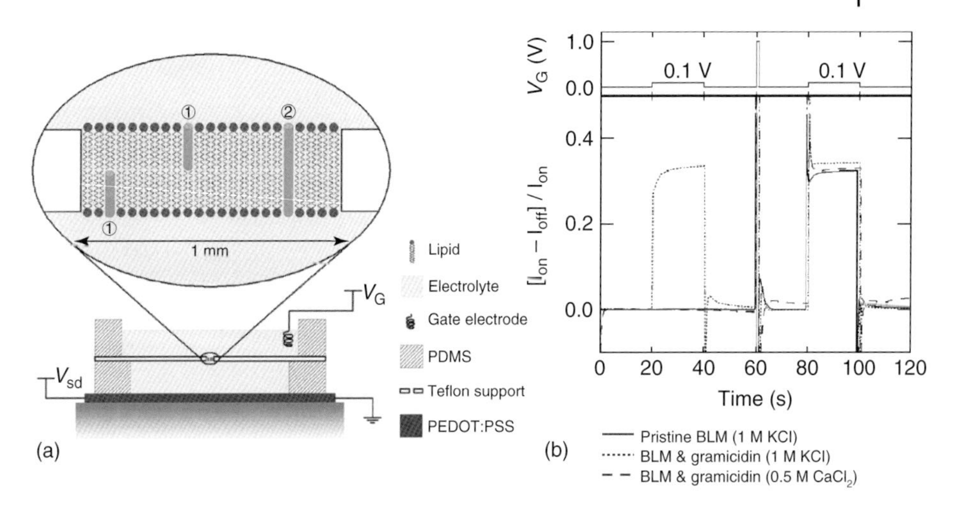

Figure 6.23 (a) Schematic cross-section of a PEDOT:PSS electrochemical transistor gated with a bilayer lipid membrane (BLM). The magnified image represents the BLM formed on a Teflon support with gramicidin, shown as ion blocking monomers (1) or ion permeable dimer (2). (b) Transient response of the device which shows the selective sensing of monovalent ions. The large V_G pulse of 1 V ruptured the BLM [91]. Reprinted with permission from Bernards, D.A., Malliaras, G.G., Toombes, G.E.S., and Gruner, S.M. Gating of an organic transistor through a bilayer lipid membrane with ion channels. *Appl. Phys. Lett.*, **89**, 053505/1–053505/3 (2006). Copyright 2006, American Institute of Physics.

ions showed negligible influence. This is a good example of valence-dependent sensing.

Water vapor is also an important analyte of interest for chemical sensors. Wrighton *et al.* employed poly(vinyl alcohol)/phosphoric acid-based solid-state electrolytes in polyaniline microelectrochemical transistors for stable and reversible sensing of relative humidity [30]. It was found that the absorption of water molecules in the electrolyte significantly increased its ionic conductivity and increased the output current. On the other hand, with lower water concentration in the electrolyte, the output current showed a pronounced decrease. On the basis of a similar mechanism, Berggren *et al.* fabricated printable humidity sensors on flexible substrates (Figure 6.24) [131]. In their application, PEDOT:PSS was printed as the active semiconductor and Nafion was adopted as the solid-state electrolyte; the ionic conductivity of Nafion depends strongly on the relative humidity level. Essentially, this prior work has shown that the higher the ionic conductivity within the Nafion film, the greater the conductivity modulation that can be achieved in the transistors.

The fact that recognition elements can be incorporated into polymers enables polymer electrochemical transistors to selectively sense different chemical analytes of interest [130, 132]. This concept has also been applied in the development of a wide variety of biochemical sensors. For example, Uchida *et al.* reported NADH (reduced form of nicotinamide adenine dinucleotide) sensors by incorporating the

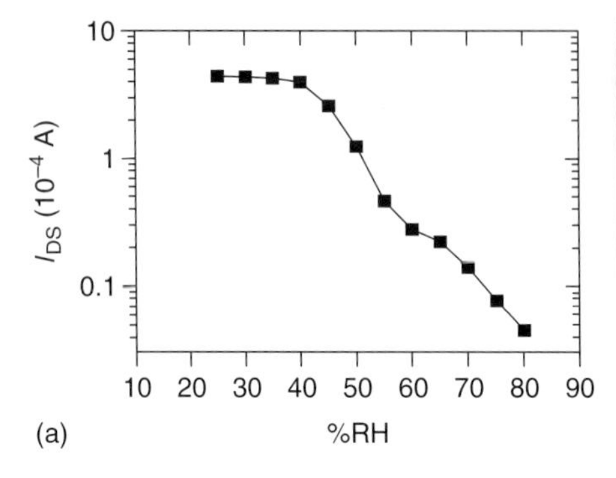

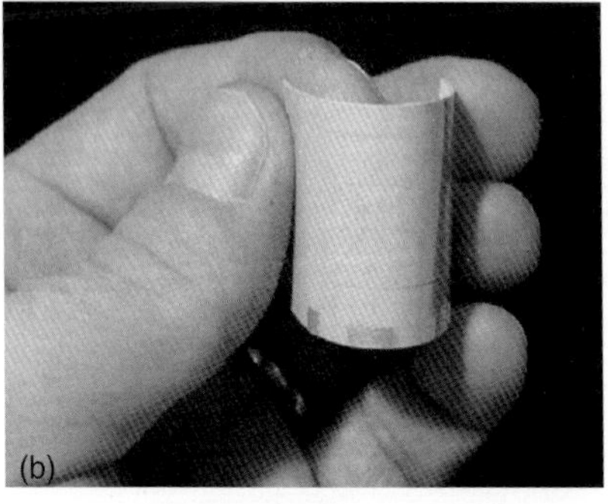

Figure 6.24 (a) Output current of a printed organic electrochemical transistor humidity sensor measured at different relative air humidities (25–80%) after the gate voltage has been applied for 15 s. (b) Image of a bendable sensor printed on a polyethylene coated paper substrate [131]. Reprinted from Nilsson, D., Kugler, T., Svensson, P.-O., and Berggren, M. An all-organic sensor-transistor based on a novel electrochemical transducer concept: printed electrochemical sensors on paper. *Sens. Actuators B*, **B86**, 193–197 (2002), with permission from Elsevier.

flavin enzyme diaphorase in polypyrrole-based electrochemical transistors [133]. Polyaniline and PEDOT:PSS transistors containing glucose oxidase have been fabricated for glucose detection by Contractor *et al.*, Bartlett *et al.*, and Malliaras *et al.*, respectively [90, 134–139]. Sensing applications for penicillin, IgG antibody, and label-free DNA electrochemical transistor sensors facilitated by various biorecognition elements (enzymes) have been demonstrated as well [140–142]. In addition, Malliaras *et al.* integrated a series of PEDOT:PSS electrochemical transistor sensors, each functionalized with a different enzyme, with a surface-directed microfluidic system [143]. In their demonstration, the aqueous analyte solution was easily distributed to every sensor through the hydrophilic pathways, and multiple metabolites in the solution such as glucose and lactate were detected simultaneously and selectively by different sensors.

6.4 Conclusions and Outlook

In this chapter, we have provided a brief overview of the operation mechanisms and applications of electrolyte-gated OTFTs. Here we conclude with a summary of advantages and limitations of these devices, as well as future prospects.

From the forgoing discussion, it is clear that a principal advantage of electrolyte-gated OTFTs is that they offer very low-voltage operation. The low-voltage switching is a direct result of the enormous capacitance of the electrolyte which exceeds the capacitance of *any other* reported gate dielectric layer. Moreover, the electrolyte layers are readily printable and the thickness of the electrolyte is not critical for achieving high capacitance – these features are potentially important for high-throughput manufacturing. Of course, the polarization response time *is* dependent on the electrolyte layer thickness, but for relatively thick 1 μm coatings of newly developed electrolyte materials (e.g., ion gels), it still should be possible to achieve 1 μs switching times, which is certainly adequate for display backplanes and many other applications.

There are of course some limitations to electrolyte-gated OTFTs. One limitation indeed relates to the low-voltage operation – that is, there are some applications (e.g., electrophoretic displays) that require TFTs to operate at 10–20 V, and this is incompatible with electrolytes as it exceeds their electrochemical stability window (±3 V). Furthermore, the use of electrolyte dielectrics requires that all materials in contact with the electrolyte are electrochemically stable (e.g., corrosion resistant). This means that only noble metals (Au, Pd, or Pt) or carbon-based conductors can be used as source, drain, and gate electrodes. Less-expensive Ag or Al electrodes are not generally compatible with electrolytes.

However, the overall prospects for electrolyte-gated OTFTs look quite promising. As noted in Section 6.3.2, electrolyte-gated OTFTs can serve as backplanes for low-voltage electrochromic displays – such displays may be ideal for integration on packaging materials, for example, as they are fully printable and require very low supply biases and thus can be powered by thin-film batteries. Biosensors are one of the most exciting prospective application areas because in this case, many of the attributes of electrolyte-gated OTFTs – sensitivity to ions, compatibility with water, inherent low-voltage operation coupled with current amplification, the option to incorporate a reference electrode, and ease of fabrication – are all major benefits. We expect that electrolyte-gated OTFTs will be increasingly developed for sensing in the coming decade. If this occurs, it will fulfill the early promise of work by Wrighton on microelectrochemical transistors in the 1980s, much of which was devoted to demonstrating sensor concepts [30, 31, 127, 128].

Finally, electrolyte gating provides extremely interesting new challenges for fundamental transport experiments on organic semiconductors (and other materials as well). The ability to "super-charge" the surface of semiconductors with high carrier densities is allowing systematic examination of the metal-insulator transition as a function of tunable carrier density, for example [51]. A kind of "holy grail" experiment would be to observe gate-induced superconductivity in organic semiconductors, a prospect that could be enhanced at the high carrier densities achievable with electrolytes [144]. Certainly, one can conclude at this juncture that in terms of both applications and fundamental experiments, the use of electrolytes to gate organic semiconductors in transistors appears to be extremely promising.

References

1. Sze, S.M. (1985) *Semiconductor Devices Physics and Technology*, John Wiley & Sons, Inc., New York.
2. Dimitrakopoulos, C.D., Purushothaman, S., Kymissis, J., Callegari, A., and Shaw, J.M. (1999) Low-voltage organic transistors on plastic comprising high-dielectric constant gate insulators. *Science (Washington)*, **283**, 822–824.

3. Majewski, L.A., Schroeder, R., and Grell, M. (2005) One volt organic transistor. *Adv. Mater. (Weinheim)*, **17**, 192–196.
4. Zirkl, M., Haase, A., Fian, A., Schoen, H., Sommer, Cm., Jakopic, G., Stadlober, B., Graz, I., Gaar, N., Schwoediauer, R., Bauer, S., and Bauer, S. (2007) Low-voltage organic thin-film transistors with high-k nanocomposite gate dielectrics for flexible electronics and optothermal sensors. *Adv. Mater. (Weinheim)*, **19**, 2241–2245.
5. Naber, R.C.G., Tanase, C., Blom, P.W.M., Gelinck, G.H., Marsman, A.W., Touwslager, F.J., Setayesh, S., and de Leeuw, D.M. (2005) High-performance solution-processed polymer ferroelectric field-effect transistors. *Nat. Mater.*, **4**, 243–248.
6. Kim, S.H., Yang, S.Y., Shin, K., Jeon, H., Lee, J.W., Hong, K.P., and Park, C.E. (2006) Low-operating-voltage pentacene field-effect transistor with a high-dielectric-constant polymeric gate dielectric. *Appl. Phys. Lett.*, **89**, 183516.
7. Haas, U., Gold, H., Haase, A., Jakopic, G., and Stadlober, B. (2007) Submicron pentacene-based organic thin film transistors on flexible substrates. *Appl. Phys. Lett.*, **91**, 043511/1–043511/3.
8. Hur, S.-H., Yoon, M.-H., Gaur, A., Shim, M., Facchetti, A., Marks, T.J., and Rogers, J.A. (2005) Organic nanodielectrics for low voltage carbon nanotube thin film transistors and complementary logic gates. *J. Am. Chem. Soc.*, **127**, 13808–13809.
9. Kang, H., Han, K.-K., Park, J.-E., and Lee, H.H. (2007) High mobility, low voltage polymer transistor. *Org. Electron.*, **8**, 460–464.
10. Klauk, H., Zschieschang, U., Pflaum, J., and Halik, M. (2007) Ultralow-power organic complementary circuits. *Nature (London)*, **445**, 745–748.
11. Yoon, M.-H., Yan, H., Facchetti, A., and Marks, T.J. (2005) Low-voltage organic field-effect transistors and inverters enabled by ultrathin cross-linked polymers as gate dielectrics. *J. Am. Chem. Soc.*, **127**, 10388–10395.
12. Roberts, M.E., Queralto, N., Mannsfeld, S.C.B., Reinecke, B.N., Knoll, W., and Bao, Z. (2009) Cross-linked polymer gate dielectric films for low-voltage organic transistors. *Chem. Mater.*, **21**, 2292.
13. Brattain, W.H. and Garrett, C.G.B. (1955) Experiments on the interface between germanium and an electrolyte. *Bell Telephone Syst. Tech. Publ. Monogr.*, **2372**, 1–48.
14. Berggren, M. and Richter-Dahlfors, A. (2007) Organic bioelectronics. *Adv. Mater. (Weinheim)*, **19**, 3201–3213.
15. Mabeck Jeffrey, T. and Malliaras George, G. (2006) Chemical and biological sensors based on organic thin-film transistors. *Anal. Bioanal. Chem.*, **384**, 343–353.
16. Leger, J.M. (2008) Organic electronics: the ions have it. *Adv. Mater. (Weinheim)*, **20**, 837–841.
17. Panzer, M.J. and Frisbie, C.D. (2008) Exploiting ionic coupling in electronic devices: electrolyte-gated organic field-effect transistors. *Adv. Mater. (Weinheim)*, **20**, 3177–3180.
18. Ortiz, R.P., Facchetti, A., and Marks, T.J. (2010) High-k organic, inorganic, and hybrid dielectrics for low-voltage organic field-effect transistors. *Chem. Rev. (Washington)*, **110**, 205–239.
19. Facchetti, A. and Marks, T.J. (2009) Self-assembled nanodielectrics (SANDs) for unconventional electronics. *Mater. Matters (Milwaukee)*, **4**, 64–67.
20. Misra, R., McCarthy, M., and Hebard, A.F. (2007) Electric field gating with ionic liquids. *Appl. Phys. Lett.*, **90**, 052905/1–052905/3.
21. Shimotani, H., Asanuma, H., Tsukazaki, A., Ohtomo, A., Kawasaki, M., and Iwasa, Y. (2007) Insulator-to-metal transition in ZnO by electric double layer gating. *Appl. Phys. Lett.*, **91**, 082106/1–082106/3.
22. Ueno, K., Nakamura, S., Shimotani, H., Ohtomo, A., Kimura, N., Nojima, T., Aoki, H., Iwasa, Y., and Kawasaki, M. (2008) Electric-field-induced superconductivity in an insulator. *Nat. Mater.*, **7**, 855–858.
23. Dhoot, A.S., Israel, C., Moya, X., Mathur, N.D., and Friend, R.H.

(2009) Large electric field effect in electrolyte-gated manganites. *Phys. Rev. Lett.*, **102**, 136402/1–136402/4.
24. Yuan, H., Shimotani, H., Tsukazaki, A., Ohtomo, A., Kawasaki, M., and Iwasa, Y. (2009) High-density carrier accumulation in ZnO field-effect transistors gated by electric double layers of ionic liquids. *Adv. Funct. Mater.*, **19**, 1046–1053.
25. Bergveld, P. (2003) Thirty years of ISFETOLOGY. What happened in the past 30 years and what may happen in the next 30 years. *Sens. Actuators B*, **88**, 1–20.
26. Bergveld, P. (1970) Development of an ion-sensitive solid-state device for neurophysiological measurements. *IEEE Trans. Bio-Med. Eng.*, **17**, 70–71.
27. Nebel, C.E., Rezek, B., Shin, D., and Watanabe, H. (2006) Surface electronic properties of H-terminated diamond in contact with adsorbates and electrolytes. *Phys. Status Solidi A*, **203**, 3273–3298.
28. Chiang, C.K., Fincher, C.R., Park, Y.W., Heeger, A.J., Shirakawa, H., Louis, E.J., Gau, S.C., and MacDiarmid, A.G. (1977) Electrical conductivity in doped polyacetylene. *Phys. Rev. Lett.*, **39**, 1098–1101.
29. White, H.S., Kittlesen, G.P., and Wrighton, M.S. (1984) Chemical derivatization of an array of three gold microelectrodes with polypyrrole: fabrication of a molecule-based transistor. *J. Am. Chem. Soc.*, **106**, 5375–5377.
30. Chao, S. and Wrighton, M.S. (1987) Characterization of a solid-state polyaniline-based transistor: water vapor dependent characteristics of a device employing a poly(vinyl alcohol)/phosphoric acid solid-state electrolyte. *J. Am. Chem. Soc.*, **109**, 6627–6631.
31. Paul, E.W., Ricco, A.J., and Wrighton, M.S. (1985) Resistance of polyaniline films as a function of electrochemical potential and the fabrication of polyaniline-based microelectronic devices. *J. Phys. Chem.*, **89**, 1441–1447.
32. Chao, S. and Wrighton, M.S. (1987) Solid-state microelectrochemistry: electrical characteristics of a solid-state microelectrochemical transistor based on poly(3-methylthiophene). *J. Am. Chem. Soc.*, **109**, 2197–2199.
33. Ofer, D. and Wrighton, M.S. (1988) Potential dependence of the conductivity of poly(3-methylthiophene) in liquid sulfur dioxide/electrolyte: a finite potential window of high conductivity. *J. Am. Chem. Soc.*, **110**, 4467–4468.
34. Ofer, D., Park, L.Y., Schrock, R.R., and Wrighton, M.S. (1991) Potential dependence of the conductivity of polyacetylene: finite potential windows of high conductivity. *Chem. Mater.*, **3**, 573–575.
35. Ofer, D., Crooks, R.M., and Wrighton, M.S. (1990) Potential dependence of the conductivity of highly oxidized polythiophenes, polypyrroles, and polyaniline: finite windows of high conductivity. *J. Am. Chem. Soc.*, **112**, 7869–7879.
36. Nilsson, D., Chen, M., Kugler, T., Remonen, T., Armgarth, M., and Berggren, M. (2002) Bi-stable and dynamic current modulation in electrochemical organic transistors. *Adv. Mater. (Weinheim)*, **14**, 51–54.
37. Chen, M., Nilsson, D., Kugler, T., Berggren, M., and Remonen, T. (2002) Electric current rectification by an all-organic electrochemical device. *Appl. Phys. Lett.*, **81**, 2011–2013.
38. Nilsson, D., Robinson, N., Berggren, M., and Forchheimer, R. (2005) Electrochemical logic circuits. *Adv. Mater. (Weinheim)*, **17**, 353–358.
39. Rani, V. and Santhanam, K.S.V. (1998) Polycarbazole-based electrochemical transistor. *J. Solid State Electrochem.*, **2**, 99–101.
40. Fenton, D.E., Parker, J.M., and Wright, P.V. (1973) Complexes of alkali metal ions with poly(ethylene oxide). *Polymer*, **14**, 589.
41. Allcock, H.R., Nelson, C.J., and Coggio, W.D. (1993) Solid polymeric electrolytes based on crosslinked MEEP-type materials. *Polym. Mater. Sci. Eng.*, **68**, 76–77.
42. Ratner, M.A. and Shriver, D.F. (1988) Ion transport in solvent-free polymers. *Chem. Rev.*, **88**, 109–124.

43. Gray, F.M. (1991) *Solid Polymer Electrolytes: Fundamentals and Technological Applications*, Wiley-VCH Verlag GmbH, New York.
44. Panzer, M.J. and Frisbie, C.D. (2006) High charge carrier densities and conductance maxima in single-crystal organic field-effect transistors with a polymer electrolyte gate dielectric. *Appl. Phys. Lett.*, **88**, 203504/1–203504/3.
45. Takeya, J., Yamada, K., Hara, K., Shigeto, K., Tsukagoshi, K., Ikehata, S., and Aoyagi, Y. (2006) High-density electrostatic carrier doping in organic single-crystal transistors with polymer gel electrolyte. *Appl. Phys. Lett.*, **88**, 112102/1–112102/3.
46. Panzer, M.J., Newman, C.R., and Frisbie, C.D. (2005) Low-voltage operation of a pentacene field-effect transistor with a polymer electrolyte gate dielectric. *Appl. Phys. Lett.*, **86**, 103503/1–103503/3.
47. Panzer, M.J. and Frisbie, C.D. (2005) Polymer electrolyte gate dielectric reveals finite windows of high conductivity in organic thin film transistors at high charge carrier densities. *J. Am. Chem. Soc.*, **127**, 6960–6961.
48. Siddons, G.P., Merchin, D., Back, J.H., Jeong, J.K., and Shim, M. (2004) Highly efficient gating and doping of carbon nanotubes with polymer electrolytes. *Nano Lett.*, **4**, 927–931.
49. Panzer, M.J. and Frisbie, C.D. (2006) High carrier density and metallic conductivity in poly(3-hexylthiophene) achieved by electrostatic charge injection. *Adv. Funct. Mater.*, **16**, 1051–1056.
50. Panzer, M.J. and Frisbie, C.D. (2007) Polymer electrolyte-gated organic field-effect transistors: low-voltage, high-current switches for organic electronics and testbeds for probing electrical transport at high charge carrier density. *J. Am. Chem. Soc.*, **129**, 6599–6607.
51. Dhoot, A.S., Yuen, J.D., Heeney, M., McCulloch, I., Moses, D., and Heeger, A.J. (2006) Beyond the metal-insulator transition in polymer electrolyte gated polymer field-effect transistors. *Proc. Natl. Acad. Sci. U.S.A.*, **103**, 11834–11837.
52. Yuen, J.D., Dhoot, A.S., Namdas, Ebinazar B., Coates, N.E., Heeney, M., McCulloch, I., Moses, D., and Heeger, A.J. (2007) Electrochemical doping in electrolyte-gated polymer transistors. *J. Am. Chem. Soc.*, **129**, 14367–14371.
53. Kaneto, K., Asano, T., and Takashima, W. (1991) Memory device using a conducting polymer and solid polymer electrolyte. *Jpn. J. Appl. Phys., Part 2*, **30**, L215–L217.
54. Taniguchi, M. and Kawai, T. (2004) Vertical electrochemical transistor based on poly(3-hexylthiophene) and cyanoethylpullulan. *Appl. Phys. Lett.*, **85**, 3298–3300.
55. Pal, B.N., Dhar, B.M., See, K.C., and Katz, H.E. (2009) Solution-deposited sodium beta-alumina gate dielectrics for low-voltage and transparent field-effect transistors. *Nat. Mater.*, **8**, 898–903.
56. Said, E., Crispin, X., Herlogsson, L., Elhag, S., Robinson, N.D., and Berggren, M. (2006) Polymer field-effect transistor gated via a poly(styrenesulfonic acid) thin film. *Appl. Phys. Lett.*, **89**, 143507/1–143507/3.
57. Herlogsson, L., Crispin, X., Robinson, N.D., Sandberg, M., Hagel, O.-J., Gustafsson, G., and Berggren, M. (2007) Low-voltage polymer field-effect transistors gated via a proton conductor. *Adv. Mater. (Weinheim)*, **19**, 97–101.
58. Said, E., Larsson, O., Berggren, M., and Crispin, X. (2008) Effects of the ionic currents in electrolyte-gated organic field-effect transistors. *Adv. Funct. Mater.*, **18**, 3529–3536.
59. Herlogsson, L., Noh, Y.-Y., Zhao, N., Crispin, X., Sirringhaus, H., and Berggren, M. (2008) Downscaling of organic field-effect transistors with a polyelectrolyte gate insulator. *Adv. Mater. (Weinheim)*, **20**, 4708–4713.
60. Sandberg, H.G.O., Baecklund, T.G., Oesterbacka, R., and Stubb, H. (2004) High-performance all-polymer

transistor utilizing a hygroscopic insulator. *Adv. Mater. (Weinheim)*, **16**, 1112–1115.

61. Backlund, T.G., Sandberg, H.G.O., Osterbacka, R., and Stubb, H. (2004) Current modulation of a hygroscopic insulator organic field-effect transistor. *Appl. Phys. Lett.*, **85**, 3887–3889.

62. Backlund, T.G., Osterbacka, R., Stubb, H., Bobacka, J., and Ivaska, A. (2005) Operating principle of polymer insulator organic thin-film transistors exposed to moisture. *J. Appl. Phys.*, **98**, 074504/1–074504/6.

63. Kaihovirta, N.J., Tobjork, D., Makela, T., and Osterbacka, R. (2008) Absence of substrate roughness effects on an all-printed organic transistor operating at one volt. *Appl. Phys. Lett.*, **93**, 053302/1–053302/3.

64. Buzzeo Marisa, C., Evans Russell, G., and Compton Richard, G. (2004) Non-haloaluminate room-temperature ionic liquids in electrochemistry--a review. *ChemPhysChem*, **5**, 1106–1120.

65. Ueki, T. and Watanabe, M. (2008) Macromolecules in ionic liquids: progress, challenges, and opportunities. *Macromolecules (Washington)*, **41**, 3739–3749.

66. Lu, W., Fadeev, A.G., Qi, B., Smela, E., Mattes, B.R., Ding, J., Spinks, G.M., Mazurkiewicz, J., Zhou, D., Wallace, G.G., MacFarlane, D.R., Forsyth, S.A., and Forsyth, M. (2002) Use of ionic liquids for pi-conjugated polymer electrochemical devices. *Science (Washington)*, **297**, 983–987.

67. Seki, S., Susan, M.A.B.H., Kaneko, T., Tokuda, H., Noda, A., and Watanabe, M. (2005) Distinct difference in ionic transport behavior in polymer electrolytes depending on the matrix polymers and incorporated salts. *J. Phys. Chem.*, **B109**, 3886–3892.

68. Klingshirn, M.A., Spear, S.K., Subramanian, R., Holbery, J.D., Huddleston, J.G., and Rogers, R.D. (2004) Gelation of ionic liquids using a cross-linked poly(ethylene glycol) gel matrix. *Chem. Mater.*, **16**, 3091–3097.

69. Marcilla, R., Alcaide, F., Sardon, H., Pomposo, J.A., Pozo-Gonzalo, C., and Mecerreyes, D. (2006) Tailor-made polymer electrolytes based upon ionic liquids and their application in all-plastic electrochromic devices. *Electrochem. Commun.*, **8**, 482–488.

70. Lee, S.W., Lee, H.J., Choi, J.H., Koh, W.G., Myoung, J.M., Hur, J.H., Park, J.J., Cho, J.H., and Jeong, U. (2010) Periodic array of polyelectrolyte-gated organic transistors from electrospun poly(3-hexylthiophene) nanofibers. *Nano Lett.* **10**, 347–351.

71. He, Y., Boswell, P.G., Buehlmann, P., and Lodge, T.P. (2007) Ion gels by self-assembly of a triblock copolymer in an ionic liquid. *J. Phys. Chem. B*, **111**, 4645–4652.

72. He, Y., Li, Z., Simone, P., and Lodge Timothy, P. (2006) Self-assembly of block copolymer micelles in an ionic liquid. *J. Am. Chem. Soc.*, **128**, 2745–2750.

73. Lodge, T.P. (2008) Materials science: a unique platform for materials design. *Science (Washington)*, **321**, 50–51.

74. Susan, M.A.B.H., Kaneko, T., Noda, A., and Watanabe, M. (2005) Ion gels prepared by in situ radical polymerization of vinyl monomers in an ionic liquid and their characterization as polymer electrolytes. *J. Am. Chem. Soc.*, **127**, 4976–4983.

75. Lee, J., Panzer, M.J., He, Y., Lodge, T.P., and Frisbie, C.D. (2007) Ion gel gated polymer thin-film transistors. *J. Am. Chem. Soc.*, **129**, 4532–4533.

76. Cho, J.H., Le, J., He, Y., Kim, B., Lodge, T.P., and Frisbie, C.D. (2008) High-capacitance ion gel gate dielectrics with faster polarization response times for organic thin film transistors. *Adv. Mater. (Weinheim)*, **20**, 686–690.

77. Cho, J.H., Lee, J., Xia, Y., Kim, B., He, Y., Renn, M.J., Lodge, T.P., and Frisbie, C.D. (2008) Printable ion-gel gate dielectrics for low-voltage polymer thin-film transistors on plastic. *Nat. Mater.*, **7**, 900–906.

78. Lee, J., Kaake, L.G., Cho, J.H., Zhu, X.Y., Lodge, T.P., and Frisbie, C.D. (2009) Ion gel-gated polymer thin-film transistors: operating mechanism and

characterization of gate dielectric capacitance, switching speed, and stability. *J. Phys. Chem. C*, **113**, 8972–8981.
79. Shimotani, H., Asanuma, H., Takeya, J., and Iwasa, Y. (2006) Electrolyte-gated charge accumulation in organic single crystals. *Appl. Phys. Lett.*, **89**, 203501/1–203501/3.
80. Uemura, T., Hirahara, R., Tominari, Y., Ono, S., Seki, S., and Takeya, J. (2008) Electronic functionalization of solid-to-liquid interfaces between organic semiconductors and ionic liquids: Realization of very high performance organic single-crystal transistors. *Appl. Phys. Lett.*, **93**, 263305/1–263305/3.
81. Ono, S., Seki, S., Hirahara, R., Tominari, Y., and Takeya, J. (2008) High-mobility, low-power, and fast-switching organic field-effect transistors with ionic liquids. *Appl. Phys. Lett.*, **92**, 103313/1–103313/3.
82. Ono, S., Miwa, K., Seki, S., and Takeya, J. (2009) A comparative study of organic single-crystal transistors gated with various ionic-liquid electrolytes. *Appl. Phys. Lett.*, **94**, 063301/1–063301/3.
83. Xia, Y., Cho, J.H., Ruden, P.P., and Frisbie, C.D. (2009) Comparison of the mobility – carrier density relation in polymer and single-crystal organic transistors employing vacuum and liquid gate dielectrics. *Adv. Mater. (Weinheim)*, **21**, 2174–2179.
84. Kaake, L.G., Paulsen, B.D., Frisbie, C.D., and Zhu, X.Y. (2010) Mixing at the charged interface of a polymer semiconductor and a polyelectrolyte dielectric. *J. Phys. Chem. Lett.*, **1**, 862–867.
85. Hamedi, M., Herlogsson, L., Crispin, X., Marcilla, R., Berggren, M., and Inganaes, O. (2009) Fiber-embedded electrolyte-gated field-effect transistors for e-textiles. *Adv. Mater. (Weinheim)*, **21**, 573–577.
86. Kaake, L.G., Zou, Y., Panzer, M.J., Frisbie, C.D., and Zhu, X.Y. (2007) Vibrational spectroscopy reveals electrostatic and electrochemical doping in organic thin film transistors gated with a polymer electrolyte dielectric. *J. Am. Chem. Soc.*, **129**, 7824–7830.
87. Tatistcheff, H.B., Fritsch-Faules, I., and Wrighton, M.S. (1993) Comparison of diffusion coefficients of electroactive species in aqueous fluid electrolytes and polyacrylate gels: step generation-collection diffusion measurements and operation of electrochemical devices. *J. Phys. Chem.*, **97**, 2732–2739.
88. Xia, Y., Cho, J., Paulsen, B., Frisbie, C.D., and Renn, M.J. (2009) Correlation of on-state conductance with referenced electrochemical potential in ion gel gated polymer transistors. *Appl. Phys. Lett.*, **94**, 013304/1–013304/3.
89. Shimotani, H., Diguet, G., and Iwasa, Y. (2005) Direct comparison of field effect and electrochemical doping in regioregular poly(3-hexylthiophene). *Appl. Phys. Lett.*, **86**, 022104/1–022104/3.
90. Zhu, Z.-T., Mabeck, J.T., Zhu, C., Cady, N.C., Batt, C.A., and Malliaras, G.G. (2004) A simple poly(3,4-ethylene dioxythiophene)/poly(styrene sulfonic acid) transistor for glucose sensing at neutral pH. *Chem. Commun. (Cambridge)*, 1556–1557.
91. Bernards, D.A., Malliaras, G.G., Toombes, G.E.S., and Gruner, S.M. (2006) Gating of an organic transistor through a bilayer lipid membrane with ion channels. *Appl. Phys. Lett.*, **89**, 053505/1–053505/3.
92. Saxena, V., Shirodkar, V., and Prakash, R. (2000) Copper(II) ion-selective microelectrochemical transistor. *J. Solid State Electrochem.*, **4**, 234–236.
93. Mannerbro, R., Ranloef, M., Robinson, N., and Forchheimer, R. (2008) Ink-jet printed electrochemical organic electronics. *Synth. Met.*, **158**, 556–560.
94. Hulea, I.N., Brom, H.B., Houtepen, A.J., Vanmaekelbergh, D., Kelly, J.J., and Meulenkamp, E.A. (2004) Wide energy-window view on the density of states and hole mobility in poly(p-phenylene vinylene). *Phys. Rev. Lett.*, **93**, 166601/1–166601/4.
95. Shimotani, H., Kanbara, T., Iwasa, Y., Tsukagoshi, K., Aoyagi, Y., and Kataura, H. (2006) Gate capacitance in electrochemical transistor of single-walled carbon nanotube. *Appl. Phys. Lett.*, **88**, 073104/1–073104/3.

96. Kruger, M., Buitelaar, M.R., Nussbaumer, T., Schonenberger, C., and Forro, L. (2001) Electrochemical carbon nanotube field-effect transistor. *Appl. Phys. Lett.*, **78**, 1291–1293.
97. Rosenblatt, S., Yaish, Y., Park, J., Gore, J., Sazonova, V., and McEuen, P.L. (2002) High performance electrolyte gated carbon nanotube transistors. *Nano Lett.*, **2**, 869–872.
98. Ozel, T., Gaur, A., Rogers, J.A., and Shim, M. (2005) Polymer electrolyte gating of carbon nanotube network transistors. *Nano Lett.*, **5**, 905–911.
99. Fukushima, T., Kosaka, A., Yamamoto, Y., Aimiya, T., Notazawa, S., Takigawa, T., Inabe, T., and Aida, T. (2006) Dramatic effect of dispersed carbon nanotubes on the mechanical and electroconductive properties of polymers derived from ionic liquids. *Small*, **2**, 554–560.
100. Roest, A.L., Kelly, J.J., Vanmaekelbergh, D., and Meulenkamp, E.A. (2002) Staircase in the electron mobility of a ZnO quantum dot assembly due to shell filling. *Phys. Rev. Lett.*, **89**, 036801.
101. Bard, A.J.F. and Larry, R. (2000) *Electrochemical Methods: Fundamentals and Applications*, John Wiley & Sons, Inc.
102. Liang, Y., Frisbie, C.D., Chang, H.-C., and Ruden, P.P. (2009) Conducting channel formation and annihilation in organic field-effect structures. *J. Appl. Phys.*, **105**, 024514/1–024514/6.
103. Xia, Y., Kalihari, V., Frisbie, C.D., Oh, N.K., and Rogers, J.A. (2007) Tetracene air-gap single-crystal field-effect transistors. *Appl. Phys. Lett.*, **90**, 162106/1–162106/3.
104. Bernards, D.A. and Malliaras, G.G. (2007) Steady-state and transient behavior of organic electrochemical transistors. *Adv. Funct. Mater.*, **17**, 3538–3544.
105. Tanase, C., Meijer, E.J., Blom, P.W.M., and de Leeuw, D.M. (2003) Unification of the hole transport in polymeric field-effect transistors and light-emitting diodes. *Phys. Rev. Lett.*, **91**, 216601/1–216601/4.
106. Hulea, I.N., Fratini, S., Xie, H., Mulder, C.L., Iossad, N.N., Rastelli, G., Ciuchi, S., and Morpurgo, A.F. (2006) Tunable Froehlich polarons in organic single-crystal transistors. *Nat. Mater.*, **5**, 982–986.
107. Houili, H., Picon, J.D., Zuppiroli, L., and Bussac, M.N. (2006) Polarization effects in the channel of an organic field-effect transistor. *J. Appl. Phys.*, **100**, 023702/1–023702/8.
108. Chabinyc, M.L., Street, R.A., and Northrup, J.E. (2007) Effects of molecular oxygen and ozone on polythiophene-based thin-film transistors. *Appl. Phys. Lett.*, **90**, 123508/1–123508/3.
109. Xia, Y., Zhang, W., Ha, M., Cho, J.H., Renn, M.J., Kim, C.H., and Frisbie, C.D. (2010) Printed sub-2 V gel-electrolyte-gated polymer transistors and circuits. *Adv. Funct. Mater.*, **20**, 587–594.
110. Rabaey, J.M. (1996) *Digital Integrated Circuits: A Design Perspective*, Prentice Hall, Upper Saddle River.
111. Herlogsson, L., Coelle, M., Tierney, S., Crispin, X., and Berggren, M. (2010) Low-voltage ring oscillators based on polyelectrolyte-gated polymer thin-film transistors. *Adv. Mater. (Weinheim)*, **22**, 72–76.
112. Hamedi, M., Forchheimer, R., and Inganas, O. (2007) Towards woven logic from organic electronic fibers. *Nat. Mater.*, **6**, 357–362.
113. Andersson, P., Forchheimer, R., Tehrani, P., and Berggren, M. (2007) Printable all-organic electrochromic active-matrix displays. *Adv. Funct. Mater.*, **17**, 3074–3082.
114. Andersson, P., Nilsson, D., Svensson, P.-O., Chen, M., Malmstrom, A., Remonen, T., Kugler, T., and Berggren, M. (2002) Active matrix displays based on all-organic electrochemical smart pixels printed on paper. *Adv. Mater. (Weinheim)*, **14**, 1460–1464.
115. Pei, Q., Yu, G., Zhang, C., Yang, Y., and Heeger, A.J. (1995) Polymer light-emitting electrochemical cells. *Science (Washington)*, **269**, 1086–1088.
116. Pei, Q., Yang, Y., Yu, G., Zhang, C., and Heeger, A.J. (1996) Polymer light-emitting electrochemical cells: in situ formation of a light-emitting

p-n junction. *J. Am. Chem. Soc.*, **118**, 3922–3929.
117. Kervella, Y., Armand, M., and Stephan, O. (2001) Organic light-emitting electrochemical cells based on polyfluorene. Investigation of the failure modes. *J. Electrochem. Soc.*, **148**, H155–H160.
118. Yang, C., Sun, Q., Qiao, J., and Li, Y. (2003) Ionic liquid doped polymer light-emitting electrochemical cells. *J. Phys. Chem. B*, **107**, 12981–12988.
119. Shin, J.-H., Xiao, S., and Edman, L. (2006) Polymer light-emitting electrochemical cells: the formation and effects of doping-induced micro shorts. *Adv. Funct. Mater.*, **16**, 949–956.
120. Shao, Y., Bazan, G.C., and Heeger, A.J. (2007) Long-lifetime polymer light-emitting electrochemical cells. *Adv. Mater. (Weinheim)*, **19**, 365–370.
121. Slinker, J.D., DeFranco, J.A., Jaquith, M.J., Silveira, W.R., Zhong, Y.-W., Moran-Mirabal, J.M., Craighead, H.G., Abruna, H.D., Marohn, J.A., and Malliaras, G.G. (2007) Direct measurement of the electric-field distribution in a light-emitting electrochemical cell. *Nat. Mater.*, **6**, 894–899.
122. Bartic, C., Palan, B., Campitelli, A., and Borghs, G. (2002) Monitoring pH with organic-based field-effect transistors. *Sens. Actuators B*, **83**, 115–122.
123. Bartic, C., Campitelli, A., and Borghs, S. (2003) Field-effect detection of chemical species with hybrid organic/inorganic transistors. *Appl. Phys. Lett.*, **82**, 475–477.
124. Loi, A., Manunza, I., and Bonfiglio, A. (2005) Flexible, organic, ion-sensitive field-effect transistor. *Appl. Phys. Lett.*, **86**, 103512/1–103512/3.
125. McQuade, D.T., Pullen, A.E., and Swager, T.M. (2000) Conjugated polymer-based chemical sensors. *Chem. Rev. (Washington)*, **100**, 2537–2574.
126. Bernards, D.A.O., Roisin, M., and Malliaras, G.G. (2008) *Organic Semiconductors in Sensor Applications*, Springer, Berlin.
127. Thackeray, J.W., White, H.S., and Wrighton, M.S. (1985) Poly(3-methylthiophene)-coated electrodes: optical and electrical properties as a function of redox potential and amplification of electrical and chemical signals using poly(3-methylthiophene)-based microelectrochemical transistors. *J. Phys. Chem.*, **89**, 5133–5140.
128. Thackeray, J.W. and Wrighton, M.S. (1986) Chemically responsive microelectrochemical devices based on platinized poly(3-methylthiophene): variation in conductivity with variation in hydrogen, oxygen, or pH in aqueous solution. *J. Phys. Chem.*, **90**, 6674–6679.
129. Gaponik, N.P., Shchukin, D.G., Kulak, A.I., and Sviridov, D.V. (1997) A polyaniline-based microelectrochemical transistor with an electrocatalytic gate. *Mendeleev Commun.*, **7**, 70–71.
130. Dabke, R.B., Singh, G.D., Dhanabalan, A., Lal, R., and Contractor, A.Q. (1997) An ion-activated molecular electronic device. *Anal. Chem.*, **69**, 724–727.
131. Nilsson, D., Kugler, T., Svensson, P.-O., and Berggren, M. (2002) An all-organic sensor-transistor based on a novel electrochemical transducer concept: printed electrochemical sensors on paper. *Sens. Actuators B*, **86**, 193–197.
132. Bartlett, P.N., Birkin, P.R., Wang, J.H., Palmisano, F., and De Benedetto, G. (1998) An enzyme switch employing direct electrochemical communication between horseradish peroxidase and a poly(aniline) film. *Anal. Chem.*, **70**, 3685–3694.
133. Matsue, T., Nishizawa, M., Sawaguchi, T., and Uchida, I. (1991) An enzyme switch sensitive to NADH. *J. Chem. Soc., Chem. Commun.*, 1029–1031.
134. Bartlett, P.N. and Birkin, P.R. (1993) Enzyme switch responsive to glucose. *Anal. Chem.*, **65**, 1118–1119.
135. Bartlett, P.N. and Birkin, P.R. (1994) A microelectrochemical enzyme transistor responsive to glucose. *Anal. Chem.*, **66**, 1552–1559.
136. Hoa, D.T., Kumar, T.N.S., Punekar, N.S., Srinivasa, R.S., Lal, R., and Contractor, A.Q. (1992) A biosensor based on conducting polymers. *Anal. Chem.*, **64**, 2645–2646.

137. Battaglini, F., Bartlett, P.N., and Wang, J.H. (2000) Covalent attachment of osmium complexes to glucose oxidase and the application of the resulting modified enzyme in an enzyme switch responsive to glucose. *Anal. Chem.*, **72**, 502–509.
138. Sangodkar, H., Sukeerthi, S., Srinivasa, R.S., Lal, R., and Contractor, A.Q. (1996) A biosensor array based on polyaniline. *Anal. Chem.*, **68**, 779–783.
139. Sukeerthi, S. and Contractor, A.Q. (1999) Molecular sensors and sensor arrays based on polyaniline microtubules. *Anal. Chem.*, **71**, 2231–2236.
140. Nishizawa, M., Matsue, T., and Uchida, I. (1992) Penicillin sensor based on a microarray electrode coated with pH-responsive polypyrrole. *Anal. Chem.*, **64**, 2642–2644.
141. Kanungo, M., Srivastava, D.N., Kumar, A., and Contractor, A.Q. (2002) Conductimetric immunosensor based on poly(3,4-ethylenedioxythiophene). *Chem. Commun. (Cambridge)*, 680–681.
142. Krishnamoorthy, K., Gokhale, R.S., Contractor, A.Q., and Kumar, A. (2004) Novel label-free DNA sensors based on poly(3,4-ethylenedioxythiophene). *Chem. Commun. (Cambridge)*, 820–821.
143. Yang, S.Y., DeFranco, J.A., Sylvester, Y.A., Gobert, T.J., Macaya, D.J., Owens, R.M., and Malliaras, G.G. (2009) Integration of a surface-directed microfluidic system with an organic electrochemical transistor array for multi-analyte biosensors. *Lab Chip*, **9**, 704–708.
144. Ahn, C.H., Bhattacharya, A., Di Ventra, M., Eckstein, J.N., Frisbie, C.D., Gershenson, M.E., Goldman, A.M., Inoue, I.H., Mannhart, J., Millis, A.J., Morpurgo, A.F., Natelson, D., and Triscone, J.-M. (2006) Electrostatic modification of novel materials. *Rev. Mod. Phys.*, **78**, 1185–1212.

Part II
Manufacturing

7
Printing Techniques for Thin-Film Electronics

Vivek Subramanian, Alejandro de la Fuente Vornbrock, Steve Molesa, Daniel Soltman, and Huai-Yuan Tseng

As discussed in other chapters, solution-processed thin-film electronics has received substantial attention as a means of realizing a range of low-cost electronic systems [1–9]. In particular, to achieve low cost, there has been substantial interest in the use of printing techniques as a means of fabricating thin-film electronic devices and circuits. To this end, in this chapter, we review the state-of-the-art in printing techniques for thin-film electronics, discuss some of the physical issues associated with printing for the same, and illustrate some of the consequences of these phenomena on the printing of thin-film electronic devices.

7.1
The Motivation for Printing of Thin-Film Electronic Devices

The widespread interest in printing for thin-film electronics fabrication is largely driven by the perceived reduction in cost that will result from the use of the same. We therefore begin by reviewing the economics of printing as applied to electronics fabrication. This is performed by contrasting the economics of conventional fabrication techniques to those of printing.

Conventionally, electronics systems are fabricated using a series of subtractive steps. A typical electronics system may contain as few as three to as many as dozens of subtractively patterned layers. A typical subtractively patterned layer would be formed as follows:

- blanket layer of material is deposited (e.g., a conductor, semiconductor, or dielectric).
- The layer is coated with photoresist.
- The photoresist is exposed using a lithographic technique.
- The photoresist is developed.
- The layer is then etched through the exposed regions of the pattern.

Organic Electronics II: More Materials and Applications, First Edition. Edited by Hagen Klauk.

- The photoresist is then stripped.
- The wafer is cleaned, and the entire process is repeated as needed.

Thus, we see that a typical subtractively patterned layer involves the use of multiple steps, each associated with a relatively high-priced tool. Typical tool costs in the modern semiconductor and display industries range from hundreds of thousands of dollars to multiple tens of millions of dollars per tool. Throughput is also limited by the need for all of these steps. On the other hand, conventional subtractive processing techniques have reached a level of maturity such that patterning of features at very high density is routinely achieved. Conventional microelectronics routinely results in the fabrication of transistors with feature sizes below 100 nm, whereas displays are often fabricated with critical feature sizes in the range of single microns.

In contrast, printing uses an additive processing paradigm. Conceptually, a printed layer would involve a single patterned deposition step, followed by a drying/sintering step. For example, to print a conductor layer using a nanoparticle ink or a conductive polymer, the steps used would be as follows:

- patterned layer is printed using an appropriately selected conductive ink.
- The ink is dried and sintered.

Thus, we see that printing potentially enables a substantial reduction in overall process complexity. In addition, many printing techniques offer extremely high throughput. Finally, overall tool cost is reduced as well, both because the cost of individual tools is expected to drop somewhat and also because the number of tools required is reduced given the reduction in overall process complexity. The precise cost advantages relative to silicon depend on process throughput and material costs, although cost advantages on the order of 1–3 orders of magnitude per unit area are expected.

On the other hand, state-of-the-art printing does not achieve the linewidth and functional density that is achievable using conventional subtractive processing techniques. Some chapters do discuss highly scaled printing techniques, but in general, there is still a substantial functional density gap between printed electronics and conventional microelectronics and display technology. This leads to an interesting paradox – while printed electronics is expected to be cheaper than conventional subtractive processing techniques per unit area, it is actually likely to be much more expensive per transistor. This is because of the much lower functional density that is achievable using printing techniques.

The conclusion that can be drawn from this, therefore, is that printed electronics in general and printing of thin-film electronic devices in particular is attractive for applications that are area constrained as opposed to function constrained. Several such applications have been identified, such as displays and RFID tags [2], and these are discussed in other chapters.

7.2
Requirements for Printing Techniques for Electronics Fabrication

Prior to reviewing the specifics of printing technology, it is worthwhile to begin by analyzing the requirements that are imposed on printing by the system needs of printed electronics. In general, these requirements can be categorized as follows:

- **Linewidth** – linewidth is a critical feature requirement for printed electronics. Linewidth often determines the performance of a printed electronics system; for example, linewidth may determine the channel length of a transistor and hence its switching speed. Linewidth may also be the key to realizing the functional system requirements of a target application; for example, linewidth may determine the achievable resolution of a display. Overall, most printed electronics systems imposed specific linewidth requirements as well as specific line-space requirements.
- **Printed film thickness** – electronic devices including both active components such as transistors and passive components such as inductors and wires typically have specific requirements in terms of the thickness of the various layers involved in fabricating the device. For example, transistors may have specific requirements for the thickness of gate dielectrics and semiconductor layers.
- **Line-edge roughness** – since patterned features are never perfectly straight edged, electronics systems typically impose a constraint in terms of required maximum line-edge roughness. For example, to realize reliable transistors, specific requirements for gate line-edge roughness typically exist.
- **Material compatibility** – of course, as is obvious, any printing technique must be compatible with the requisite material needed to form the layer in question.
- **Throughput** – to meet the cost and manufacturing requirements of the intended application, printing techniques for fabrication of thin-film electronics will typically have required minimum throughputs.
- **Layer-to-layer registration** – since the overlap between the layers often introduces parasitic capacitance elements into a circuit, there is often the requirement in terms of maximum allowed overlap, that is, there is a requirement in terms of layer-to-layer registration accuracy.

Having introduced the various parameters of interest to be used while evaluating a printing technique for electronics fabrication, it is now possible to introduce the various printing techniques that have been considered for printed electronics to date.

7.3
A Survey of Printing Techniques for Printed Electronics

Several well-known printing techniques have been demonstrated for the production of printed electronics components, including screen printing, flexographic printing, gravure printing, offset printing, and ink-jet printing [1, 2, 10–12]. To date, the vast majority of printed electronics demonstrations have been in the area of printed wiring boards, membrane switches, and printed passive components for applications such as RFID tags and touch screens. In these applications, printing has been deployed both in research and in production. In these applications, screen and flexo have been extensively demonstrated. Ink-jet has been used in conjunction with other techniques as well. Here, we will review these techniques, specifically focusing on the capabilities of and limitations imposed by the same.

7.3.1
Screen Printing

Screen printing is extensively used in the printed circuit board industry, so it represents a natural migration path for printed electronics development [13]. Screen printing typically makes use of pastelike inks, with very high viscosities. These inks are deposited onto the substrate by pushing the ink through a patterned screen; the high viscosity is required to ensure pattern fidelity by minimizing flowing of the ink on the substrate. Typical resolution of screen printing is >50 μm, although sub-30 μm screen-printing techniques have been described. A typical screen-printing process is shown schematically below (Figure 7.1).

The main consequences of screen printing arise from its viscosity requirements and also from its resolution limits. The high viscosity requirements of screen printing typically necessitate the use of inks with binders within the formulation. These binders help increase the viscosity of the inks. Therefore, a typical conductive ink formulated for screen printing consists of conductive particles or flakes interspersed with a binder material such as polymer within a solvent base. After screen printing and subsequent drying, the resulting pattern typically consists of

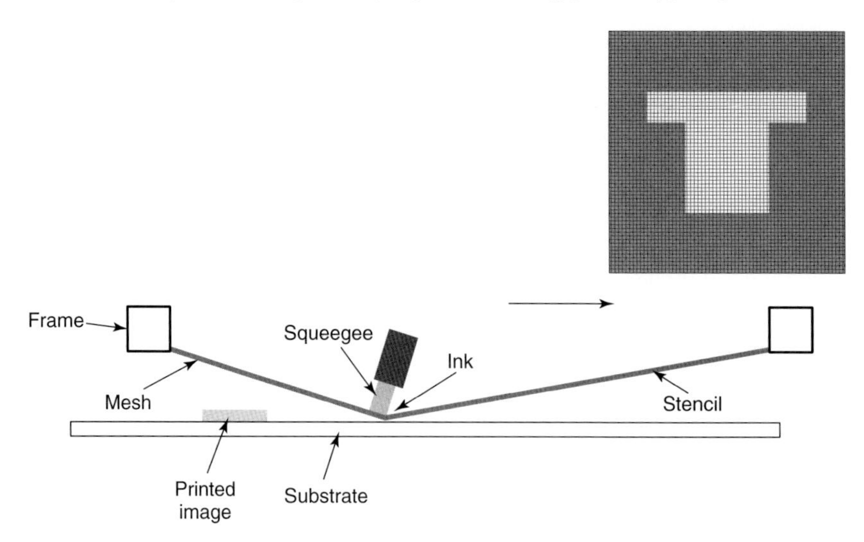

Figure 7.1 Schematic representation of screen printing. Notes: (i) High viscosity is required to prevent bleeding. (ii) Resolution limits are also set by screen mesh size.

metallic flakes and particles with interspersed binder material. Since the binder material is relatively nonconductive, the overall resistivity of screen-printed material is typically many times larger than an equivalent film of pure metal; this necessitates the use of thicker/wider features to minimize resistance.

The resolution limits of screen printing also have an impact, since they impact the ability produce high-resolution circuits and devices. Furthermore, given the nature of screen inks as discussed above, such inks typically result in films with relatively large roughness. This, in turn, impacts the viability of screen printing for realization of thin-film devices, since roughness typically limits the ability to use overlying thin films. As a result, while screen printing is generally important for the printing of printed circuits boards, and so on, for the realization of thin-film devices, as is the subject of this chapter, screen printing is arguably less important. Therefore, screen printing will not be discussed in further detail here.

7.3.2
Gravure/Flexographic/Offset Printing

Gravure and flexographic printing are somewhat related to each other. Both techniques make use of a drum with prepatterned features (Figure 7.2). Ink is deposited into the drum such that it lies within the wells produced by the features but not in the elevated regions of the drum (i.e., the ink lies in the depressed regions of the pattern, but not in the elevated "field" regions of the pattern) [14, 15]. The drum is pressed against the substrate, causing most of the ink in the wells to transfer to the substrate. The main difference between gravure and flexographic printing is the choice of drum material. Gravure makes use of a hard metallic drum, whereas flexographic printing makes use of a somewhat compliant, rubberlike drum. This results in different viscosity and surface wetting requirements for the ink, and also results in slightly different pattern fidelity characteristics. Offset printing adds an intermediate step – the ink from the master drum is transferred to an intermediate drum, which, in turn, transfers the ink to the substrate [11]. This typically results in better drum reusability; however, the double transfer typically necessitates the use of higher viscosity inks.

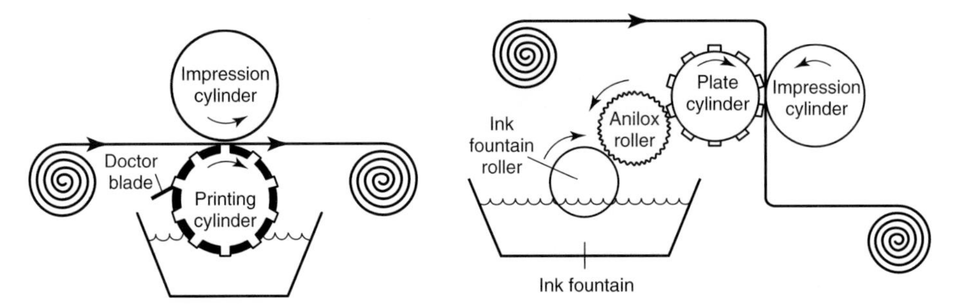

Figure 7.2 Schematic representations of (a) rotary gravure system and (b) flexographic system.

Ink viscosities required for gravure and flexo typically necessitate the use of binders, similar to those used for screen-printing inks. However, since the viscosity requirements for flexo and gravure are less than that required for screen printing, binder content can usually be reduced, which allows for the realization of features with lower resistivity. In addition, higher resolution gravure features have been reported compared to screen printing. The overall consequences are similar to those that exist for screen printing. For printed electronics, gravure does offer some advantages. First, since gravure has excellent pattern fidelity, it allows for the realization of fine lines and fine spaces in printed electronics patterns. In addition, since film thickness control is very good, assuming surface roughness issues with the inks are dealt with, it is possible to realize high-quality thin films suitable for use in thin-film electronic devices using gravure printing.

The gravure roll used to pattern in this printing technique is typically metallic and has patterns etched or engraved into its surface. These patterns are inked, and any excess ink that is in the surrounding field regions is subsequently wiped off using a doctor blade. The roll is then contacted to the substrate, resulting in transfer of ink from the patterned wells to the substrate (Figure 7.3) [14].

The wiping process is critical, since the doctor blade must remove essentially all ink from the nonpatterned areas of the roll. This places specific requirements on the quality and properties of the surface of the roll. To satisfy these requirements, gravure rolls are commonly made out of chrome-coated copper. Most rolls are formed using a steel or aluminum core, with an overlying layer of electroplated copper, which can be easily machined and patterned, and then coated with a hard chromium layer for improved wear resistance.

The wells are commonly classified into two families – intaglio and gravure. In intaglio, the pattern to be printed is etched or engraved directly into the roll. For example, if a line is to be printed, then a thin channel is patterned into the roll. Likewise, if a solid-filled square is to be printed, then a square is etched into the roll. In gravure, patterns are broken up into pixels, that is, a typical pattern is

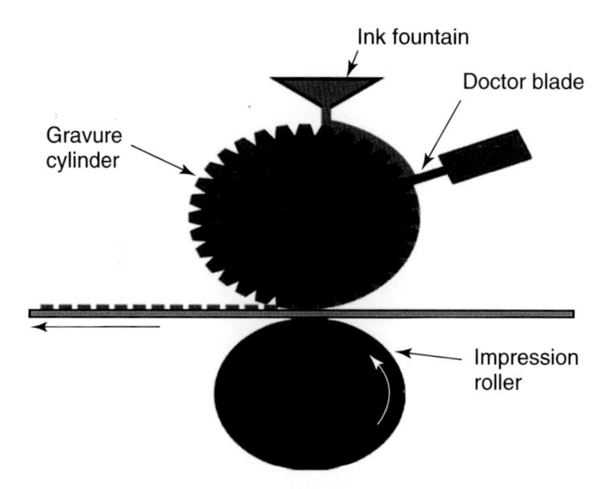

Figure 7.3 The gravure printing process.

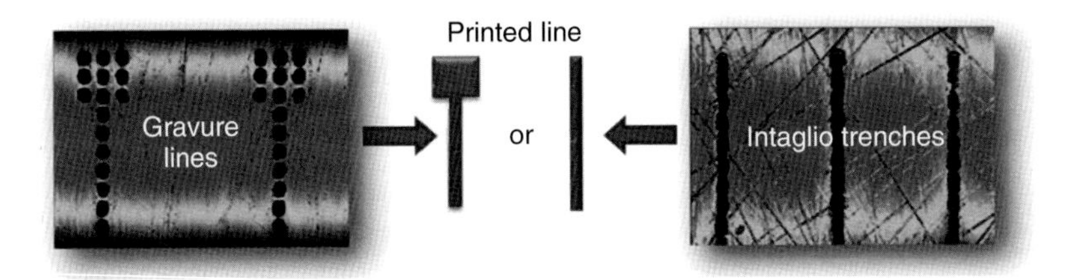

Figure 7.4 Gravure versus intaglio.

formed from a series of smaller unit cells. The cells are separated from each other by a space sufficient to prevent the cells from accidentally merging during roll fabrication (Figure 7.4).

Conceptually, intaglio would be expected to be the more desirable printing method, since there is a one-to-one correspondence between the pattern on the roll and the desired printed image. Intaglio can transfer more ink per unit area than gravure, and can make smoother line edges, but it suffers from a variety of drawbacks. Because intaglio patterns are formed using a single well, controlling the amount of ink deposited is complicated. The lack of a cell wall causes a deflection in the doctor blade when crossing large patterned areas, reducing its ability to wipe the nonimage areas. Further, fluid inside intaglio patterns can redistribute itself during the print, causing print inconsistencies. Thus, intaglio has been largely replaced by gravure, and is only considered attractive for the printing of very small patterns in printed electronics.

In gravure, the ink is metered by the size of the individual cells. Large-area and small-area patterns alike can be printed with the same thickness of ink because of this metering. Further tonal differences in an image can be easily implemented by modulating the size of the wells, thus reducing the thickness of ink deposited. A drawback, however, is that gravure typically will have worse line-edge roughness than intaglio, for obvious reasons. In practice, both types of patterns can be made on a single roll taking advantage of each cell type, where most appropriate.

A key issue is the patterning of the wells in a gravure cylinder. Two techniques are commonly used – etching and engraving. Patterning by wet etching uses a ferric chloride solution to isotropically etch the copper, which has been patterned with an etch resist. The traditional resist consists of a pigment paper with a light-sensitive gelatin layer. This resist is patterned off the roll by exposing light through a photomask in a copying frame. The patterned resist is then laminated onto the roll and the paper backing is removed. More modern systems use a laser to pattern a resist directly on the roll. This is known as the *indirect laser method.*

Electromechanical engraving uses a diamond stylus to cut wells into the copper layer. As the cutting operation is performed, the gravure roll rotates as the diamond stylus is pushed into and out of the surface, cutting the cell. The rate of approach and withdrawal can be adjusted to control cell depth in a continuously variable manner (Figure 7.5).

The smoothness of the gravure cylinder is also a concern. Therefore, polishing is extensively used in the cylinder preparation process. For example, polishing is used

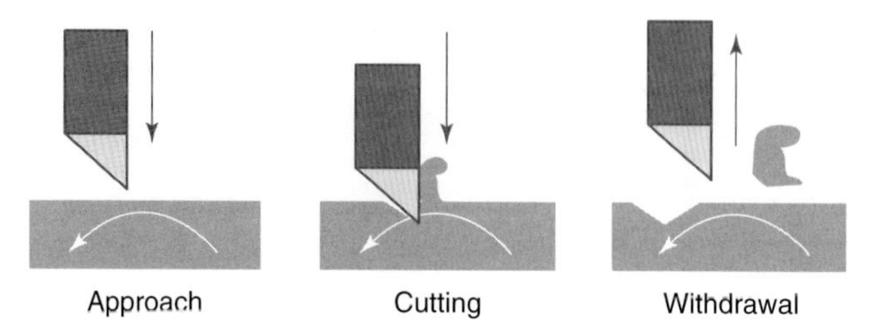

Figure 7.5 Electromechanical engraving process.

to smooth any imperfections in the electroplated copper film, allowing for a faithful reproduction of the intended pattern. Polishing is also used to provide a controlled surface roughness on the chrome that allows the doctor blade to maintain some lubrication in the nonimage areas, yet prevents any significant amount of ink from transferring to the surface.

The overall gravure printing process can be separated into two parts, cell emptying and drop spreading. Both play important roles in defining the final printed features. The ink within the wells first makes contact with the substrate, then transfers to the substrate as the cell rotates off the substrate, and finally spreads. Clearly, the final width and thickness of the drop is dependent on the cell width and depth (Figure 7.6).

During the transfer process, the cell does not completely empty, and the exact amount of ink that is transferred from the cell to the substrate depends on the rheology of the fluid, the shape of the cell, and the ink–surface interactions. Smaller cells tend to empty proportionally less ink than larger cells because of the larger surface-to-volume ratio in the smaller cells, resulting in relatively larger adhesive forces between the cell and the fluid. Also, higher aspect ratio wells require ink to be pulled further from a narrower body of fluid onto the substrate, if equal volumes of ink are to be transferred compared to a lower aspect ratio cell. Higher surface energy substrates will tend to adhere better to the ink and pull it out of the well better than low surface energy inks. The print speed is also important in determining how much ink is removed from the cell, as the speed determines how much sheer is applied to the fluid. Higher viscosity inks will have a greater resistance to sheer and will thus empty less at high speeds.

After leaving the cell, the drop will begin to spread on the substrate, thus minimizing its total surface energy. How far the drop spreads is determined by the surface energies of the three intersecting phases, the solid and liquid, the solid

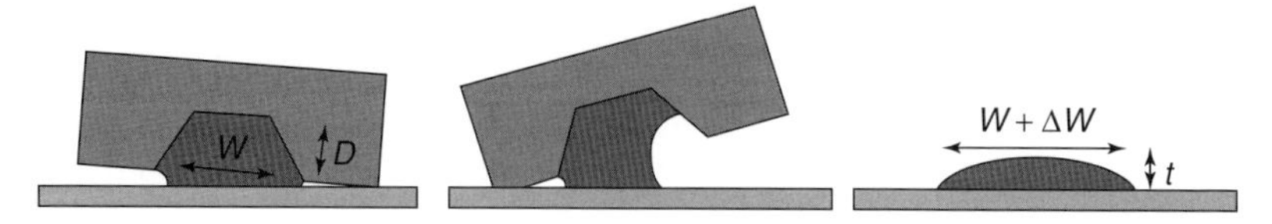

Figure 7.6 Schematic of the print process as ink contacts the substrate, is removed from the cell, and spreads.

and air, and the liquid and air. The issue of droplet spreading will be discussed in more detail in a later section of this chapter, since it is an important issue affecting numerous printing techniques.

Line formation occurs when drops are deposited close together and thus coalesce as they spread on the substrate. Line formation has been well studied for ink-jet printing, and will be discussed later in this chapter, since analogous phenomena apply to gravure-printed lines as well.

Viscosity plays another important role in achieving good lines. If the ink is too viscous, drops will not flow smoothly into each other and the resulting lines will have poor width and thickness uniformity. On the other hand, if the ink viscosity is too low the ink will spread too far, degrading achievable linewidth.

7.3.3 Ink-jet Printing

Ink-jet printing is arguably the most important printing technique for printed electronics today. Of the numerous demonstrations of printed devices that have been described in the literature, the vast majority have been fabricated using ink-jet printing because of its excellent material compatibility and ease of use [1, 9, 12, 16–21]. Ink-jet printing makes use of a drop-by-drop process to deposit ink into a substrate through a nozzle. The droplets are typically generated using either thermal bubble techniques or piezoelectric transducers to squeeze out a drop from a reservoir. For most electronics applications, the latter technique is used to maximize material compatibility. Both techniques have certain aspects in common – in particular, both techniques used the actuation means (either a thermal bubble or a piezo constriction) to force the droplets through a nozzle in a repeated manner.

A thermal dispenser uses a heater in the form of a resister to superheat the ink to its spinodalian limit, 312 °C for water, whereby the ink boils and a bubble is created. The bubble is created in the time span of microseconds and causes pressures that cause droplet ejection (Figure 7.7). Owing to the relative simplicity of heads using such designs, highly scaled heads have been realized using thermal bubble processes.

However, the drawback of heads that use thermal ink-jetting is that the requisite heating step poses compatibility concerns for thin-film electronic materials, where the inks are usually temperature sensitive.

Piezoelectric heads are actuated using a piezoelectric crystal. By applying a potential across the crystal, the crystal deforms, and, if properly oriented with respect to an ink-containing chamber of a dispenser, creates a pressure wave that expels ink droplets (Figure 7.8). Unlike thermal ink-jet heads, therefore, piezo heads do not suffer from ink degradation concerns because of heating.

Ink-jet has several advantages and disadvantages for printed electronics fabrication. First, ink-jet typically allows the use of lower viscosity inks than any other printing technology. Therefore, it is possible to produce ink-jet-compatible inks with no binder materials. This allows for the realization of higher quality

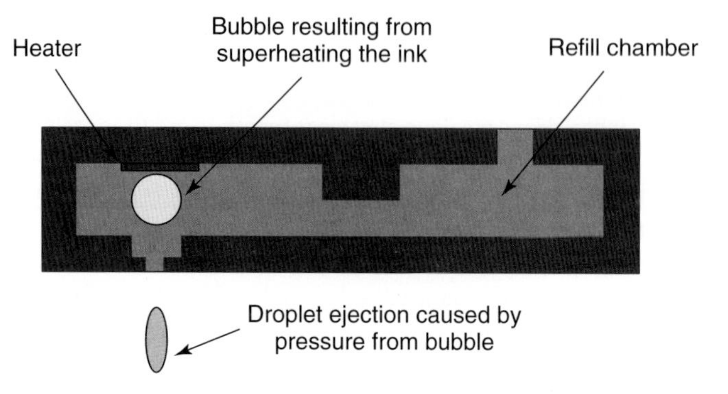

Figure 7.7 Cross-section of a typical bubble jet dispenser showing how droplet generation occurs.

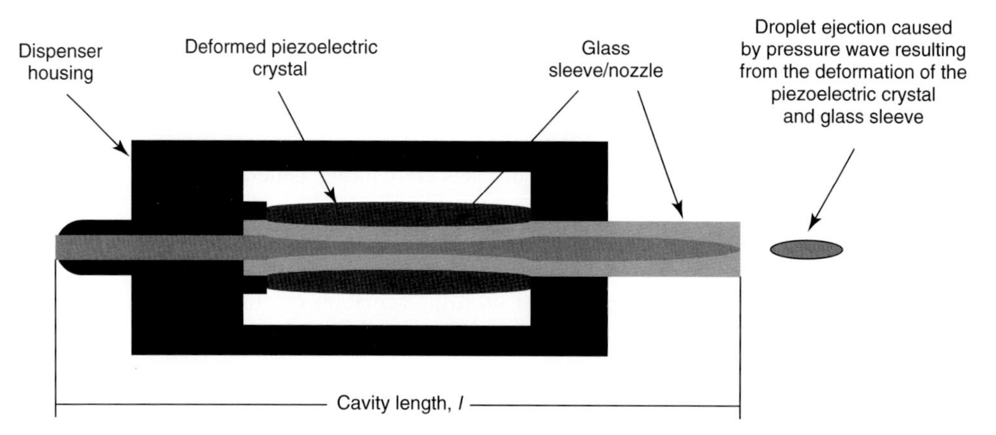

Figure 7.8 Cross-section of a piezoelectric ink-jet dispenser showing how droplet ejection occurs.

ink-jet printed patterns. Second, being a digital fabrication technique, ink-jet allows for on-the-fly correction as well as easy pattern changes. Unfortunately, ink-jet has several drawbacks. First, the process stability of ink-jet is generally a concern, although this issue is being addressed through the production of advanced heads specifically designed for electronics applications. Second, since ink-jet uses a drop-by-drop approach, it typically requires several passes to produce relatively thick films for applications that require thicker film buildup. This, in turn, reduces process throughput. Third, since ink-jet tends to be a relatively slow technique, it is necessary to use large arrays of heads to produce high-throughput ink-jetting systems; the manufacturability and process stability of such large arrays of heads is still uncertain for electronics applications, although page-wide heads are already in use in the graphic arts industry.

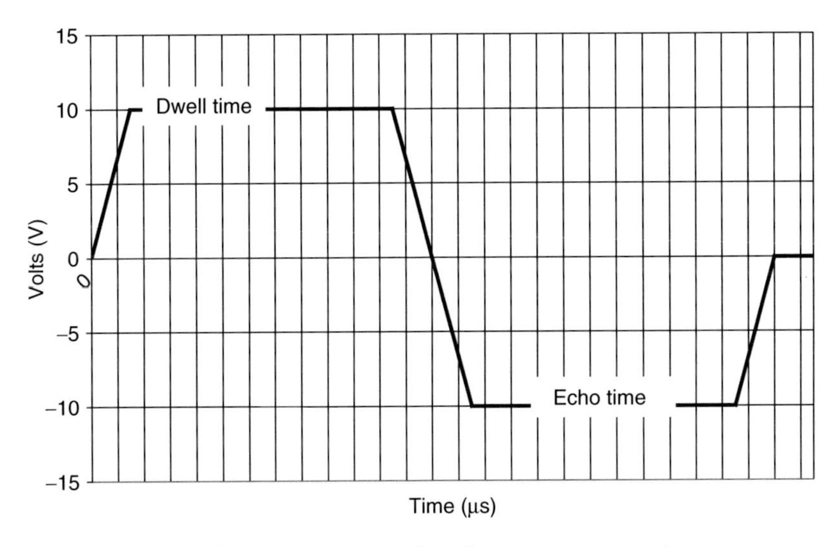

Figure 7.9 A typical bipolar actuation waveform for a piezo-ink-jet head.

Most piezo-ink-jet dispensers use either a unipolar or bipolar voltage waveform to ink-jet the material (Figure 7.9). In recent years, bipolar waveforms have become more common. Conceptually, the idea is as follows:

- The positive half of the waveform causes the piezo membrane to deflect, distorting the meniscus at the nozzle such that a hemispherical bulge of fluid hangs from the nozzle.
- During the negative half of the waveform, the piezo membrane is pulled back. This causes the ink to attempt to flow back into the nozzle. Since the nozzle is constricted, a neck will develop in the ink, ultimately pinching off a droplet and ejecting it from the nozzle.
- The ink then equilibrates, and the process is repeated for future pulses.

7.4 Pattern Formation During Printing

As is apparent from the above sections, printing techniques for the realization of thin-film electronics typically made use of droplet-based pattern formation, whether that drop is deposited via an ink-jet printer or by the emptying of a gravure pattern. To this end, we end our discussion of printing of thin-film electronics by reviewing the physics and mechanisms of pattern formation from discrete droplets.

For printed electronics, ideal lines should be smooth, even, narrow, and straight. However, depending on printing conditions, it is possible for the resulting patterns to deviate substantially from this desired behavior [22–25]. A few principal behaviors emerge when examining printed patterns across a variety of drop spacings, delay

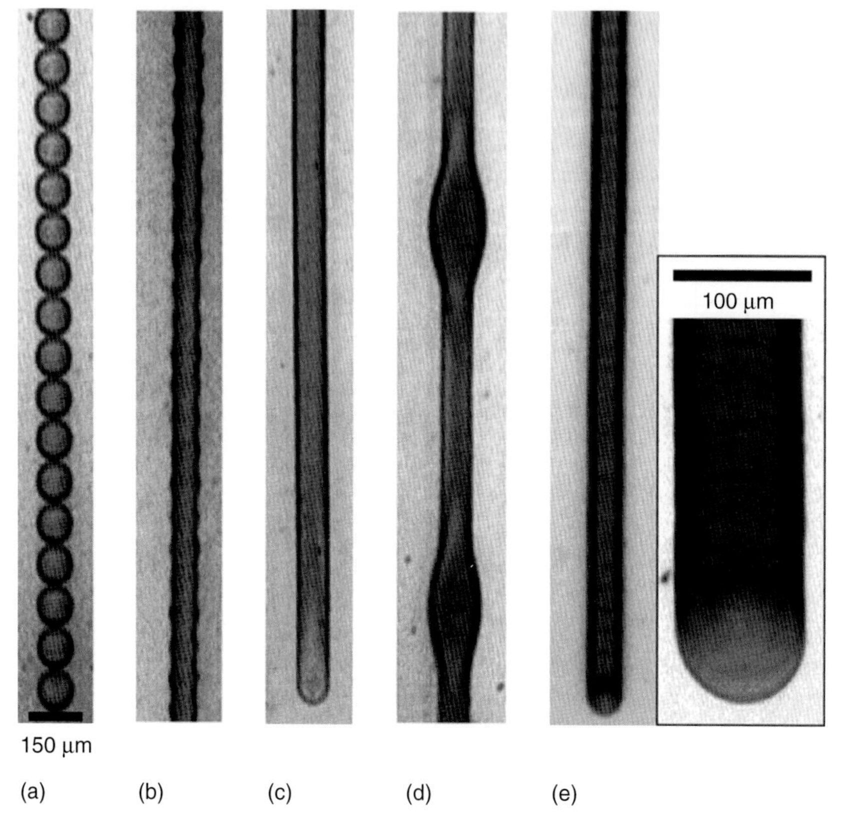

Figure 7.10 Examples of principal printed line behaviors: (a) individual drops; (b) scalloped; (c) uniform; (d) bulging; and (e) stacked coins. Drop spacing decreases from left to right.

periods between deposition of adjacent drops and substrate temperatures. We label these as individual drops, scalloped line, uniform line, bulging line, and stacked coins (Figure 7.10).

The origins of the observed behaviors relate to the way fluids flow as the droplets dry. When a drop is deposited on a nonabsorbent substrate, the spherical drop expands to a determinable hemispherical radius. It often is then pinned at this radius. As the solvent evaporates from the drop, there is faster evaporation at the edges of the drop than at the center. Consequently, a convective flow develops, pulling material toward the edges of the drop. In isolated drops, this results in the commonly observed "coffee ring," as has been discussed by numerous authors. The same flow is also responsible for the range of behaviors observed above. If drops are printed too far apart, more than twice a drop's radius, then isolated drops land and dry. As drop spacing decreases, isolated drops overlap and merge but retain individual rounded contact lines, and a scalloped pattern emerges. Further decreasing the drop spacing will eliminate the scalloping and lead to a smooth, straight line. These lines have a uniform smooth edge and top. Printing drops even closer together leads to discreet bulging along the line's length, separated by regions of uniform narrow line (Figure 7.10d). These bulges tend to form periodically and also at the beginning of the line. Essentially, rapid addition of additional fluid via printing of subsequent closely spaced drops exceeds a bead's equilibrium contact angle and discreet regions of outflow result, leading to rounded bulges in the dried feature. If the substrate temperature increases such that the evaporation time of a single drop is less than the drop jetting period, then each landing drop will dry individually regardless of overlap, leading to what looks like offset stacked coins. At a given substrate temperature, increasing drop delay will affect the onset of the stacked coin behavior.

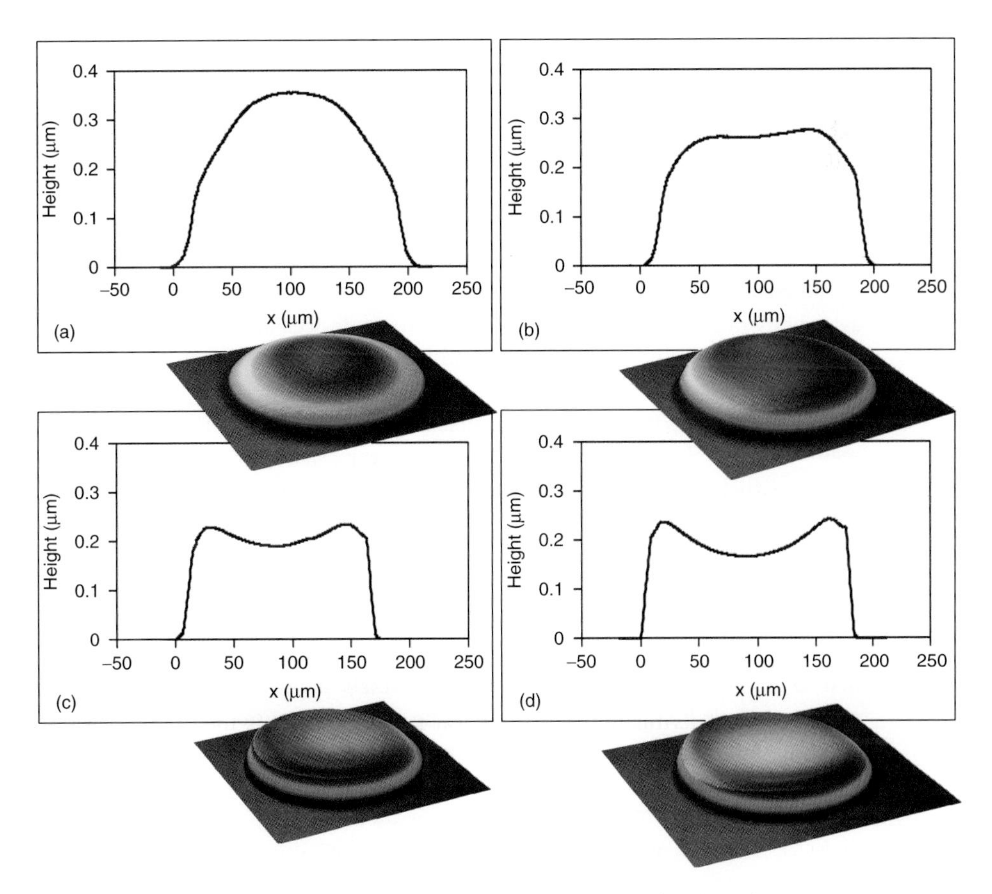

Figure 7.11 Cross-sectional profile of uniform lines printed at (a) 17 °C; (b) 30 °C; (c) 45 °C; and (d) 60 °C using mechanical stylus profiler. (Delay and spacing are adjusted to be appropriate for each temperature.)

Through appropriate optimization, it is possible to print a smooth, narrow line with an even edge. It is possible to avoid bulging by slowing down the drop frequency until the advancing contact angle is never exceeded, but is not so slow that drops dry within the period of one or two drops landing, avoiding stacked coins. Correspondingly, at the other limit, the drop spacing is made low enough to avoid scalloping.

The manifestation of the "coffee ring" is also visible in printed lines, based on the evaporation-induced convective fluxes within the bead. These can be controlled, for example, by varying substrate temperature (Figure 7.11). They can also be controlled by altering the solvent evaporation rate or breaking up the convective flow using a multisolvent system.

7.5
Printed Device Considerations

In previous sections, we have discussed the reasons to consider printing, have discussed printing technology in general, and have examined the issues associated with formation of patterns for well-defined electronic structures. We end this chapter by reviewing some achievements in printed electronics to illustrate the power of printing for this application. Further details on the application of printing to the realization of electronic devices are available in other chapters.

The basic need for realizing any printed electronics system is, of course, the availability of suitable "inks." To print electronic devices and circuits, a range of printable semiconductor, conductor, and dielectric inks are required. In recent years, two main classes of materials have become dominant – inorganic structures are realized using nanoparticle precursors [26, 27] and organic structures are realized using soluble organic inks [1, 9, 12, 16, 28].

Nanoparticles are a particularly attractive material system for printed electronics for two reasons. First, it is possible to realize soluble and, therefore, printable forms of a range of materials using nanoparticles, including a range of semiconductors, dielectrics, and conductors. Second, through appropriate design of the nanoparticles, it is possible to produce inks that may be printed and subsequently sintered to produce very high-quality polycrystalline thin films. This latter behavior is due to the fact that nanoparticles can be designed to show dramatically depressed melting points relative to their bulk counterparts, enabling low-temperature, plastic-compatible annealing of nanoparticle films to realize high-quality thin films with good grain structure and resulting high electrical performance.

Using the aforementioned nanoparticle conductors in conjunction with polymer dielectrics, a range of printed passive components including inductors, capacitors, multilevel interconnects, and so on, have been realized [29]. Owing to the high conductivity, very good performance is achievable; high-frequency performance is limited by the inability to build up thick films and by dispersion that manifests itself in some polymer dielectrics at high frequency (Figure 7.12).

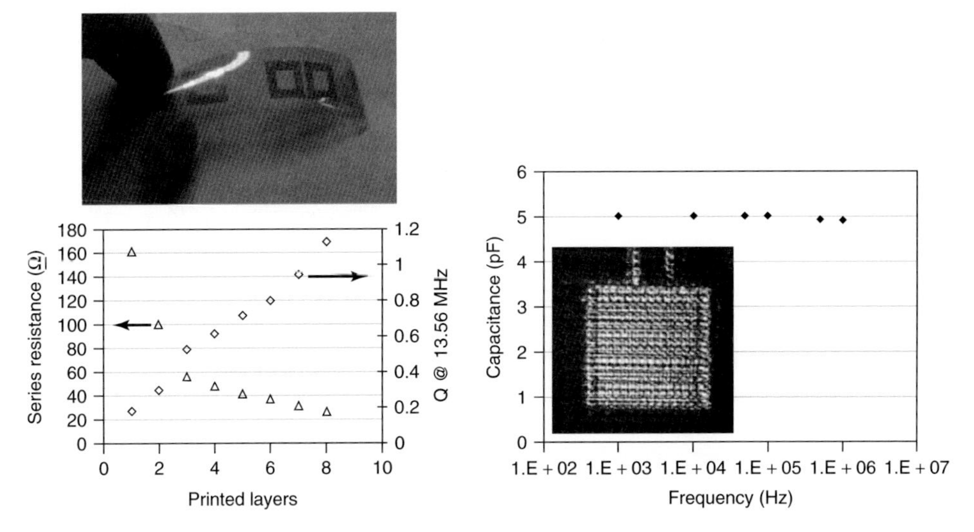

Figure 7.12 (Left) Printed inductors achieved using ink-jet printing; even higher performance (Q ~ 10) is achievable using thicker buildup of films via multipass ink-jet, gravure printing, or printing followed by plating. (Right) Printed capacitor formed using polymer dielectric and nanoparticle electrodes. There is little dispersion up to 13.56 MHz.

These same materials can be utilized to realize printed transistors. Over the years, several groups have realized printed transistors, some of which are summarized in other chapters [1, 12, 16–18]. Conceptually, the idea is to build up multilayer devices using individual layers that are printed using the techniques described above. Clearly, it is necessary to have excellent control of the pattern quality parameters discussed above to realize such devices. To illustrate this, we conclude by examining a prototypical printed transistor process flow to highlight how printing impacts the realization of printed electronics systems and devices.

A typical printed transistor requires the printing of at least three or four layers including the gate, the gate dielectric, the source and drain electrodes, and the semiconductor (Figure 7.13). Conventionally, layer-to-layer alignment is achieved using the registration of the printing system; this results in a relatively large overlap capacitance in printed electronic devices. More recently, however, there have been demonstrations of self-alignment between layers by exploiting wetting characteristics of the inks, and so on [19, 20, 30]. For example, by utilizing wetting-based self-assembly, in which the source/drain lines are pinned to the corners of an underlying gate stack, transistors with very low overlap capacitance, and hence, relatively fast switching speed have been reported. By controlling the roll-off characteristics, the overlap capacitance can be minimized. Clearly, this attests to the importance of an understanding of both printing in general and fluid behavior in particular (Figure 7.14).

Looking at the aforementioned printed transistor process flow, it is now clearly apparent why line morphology and pattern fidelity is so important. Since multiple layers are deposited on top of each other, it is necessary to have excellent control of

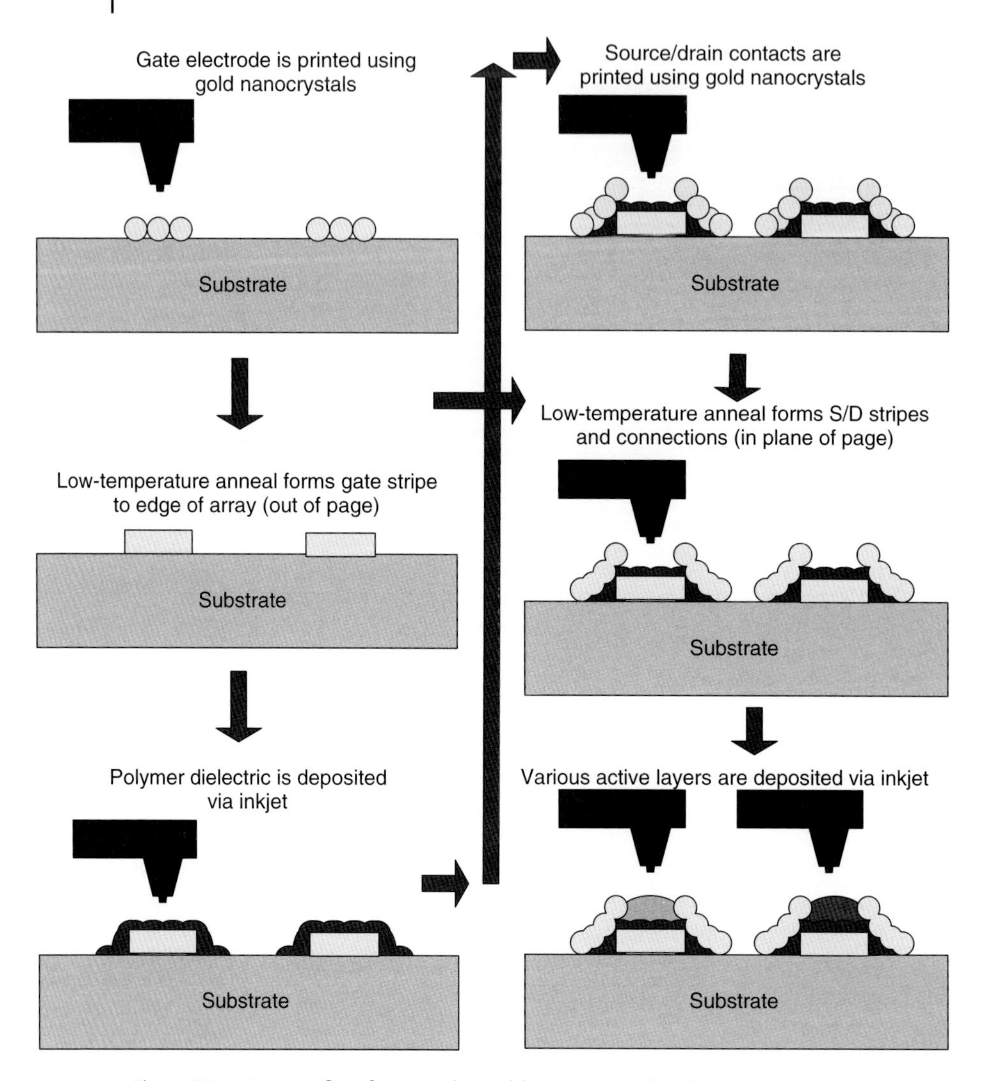

Figure 7.13 Process flow for an archetypal bottom-gated thin-film transistor.

line-edge roughness, film thickness, and overall pattern quality. The techniques to control the bulging, scalloping, and linewidth and thickness, as discussed above, clearly influence the final device characteristics. Using these approaches, various groups have been able to demonstrate devices with performance approaching or even exceeding that of amorphous silicon, by exploiting the simplicity and additive process capabilities of printing techniques, including ink-jet printing and gravure printing, as discussed in this chapter. Overall, therefore, we see how the field of printing, including its physics, technology, rheology, materials science, and device integration, can be used to drive the development of printed electronics. More details on the demonstration of printed devices are provided in other chapters.

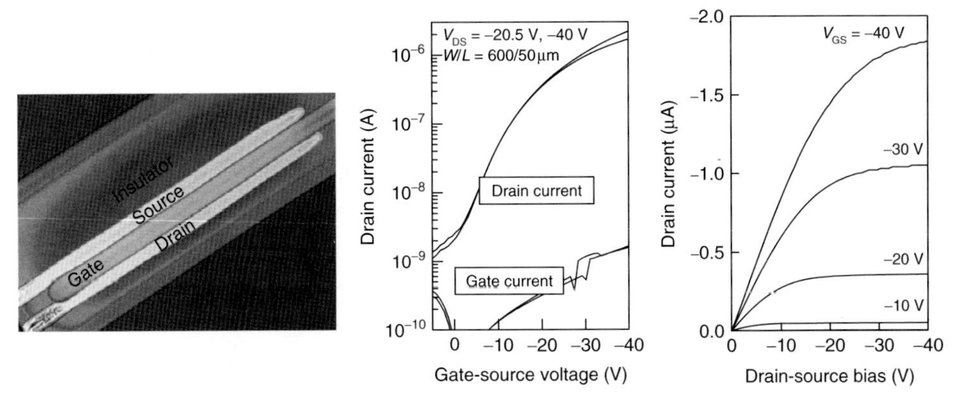

Figure 7.14 (a) Optical micrograph and (b,c) electrical characteristics of a printed self-aligned transistor. The source/drain electrodes roll off the gate stack to achieve self-alignment. All layers are printed using ink-jet printing.

References

1. Kawase, T., Shimoda, T., Newsome, C., Sirringhaus, H., and Friend, R.H. (2003) Inkjet printing of polymer thin film transistors. *Thin Solid Films*, **438**, 279–287.
2. Subramanian, V., Frechet, J.M., Chang, P.C., Huang, D.C., Lee, J.B., Molesa, S.E., Murphy, A.R., Redinger, D.R., and Volkman, S.K. (2005) Progress toward development of all-printed RFID tags: materials, processes, and devices. *Proc. IEEE*, **93**, 1330–1338.
3. Cantatore, E., Geuns, T.C.T., Gelinck, G.H., van Veenendaal, E., Gruijthuijsen, A.F.A., Schrijnemakers, L., Drews, S., and De Leeuw, D.M. (2007) A 13.56-MHz RFID system based on organic transponders. *IEEE J. Solid-State Circuits*, **42**, 84–92.
4. Crone, B., Dodabalapur, A., Gelperin, A., Torsi, L., Katz, H.E., Lovinger, A.J., and Bao, Z. (2001) Electronic sensing of vapors with organic transistors. *Appl. Phys. Lett.*, **78**, 2229.
5. Liao, F., Chen, C., and Subramanian, V. (2005) Organic TFTs as gas sensors for electronic nose applications. *Sens. Actuators B Chem.*, **107**, 849–855.
6. Someya, T., Kato, Y., Sekitani, T., Iba, S., Noguchi, Y., Murase, Y., Kawaguchi, H., and Sakurai, T. (2005) Conformable, flexible, large-area networks of pressure and thermal sensors with organic transistor active matrixes. *Proc. Natl. Acad. Sci. U.S.A.*, **102**, 12321.
7. Park, S.K., Jackson, T.N., Anthony, J.E., and Mourey, D.A. (2007) High mobility solution processed 6,13-bis(triisopropyl-silylethynyl) pentacene organic thin film transistors. *Appl. Phys. Lett.*, **91**, 063514.
8. Park, S.K., Anthony, J.E., and Jackson, T.N. (2007) Solution-processed TIPS-pentacene organic thin-film-transistor circuits. *IEEE Electron Device Lett.*, **28**, 877–879.
9. Paul, K.E., Wong, W.S., Ready, S.E., and Street, R.A. (2003) Additive jet printing of polymer thin-film transistors. *Appl. Phys. Lett.*, **83**, 2070.
10. Jung, M., Kim, J., Noh, J., Lim, N., Lim, C., Lee, G., Kim, J., Kang, H., Jung, K., Leonard, A., Tour, J., and Cho, G. (2010) All-printed and roll-to-roll-printable 13.56-MHz-operated 1-bit RF tag on plastic foils. *IEEE Trans. Electron Devices.*, **57**, 571–580.
11. Zielke, D., Hübler, A.C., Hahn, U., Brandt, N., Bartzsch, M., Fügmann, U.,

Fischer, T., Veres, J., and Ogier, S. (2005) Polymer-based organic field-effect transistor using offset printed source/drain structures. *Appl. Phys. Lett.*, **87**, 123508.

12. Subramanian, V., Chang, P.C., Lee, J.B., Molesa, S.E., and Volkman, S.K. (2005) Printed organic transistors for ultra-low-cost RFID applications. *IEEE Trans. Compon. Packag. Technol.*, **28**, 742.
13. Ong, N.S. (1995) Manufacturing cost estimation for PCB assembly: an activity-based approach. *Int. J. Prod. Econ.*, **38**, 159–172.
14. Sung, D., de la Fuente Vornbrock, A., and Subramanian, V. (2010) Scaling and optimization of gravure-printed silver nanoparticle lines for printed electronics. *IEEE Trans. Compon. Packag. Technol.*, **33**, 105–114.
15. Pudas, M., Halonen, N., Granat, P., and Vähäkangas, J. (2005) Gravure printing of conductive particulate polymer inks on flexible substrates. *Prog. Org. Coat.*, **54**, 310–316.
16. Sirringhaus, H., Kawase, T., Friend, R.H., Shimoda, T., Inbasekaran, M., Wu, W., and Woo, E.P. (2000) High-resolution inkjet printing of all-polymer transistor circuits. *Science*, **290**, 2123.
17. Molesa, S.E., Volkman, S.K., Redinger, D.R., Vornbrock, A.F., and Subramanian, V. (2004) A high-performance all-inkjetted organic transistor technology. IEEE International Electron Devices Meeting, 2004. IEDM Technical Digest, pp. 1072–1074.
18. Molesa, S.E., de la Fuente Vornbrock, A., Chang,P.C., and Subramanian, V. (2005) Low-voltage inkjetted organic transistors for printed RFID and display applications. IEEE International Electron Devices Meeting, 2005. IEDM Technical Digest, pp. 109–112.
19. Noh, Y.Y., Zhao, N., Caironi, M., and Sirringhaus, H. (2007) Downscaling of self-aligned, all-printed polymer thin-film transistors. *Nat. Nanotechnol.*, **2**, 784–789.
20. Sele, C.W., von Werne, T., Friend, R.H., and Sirringhaus, H. (2005) Lithography-free, self-aligned inkjet printing with sub-hundred-nanometer resolution. *Adv. Mater.*, **17**, 997–1000.
21. Calvert, P. (2001) Inkjet printing for materials and devices. *Chem. Mater.*, **13**, 3299–3305.
22. Soltman, D. and Subramanian, V. (2008) Inkjet-printed line morphologies and temperature control of the coffee ring effect. *Langmuir*, **24**, 2224–2231.
23. de Gans, B.J. and Schubert, U.S. (2004) Inkjet printing of well-defined polymer dots and arrays. *Langmuir*, **20**, 7789–7793.
24. Deegan, R.D., Bakajin, O., Dupont, T.F., Huber, G., Nagel, S.R., and Witten, T.A. (1997) Capillary flow as the cause of ring stains from dried liquid drops. *Nature*, **389**, 827–828.
25. Duineveld, P.C. (2003) The stability of ink-jet printed lines of liquid with zero receding contact angle on a homogeneous substrate. *J. Fluid Mech.*, **477**, 175–200.
26. Huang, D., Liao, F., Molesa, S., Redinger, D., and Subramanian, V. (2003) Plastic-compatible low resistance printable gold nanoparticle conductors for flexible electronics. *J. Electrochem. Soc.*, **150**, G412–G417.
27. Volkman, S.K., Molesa, S.E., Lee, J.B., Mattis, B.A., de la Fuente Vornbrock, A., Backhishev, T., and Subramanian, V. (2004) A novel transparent air-stable printable n-type semiconductor technology using ZnO nanoparticles. 2004 IEEE International Electron Devices Meeting Technical Digest, pp. 769–772.
28. Afzali, A., Dimitrakopoulos, C.D., and Breen, T.L. (2002) High-performance, solution-processed organic thin film transistors from a novel pentacene precursor. *J. Am. Chem. Soc.*, **124**, 8812–8813.
29. Redinger, D., Molesa, S., Yin, S., Farschi, R., and Subramanian, V. (2004) An ink-jet-deposited passive component process for RFID. *IEEE Trans. Electron Devices.*, **51**, 1978–1983.
30. Tseng, H. and Subramanian, V. (2009) All inkjet printed self-aligned transistors and circuits applications. Electron Devices Meeting (IEDM), 2009 IEEE International, pp. 1–4.

8
Picoliter and Subfemtoliter Ink-jet Technologies for Organic Transistors

Tsuyoshi Sekitani and Takao Someya

Realization of a sustainable society will require the development of industrial manufacturing processes that have minimal impact on the environment. From this viewpoint, the emerging field of printed electronics should attract considerable attention: it has the potential to drastically reduce the ecological footprint, or the energy consumed, in manufacturing electronic products. In addition, developments in printing are allowing an ever-greater economization in the use of inks, thus minimizing material wastage in comparison to conventional manufacturing techniques. However, a major obstacle to the development of large-area printed electronics such as large-area displays and sensors has been the fundamental compromise between manufacturing efficiency, transistor performance, and power consumption. Past improvements in manufacturing efficiency achieved through the use of printing techniques have come at the inevitable low device performance and large power consumption, while attempts to improve performance or reduce power have led to higher process temperatures and increased manufacturing cost. This review describes how this fundamental trade-off is avoided with picoliter and subfemtoliter ink-jet printing, allowing the creation of high-performance organic transistors and low-power, high-speed complementary circuits. Furthermore, the future prospects of printed large-area electronics are discussed and a picture is presented of the nature of the material science, electronic, and mechanical engineering requirements for the next-generation manufacturing processes.

8.1
Introduction

Digital fabrication with ink-jet technology is expected to become an important industrial manufacturing process since it can be used to pattern high-purity electronic functional materials without the need for original printing masks [1–9], thus reducing manufacturing costs and turnaround time. Ink-jet technology has recently found wide adoption in the mass production of color filters for liquid crystal displays [10–12] and electrical wirings on flexible substrates. This further

Organic Electronics II: More Materials and Applications, First Edition. Edited by Hagen Klauk.

indicates that printing may soon emerge as the dominant paradigm in electronic device manufacturing.

However, there still exists a rather wide gap between the resolution allowed by current design rules for active devices such as transistors and the typical resolution of conventional ink-jet printers. For example, 32 nm single-crystal silicon-based processors are currently being manufactured [13], while amorphous silicon allows for a channel length of only a few microns. However, an ink-jet head typically maintains a picoliter-order discharge volume, which creates dots with diameters of 30–50 μm on regular paper. The minimum size of the droplet ejected from an ink-jet head determines the printing resolution, and it is, in principle, limited by surface tension. Although this makes it difficult to control resolution at subpicoliter volumes, the on-demand customizability provided by this technology shows its significant potential.

This review focuses on the manufacture of organic transistors with ink-jet technology. First, we describe how high-performance organic thin-film transistors (TFTs) with top-contact geometry can be directly patterned using picoliter ink-jet printing on high-performance organic semiconductors. We show that by supplying carefully controlled waveforms to the piezoelectric actuators in ink-jet nozzles, we can accurately produce 1.4–17 pl. ink droplets containing silver (Ag) nanoparticles. We found that the TFT characteristics improve significantly when the ink volume is less than 3 pl. Taking full advantage of the technique, we describe the manufacture of a 33 cm diagonal large-area organic TFT active matrix, which was integrated with a pressure-sensitive rubber sheet. This formed the basis of a method for manufacturing large-area pressure sensor sheets that can monitor the spatial distribution of pressure. Second, we show the feasibility of employing ink-jet technology to create subfemtoliter droplets, the smallest droplets ever produced, with electrical functional inks. We then describe how the source/drain contacts of p- and n-channel organic TFTs can be prepared by subfemtoliter ink-jet printing of Ag nanoparticles directly on the surface of the organic semiconductor layers, without the need for any photolithographic prepatterning [14, 15] or surface pretreatment [16–20]. This allows us to prepare top-contact TFTs with a channel length of 1 μm. Because only a very small amount of organic solvent is dispensed during subfemtoliter ink-jet printing and the Ag nanoparticle sintering temperature after printing is low (130 °C), the morphology of the organic semiconductors is not altered. For bottom-contact TFTs, the contacts are defined before the deposition of the organic semiconductor, and hence the channel length can be very small despite the use of photolithography or electron-beam lithography. However, top-contact TFTs benefit from significantly lower contact resistance [21], but require a contact patterning technique that does not harm the organic semiconductor [22]. Thus, with subfemtoliter ink-jet printing, TFTs with both a short channel length (1 μm) and small contact resistance (5 kΩ·cm) and parasitic capacitance can be produced for the first time.

Some of the features of our subfemtoliter ink-jet system are similar to those of the electrohydrodynamic jet system recently reported by Rojers *et al.* [23], but an important difference is that our system is utilized to print metal nanoparticle inks directly onto the surface of high-mobility organic semiconductor films.

All the materials and manufacturing processes described in this review, including the electrical functional materials (inks) and picoliter and subfemtoliter ink-jets, are compatible with a large-area, cost-effective manufacturing.

The remainder of this review is organized as follows: Section 8.2 describes the Ag nanoparticle ink used and Section 8.3 describes the methods of picoliter and subfemtoliter printing. Section 8.4 describes the manufacturing processes and electrical characteristics of the organic TFTs, while Section 8.5 discusses the merits of the new methods and the future prospects of large-area printed electronics.

8.2 Silver Nanoparticle Ink

The "ink" for the electrodes consists of monodispersed Ag nanoparticles with a diameter of 2–3 nm that are functionalized with a proprietary dispersing agent and suspended in tetradecane, a nonpolar organic solvent, as shown in Figure 8.1a–c (Harima Chemical Co. Ltd.; NPS-J-HP, viscosity: 10 mPa·s). The Ag content by volume is 10%.

As shown in Figure 8.1d, the resistance (measured with a four-point probe) systematically decreases with increasing sintering temperature. Sintering at 180 °C produces resistivity of 5.1 μΩ cm, which is approximately three times larger than that of pure Ag (1.6 μΩ cm) and sufficiently small for electrical circuits.

Figure 8.1e shows cross-sectional micrographs of a 2 μm thick Ag electrode sintered at different temperatures for 1 h. The samples were processed by focused ion beam (FIB) and observed under a transmission electron microscope (TEM). The electrode sintered at 180 °C is porous, but Ag crystallizes at temperatures higher than 200 °C, which produces low resistivity. Sintering of the ink-jet printed lines at 220 °C for 1 h in nitrogen or air removes the dispersing agent and fuses the nanoparticles into a homogeneous metallic line with large electrical conductivity (Figure 8.2) [24, 25].

8.3 Ink-jet Technologies with Pico- and Subfemtoliter Accuracies

8.3.1 Picoliter Ink-jet Printing

The picoliter ink-jet printer fabricated by RICHO Printing Systems has a printing head with 192 piezoelectrically driven nozzles of 25 μm diameter. The stage size is 300 × 300 mm^2. Typically, this picoliter ink-jet printer can create a pattern of dots with 40 μm diameter and lines with 50 μm width. The dot size and line width can be controlled by changing conditions during ink ejection. As shown in Figure 8.2c, a single 25 V pulse applied to the piezoelectric elements creates a droplet of volume

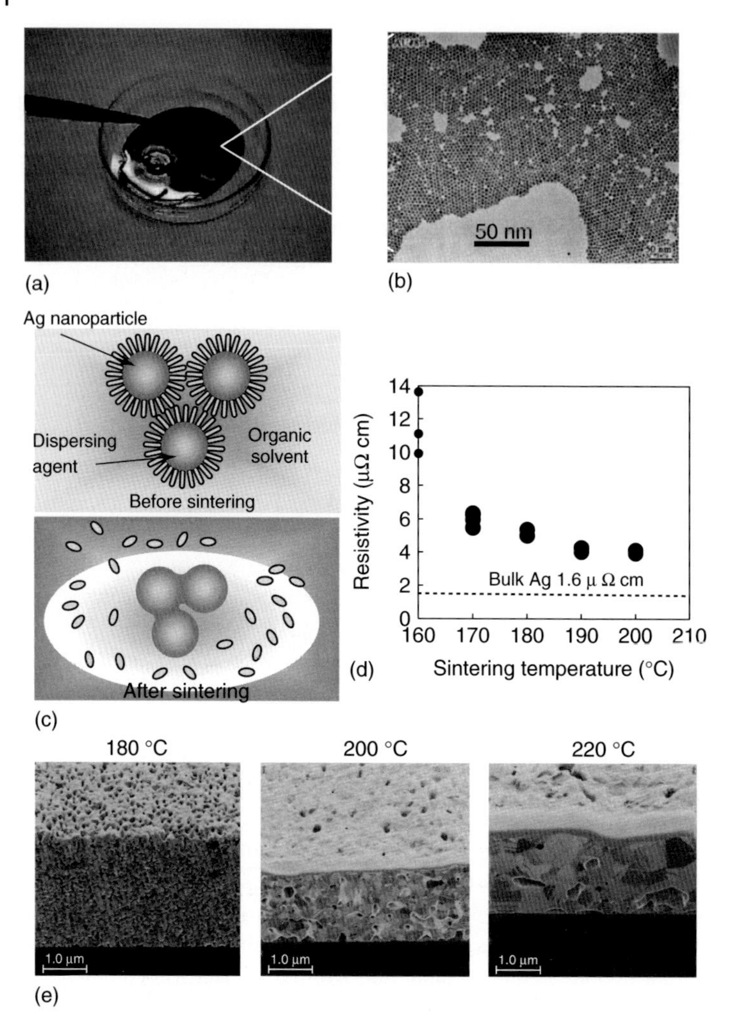

Figure 8.1 (a) Silver (Ag) nanoparticle ink and (b) magnified image of Ag nanoparticle observed using transmission electron microscope (TEM). (c) Schematic illustration of Ag nanoparticle ink before and after sintering. Sintering removes the dispersing agent and organic solvents and fuses nanoparticles into homogeneous metal. (d) Resistivity of printed Ag nanoparticle electrodes as a function of sintering temperature. (e) Cross-sectional pictures of printed Ag nanoparticles sintered at 180, 200, and 220 °C. Samples prepared using focused ion beam (FIB) and observed using scanning electron microscope (SEM).

17 pl. But with two pulses, the droplet volume can be reduced further. Reducing the voltage applied to the piezoelectric elements also significantly decreased the volume, as shown in Figure 8.2d,e. The minimum volume can be as small as 1.4 pl. When the voltage of the two-pulse wave is increased, two ink droplets are ejected simultaneously. Thus, a single-pulse wave is applied when a volume exceeding 10 pl is required.

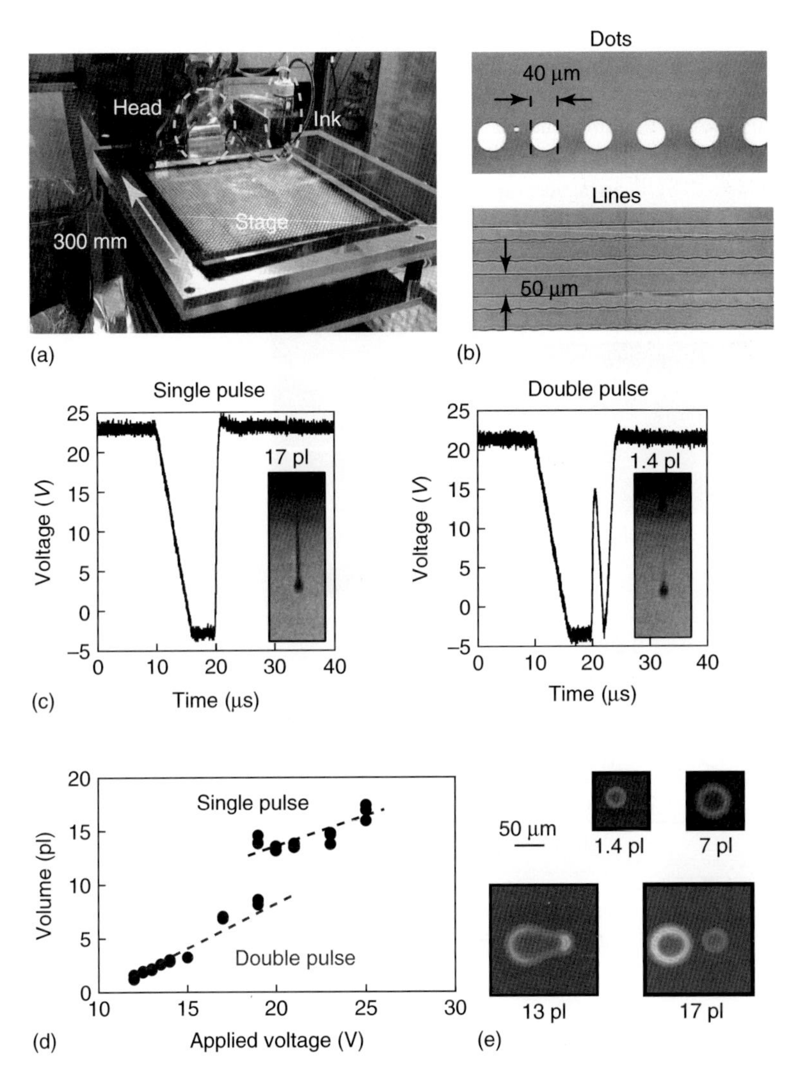

Figure 8.2 (a) A picture of picoliter ink-jet system. (b) Dots and lines of Ag nanoparticle inks patterned on an organic semiconductor (pentacene). (c) Voltage applied to the piezoelectric actuators of the ink-jet nozzle. Single- and double-pulse waveforms are used to control droplet volume. Picture of droplets with ejected volumes of 17 and 1.4 pl. (d) Volume of Ag ink droplets as a function of the applied voltage. The black and gray plots are the volumes obtained with single-pulse and double-pulse waveforms, respectively. The two waveforms used in this experiment are shown in the inset. (e) Atomic force microscopic (AFM) images of the ink-jet-patterned Ag dots. From these images, we estimate that the volume of the ink droplets varies from 1.4 to 17 pl when the applied voltage is changed from 12 to 25 V. Adapted from [30]. Copyright, American Institute of Physics, 2008.

8.3.2
Subfemtoliter Ink-jet

It is well known that organic semiconductor molecules are damaged by the organic solvents used in inks. For example, the morphology of pentacene is easily collapsed by even a very small amount of organic solvent, as shown in Figure 8.3a. Reduction of droplet volume is the most effective solution because the large surface ratio after impacting of substrate leads to faster volatilization

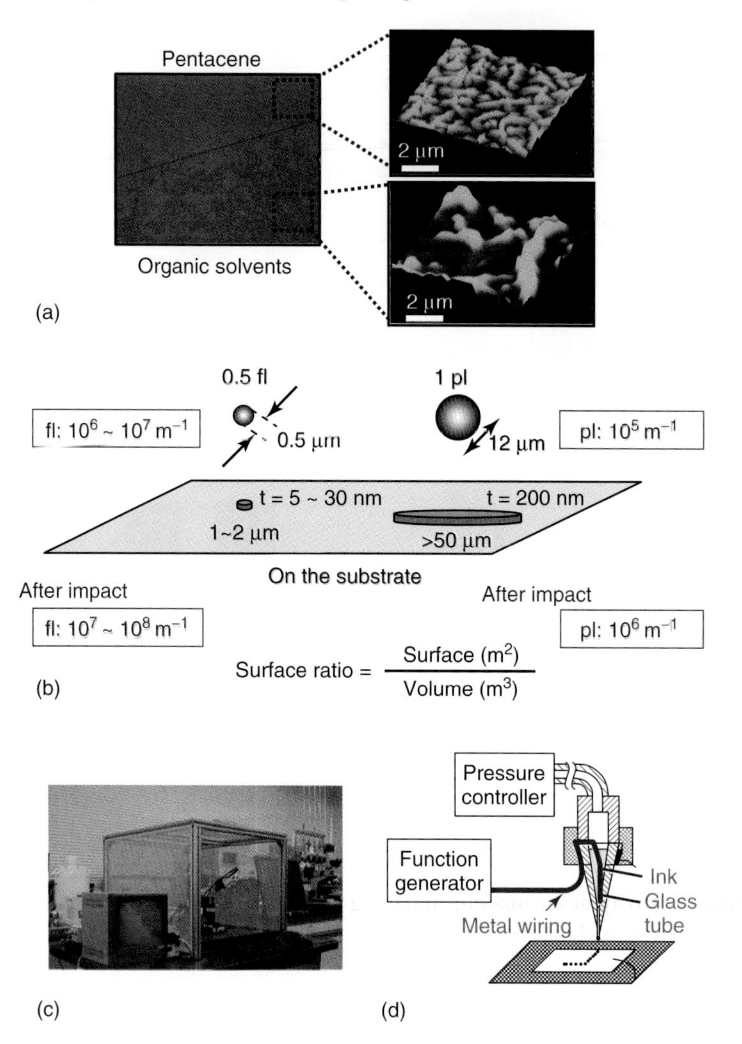

Figure 8.3 (a) Micrograph and AFM image showing morphology of organic semiconductor, pentacene. Morphology is damaged even with small amount of organic solvent. (b) A schematic illustration of estimated surface ratio (m^{-1}) at 0.5 fl and 1 pl. The surface ratio of 0.5 fl droplets is 1 order of magnitude larger than that of 1 pl, thus resulting in faster volatilization of organic solvents in inks. (c) Photograph of subfemtoliter ink-jet system.

of the solvent (Figure 8.3b). Furthermore, the organic solvent evaporates quickly even before the drop impacts the substrate, thus reducing the damage to the organic semiconductor. We employed a subfemtoliter ink-jet printer that achieves the smallest droplet volume produced thus far (0.5–0.7 fl) [26, 27].

8.3.2.1 Ejection Mechanism

Figure 8.3c shows a photograph of the subfemtoliter ink-jet printer. The nozzle of the subfemtoliter ink-jet system is manufactured from a very fine capillary glass tube with a tip diameter of less than 1 μm. The inside of the capillary tube is hydrophilic, whereas the outside is hydrophobic. A pressure controller is utilized to control the pressure inside the tube. A fine, electrically conducting wire is located inside the nozzle to charge the ink and electric pulses of less than 200 V are applied to eject droplets with subfemtoliter volume. The ejected droplets travel straight through the electric flux line. The accuracy of the mechanical *xy*-stage is better than 1 μm. The substrate is at room temperature.

8.3.2.2 Subfemtoliter Droplets on Organic Semiconductors

A subfemtoliter ink-jet printer is employed to deposit narrow metal lines with 1 μm accuracy directly on top of the pentacene and $F_{16}CuPc$ layers.

The performance of the subfemtoliter ink-jet system is illustrated in Figure 8.4 The Ag nanoparticle droplets dispensed by the ink-jet nozzle have a volume of 0.7 ± 0.2 fl and a diameter of less than 1 μm in air. The diameter of the Ag dots thus created on the pentacene surface can be controlled between 2 and 6 μm, and the thickness of the dots is 30 nm. Figure 8.4a,b shows optical microscopic and atomic force microscopic (AFM) images of a pentacene film with an array of printed Ag dots after sintering with an average diameter of 2 ± 0.5 μm. The distinct morphology of the pentacene film and the excellent uniformity of the ink-jet-printed metal dots are clearly seen. Figure 8.4c,d confirms that the ink-jet-printed metal lines with line widths as small as 2 μm are uniform and continuous. Figure 8.4c also shows that lines of width 1 μm can be created as well, although their uniformity is insufficient to guarantee metallic connectivity over long distances. Lines of width 2 μm, on the other hand, can be printed uniformly and reproducibly over large areas, as shown in Figure 8.4d.

To control the thickness and electrical conductance of the metal lines, multiple ink-jet passes are used. Figure 8.4e,f illustrates the evolution of the morphology and electrical resistivity of an ink-jet-printed Ag line with increasing number of passes. The effective thickness of the line increases linearly with the number of passes, from 30 nm after a single pass to 600 nm after 20 passes. After sintering (at 130 °C for 1 h in nitrogen), the line resistivity was measured at room temperature in air. Figure 8.4f shows that 10 passes are sufficient to obtain continuous lines with a resistivity of less than 25 μΩ·cm, which is sufficiently low for practical applications.

Compared with picoliter ink-jet printing, subfemtoliter printing significantly reduces the sintering temperature required to achieve near-bulk-metal resistivity.

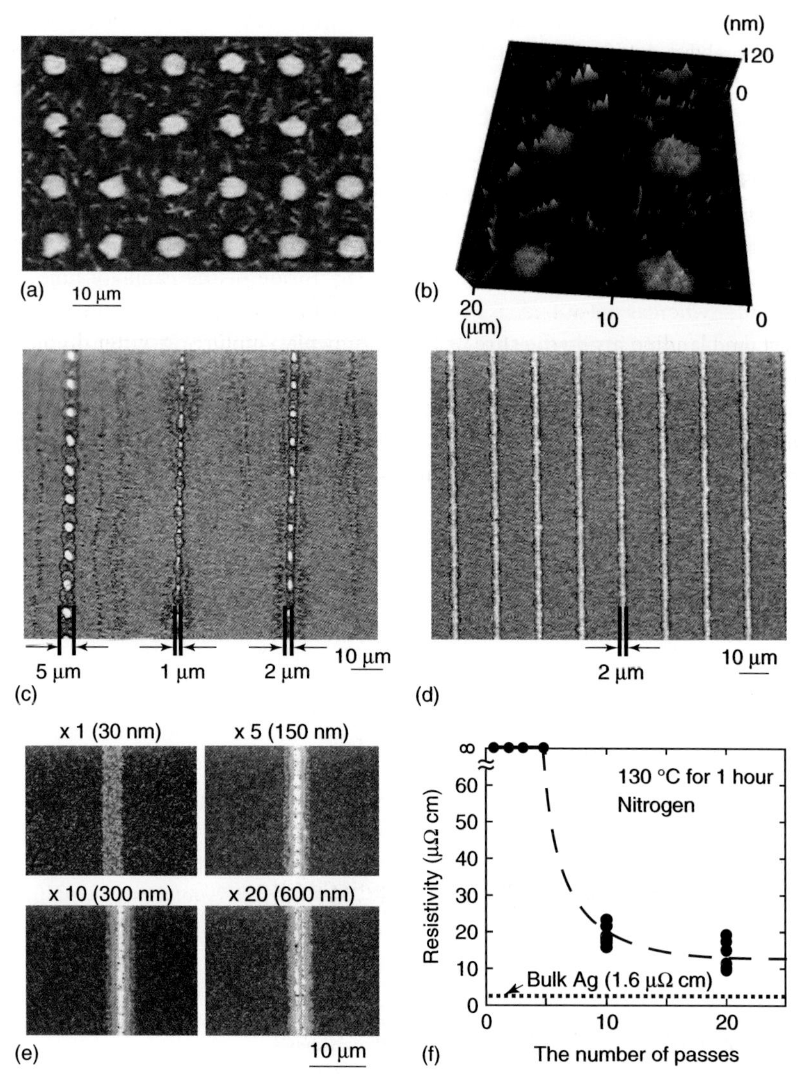

Figure 8.4 (a) Optical microscopic and (b) atomic force microscopic (AFM) image of fine dots of Ag nanoparticles deposited by subfemtoliter ink-jet printing on the surface of a thin pentacene film after sintering at 130 °C. The diameter of the dots is approximately 2 μm and thickness is 30 nm. The dots were formed with a single printing pass. (c) Optical microscope image of ink-jet-printed Ag lines after a single printing pass before sintering. Line widths of between 1 and 5 μm were obtained in a controlled manner by adjusting the electric field inside the ink-jet nozzle. (d) Optical microscopic image of ink-jet-printed Ag lines after a single printing pass and after sintering at 130 °C. Lines of width down to 2 μm are uniform and continuous over large areas. (e) Optical microscopic image showing the effect of multiple-pass printing on the evolution of the morphology of an ink-jet-printed Ag line after sintering. The effective thickness of the line increases from 30 nm after a single pass to 600 nm after 20 passes. (f) Evolution of the electrical resistivity of ink-jet-printed Ag lines with the number of passes after sintering at 130 °C for 1 h in nitrogen. Adapted from [43]. Copyright, National Academy of Sciences, 2008.

Low sintering temperatures are important for top-contact devices, since temperatures above ~150 °C will irreversibly damage the organic semiconductors [28, 29]. Using the same Ag nanoparticle ink, but employing a conventional picoliter ink-jet system, we obtained a resistivity of 56 ± 8 μΩ cm after sintering at 130 °C, and a temperature well above 150 °C was required to obtain a resistivity below 25 μΩ·cm. With subfemtoliter printing, a sintering temperature of 130 °C for 1 h was sufficient to obtain a resistivity of 25 μΩ·cm.

The landing accuracy of the ejected droplets is one of the most important factors for ink-jet technologies because it limits printing resolutions. Figure 8.5a shows the typical droplet volumes (1 pl and 0.5 fl), diameters before and after impact, and landing accuracy estimated from Brownian motion. Picoliter droplets of 12 μm diameter have a variation in landing accuracy of about 2 μm, which is small enough compared with that of 50 μm dots after impact. However, the landing accuracy greatly reduces with decreasing droplet volume because of fluctuations that originate from Brownian motion and is estimated to be more than 5 μm for subfemtoliter droplets. However, as shown in Figure 8.12b, 8.2 μm wide lines are

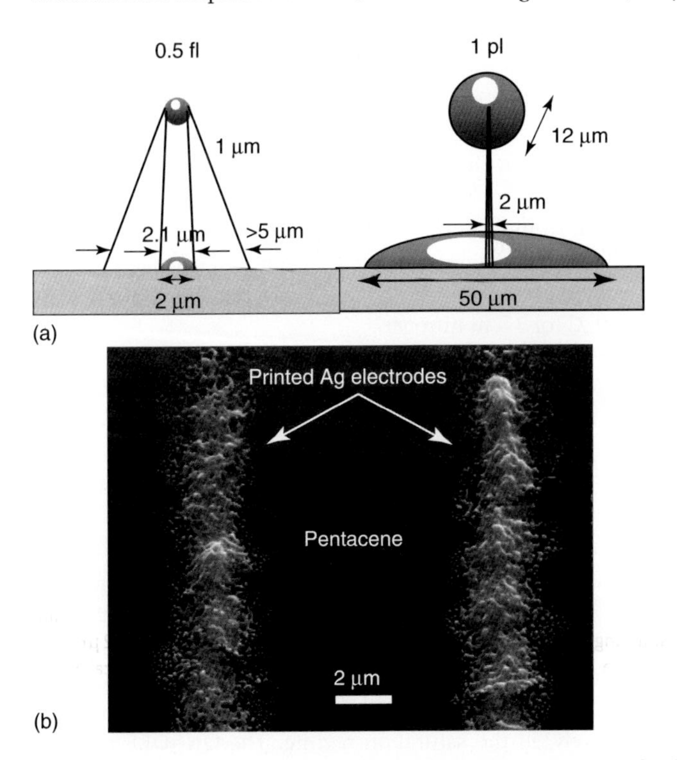

Figure 8.5 (a) Illustration of typical droplet volumes (1 pl and 0.5 fl), diameters before and after impact, and landing accuracy estimated from Brownian motion. Picoliter droplets of 12 μm diameter have a variation in landing accuracy of about 2 μm, while the uncertainty in landing position of subfemtoliter droplets is estimated to be more than 5 μm. (b) SEM image of printed Ag nanoparticle electrodes with 2 μm width that are uniformly patterned on pentacene.

uniformly patterned on pentacene. The actual landing accuracy of droplets ejected from the subfemtoliter ink-jet printer is less than 500 nm. This is quite reasonable because charged droplets travel straight though the electric flux lines from the ink-jet nozzle to the substrate.

8.4 Manufacturing Processes and Electrical Characteristics of Organic Transistors

8.4.1 Organic Transistors with Source/Drain Electrodes Printed Using Picoliter Ink-jet

8.4.1.1 Transistor Characteristics with Changing Droplet Volume

Organic transistors were fabricated using source/drain electrodes printed by varying the volume of Ag ink from 1.4 to 17 pl [30]. Figure 8.6a shows a cross-sectional view of the organic TFT with top-contact geometry. A gate electrode comprising a 5 nm thick Cr layer and 50 nm thick Au layers was evaporated through a shadow mask on a 75 μm thick polyimide film substrate. A polyimide precursor (Kemitite CT4112, Kyocera Chemical) was spin-coated and cured at 180 °C for 1 h in nitrogen to form a 450 nm thick gate dielectric layer [31]. For the organic semiconducting channel, pentacene was used for this experiment [32–38]. A 50 nm thick pentacene channel layer was deposited on the dielectric layer by thermal evaporation using a shadow mask. Finally, Ag nanoparticles were printed by the ink-jet technique at a substrate temperature of 90 °C to form the source/drain electrodes. In this study, the voltage applied to the piezoelectric elements of picoliter ink-jet was 12–25 V. The Ag ink droplets were deposited with a spacing of 50 μm. The patterned Ag lines were sintered at 130 °C for 3 h in nitrogen.

Micrographs of the fabricated device are shown in Figure 8.6b,c. The Ag electrodes are wider on the pentacene film than on the polyimide dielectric layer because the pentacene film has higher surface energy. For the transistors fabricated with droplet volumes of 1.4 and 17 pl (which, in this chapter, are simply referred to as 1.4 pl TFTs and 17 pl TFTs, respectively), the line widths of the electrodes are 60 and 130 μm, respectively.

Figure 8.6b,c shows the source–drain current (I_{DS}) of the 1.4 and 17 pl TFTs, respectively, as a function of the source–drain voltage (V_{DS}). The gate voltage (V_{GS}) was changed from 20 to −40 V in steps of 10 V. The 1.4 pl TFT exhibits excellent characteristics in the linear regime. This clearly indicates that the contact resistance can be suppressed by reducing the volume of the ink droplets. The transfer curves of the devices are shown. The mobilities of the 1.4 and 17 pl TFTs are 0.3 and $0.09\,cm^2\,V^{-1}\,s^{-1}$, respectively, in the saturation regime. The ON–OFF ratios of the two transistors are independent of the volume and have a value of 10^6 each. These characteristics are comparable to those of organic transistors comprising Au source/drain electrodes fabricated by thermal evaporation.

Figure 8.7a shows the mobility as a function of the volume of the Ag ink droplets used for manufacturing these devices. The mobility increases significantly when

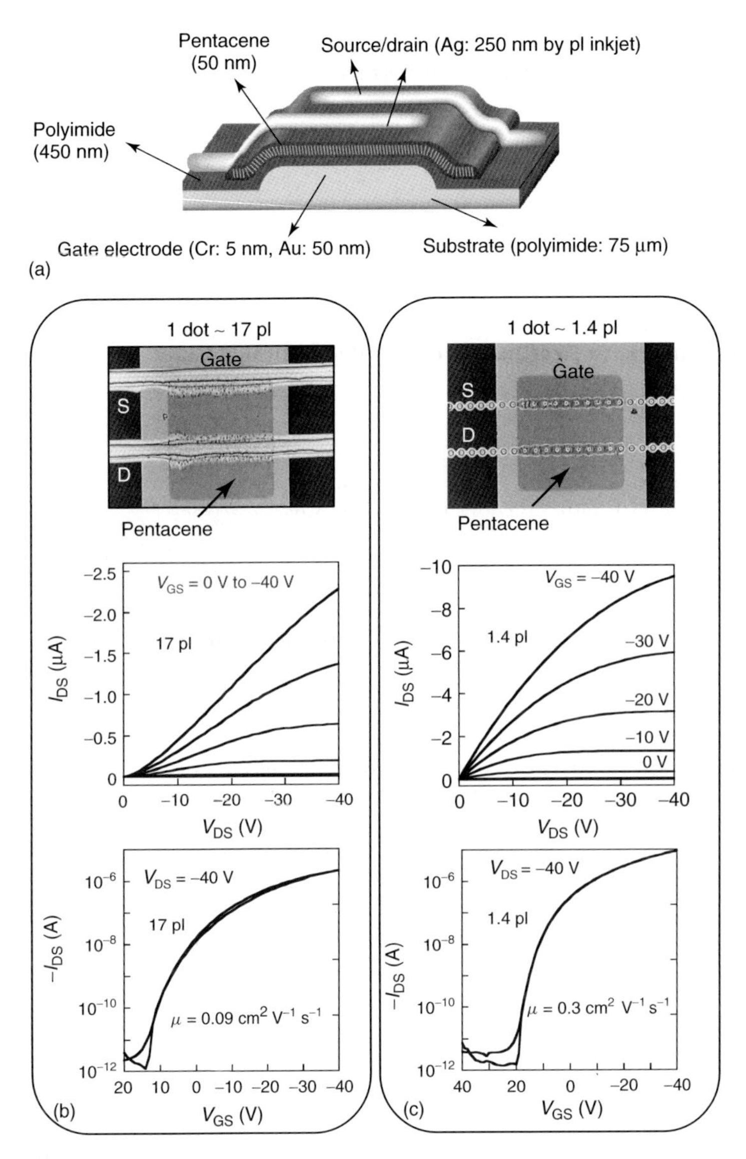

Figure 8.6 (a) A schematic cross-sectional illustration of organic transistors, in which the gate and organic semiconductor layers (50 nm thick pentacene) are patterned by vacuum evaporation, and the 450 nm thick polyimide dielectric layer is spin-coated and cured at 180 °C for 1 h. Ag nanoparticles are printed on the pentacene layer by picoliter ink-jet printing. (b,c) Micrograph of the manufactured organic transistors with different printed source (S)/drain (D) electrodes. Droplet volumes used for these transistors are 17 and 1.4 pl, respectively. Transistor characteristics of the manufactured organic TFTs are also shown for comparison. Source–drain current (I_{DS}) of the 1.4 pl TFT and the 17 pl TFT are also shown as a function of the source–drain voltage (V_{DS}). The gate voltage (V_{GS}) is varied from 20 to −40 V in steps of 10 V. The transfer curves of the devices are shown. (V_{GS} is varied from 40 to −40 V with V_{DS} = −40 V). Adapted from [30]. Copyright, American Institute of Physics, 2008.

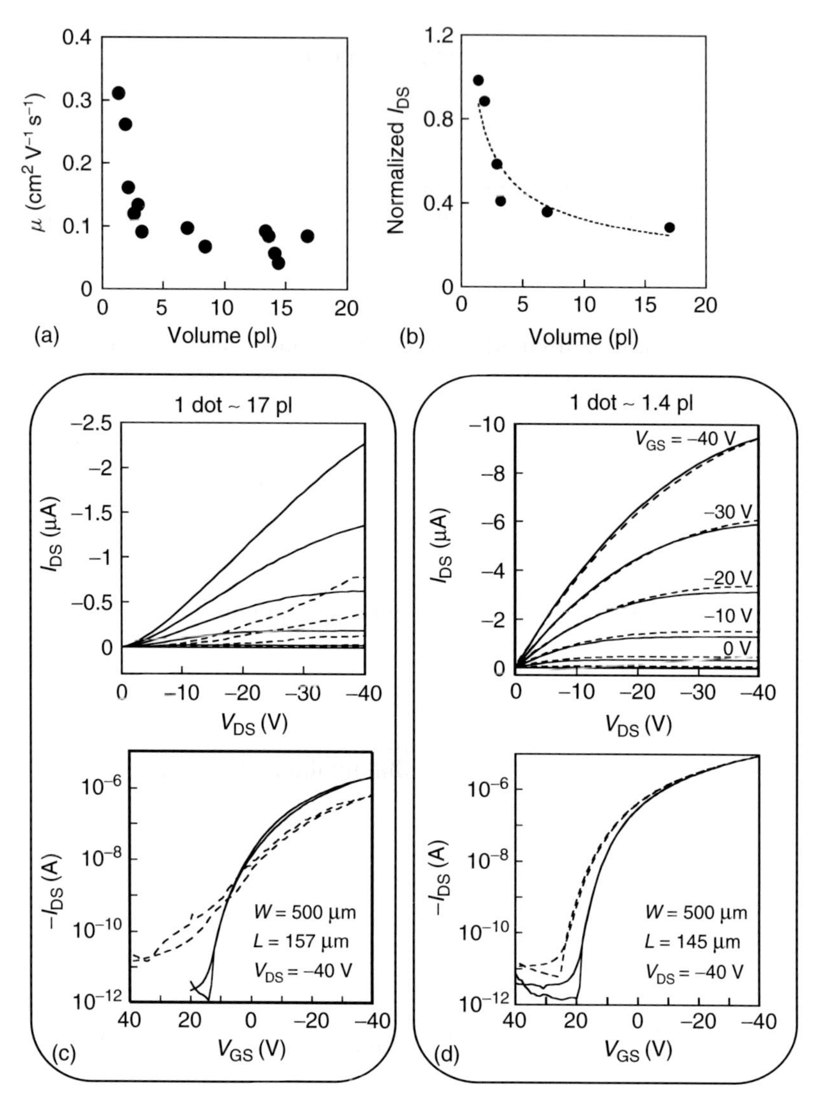

Figure 8.7 (a) Mobility of organic transistors with printed source/drain electrodes. Droplet volumes are changed from 1.4 to 17 pl. (b) Normalized channel currents (I_{DS}) as a function of droplet volume. It is normalized with the value measured as soon as the device is manufactured. (c,d) Long-term stability of organic transistors with printed source/drain electrodes. Solid lines represent transistor characteristics of the manufactured organic TFTs with 1.4 and 17 pl printed source/drain electrodes, which are replots of Figure 8.6b,c. The dashed lines in each graph represent the transistor characteristics after three weeks. Changes in transistor performances are smaller with decreasing droplet volume. Adapted from [30]. Copyright, American Institute of Physics, 2008.

the volume is less than 3 pl and reaches $0.3\,cm^2\,V^{-1}\,s^{-1}$ at 1.4 pl. This shows that the ink volume significantly affects the transistor characteristics.

In order to evaluate long-time stability, the transistors were placed in a nitrogen environment for three weeks. The device characteristics are shown by the dashed lines in Figure 8.7c,d. As shown in this figure, the ON-state current of the 17 pl TFT decreases, whereas its OFF-state current increases; the 1.4 pl TFT does not exhibit any significant change. The changes in the ON-state current measured after three weeks are shown in Figure 8.7b as a function of the volume of the ink droplets. The ON-state current was normalized with the data measured as soon as the devices were manufactured. The ON-state current decreases significantly when the volume of the droplets decreases. This clearly demonstrates that for the printed organic transistors, a reduction in the ink volume to near 1 pl is crucial to obtain high mobility ($\sim 0.3\,cm^2\,V^{-1}\,s^{-1}$) as well as good stability over many weeks.

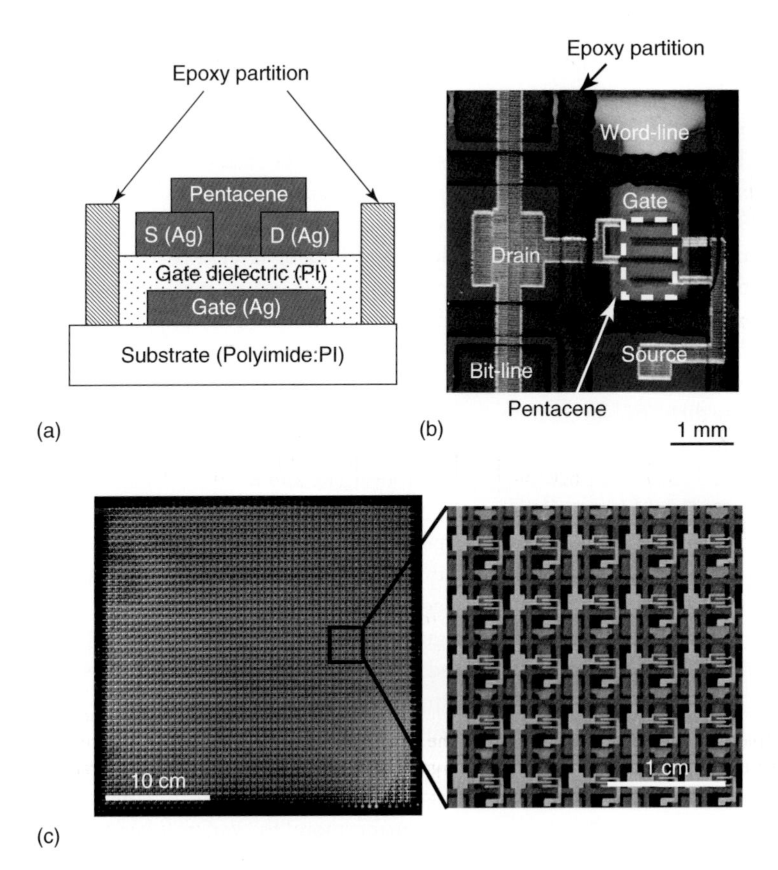

Figure 8.8 (a,b) A cross-sectional illustration and micrograph of printed organic transistor. All the electrodes including gate electrodes are formed using picoliter ink-jet, while pentacene is formed using vacuum evaporation. (c) A picture of active matrix comprising 2025 printed organic transistors on plastic. The sheet size is $300 \times 300\,mm^2$, and the periodicity is 5.04 mm. Adapted from [39, 40]. Copyright, American Institute of Physics, 2006, 2007.

8.4.1.2 **Printed Organic Transistor Active Matrix Using Picoliter Ink-jet**

In order to demonstrate the feasibility of the method, a printed organic transistor active matrix was manufactured and tested. A cross-sectional illustration and photograph of the device are presented in Figure 8.8. The picoliter ink-jet printing system was equipped with two kinds of heads: one employing Ag nanoparticles, and the other, polyimide precursors [31]. First, Ag nanoparticles [39, 40] were patterned using a 75 μm thick polyimide film to form 2025 gate electrodes and 45 word-lines. During the ink-jet process, the substrate temperature was maintained at 40 °C. After printing, the base films were maintained at 180 °C for 1 h in nitrogen to sinter the Ag nanoparticles.

To improve the uniformity of the thickness of the ink-jet-printed polyimide layers, a partition of epoxy resin was formed using a screen printing machine (MT-550, Micro-tec Co., Ltd) before patterning the polyimide precursor. The epoxy resin was stiffened at 150 °C for 30 min in ambient air. Each epoxy partition was square shaped and had a height of 5 μm. The inner and outer sizes of the partition were 2×2 and $2.5 \times 2.5\ mm^2$, respectively.

A solution containing polyimide precursors (KEMITITE CT4112, Kyocera Chemical) was diluted with *N*-methyl-2-pyrrolidone (NMP), and its viscosity was set to 11 mPa·s. The polyimide precursor ink was filled up into the inside of the epoxy partition by the ink-jet printer. Precautions were taken to prevent ink overflow from the partition [39, 40]. The patterned polyimide precursors were cross-linked at 180 °C for 1 h in nitrogen, and a 1 μm thick polyimide insulator layer was formed. The dielectric strength voltage and surface smoothness (0.2 nm in root-mean-square (RMS)) of the obtained layer are almost comparable with those obtained using spin coating, as shown in Figure 8.9.

Next, the above-mentioned Ag nanoparticles were patterned by the ink-jet on the polyimide gate dielectric layers to form source, drain electrodes, and bit-lines. The substrate temperature in the ink-jet process was maintained at 80 °C to avoid ink lappets and produce a fine pattern of Ag electrodes. The applied nanoparticles were sintered at 180 °C for 1 h in ambient air.

A 50 nm thick pentacene layer was deposited as the channel layer by vacuum sublimation. The channel width was 2.5 mm, while its length was systematically changed from 60 to 200 μm. The TFTs used in the pressure sensor matrices had a channel length of 150 μm, as shown in Figure 8.8b.

When these insulator layers are employed in organic transistors, the homogeneity of the film thickness is very crucial in minimizing variation in transistor performance. As mentioned earlier, we first defined the size of the device by preparing an epoxy partition [39, 40]. The polyimide film showed singular thickening near the circumference of the partition, although the variation in film thickness was ±7% in the central parts ($1 \times 1\ mm^2$ size). Furthermore, the variation among the partitions was ±6%. In this manner, a uniform device was manufactured with the partition, which is crucial for realizing reliable organic transistor active matrices.

We measured the characteristics of the pentacene transistors manufactured by the above processes. The source–drain current (I_{DS}) of the transistors was measured as a function of the source–drain voltage (V_{DS}), as shown in Figure 8.10a. The gate

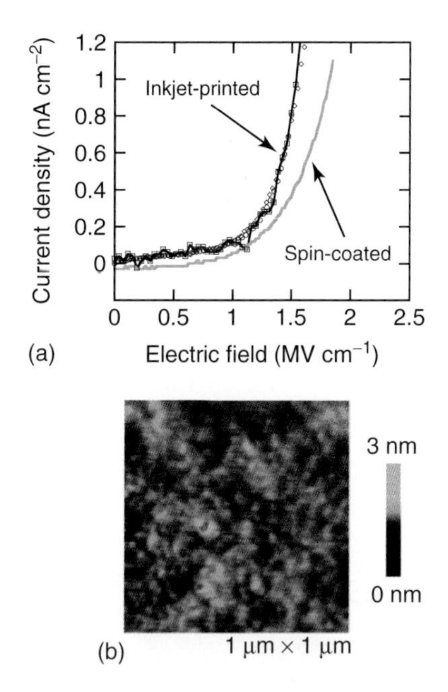

Figure 8.9 (a) Current density as a function of electric field. The breakdown test is performed on a capacitor structure formed using ink-jet (300 nm thick printed Ag electrode/1 μm thick printed polyimide gate dielectric layer/300 nm thick printed Ag-printed electrode). A capacitor structure formed using conventional process (vacuum evaporation and spin-coating) is also evaluated for comparison. (b) AFM image of the surface of printed polyimide. RMS value is approximately 0.2–0.3 nm, which is comparable to that formed using conventional processes.The scan size is 1 μm×1 μm. Adapted from [39, 40]. Copyright, American Institute of Physics, 2006, 2007.

voltage (V_{GS}) was varied from 0 to −100 V in steps of −20 V. Figure 8.10b shows the corresponding transfer curves of the same device when V_{GS} was varied from 40 to −100 V and V_{DS} was set at −100 V. The hysteresis was very small. In Figure 8.10b, the ON–OFF ratio exceeds 10^6, showing that the leakage current through the ink-jet polyimide layer is sufficiently low. The mobility in the saturation regime was 0.7 cm^2 V^{-1} s^{-1}, indicating excellent transistor characteristics.

8.4.1.3 A Large-Area Pressure Sensor Sheet

After manufacturing the organic transistor active matrix, the device was coated with 5 μm thick parylene passivation layers. A CO_2-laser drilling machine was used to make via-holes through the parylene on the source electrodes, and then Au pads were deposited to connect to each electrode through the via-holes. Finally, pressure-sensitive rubber and polyethylene naphthalate (PEN) films with Cu electrodes were laminated with the active matrix. A cross-sectional illustration, photograph, and circuit diagram are shown in Figure 8.11a–c.

Figure 8.11d shows the transfer characteristics of one pressure sensor cell with and without the application of pressure. Under pressure, the resistance of the

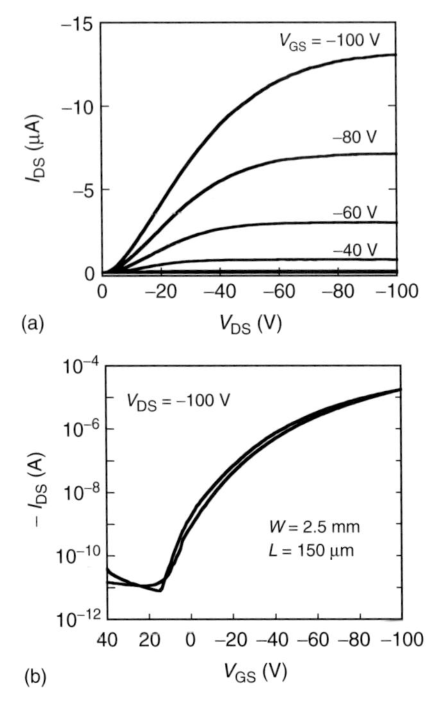

Figure 8.10 (a) The source–drain current (I_{DS}) of the TFTs as a function of the source–drain voltage (V_{DS}). The gate voltage (V_{GS}) is varied from 0 to −100 V in steps of −20 V. (b) Transfer curves of the same device (V_{GS} is varied from 40 to −100 V with V_{DS} = −100 V). Channel width and length were 2500 and 150 μm, respectively. Adapted from [39, 40]. Copyright, American Institute of Physics, 2006, 2007.

pressure-sensitive rubber decreased and then the organic TFTs were activated because of the application of the power supply voltage V_{DD}. The principle of operation of the pressure sensors is explained in Ref. [39, 40]. When pressure was applied with rubber objects, we clearly obtained the spatial distribution of the pressure, as shown in Figure 8.12, showing the feasibility of the printed organic transistor active matrix as a read-out circuit for sensor applications.

8.4.2 Organic Transistors with Source/Drain Electrodes Printed Using Subfemtoliter Ink-jet

8.4.2.1 TFTs on Polyimide Gate Dielectric

Organic transistors were fabricated using source/drain electrodes printed by a subfemtoliter ink-jet printer with 0.5 fl accuracy. Figure 8.13a shows the cross-sectional view of the organic TFT with top-contact geometry where the device structure is almost the same as that in Figure 8.6a; however, top-contact source/drain electrodes are formed using Ag nanoparticles patterned by a subfemtoliter ink-jet printer. Figure 8.13b shows the top view of the organic transistor. Two Ag nanoparticle lines, each 2 μm wide and 800 μm long, are clearly patterned on the pentacene

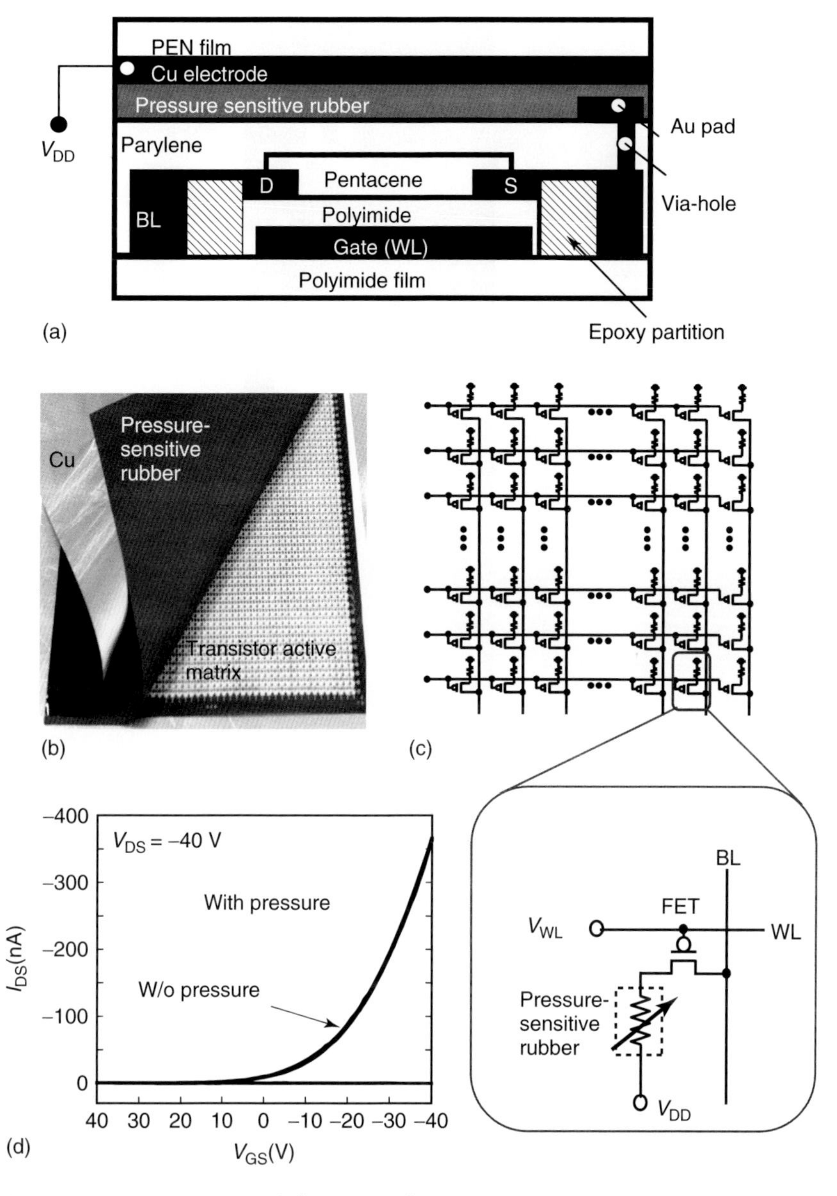

Figure 8.11 (a) Cross-sectional illustration of a pressure sensor. V_{DD} and V_{WL} represent power supply voltage and voltage for word-lines, respectively. WL and BL represent word-lines and bit-lines, respectively. Gate (G), source (S), and drain electrodes (D) were manufactured using ink-jet-printed nanoparticles. Gate dielectric layers were also manufactured using ink-jet-printed polyimide. Epoxy partitions were made using screen printing systems. (b) An exploded picture of a pressure sensor, comprising an organic transistor active matrix, a pressure-sensitive rubber, and a PEN film with Cu electrode. (c) Circuit diagrams of the pressure sensor and of a stand-alone pressure sensor cell. (d) Transfer characteristics of the transistor with and without (W/o) the application of pressure (30 kPa). Adapted from [39, 40]. Copyright, American Institute of Physics, 2006, 2007.

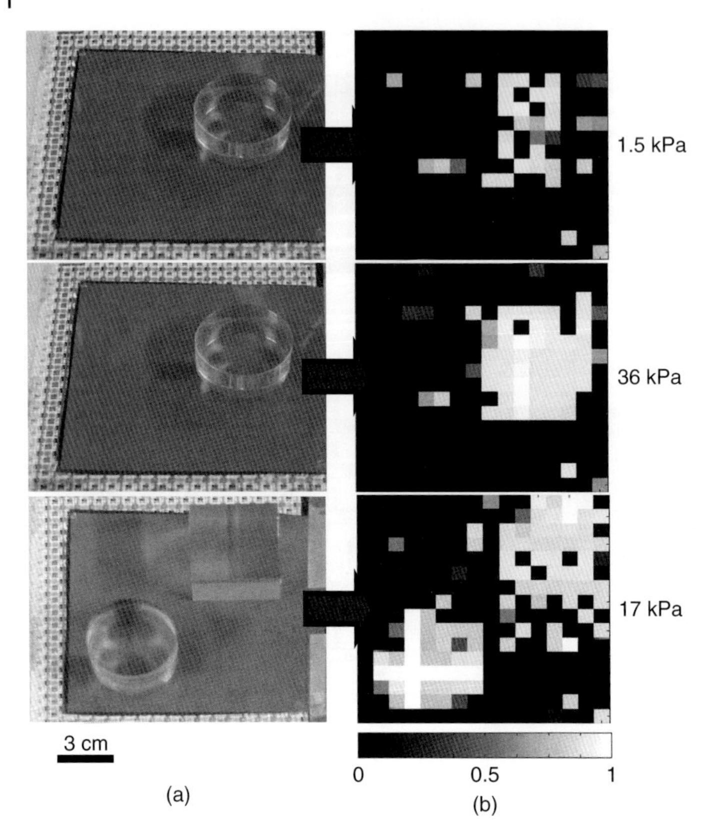

Figure 8.12 Pictures of rubber objects for applying pressures (a) and spatial distributions of pressures read out by organic TFT active matrix (b). Applied pressures are also shown for each experiment. Adapted from [39, 40]. Copyright, American Institute of Physics, 2006, 2007.

layer. The channel length and width of the organic transistor are 5 and 500 μm, respectively.

Figure 8.13c shows the characteristics of the organic transistor. Under an operation voltage of −40 V, the mobility was $0.7\,\mathrm{cm^2\,V^{-1}\,s^{-1}}$ and the ON–OFF ratio exceeded 10^7. These characteristics are almost identical to those of the device formed using vacuum evaporation. Thus, printed source and drain electrodes do not damage the organic semiconductors.

8.4.2.2 TFTs with Self-Assembled Monolayer as a Very Thin Gate Dielectric

Self-assembled monolayers (SAMs) can be used as very thin gate dielectrics for organic transistors to reduce the operation voltage [41, 42]. We have manufactured organic transistors with subfemtoliter source/drain electrodes on a SAM gate dielectric layer [43]. A cross-sectional illustration of the fabricated organic transistors is shown in Figure 8.14a. The base film (substrate) of PEN or SiO_2 is covered with a 20 nm thick Al layer with excellent surface flatness by vacuum evaporation.

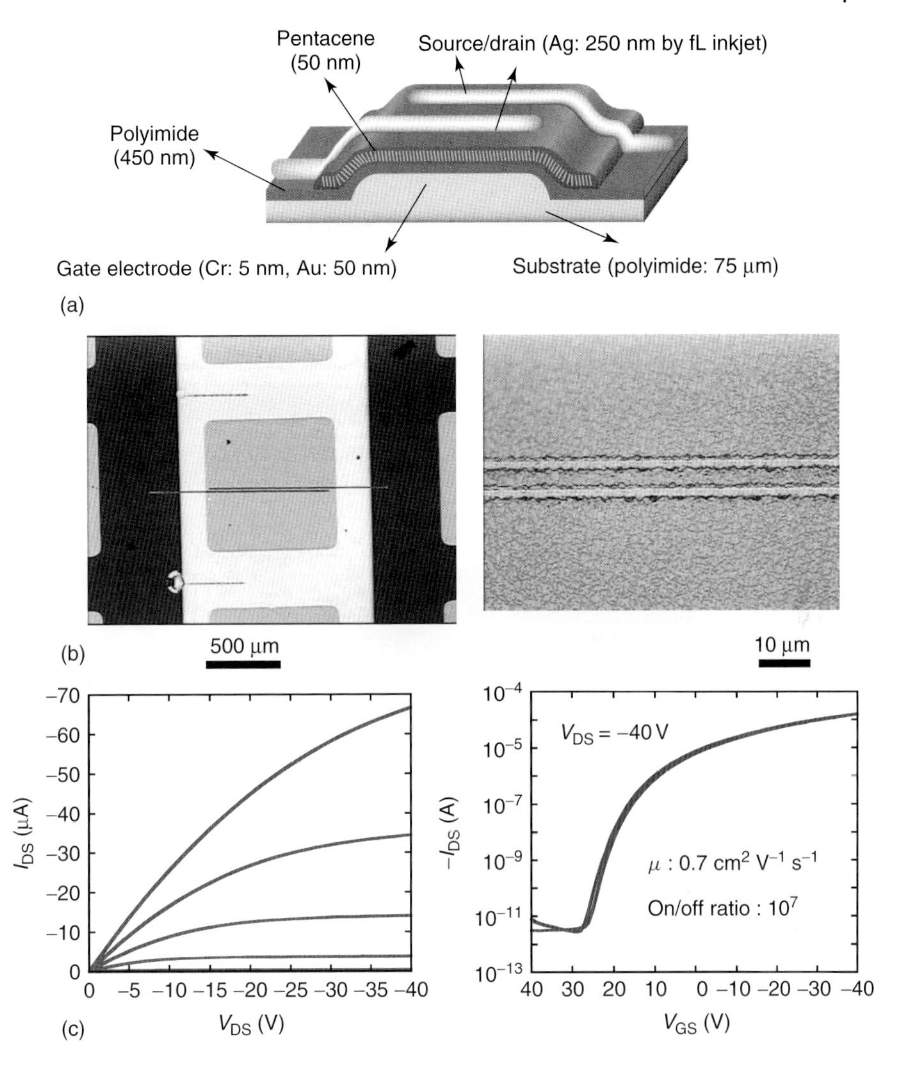

Figure 8.13 (a) A schematic cross-sectional illustration of organic transistors. Source/drain electrodes are formed using subfemtoliter ink-jet, while other layers are formed using vacuum evaporation and spin-coating. (b) A micrograph and magnified view of printed organic transistors. Line width and length are 2 and 800 μm, respectively, while transistor channel length and width are 5 and 400 μm, respectively. (c) Transistor characteristics of the organic transistor.

The Al layer is then activated by exposure to oxygen plasma in a parallel-plate reactor operating at 13.56 MHz. A dense SAM is formed on the Al surface as a 2.5 nm thick gate dielectric layer. The SAM is created in a one-step process without chemical conversion. The SAM manufacture process is described in detail in Ref. [41, 42]. Purified pentacene is then deposited by vacuum evaporation at a substrate temperature of 60 °C to form a 50 nm thick semiconductor layer.

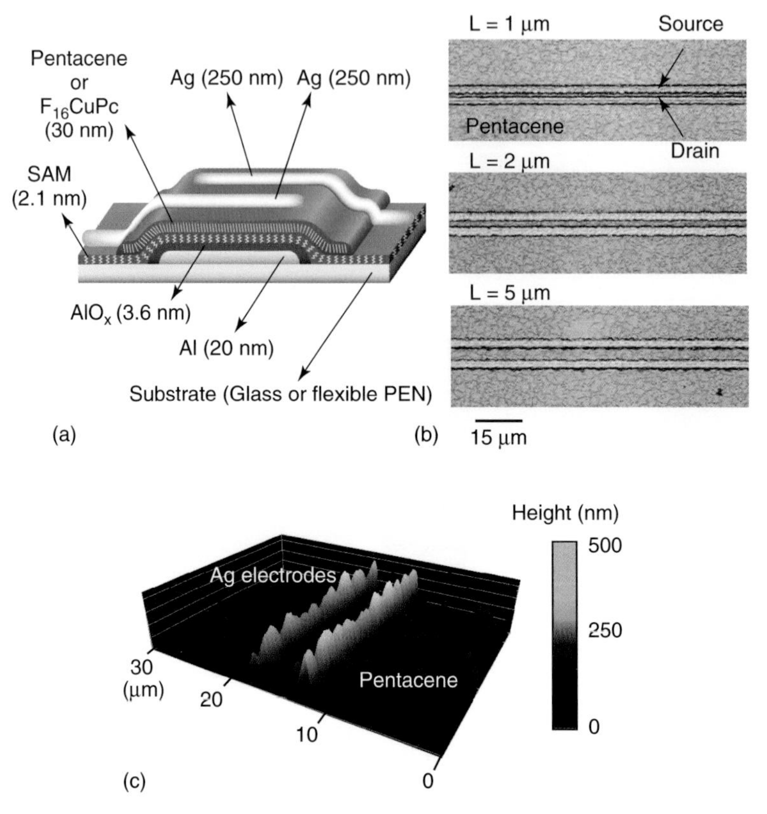

Figure 8.14 (a) Schematic cross-section of the organic thin-film transistors with patterned Al gates, ultra-thin gate dielectric, vacuum-deposited organic semiconductor, and subfemtoliter ink-jet-printed Ag nanoparticle source/drain contacts. (b) Optical microscopic images of pentacene TFTs with channel length of 1, 2, and 5 μm after sintering (the line width of the ink-jet-printed contact lines is 5 μm). (c) AFM image of a pentacene TFT with a channel length of 5 μm and a contact line width of 2 μm. Adapted from [43]. Copyright, National Academy of Sciences, 2008.

The distance between the source and drain electrodes (channel length) can be controlled between 1 and 100 μm while maintaining a channel width of 300 μm (Figure 8.14b). Such fine and direct printing of Ag nanoparticles on organic semiconductors has not been previously achieved. Sintering at 130 °C for 1 h removes the dispersing agents and fuses the dispersed Ag nanoparticles. Figure 8.14c shows the AFM image of a TFT with a channel length of 5 μm.

The DC electrical characteristics of the TFTs were measured in air using a semiconductor parameter analyzer (Agilent 4156C). Figure 8.15a,b shows the current–voltage characteristics of a p-channel pentacene TFT with a channel length of 1 μm and width of 300 μm. Owing to the large capacitance of the thin gate dielectric (0.7 μF cm^{-2}), the TFTs show excellent linear and saturation characteristics at gate–source and drain–source voltages of 3 V. Despite the short

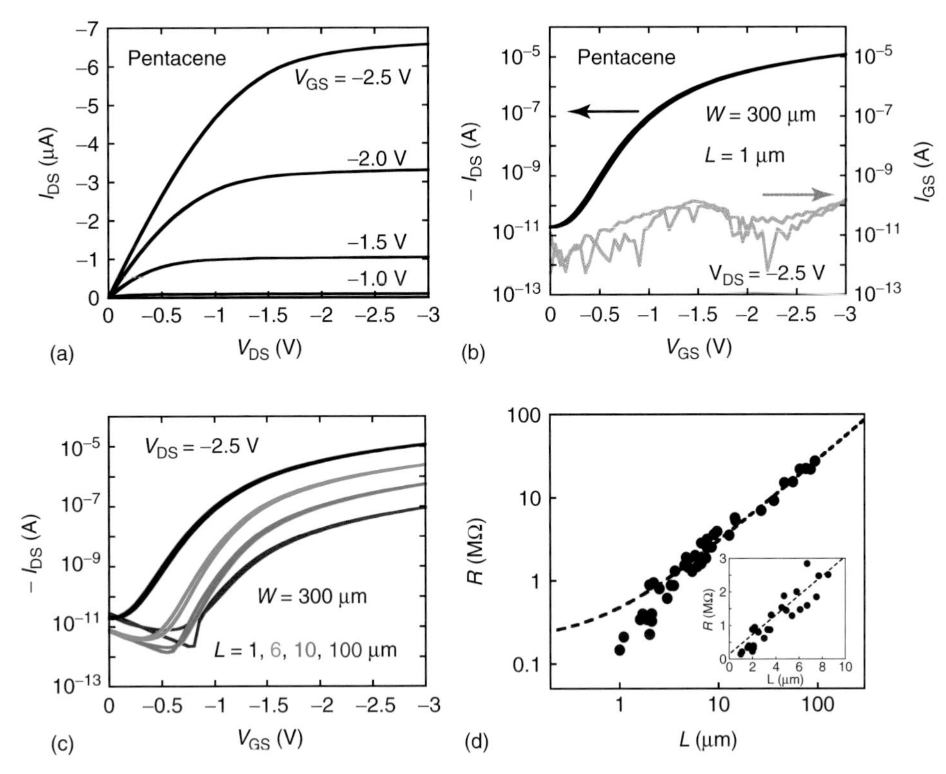

Figure 8.15 (a) Output and (b) transfer characteristics of a p-channel pentacene TFT with a channel length of 1 μm and a channel width of 300 μm. The measurements were carried out in air. (c) Transfer characteristics of pentacene TFTs with channel lengths of 1, 6, 10, and 100 μm, showing the scaling of the maximum drain current with channel length (channel width is 300 μm for all devices). The measurements were taken on TFTs prepared without substrate heating during the pentacene deposition. (d) The resistance of pentacene TFTs in the saturation regime scales linearly with channel length, as predicted by transistor theory. By extrapolating the linear fit to a channel length of zero, a contact resistance of 5 kΩ·cm is extracted. Adapted from [43]. Copyright, National Academy of Sciences, 2008.

channel length (1 μm), the OFF-state current of $V_{GS} = 0\,\text{V}$ is less than 20 pA and the ON–OFF current ratio is about 10^6. Despite the thin room-temperature gate dielectric, the maximum gate current at $V_{GS} = -3\,\text{V}$ is only about 100 pA. Figure 8.15c,d summarizes the results obtained for pentacene TFTs with channel lengths ranging from 1 to 100 μm and a channel width of 300 μm. The TFT resistance in the saturation regime scales linearly with channel length over the entire range, as predicted by transistor theory. By extrapolating the linear fit to a channel length of zero, an effective contact resistance of 5 kΩ·cm is extracted. This is smaller than the contact resistance of pentacene TFTs with bottom contacts [21] and identical to the contact resistance of pentacene TFTs with evaporated Ag top contacts [44], which confirms the high quality of the interface between the organic semiconductor and the ink-jet-printed contacts. Depending on the pentacene deposition conditions, the field-effect mobility of the TFTs is between 0.03 cm^2 (substrates not heated during pentacene deposition) and 0.3 $cm^2\,V^{-1}\,s^{-1}$ (pentacene deposited onto substrates held at a temperature of 60 °C during vacuum deposition).

The present subfemtoliter ink-jet technologies can be applied not only to p-type organic semiconductors such as pentacene but also to n-type organic semiconductors such as $F_{16}CuPc$. Figure 8.16a,b shows the characteristics of n-type organic transistors with $F_{16}CuPc$ in air. The manufacturing process is identical to that of organic transistors with pentacene described above; this, however, does not hold true for all organic semiconductors. The transistors exhibit a mobility of 0.02 $cm^2\,V^{-1}\,s^{-1}$ in air, V_{th} of −0.32 V, and an ON–OFF current ratio in excess of 10^4.

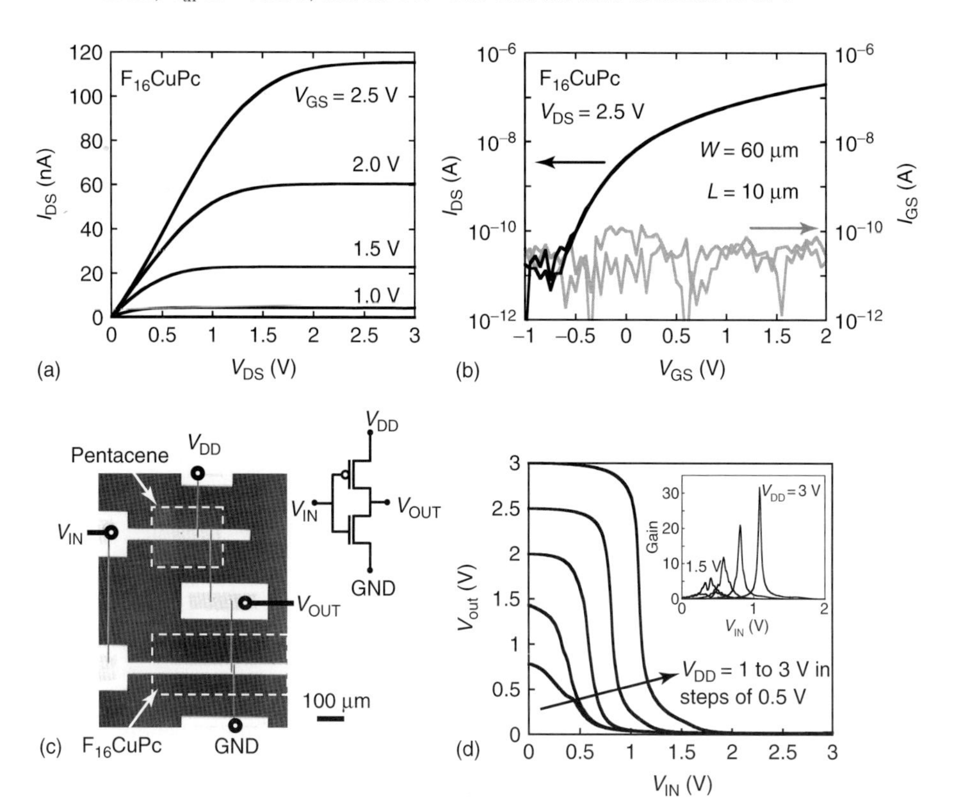

Figure 8.16 (a) Output and (b) transfer characteristics of an n-channel $F_{16}CuPc$ TFT with a channel length of 10 µm and a channel width of 60 µm. The measurements were carried out in air. (c) Optical microscope image and (d) transfer characteristics of an organic complementary inverter. The pentacene TFT has a channel length of 50 µm, the $F_{16}CuPc$ TFT has a channel length of 5 µm, and both TFTs have a channel width of 60 µm. The inverter operates with supply voltages between 1.5 and 3 V and with a small-signal gain greater than 10. Adapted from [43]. Copyright, National Academy of Sciences, 2008.

These characteristics are comparable to those of organic transistors manufactured using vacuum evaporation. This is consistent with our previous work [28, 29] and with our observation that the morphology of the pentacene films (examined by AFM) does not change during sintering at 130 °C, although others have reported changes in pentacene morphology and mobility degradation during annealing [45, 46]. The reason for this discrepancy is unknown, but possibly related to the different gate dielectric materials, that is, organic dielectrics (polyimide or SAM) versus inorganic dielectrics (SiO_2).

An additional benefit of the subfemtoliter ink-jet device process described here is that small gate-to-source and gate-to-drain overlap areas are easily obtained even without precise alignment of the contacts with respect to the gate electrode and without the use of self-alignment schemes [16, 17, 47–49]. For example, the pentacene TFT shown in Figure 8.15a,b has a parasitic capacitance of only 6 pF. With a carrier mobility of $0.3\,cm^2\,V^{-1}\,s^{-1}$, this translates into a cutoff frequency of above 2 MHz – approximately 5 orders of magnitude faster than the best result reported to date for printed organic circuits [50].

Next, we fabricated organic complementary inverters using p-channel pentacene transistors, n-channel $F_{16}CuPc$ transistors, patterned Al gates, a SAM-based gate dielectric, and ink-jet-printed source/drain contacts. A photograph and the electrical transfer characteristics of such an inverter are shown in Figure 8.16c,d. The pentacene TFT has a channel length of 50 μm, the $F_{16}CuPc$ transistors have a channel length of 5 μm, and both transistors have a channel width of 60 μm. The difference in channel length is necessary to achieve similar drain currents for both transistors despite the significant difference in carrier mobility ($0.1\,cm^2\,V^{-1}\,s^{-1}$ for the pentacene TFT and $0.02\,cm^2\,V^{-1}\,s^{-1}$ for the $F_{16}CuPc$ TFT). The inverter operates with supply voltages between 1.5 and 3 V and with a small-signal gain greater than 10. From a circuit design perspective, complementary circuits have several advantages over circuits based on a single carrier type, including greater noise margin, lower power consumption, and faster switching [51].

8.5 Discussion and Future Prospects of Large-Area Printed Electronics

With the manufacturing process described in this work, all three geometrical parameters that contribute to the high-frequency performance of organic transistors are scaled simultaneously and to a degree unprecedented in printed electronics. First, the channel length of the transistors is reduced to 1 μm, which is the shortest reported for organic transistors with patterned gates and printed source/drain contacts. Importantly, this short channel length is obtained along with low contact resistance since a top-contact structure is employed. Second, the gate dielectric thickness is reduced to a few nanometers without introducing prohibitive gate leakage. This means that the transistors can be operated with low voltages (3 V), while it also assures that the transistors have very small OFF-state currents (~20 pA), large ON–OFF current ratio (~10^6), and excellent drain current saturation even at small drain–source voltages. Finally, the parasitic capacitance of the transistors is reduced to a few picofarads since the printed source/drain contacts have line widths of only 2 μm. This extraordinary miniaturization of the key transistor parameters was achieved within the boundaries of a low-temperature, flexible-substrate-compatible, low cost, and large-area manufacturing approach.

The fact that the source/drain contacts can be ink-jet printed on top of the pentacene and $F_{16}CuPc$ surfaces without solvent damage is a direct consequence of the small droplet volume. Furthermore, the small size and large surface area of the droplets on the surface (thickness, ~30 nm) reduces the temperature required to remove the dispersing agent and fuse the nanoparticles into a metallic line with large electrical conductivity (25 μΩ·cm at 130 °C). Printing the same ink with a picoliter ink-jet system, we obtained droplets on the surface with a diameter of 50 μm and a thickness of 150 nm, which required sintering at 150 °C, although this caused significant changes to the morphology of the organic semiconductor film.

It is very interesting to compare the transistor performances shown in Figures 8.6 and 8.13 for the picoliter- and subfemtoliter-printed electrodes, respectively. The mobility of the subfemtoliter transistors are much higher than those of the picoliter transistors. This further shows that the subfemtoliter droplets do not damage the organic semiconductors.

Although other nonlithographic patterning techniques have been employed to manufacture organic transistors with micron and submicron dimensions [18–20, 52], these methods usually require prepatterning processes based on original patterning forms such as photolithography masks, or they work only with low mobility semiconductors. Here, we described a method for ultrafine patterning with 1 μm resolution on high-mobility p- and n-channel organic semiconductors without the need for any prepatterning processes or original patterning forms.

The main limitation of the subfemtoliter ink-jet system described here is the relatively small throughput, which is a consequence of the time required to cover large areas with multiple printing passes. Therefore, we believe that future manufacturing schemes for large-area printed electronic systems will benefit from utilizing high-resolution subfemtoliter ink-jet printing solely for the definition of critical device features such as source/drain contacts for short-channel transistors, while employing high-throughput, low-resolution printing techniques such as picoliter ink-jet printing, screen printing, and gravure printing to define large-area features that require lower accuracy, such as bit-lines, word-lines, and interconnects.

Picoliter ink-jet and screen printing have high throughput and very short tact time, although the minimum printing size is 20–50 μm. Hence, they have great advantages for manufacturing low-speed, cost-effective circuits or electrical wirings such as word-lines and bit-lines. Even though the single nozzle does not achieve a high throughput, the minimum printing size is 1 μm and it provides very fast, high-performance circuits. We believe these printing systems compliment each other and open up a new era of printed electronics.

Acknowledgments

This study was partially supported by JST/CREST, the Grant in Aid for Scientific Research (KAKENHI; WAKATE S), NEDO, and the Special Coordination Funds for Promoting and Technology. The high-quality Ag nanoparticles were supplied by Dr. Y. Matsuba and Dr. H. Saito (Harima Chemical Corp, Japan). The authors acknowledge Dr. K. Murata (AIST) and D. Mori (SIJ technologies) for their technical support on the high-definition ink-jet process. We thank Dr. Hagen Klauk and Dr. Ute Zschieschang, (Max Planck Institute for Solid State Research) for fabricating self-assembled monolayer and Dr. Y. Noguchi and T. Yokota (University of Tokyo) for the support of printing technologies. We also thank Kyocera Chemical for manufacturing the solution of polyimide insulators and Daisankasei Co., Ltd. for high-purity parylene (diX-SR).

References

1. Service, R.F. (2004) *Science*, **304**, 675.
2. Katz, H.E. (2004) *Chem. Mater.*, **16**, 4748.
3. Rogers, J.A. (2001) *Science*, **291**, 1502.
4. Hamedi, M., Forchheimer, R., and Inganäs, O. (2007) *Nat. Mater.*, **6**, 357.
5. Garnier, F., Hajlaoui, R., Yassar, A., and Srivastava, P. (1994) *Science*, **265**, 1684.
6. Rogers, J.A., Bao, Z.N., Baldwin, K., Dodabalapur, A., Crone, B., Raju, V.R., Kuck, V., Katz, H., Amundson, K., Ewing, J., and Drzaic, P. (2001) *Proc. Natl. Acad. Sci. U.S.A.*, **98**, 4835.
7. Sekitani, T., Takamiya, M., Noguchi, Y., Nakano, S., Kato, Y., Sakurai, T., and Someya, T. (2007) *Nat. Mater.*, **6**, 413–417.
8. Calvert, P. (2001) *Chem. Mater.*, **13**, 3299.
9. de Gans, B.J., Duineveld, P.C., and Schubert, U.S. (2004) *Adv. Mater.*, **16**, 203.
10. Shimoda, T. *et al.* (1999) *Proc. Soc. Inf. Display*, **99**, 376.
11. Bharathan, J. and Yang, Y. (1998) *Appl. Phys. Lett.*, **72**, 2660.
12. Sirringhause, H. *et al.* (2000) *Appl. Phys. Lett.*, **77**, 406.
13. Bai, P.P. *et al.* (2004) International Electron Device Meeting 2004 - IEDM Technical Digest, p. 657.
14. Sirringhaus, H., Kawase, T., Friend, R.H., Shimoda, T., Inbasekaran, M., Wu, W., and Woo, E.P. (2000) *Science*, **290**, 2123–2126.
15. Wang, J.Z., Zheng, Z.H., Li, H.W., Huck, W.T.S., and Sirringhaus, H. (2004) *Nat. Mater.*, **3**, 171–176.
16. Stutzmann, N., Friend, R.H., and Sirringhaus, H. (2003) *Science*, **299**, 1881–1884.
17. Sele, C.W., Werne, T., Friend, R.H., and Sirringhaus, H. (2005) *Adv. Mater.*, **17**, 997–1001.
18. Loo, Y.L., Willett, R.L., Baldwin, K.W., and Rogers, J.A. (2002) *Appl. Phys. Lett.*, **81**, 562.
19. Chabinyc, M.L., Wong, W.S., Arias, A.C., Ready, S., Lujan, R.A., Daniel, J.H., Krusor, B., Apte, R.B., Salleo, A., and Street, R.A. (2005) *Proc. Inst. Electr. Electron. Eng.*, **93**, 1491.
20. Blanchet, G.B., Loo, Y.L., Rogers, J.A., Gao, F., and Fincher, C.R. (2003) *Appl. Phys. Lett.*, **82**, 463.
21. Necliudov, P.V., Shur, M.S., Gundlach, D.J., and Jackson, T.N. (2003) *Solid-State Electron.*, **47**, 259–262.
22. Gundlach, D.J., Jackson, T.N., Schlom, D.G., and Nelson, S.F. (1999) *Appl. Phys. Lett.*, **74**, 3302–3304.
23. Park, J.U., Hardy, M., Kang, S.J., Barton, K., Adair, K., Mukhopadhyay, D.K., Lee, C.Y., Strano, M.S., Alleyne, A.G., Georgiadis, J.G., Ferreira, P.M., and Rogers, J.A. (2007) *Nat. Mater.*, **6**, 782–789.

24. Saito, H. and Matsuba, Y. (2006) Proceedings of the 39th International Symposium on Microelectronics, San Diego, October 8–12, 2006, pp. 470–477.
25. Saito, H., Ueda, M., Oyama, K., and Matsuba, Y. (2005) Proceedings of the International Conference on Electronics Packaging, Tokyo, April 13–15, 2005, pp. 259–262.
26. Murata, K., Matsumoto, J., Tezuka, A., Matsuba, Y., and Yokoyama, H. (2005) *Microsyst. Technol.*, **12**, 2–7.
27. Murata, K. (2003) Proceedings of the 2003 International Conference on MEMS, NANO, Smart Systems (ICMENS'03, Bannf, Canada, July 20–23, 2003), ISBN: 0-7695-1947-49, pp. 346–349.
28. Sekitani, T., Iba, S., Kato, Y., and Someya, T. (2004) *Appl. Phys. Lett.*, **85**, 3902–3904.
29. Sekitani, T., Someya, T., and Sakurai, T. (2006) *J. Appl. Phys.*, **100**, 024513.
30. Noguchi, Y., Sekitani, T., Yokota, T., and Someya, T. (2008) *Appl. Phys. Lett.*, **93**, 043303.
31. Kato, Y., Iba, S., Teramoto, R., Sekitani, T., Someya, T., Kawaguchi, H., and Sakurai, T. (2004) *Appl. Phys. Lett.*, **84**, 3789.
32. Dimitrakopoulos, C.D. and Malenfant, P.R.L. (2002) *Adv. Mater.*, **14**, 99.
33. Dimitrakopoulos, C.D., Purushothaman, S., Kymissis, J., Callegari, A., and Shaw, J.M. (1999) *Science*, **283**, 822.
34. Klauk, H., Gundlach, D.J., Nichols, J.A., and Jackson, T.N. (1999) *IEEE Trans. Electron Devices*, **46**, 1258.
35. Bao, Z., Lovinger, A., and Brown, J. (1998) *J. Am. Chem. Soc.*, **120**, 207–208.
36. Bao, Z., Dodabalapur, A., and Lovinger, A.J. (1996) *Appl. Phys. Lett.*, **69**, 4108.
37. Chang, P.C., Molesa, S.E., Murphy, A.R., Frechet, J.M.J., and Subramanian, V. (2006) *IEEE Trans. Electron Devices*, **53**, 594.
38. Dodabalapur, A., Torsi, L., and Katz, H.E. (1995) *Science*, **268**, 270.
39. Noguchi, Y., Sekitani, T., and Someya, T. (2006) *Appl. Phys. Lett.*, **89**, 253507.
40. Noguchi, Y., Sekitani, T., and Someya, T. (2007) *Appl. Phys. Lett.*, **91**, 133502.
41. Klauk, H., Zschieschang, U., Pflaum, J., and Halik, M. (2007) *Nature*, **445**, 745.
42. Halik, M., Klauk, H., Zschieschang, U., Schmid, G., Dehm, C., Schutz, M., Maisch, S., Effenberger, F., Brunnbauer, M., and Stellacci, F. (2004) *Nature*, **431**, 963.
43. Sekitani, T., Noguchi, Y., Zschieschang, U., Klauk, H., and Someya, T. (2008) *Proc. Natl. Acad. Sci. U.S.A.*, **105**, 4976–4980.
44. Pesavento, P.V., Chesterfield, R.J., Newman, C.R., and Frisbie, C.D. (2004) *J. Appl. Phys.*, **96**, 7312–7324.
45. Kang, S.J., Noh, M., Park, D.S., Kim, H.J., Whang, C.N., and Chang, C.H. (2004) *J. Appl. Phys.*, **95**, 2293–2296.
46. Ye, R., Baba, M., Suzuki, K., Ohishi, Y., and Mori, K. (2003) *Jpn. J. Appl. Phys.*, **42**, 4473–4475.
47. Ando, M., Kawasaki, M., Imazeki, S., Sasaki, H., and Kamata, T. (2004) *Appl. Phys. Lett.*, **85**, 1849.
48. Lu, J.P., Mei, P., Rahn, J., Ho, J., Wang, Y., Boyce, J.B., and Street, R.A. (2000) *J. Non-Cryst. Solids*, **266**, 1294–1298.
49. Nagai, T., Naka, S., Okada, H., and Onnagawa, H. (2007) *Jpn. J. Appl. Phys.*, **46**, 2666–2668.
50. Hübler, A.C., Doetz, F., Kempa, H., Katz, H.E., Bartzsch, M., Brandt, N., Hennig, I., Fuegmann, U., Vaidyanathan, S., Granstrom, J., Liu, S., Sydorenko, A., Zillger, T., Schmidt, G., Preissler, K., Reichmanis, E., Eckerle, P., Richter, F., Fischer, T., and Hahn, U. (2007) *Org. Electron.*, **8**, 480–486.
51. Crone, B.K., Dodabalapur, A., Sarpeshkar, R., Filas, R.W., Lin, Y.Y., Bao, Z., O'Neill, J.H., Li, W., and Katz, H.E. (2001) *J. Appl. Phys.*, **89**, 5125–5132.
52. Noh, Y.Y., Zhao, N., Caironi, M., and Sirringhaus, H. (2007) *Nat. Nanotechnol.*, **2**, 784–789.

9
Ink-jet Printing of Downscaled Organic Electronic Devices

Mario Caironi, Enrico Gili, and Henning Sirringhaus

9.1
Introduction

Ink-jet printing is a very promising technique for the large-scale production of low-cost, large-area, and flexible electronic circuits. Additive printing offers, in fact, advantages in terms of process simplification, layer-to-layer alignment, compatibility with large-area processing, and potential cost reduction compared to subtractive lithographic patterning. Since there is a large number of organic semiconductors that can be processed from solution, this technique is well suited to the field of organic electronics. Despite not being as high throughput as other graphical art printing techniques [1], ink-jet printing, which is a noncontact technique, offers high flexibility and is compatible with a large range of functional materials. In some cases, such as organic light emitting diodes (OLEDs), it was successfully adopted to realize prototypal RGB displays, which are in an advanced stage of development. In other cases, such as organic thin film transistors (OTFTs), it is helping to transfer this technology from laboratory demonstrations to industrial fabrication processes. What is currently limiting the development in such applications is the relatively low resolution of standard tools that do not allow feature sizes to be smaller than 10 μm [2]. Despite the synthesis of various organic semiconductors with a mobility exceeding the one of amorphous silicon, this is reflected in low circuit switching speeds and high-operating voltages, precluding a widespread integration of organic electronics in many applications where downscaling is required.

The highest operational frequency of a transistor is described by its transition frequency f_T, which under certain assumptions scales with the channel length L according to the relationship

$$f_T \propto \frac{1}{L^2} \tag{9.1}$$

It is therefore evident that a reduction in the lowest printable feature size is a key issue in the development of printed downscaled transistors. Moreover, as it will be better detailed in Section 9.4, to retain the correct transistor behavior, other

Organic Electronics II: More Materials and Applications, First Edition. Edited by Hagen Klauk.

parameters have to be correspondingly downscaled, as the dielectric thickness and the parasitic capacitance.

It is therefore mandatory to develop new techniques and tools that could allow to go beyond the resolution offered by the standard mass production instruments. Indeed, there are several examples where high-resolution printing was demonstrated, both in terms of reduced printed linewidths and higher droplets placement accuracy. These demonstrations open up new possibilities for the downscaling of organic devices. However, it is important that such approaches are developed compatibly with the requirements of large-area and flexible electronics manufacturing. Moreover, technological challenges have to be faced in order to reach the high level of printing uniformity and yield, which are a basic prerequisite for the adoption of these direct-write techniques in real fabrication processes.

In this chapter, we aim to give an overview on the recent advancements in high-resolution printing of downscaled organic devices, with a focus on OTFTs. By providing examples that are present in the literature, we describe the technologies involved and point out the issues that have already been overcome and we give an overview of the areas in which progress is needed. Different approaches are reported and analyzed in terms of performance improvement and compatibility with low-cost manufacturing.

9.2 Ink-Jet Printing: Technologies, Tools, and Materials

9.2.1 Principle of Operation of Ink-Jet Printers

Since the observations by Lord Rayleigh in 1878 that a liquid stream has the tendency to break up into individual droplets [3], several ink-jet printing technologies have been developed. Although ink-jet printing has been historically extremely successful in the graphics market, several other applications have more recently been developed in the fields as diverse as electronics and biology. These novel applications are based on the idea of using this technology to position functional inks on a substrate. These inks can then be transformed, usually by a thermally activated process, in functional materials with structural, electrical, optical, chemical, or biological properties.

Ink-jet printing technology is very attractive because of its unique characteristics and flexibility. First, it is a digital technology that does not require a physical template in order to pattern a material on a substrate. Second, it is an additive technique which only deposits the material of choice where it is needed. For this reason, ink-jet printing typically has a lower consumption of active materials and a lower production of waste compared to conventional additive technologies. Third, ink-jet printers typically do not need a special environment for their operation. In particular, they do not require to operate under vacuum, unlike most conventional deposition systems. This leads to considerably lower equipment cost compared with conventional microfabrication techniques. Finally, this technique allows to deposit a wide range of materials, usually in the form of solutions or suspensions in solvents.

Ink-jet printers are basically dispensers that propel droplets of a fluid toward a substrate. This can be achieved in two different ways: continuous and drop on demand (DOD) printing [4]. In continuous ink-jet printers, a stream of droplets is ejected from a nozzle. The fluid is then either steered on the substrate to deposit a pattern or recirculated. On the other hand, DOD printers have an actuator that triggers the ejection of a droplet from a nozzle only when this is needed to create a pattern on a substrate. Figure 9.1 shows a schematic comparison between a continuous ink-jet printer and a piezoelectric DOD printer. In the following sections, we review two continuous printing technologies (continuous ink-jet printing and aerosol jet printing) and five DOD printing technologies (Piezoelectric ink-jet printing, acoustic ink-jet printing, electrohydrodynamic-jet printing, and thermal ink-jet printing). Table 9.1 compares the drop volume and diameter achievable with the printing technologies most commonly used to print functional materials.

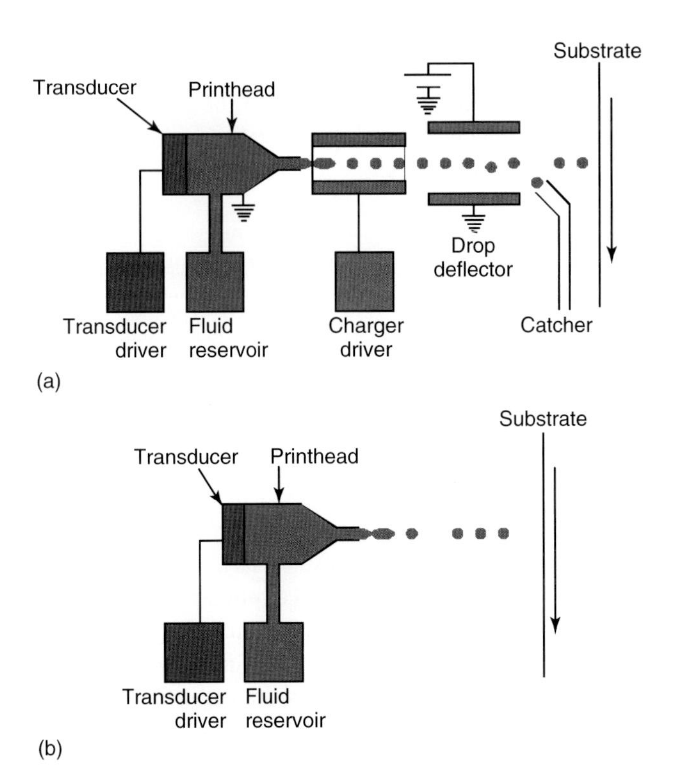

Figure 9.1 Operating principle of (a) continuous ink-jet printer and of a (b) piezoelectric drop on demand ink-jet printer.

Table 9.1 Comparison of the drop volume and diameter achievable with the different drop on demand piezoelectric ink-jet technologies most commonly used to print functional materials [5–8].

Technology	Minimum drop volume (fl)	Minimum drop diameter (μm)	Multiple nozzle printheads availability	Viscosity range (mPa s)
Piezoelectric	1000	12	Yes	0.5–40
Acoustic	65	5	No	N/A
Electrohydrodynamic	0.7	1	No	N/A
Aerosol jet printing	N/A (Jet)	5 (Jet)	Yes	0.7–2500

9.2.2 Continuous Ink-Jet Printing Technologies

Continuous ink-jet printing systems are usually very reliable, because they eject a continuous stream of fluid from an orifice. This prevents clogging of the nozzles. On the other hand, in order to produce a pattern, they require a system to select the fluid that needs to be deposited on the substrate from the fluid that needs to be recirculated. Moreover, continuous printers are typically considerably more complex and expensive than the DOD printers.

9.2.2.1 Continuous Ink-Jet Printing

In continuous ink-jet printers (Figure 9.1a), a stream of liquid is continuously projected from an orifice. A piezoelectric crystal creates an acoustic wave, as it vibrates within the cavity and causes the stream of liquid to break into droplets at regular intervals, due to Rayleigh instability [3]. During their formation, an electric charge is imparted to the drops by an electrode. Finally, a second pair of electrode steers the charged droplets by means of an electric field. The droplets can either be directed onto the substrate or steered toward a catcher, which allows to recirculate the unused liquid.

Continuous ink-jet printing is the system of choice for high-throughput, low-resolution industrial graphics printing systems. In fact, the high-velocity ($\sim$50 m s^{-1}) of the ink droplets allows for a long distance between print head and substrate and for a high printing speed. Moreover, as the liquid is ejected continuously from the print head, this system is not susceptible to nozzle clogging. On the other hand, continuous ink-jet printing has a few disadvantages that hinder its application for printing functional materials. First, the system has poor flexibility, as it is usually designed for only one type of ink. Second, continuous ink-jet system usually has stringent requirements on ink viscosity. Third, the liquid of choice needs to be suitable for electrostatic charging.

9.2.2.2 **Aerosol Jet Printing**

In order to enable direct-write patterning of high-viscosity fluids, a contact-less printing technology called *aerosol jet printing* has been developed [9, 10]. This system is composed of an atomizer that produces an aerosol, which is then focused and directed toward a substrate. The aerosol stream can be shuttered mechanically in order to selectively deposit the ink on a substrate.

The wide range of viscosity handled by the system (0.7–2500 mPa s [8]) is enabled by the use of different types of atomizers. For fluids with viscosity between 0.7 and 10 mPa s, an ultrasonic transducer nebulizes the fluid with high-frequency pressure waves. On the other hand, higher viscosity fluids (10–2500 mPa s) require a pneumatic atomizer, in which a high-velocity gas stream is used to shear the liquid stream into droplets. For materials with viscosity higher than 1000 mPa s, a heater/stirrer is also required. The aerosol stream is then fed to a nozzle in which the aerosol is completely surrounded by a focusing gas. The droplets therefore do not touch the inner walls of the nozzle, eliminating clogging and allowing to focus the aerosol to a jet diameter which is smaller than that of the nozzle.

The aerosol jet printing technology has the main advantage of allowing to print fluids over a very wide range of viscosities. This enables conformal printing on patterned substrates. High-viscosity materials also allow to pattern thicker films in a single pass, thanks to the lower dilution of the active material. Moreover, this system enables to print features smaller than the size of the orifice from which the jet is ejected. Feature sizes down to 10 μm [8] can be produced without downscaling the orifice of the nozzle, therefore preventing clogging. The system also produces a collimated beam of small droplets instead of one larger droplet as in conventional ink-jet systems. For this reason, the system can print from a distance of up to 5 mm [8] from the substrate, which enables printing on curved or irregular surfaces. Finally, aerosol jet printing systems are available in both single-nozzle and multinozzle versions, and are therefore suitable for industrial applications.

One drawback of this technology is the requirement of separate systems for the atomization and the delivery of the fluid. This is likely to increase the complexity and cost of these dispensers. Another disadvantage is the use of a mechanical shutter to interrupt the stream and pattern the material. The response time of the shutter is limited to a minimum of 2 ms [8], and this could affect the throughput of the system.

9.2.3
DOD Ink-Jet Printing Technologies

DOD ink-jet printing is an ideal technique for printing functional materials. First, it is highly flexible, as fluids can be easily interchanged. Second, high-resolution DOD systems are commercially available. Third, the use of actuators driven by an electrical signal allows to jet a wide range of inks by adapting the shape and intensity of the electrical pulse to their characteristics. Nevertheless, DOD printing has a major drawback due to the intermittent creation of pressure waves in the printhead cavity. For this reason, DOD ink-jet printheads are more susceptible to clogging than continuous printing systems.

An important figure of merit of a DOD printhead is the printing resolution, which is dependent on the minimum drop size that can be achieved (Table 9.1). Electrohydrodynamic printheads allow to obtain the smallest drops and the highest resolution reported to date. Nevertheless, only piezoelectric printheads have been scaled up to multinozzle dispensing systems, which have sufficient throughput for applications in manufacturing.

9.2.3.1 **Thermal Ink-Jet Printing**

In thermal ink-jet printers the transducer is composed of a heating element that is electrically activated by a pulse of current. When the transducer is heated, the liquid contained in the printhead cavity is locally heated, and it evaporates forming a bubble, which generates a pressure wave in the cavity. Although thermal DOD ink-jet printing of nonaqueous fluids has been reported [11], the use of this technology to print functional materials has proved challenging. In fact, in conventional graphics thermal ink-jet printing, additives are used to modify the properties of an ink and to optimize it for jetting. This is not always possible when processing functional inks, because the interaction of the additives with the functional material has to be taken into consideration.

9.2.3.2 **Piezoelectric Ink-Jet Printing**

In piezoelectric ink-jet printers (Figure 9.1b), a transducer driven by an electrical signal generates a pressure wave in the cavity of the printhead, which is full of ink. The transducer only generates a signal when a drop needs to be ejected to print a dot on a substrate. The pressure wave has to be optimized to eject only one drop for every electrical pulse applied to the transducer. Once ejected, the droplet is deposited onto the substrate. The minimum volume of the drops that can be generated is mainly limited by the nozzle size. Piezoelectric printheads generating drops with calibrated volume as small as 10 pl [2] (26 μm drop diameter) are used in manufacturing, while 1 pl (12 μm drop diameter) dispensers are used in research and development.

Piezoelectric printheads incorporate a transducer composed of a piezoelectric material attached to the nozzle, connected with a signal generator. When a voltage is applied, the piezoelectric material changes shape or size, thus reducing or increasing the cross-section of the cavity full of ink. This generates a pressure wave in the fluid forcing a droplet of ink from the nozzle. An example of the cycle leading to the ejection of a drop in a piezoelectric printhead is schematically shown in Figure 9.2 and can be divided into four steps. For each step, Figure 9.2 shows a schematic picture of the ink cavity and the nozzle. In the inset, a typical pulse is shown in the $V(t)$ diagram.

When the printhead is inactive, or before the start of the pulse, a small voltage is applied to the piezoelectric element to keep it slightly deflected (Figure 9.2a). When the pulse starts (Phase 1), a step signal is applied to the piezoelectric element to relax it and expand the ink cavity (Figure 9.2b). This generates a negative pressure

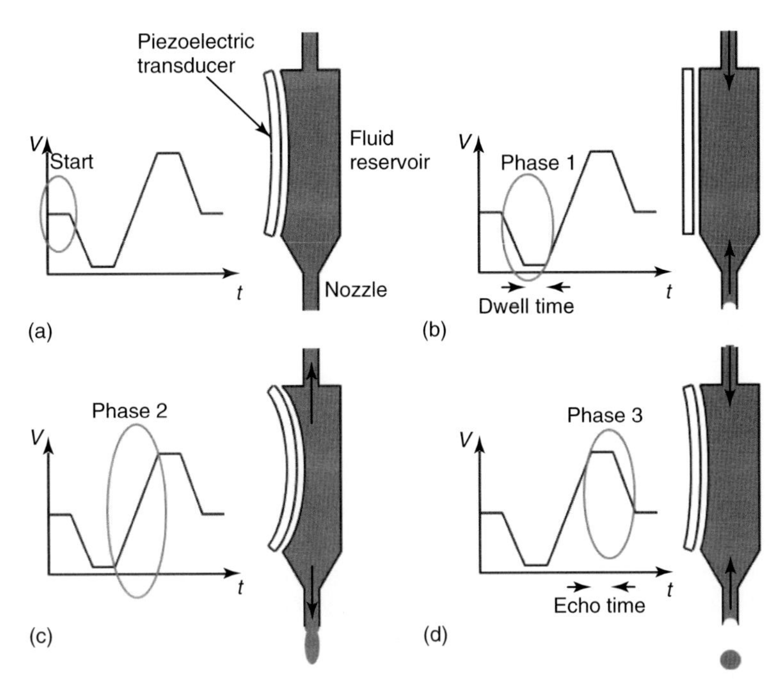

Figure 9.2 Schematic example of a typical drop ejection cycle in a drop on demand piezoelectric printhead. (a) Initial printhead configuration; (b) Phase 1 (expansion): ink is sucked into the nozzle from the reservoir; (c) Phase 2 (compression): a droplet is ejected; (d) Phase 3 (expansion): additional phase added to eliminate satellite drops. V is the voltage applied to the piezoelectric element and t is time.

(expansion) wave which moves both toward the supply and the orifice ends of the cavity. At the supply end of the cavity, the pressure wave is reflected as a positive (compression) wave. During this phase, fluid is pulled into the chamber through the inlet. The voltage then stays constant for the duration of the "dwell time."

During Phase 2 of the pulse, the piezoelectric element bends inward and compresses the fluid in the cavity (Figure 9.2c). This generates a compression wave that, if properly timed, can add to the reflected compression wave from Phase 1, thus generating a reinforced pressure wave. When this wave reaches the orifice, the acoustic energy is converted into kinetic energy which causes the fluid to start emerging from the orifice at high velocity. When a negative pressure wave, reflected at the supply end of the cavity, reaches the orifice, the fluid is pulled back and a drop is formed.

An additional phase can be added to the pulse to cancel the residual acoustic waves propagating in the tube (Figure 9.2d). After the signal is kept constant for the duration of the "echo time," it is brought back to its original value, causing a second expansion of the ink cavity. If properly timed, the cavity expansion removes any residual wave propagating in the cavity, avoiding satellites drops to be ejected from the nozzle. Phase 3 is an optional feature of the signal pulse, which can be used to optimize the jetting characteristics of a piezoelectric printhead.

The signal applied to a piezoelectric printhead needs to be adjusted for every fluid that is introduced in the jetting cavity. This is usually done empirically, but the suitability of a fluid to be used for a particular printhead should be established before attempting to jet it. The main physical parameters that affect the jetting performance of a fluid are the following:

1) Viscosity
2) Density
3) Surface tension
4) Speed of sound in the fluid, and consequently speed of a pressure wave inside the fluid cavity
5) Boiling temperature, and consequently volatility of the fluid at the nozzle orifice, where evaporation occurs

The design of a printhead and the choice of a particular ink for a piezoelectric printing system involves several factors which should all be taken into account. First, the ink should be formulated within a narrow range of viscosity and surface tension, optimized for the printhead of choice. The general requirements for a fluid to be used in a piezoelectric dispenser are a viscosity of 0.5–40 mPa s (Newtonian fluids) and a surface tension of 20–70 $\mathrm{mN\,m^{-1}}$. If a fluid containing nanoparticles is used, the diameter of the nanoparticles should be less than 5% of the diameter of the nozzle to avoid jetting instability [12]. Any additive added to modify the physical properties of the ink should be carefully assessed to ensure that it will not affect negatively the performance of the functional material after deposition. Second, no chemical interaction should occur between the ink and the printhead, to prevent structural failure or malfunction. Third, a reliable and stable jet should be achieved before an ink can be used for printing. In particular, the presence of satellite drops may affect the printing resolution. Moreover, nozzles can become clogged due to the presence of air bubbles in the ink, evaporation of ink at the nozzle plate or presence of particles of size similar to the diameter of the nozzle. Fourth, if a multinozzle printhead is to be used successfully, the direction and speed of the drops ejected has to be highly uniform in order to ensure an accurate positioning of a large number of drops in well-defined substrate locations.

9.2.3.3 **Acoustic Ink-Jet Printing**

One of the main drawbacks of piezoelectric printheads is their tendency to become clogged. This is particularly critical when attempting to deposit inks composed of particles suspensions in solvents with low-boiling temperature. In this case, nozzles can become clogged due to evaporation of solvent on the nozzle plate. An alternative solution to overcome these limitations is acoustic ink-jet printing [6]. This system has a nozzleless printhead, in which an acoustic beam is focused by a high-frequency acoustic lens on a free liquid surface. The acoustic wave is created by a high-frequency transducer attached to the back of the lens. The lens focuses the acoustic energy, forming a pressure wave that overcomes the surface tension of the liquid standing over the lens and expels a drop from the liquid surface.

An acoustic lens focuses sound in much the same way that an optical lens focuses light. Snell's law describes the refraction of sound as it passes through an interface between two materials of differing sound speed. An acoustic lens provides the appropriate material thicknesses to focus a parallel wavefront of sound to a single focal point. The liquid surface is at the focal plane of the acoustic beam, which is confined to a spot with a diameter of the order of the acoustic wavelength. The acoustic energy transmitted by the beam to the liquid surface causes a mound or liquid to rise and, consequently, a single droplet to be expelled. Stable formation of drops with volume as small as 65 fl (5 μm diameter) has been reported [6].

The nozzleless design of acoustic ink-jet printheads minimizes clogging, improving jetting reliability. OTFTs fabricated using a polymeric semiconducting layer printed by acoustic ink-jet printing have been reported [13]. Nevertheless, the scalability of this technology to multiejector printheads, necessary for applications in manufacturing, has not been demonstrated to date.

9.2.3.4 Electrohydrodynamic-Jet (e-Jet) Printing

While thermal and piezoelectric printers use acoustic energy to destabilize the fluid contained in a nozzle, electrohydrodynamic-jet (e-jet) printers achieve the same result using an electric field. In this approach, a conductive liquid, when subjected to an intense electric field, is disrupted into a spray of charged droplets. This phenomenon has been used in a wide range of applications including paint spraying, electrostatic printing, electrostatic emulsification, and fuel atomization in combustion systems [14]. This technique has been found to be particularly attractive because it allows to decouple the diameter of the nozzle from the size of the drop ejected. On the other hand, in piezoelectric ink-jet printers, the nozzle size is approximately half the size of the ejected droplet. In one application, e-jet printing has been used to print living cells using nozzles large enough in order to avoid clogging [15]. In this experiment, 500 μm wide nozzles were used to print droplets as small as 17 μm.

This technique has also been used to print functional materials for electronic applications, such as nanoparticle inks, conductive polymers, and carbon nanotubes suspensions. Park *et al.* [16] describe a setup in which the external surface of very fine glass capillaries is coated with gold. A self-assembled monolayer (SAM) is subsequently deposited on the gold surface, which becomes hydrophobic, while the internal surface of the capillary remains hydrophilic. A pressure controller is used to control the ink pressure inside the tube. A DC voltage is applied between the gold coating on the nozzle and a substrate, which rests on an electrically grounded metal plate. The voltage generates an electric field that causes mobile ions in the ink to accumulate near the surface of the nozzle, creating a flow of ink that, if correctly optimized, can generate a very small drop of ink. The droplet travels from the nozzle to the substrate, where its final location is controlled by a mechanical x-y stage. With this method, features as small as 0.5 μm could be printed with a polyurethane ink using nozzles with a diameter of 0.5 μm. Using a similar setup, Murata *et al.* [17] printed a silver nanoparticle ink to pattern conductive lines as narrow as 1 μm using an electrohydrodynamic printer capable to eject drops with volume as low as 0.7 fl, about three orders of magnitude smaller than achievable with piezoelectric ink-jet printing systems.

In order to improve the control on the electrohydrodynamic jet, an alternative setup was developed by Lee *et al.* [18]. In this approach, a guide ring was introduced between the conductive nozzle and the substrate to focus the jet. Moreover, the ground electrode used to create the electric field was composed by a pin-type electrode located below the substrate. This allowed to increase the axial electric field but not the radial field, therefore reducing jet breakup. This approach may also be useful to develop multinozzle electrohydrodynamic printing systems in the future in order to improve the throughput of this technique. In fact, the low throughput is the main drawback of this technology, especially in the case of the high-resolution systems. In particular, several passes are necessary to print high-resolution metallic lines with conductivity high enough for electronic applications [17]. Another drawback of the technique is that the printed droplets have a substantial charge as a result of the application of the electric field. This may lead to unwanted consequences in device performance, especially when printing semiconductive or insulating layers.

9.2.4
Conductive Inks for Ink-Jet Printing of Electrodes and Interconnections

Ink-jet printing is a very attractive technique to deposit functional materials for electronics because complete devices can be printed using only one tool, with no need of high-vacuum systems or large clean rooms. As discussed before, many different functional inks have been developed for a wide range of applications. In this paragraph, we concentrate on the use of highly conducting inks for ink-jet printing of very low resistive contacts and tracks needed to fabricate organic electronic devices and interconnections. Table 9.2 compares the different approaches compatible with low-temperature processing to print metallic tracks with high-electrical conductivity.

One of the interconnection technologies that is fully compatible with ink-jet printing is based on organic conductors. The most widely used printable organic conducting material is a water-based ink of the conducting polymer poly(3,4-ethylenedioxy thiophene) doped with polystyrene sulfonic acid (PEDOT:PSS), with a conductivity range of 30–150 $\mathrm{S\,cm^{-1}}$ (e.g., Clevios conductive polymers, produced by H. C. Starck GmbH) [19]. Given the large interconnection distances and the small cross-section of the conducting tracks needed for many potential printed electronic applications, the use of organic conductors for interconnects results in unacceptably high-parasitic resistance and therefore low-switching speed.

Recently, a new class of inks containing metallic nanoparticles, suitable for ink jetting, has received much interest. Several companies now offer inks containing gold or silver nanoparticles. These two materials are preferred for ink-jet applications because they do not grow insulating oxides when deposited in the presence of oxygen. This simplifies the processing considerably, as enclosing an ink-jet printer in a oxygen-free atmosphere introduces considerable technical challenges.

Table 9.2 Comparison of the different approaches compatible with low-temperature processing to print conductive electrodes and interconnections [19–24].

Material	Printed line conductivity ($S\ cm^{-1}$)	Annealing temperature (°C)	Bulk material conductivity ($S\ cm^{-1}$)
Aqueous dispersion of PEDOT/PSS (Clevios)	30–150	100	N/A
Gold nanoparticles (annealed)	3.3×10^5	150	4.5×10^5
Gold nanoparticles (laser sintering)	1.7×10^5	N/A	4.5×10^5
Silver nanoparticles (microwave sintering)	2.3×10^5	N/A	6.2×10^5
Aluminum nanoparticles (electro-magnetic sintering)	2.3×10^5	N/A	5.9×10^5
Organic complex-based silver (TEC-I)	4.5×10^5	130	6.2×10^5
Silver seed plus electroless copper plating	5.5×10^5	N/A	5.9×10^5

The melting temperature of nanometer-sized metallic particles has been found to be dependent on the size of the particles [25]. In particular, particles with a diameter lower than 5 nm exhibit a marked drop in melting temperature that in some cases makes them suitable for processing on flexible substrates. In order to be diluted in a solvent suitable for ink-jet printing without forming larger aggregates, the metallic nanoparticles need to be encapsulated in organic carbon chains (e.g., alkylthiols), which makes the particles soluble in common nonpolar solvents such as xylene. By optimizing the chain length of the encapsulant carbon chain, Huang *et al.* [20] demonstrated ink-jet printed gold lines with a conductivity $\sigma = 3.3 \times 10^5$ $S\ cm^{-1}$ after annealing at 150 °C, compared with a value for bulk gold of 4.5×10^5 $S\ cm^{-1}$.

An alternative approach to lower the processing temperature even further is laser curing of ink-jet printed metallic nanoparticles [26]. In this approach, ink-jet printed gold nanoparticles were sintered by irradiation with a laser beam of suitable wavelength to deposit enough thermal energy to induce highly localized melting of the nanoparticles. Ink-jet printed lines with $\sigma = 1.7 \times 10^5\ S\ cm^{-1}$ were fabricated using this approach [21]. Local laser heating allowed to reduce the heat-affected zone, lowering the processing temperature considerably.

Both thermal and laser annealing of printed nanoparticle inks suffer from limited throughput, due to the anneal time (of the order of 1 h) in the case of thermal processing, and to the low writing speed in the case of laser processing. For this reason, the annealing of printed silver nanoparticle lines by microwave sintering has been investigated [27]. Metallic nanoparticles are found to absorb microwave radiation efficiently, leading to sintering of a metallic film. Electrodes with a conductivity of 2.3×10^5 S cm^{-1} have been obtained with an exposure to microwave radiation for about 1 min. Nevertheless, the capability of the material to absorb microwave radiation is found to depend on the surface area of the electrodes, potentially leading to conductivity variations of different devices within the same electronic circuit.

Although gold and silver nanoparticle inks are currently used in research and in many industrial applications owing to their chemical stability and high conductivity, the growing price of silver may be a concern for its future applications in printed electronics. For this reason, recently several research organizations have tried to develop alternative printable inks based on low-cost materials. With this objective in mind, copper is an excellent candidate because of its low cost and high-electrical conductivity. Nevertheless, the development of printable copper inks has proved very difficult, mainly because copper is not stable in air, but has a strong tendency to oxidize. In particular, pure nonnoble metal nanopowders show very high reactivity in oxidizing atmosphere, because of their very high specific area. Moreover, high temperatures are typically required to anneal copper inks, which prevents their use on flexible polymeric substrates. In an attempt to solve these problems, precursor-based inks have been used to print copper conductive lines [28]. Nevertheless, this approach requires annealing temperatures of the order of 200 °C in a reducing atmosphere. This requires sophisticated equipment and therefore is not compatible with low-cost device fabrication. In a promising approach to develop copper inks compatible with ink-jet printing, a capping polymer has been used to synthesize air-stable copper nanoparticles [29, 30], requiring temperatures of more than 250 °C in order to obtain bulklike conductivity. In order to reduce processing temperatures, several companies have developed copper inks, which can be annealed at temperatures compatible with polymeric substrates. For example, Novacentrix has developed copper nanoparticle inks suitable for ink-jet printing, which can be sintered within microseconds using electromagnetic irradiation. This approach allows to produce copper films with a conductivity only three times higher than bulk copper and at process temperatures compatible with flexible substrates such as polyethylene terephthalate (PET) [22].

Printable inks composed of metal-containing organic complexes dissolved in a solvent were developed as an alternative to nanoparticle metallic inks. An example of this class of inks are the TEC-IJ transparent silver inks, produced by InkTec Corporation. In this case, the ink does not contain nanoparticles, but instead is composed of a silver-containing organic complex dissolved in an anisole-based solvent. During annealing, at first the solvent contained in the ink evaporates. In this phase, nucleation and growth of silver nanoparticles is observed. This is followed by a sintering step, when the silver nanoparticles join together to form a continuous metal line. With this approach, ink-jet printed lines with $\sigma = 4.5 \times 10^5$ S cm^{-1} were obtained after annealing at 130 °C, compared with a value for bulk silver of 6.2×10^5 S cm^{-1} [23]. This is the lowest sintering temperature achieved by thermal annealing of metallic inks demonstrated to date, and it is fully compatible with several commercially available flexible polymeric substrate materials, such as PET and polyethylene naphthalate (PEN) [31]. Moreover, the absence of nanoparticles facilitates the jetting of this class of inks. Nanoparticle inks are more likely to cause nozzle clogging than organic complex-based inks, because the nanoparticles can form large aggregates at the nozzle orifice due to evaporation of solvent on the nozzle plate.

An alternative approach to pattern metal tracks with very high conductivity is a printing plus electroless plating process. In this method, a seed layer is ink-jet printed, followed by electroless deposition of a thick layer of the metal required by the application. This technique is very flexible, as different metals can be plated using the same seed layer. For example, copper, silver, gold, nickel, and cobalt can all be deposited using this technique. Moreover the conductivity of the film is very close to that of bulk material [24]. The possibility of growing thick metal layers without need of multiple printing passes allows to tune the line resistivity without affecting the throughput. Finally, electroless plating is a low-temperature technique, compatible with flexible substrates. The process temperature only depends on the technique chosen to cure the seed layers. Nevertheless, the resolution that can be achieved with this technique is limited. Therefore, it is used mainly in applications in which low resolution and high conductivity metal tracks are needed, for example, to fabricate antennas for radio frequency identification tags (RFIDs).

9.2.5 Ink-Jet Printing of Organic Electronic Devices

In the last 10 years, a growing number of applications has been found to be suitable for ink-jet printing of functional materials. In each of these, inks with a wide range of compositions are used to deposit materials with specific electrical, structural, optical, chemical, or biological properties. The chosen applications benefit from the possibility of depositing small, controlled quantities of materials in precise locations on a substrate and at a low temperature. The low cost of the technique and its extreme flexibility (every run can be tailored to a specific application, as no template is needed) make it even more appealing to an extremely varied range of technological sectors. In this section, we review some of the applications of ink-jet technology to fabricate organic electronic devices with the aim to point out case by case which of the characteristics of this deposition technology is advantageous. A recent progress report with the latest developments in the field can be found in [32].

The most advanced industrial application of ink-jet printing in organic electronics is the fabrication of OLEDs for the display market. Another application that is on the verge of becoming mainstream is the use of ink-jet printing to fabricate OTFTs to be integrated in active matrix backplanes, flexible sensors, and RFIDs. The suitability of the technology to fabricate organic photovoltaic cells and other devices, such as sensors and memories, has also been investigated.

9.2.5.1 Fabrication of OLEDs by Ink-Jet Printing

Within the organic display industry, ink-jet printing has been found to be particularly suitable for the fabrication of OLEDs [33]. In order to fabricate these devices, the necessity to deliver light emitting polymers to each pixel of the display makes the use of ink-jet printing very attractive, since it is feasible to deliver a finely controlled amount of polymer on a precise position of the substrate. Moreover ink-jet printing allows a combinatorial approach to the development of materials for OLEDs, which involves time-consuming devices and materials optimization [34]. Early studies focused on the development of the deposition of electroluminescent organic materials in discrete device elements [35], while soon the patterning of multicolor displays was demonstrated, where ink-jet printing was combined with more standard deposition techniques [36]. The technology is currently adopted to deliver all the different semiconductors that define the three primary colors of each pixel for full-color displays. One of the main requirements for this application is to deposit the polymer with a precise placement of the droplets on the substrate. In order to achieve this, a process based on a surface-energy pattern has been developed to compensate for any imprecision in the alignment of the printhead with the substrate and for variations of drop directionality in multinozzle printers [37]. In this approach, the substrate is prepatterned with polyimide wells where the droplets need to be placed. Two successive plasma treatments modify the surface energy of the substrate, making it hydrophilic, and of the polyimide walls of the wells, which become hydrophobic. When a drop reaches the substrate, if misaligned, it could overlap the walls of a well. As far as the misalignment is not excessive, in this case the drop dewets from the polyimide walls and flows inside the well. This technique provides an accuracy for the placement of the droplets on the order of 1 μm.

9.2.5.2 Fabrication of Organic Thin Film Transistors by Ink-Jet Printing

The fabrication of solution processed TFTs (thin film transistors) has recently become a viable technology, as the mobility of FETs with polymer semiconductors has reached or exceeded that of devices using an amorphous silicon active layer ($\mu = 0.5–1\,\mathrm{cm^2\,V^{-1}\,s^{-1}}$) [38]. In the past, only p-type semiconducting polymers achieved these performances. Nevertheless, recently n-type semiconducting polymers with electron mobilities up to $0.85\,\mathrm{cm^2\,V^{-1}\,s^{-1}}$ have become available, paving the way to the development of all-printed organic CMOS circuits [39]. A wide range of applications has therefore become possible. Manufacturing integrated circuits by ink-jet printing offers several advantages, such as

1) low process cost due to process simplification, no need to fabricate templates and reduced material consumption due to the additive approach,
2) lower equipment investment if all-printed technologies can implemented,
3) low-temperature processes that allow the fabrication of circuits on flexible displays,
4) a high-throughput technology compatible with roll-to-roll and wide area processing,

5) a more environmental friendly technology compared with conventional subtractive processes, where large amounts of waste solvents are needed to remove unwanted materials.

The most obvious application of ink-jet printed TFTs is to fabricate organic-active matrix backplanes for flexible displays [40, 41]. The low frequency operation of these applications does not require very fast switching devices, and therefore, it is compatible with solution processed polymeric semiconductors. Nevertheless, the necessity of defining devices with sub-10 μm channel length is very demanding in terms of the resolution of the systems. Moreover, the necessity of printing different materials, to be used for the active layer, dielectric, electrodes, and interconnects of the circuits, requires a printhead compatible with a broad range of different solvents.

As we have seen in Section 9.2.4, a lot of approaches have been developed to allow the printing of suitable highly conductive electrodes. The development of high-resolution printing techniques that allow to define downscaled channel lengths, needed to enhance the TFT switching speed in more demanding applications, is covered in Section 9.3.

While less developed ink-jet printing of organic semiconductors is of course a central aspect for the development of fully printed devices. Here the main issue, when adopting ink-jet printing, is to control the morphology and microcrystalline structure of the printed layer, since these have a strong effect on the charge transport and therefore on the device final mobility [38]. Some examples can be found in the literature, especially in the case of small molecule-based inks [42, 43]. In particular, for 6,13-bis((triisopropylsilylethynyl) pentacene (TIPS-p), it was shown that it is possible to suppress coffee-stain effects by a proper formulation of the ink. By adding high-boiling solvents, such as dodecane, to a chlorobenzene TIPS-p solution reduces the evaporation rate of the ink and allows the formation of a recirculating Marangoni flow, which opposes the convective flow, which is the primary cause of the coffee-stain effect. Highly ordered ink-jet printed polycrystalline TIPS-p layers could be fabricated using this optimized formulation, showing a field-effect mobility higher than $\mu = 0.1\,\mathrm{cm^2\,V^{-1}\,s^{-1}}$ [44]. In a similar way, control of the surface energy of the substrate, obtained by growing SAMs, is found to have an important role in the final degree of order for printed TIPS-p. Pinning of the contact line on higher energy surfaces favors the nucleation of crystals from the periphery toward the center of the droplets, as opposed to the formation of randomly oriented agglomerates in the case of lower surface-energy substrates [45]. Ink-jet printing of semiconducting polymers has also been investigated, with examples of regio-regular polythiophenes deposited by aerosol printing and piezoelectric ink-jet printing of poly(2,5-bis(3-alkylthiophen-2-yl)thieno [3,2-b] thiophene) (pBTTT) to fabricate complementary logic inverters [46].

In most of the studies, the dielectric material is deposited by conventional lithography or spun from solution. Direct printing of it is very challenging given the usually high viscosity of the ink needed to achieve a good insulation and low-leakage currents. Despite the presence of a few examples related to specific systems [26, 47, 48], this can be regarded as an open issue for ink-jet printing and it is still debated whether it will be the technique of choice for this process step, rather than a more standard printing technique. However low-voltage, fully printed TFTs were recently demonstrated, in which the adopted polymer dielectric belongs to the particular class of ion gels. This case will be reviewed in Section 9.4.2.1.

Another area in which there has been a lot of interest in ink-jet processing is the fabrication of all-printed organic RFIDs [49]. This is a very competitive market where a new technology could only compete with the mainstream silicon CMOS circuits by introducing extreme cost savings. The most likely frequency band in which solution processed RFID tags could drive the fabrication costs down is the high-frequency band at 13.56 MHz. In this frequency range, the coil can be integrated in a planar process and the systems are only moderately susceptible to the interference from metals and liquids. Therefore 13.56 MHz all-printed RFIDs are a very promising technology for use in RF barcodes for item-level tagging.

9.2.5.3 **Fabrication of Organic Photovoltaic Cells by Ink-Jet Printing**

It is widely accepted that the widespread adoption of solar cells is still limited by high fabrication costs. The use of solution-processable organic semiconductors has been seen as a possible solution toward low-cost, environmentally friendly fabrication of large solar modules. The most efficient technology to date is based on a polymer:fullerene blend heterojunction, whose laboratory fabrication usually involves the deposition of the active layer by standard techniques such as spin coating and doctor blade. More cost-effective fabrication tools are needed in order to develop industrial applications, and printing is regarded as the most promising option. Ink-jet printing offers a limited throughput with respect to other graphical arts techniques such as gravure or screen printing [50, 51], but these techniques require a complicated master preparation. As ink-jet printing is a contact-less technology, it is compatible with a wide range of substrates, and furthermore, it also offers the possibility to define with the same tool all the functional layers needed. For example, ink-jet printing allows to pattern directly the conductive lines needed to interconnect different modules. So far the research has been limited to the deposition and optimization of the photoactive blend layer, which is the most crucial for the photoconversion efficiency. Development here is at a very early stage and the technology adopted has preferably been piezoelectric. Several examples have been reported in the literature for blends of poly(3-hexylthiophene) (P3HT) and [6,6]-phenyl C61 butyric acid methyl ester (PCBM), where it was shown that optimum conditions in terms of material properties and solvent characteristics are very peculiar to the printing technique, with results which are very far from what has been reported in the literature in the last decade for films deposited by spin coating and doctor blade. It was reported that a mixture of high-boiling solvents, which prevent clogging of the head, and lower boiling point solvents, with higher vapor pressure, needed to increase the drying rate of the ink on the substrate, can lead to good intermixing of the polymer and the fullerene in the blend, leading to increased efficiency [52]. Control of the gelation of the blend is crucial too, and to this extent regioregulairity of the P3HT must be reoptimized accordingly [53]. Another important factor is the possibility to cover large areas with uniform, smooth, pin-holes free photoactive films, and research in this direction showed promising results [54].

9.2.5.4 Other Organic Devices

A final application that has recently received a lot of attention in the field of ink-jet printed organic electronics is the fabrication of chemical and mechanical sensors on flexible substrates. A remarkable example of this area of application is the work presented by Someya *et al.* [55] on flexible pressure sensors for the fabrication of E-skins. In this approach, an active matrix composed of organic TFTs is fabricated by ink-jet printing of silver nanoparticle inks and polyimide precursors on a flexible substrate. The active matrix circuitry can sense the pressure applied on a pressure-sensitive rubber layer, which is deposited on the whole surface of the sensor. The flexibility of the circuitry could allow for this sensor to be used in a large-area detection system.

Other organic devices could benefit from the fabrication by means of ink-jet printing. For example, the technology could be employed to fabricate photodetectors [56], providing the same advantages as in solar cells. Moreover direct writing of light detecting elements in the form of arrays would be extremely simplified by ink-jet printing [57].

Recently, many examples of memory diodes based on organic materials have been proposed [58]. These are usually vertical structures where one or more layers of an active material is sandwiched between two electrodes. In this case, there is no doubt that the possibility to pattern large arrays of memory elements at a low cost would be a great advantage. An example of arrays of very low power, memory fuses where the metal lines were printed by piezoelectric ink-jet printing of nanoparticles-based metal inks has been recently demonstrated. Especially for this application, the possibility to downscale the area-per-bit through an increased lateral resolution would be very important to achieve denser memory chips. Possible techniques that can be adopted to reduce the printed linewidth will be reviewed in the next section.

9.3 High-Resolution Printing of Highly Conductive Electrodes

The limited lateral resolution achievable with standard ink-jet printing tools, which in most cases is not better 10–100 μm [2, 59], represents a strong limitation in terms of device scalability and integration density. Several approaches have been proposed in order to achieve a higher resolution, especially in the definition of highly conductive contacts that could allow, for example, the shortening of channel lengths in TFTs, the definition of very dense passive and active cross matrixes for displays, memory modules, and in general for sandwich structured devices.

The most obvious approach in order to achieve better feature sizes is to narrow the diameter of the nozzle orifice. This has represented a technological challenge and comes at the cost of a reduced throughput. In the case of e-jet printing, it was demonstrated to be a viable option, with examples of nozzle internal diameters as small as a few hundreds of nanometers that allowed the definition of conductive tracks with a similar critical dimension [16]. On the contrary, in the case of piezoelectric DOD printheads, this solution is not possible because a nozzle internal diameter below 10 μm would severely limit the choice of printable inks due to the increased risk of clogging. In order to overcome this limitation, different advanced techniques, most of which rely on the fine control of the spreading of the ink on the substrate, were devised.

In general, the techniques adopted to achieve a higher resolution patterning with ink-jet printing can be roughly divided in two approaches.

1) **Ink-Jet Printing of Narrow Linewidths**
   These comprise the reduction of the dispensed ink volume to the subfemtoliter range in the case of e-jet; in the case of piezoelectric printing, this can be achieved through the control of the ink spreading with topological modifications of the substrate or with sintering of the dispensed ink with a focused laser beam.
2) **High-Resolution Printing Assisted by Surface-Energy Patterns**
   In this case, conventional DOD printheads are used and therefore the linewidth is not reduced. Instead, the droplet placing accuracy is strongly enhanced by the presence on the substrate of areas with different wettability. This allows to control the relative position of one printed pattern with respect to another beyond the resolution of the printing technique. In this category fall, the techniques that rely on a prepatterning of the substrates through standard lithography or microcontact printing, for the creation of the surface-energy patterns. An evolution is represented by self-aligned printing (SAP), a complete bottom-up process, where no prepatterning is necessary.

These two approaches have been so far demonstrated independently in the literature, but there is no reason to consider them as alternative, since an integration of both could greatly enhance the resolution achievable with ink-jet printing techniques.

Various examples of resolution enhancing techniques are reviewed in the following, focusing on high-resolution printing of highly conductive tracks and, in particular, electrodes. We also describe an example of a highly reliable printing process of nanoscale contact arrays.

9.3.1 Ink-Jet Printing of Narrow Linewidths

Although in theory the narrowest line printable by DOD piezoelectric ink-jet printing is of the order of ≈20 μm, in practice the typical linewidth achievable with standard tools and processes is of the order of ≈50–100 μm or higher

[2, 59]. Reliable printing of conductive tracks with reduced width would of course be advantageous for an improved lateral resolution.

As mentioned above, physical factors limit the miniaturization of piezo-driven nozzles. However high-resolution printheads are currently available: these, thanks to a reduced nozzle orifice allow to reduce the ejected ink volume to 1 pl, equivalent to a droplet diameter of 12 μm, for a relatively large range of printable materials. This currently represents the limit for piezo-driven printheads. Limiting the droplet volume is, however, not enough, since an important factor affecting the final width of the printed line is given by the spreading of the ink on the substrate, determined by the surface tension of the former relatively to the surface energy of the latter. Careful selection of substrates with proper surface energy is very important, and in this case polyarylates foils offer intermediate surface energies, which can guarantee enough wettability while limiting the spreading of the ink (Figure 9.3a). By adopting these substrates, by raising the substrate temperature above room temperature in order to induce a faster evaporation of the solvents and limiting the ink spreading, and thanks to a reduced nozzle diameter, 40 μm wide silver conductive tracks were demonstrated [60]. Similar results could be obtained by optimizing the dot-to-dot spacing and the solvent composition of a silver ink printed on hexamethyldisilazane glass substrates [59].

Such a resolution represents an improvement for printing without the assistance of any kind of substrate prepatterning, but still is very limiting for the down scalability of devices. Much narrower silver tracks were demonstrated by Schubert *et al.* thanks to hot-embossed grooves in a plastic substrate guiding the spreading of printed droplets (Figure 9.3b). The ink, printed on a limited portion of millimeters to centimeters long and micrometers-wide grooves, fill the channels thanks to capillary forces with a very uniform width, in the range of 5–15 μm.

Another proposed approach in order to reduce the linewidth of conductive tracks is based on laser patterning, introduced in Section 9.2.4. The ink is sintered by irradiation with a very narrow focused laser beam to improve the resolution of the ink-jet printed lines. By washing away the unsintered nanoparticles after laser irradiation, lines as narrow as 1–2 μm can be fabricated [26] (Figure 9.3c).

A similar feature size can be directly obtained with a single printing step, without the need to resort to a second patterning process, if high-resolution e-jet technology is adopted. This technology (introduced in Section 9.2.3.4) offers a clear advantage with respect to piezoelectric DOD printers as it allows to narrow the nozzle diameter, in order to reduce the drop volume. In this case, the presence of a high-electric field allows formation of the jet with a relatively wide range of ink viscosities and surface tensions even when the internal diameter of the nozzle reaches the submicrometer range. With this approach, subfemtoliter droplets can be generated, the smallest volume reported in the literature to date. Park *et al.* [16] reported printing lines of polyethyleneglycol with a minimum width of 700 nm. They also reported using e-jet to print a range of inks including PEDOT:PSS and a carbon nanotubes ink. They also demonstrated fine gold patterning by printing a polyurethane ink acting as a etch resist for the metal, while direct printing of silver tracks from a commercial

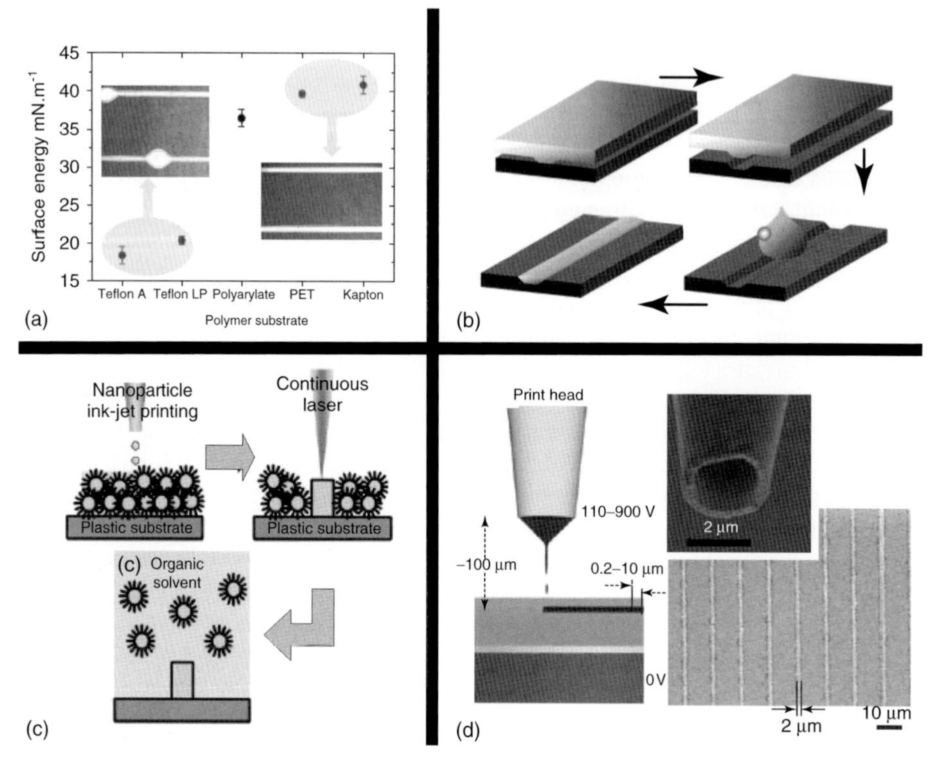

Figure 9.3 (a) Surface-energy values for different polymer substrates and, in the insets, their effect on printed silver tracks. Reproduced with permission from [60]. (b) Processing steps for the printing of narrow lines within hot-embossed grooves. Reproduced with permission from [61]. (c) Narrow gold lines pattering by sintering with a focused laser beam. Adapted with permission from [26]. (d) Sketch of an e-jet printing system, with a SEM image of a gold-coated glass microcapillary nozzle (adapted with permission from [16]) and an optical microscope image of e-jet printed Ag lines (adapted with permission from [7]).

nanoparticles-based ink was actually demonstrated with a similar experimental setup by Sekitani *et al.* [7] from the University of Tokyo. Drops with a volume of 0.7 fl and a diameter of 1 μm were obtained, yielding lines as narrow as 2 μm after 10 printing passes, made necessary in order to obtain continuous lines with a resistivity low enough for practical applications (Figure 9.3d). High-resolution e-jet printing is therefore a very interesting technology for an improved printing resolution, even though it shows some drawbacks such as the limited throughput and the fact that it is much less developed than standard printing piezoelectric DOD tools: to date this system has not been scaled to multinozzle operation, which would be necessary for its application in manufacturing high-performance printed circuits.

9.3.2
Ink-Jet Printing Assisted by Surface-Energy Patterns

An improvement of the resolution in the definition of narrow gaps between conductive lines, and therefore potential channels for scaled planar electronic devices, such as transistors and detectors, can be obtained with standard piezoelectric printers, without drastically reducing the droplets volume and therefore maintaining the throughput and the process reliability. This can be achieved by precisely controlling the flow and the spreading of the ink on the substrate, with prepatterned areas characterized by different surface energy, thus substantially enhancing the droplets placement accuracy. In the case of a water-based ink, if a pattern containing hydrophobic and hydrophilic areas is defined, the droplets landing on the substrate will be subjected to a driving force that pushes them away from hydrophobic areas toward hydrophilic areas, leading to a confinement of the ink (Section 9.2.5.1). Furthermore the contact angle of the ink on both areas can be controlled, thus, allowing to limit the spreading of the ink. In practice, the process and its resolution are complicated by many factors such as the surface tension of the ink and its viscosity, the receding contact angle (which if too small can induce a pinning of the fluid) and the drying rate (which can dynamically influence the other parameters). Notwithstanding the complexity of the mechanism, the viability of surface-energy patterns assisted ink-jet printing in achieving high-resolution printing of channel lengths as short as 5 μm was successfully demonstrated [62]. In this work, a hydrophobic polyimide pattern was defined by means of standard photolithography on a hydrophilic glass substrate. After that, PEDOT:PSS contacts were printed with high resolution, thanks to the confinement offered by the polyimide structures.

The same concept was then developed on substrates patterned by means of processes more appealing from the point of view of manufacturing, such as soft lithographic techniques [40] and digital direct-write laser patterning [63]. These provided a viable tool to develop ink-jet printed backplanes for displays [1], where a micron-scale resolution is sufficient. To address more demanding applications, such as the fabrication of logic circuitry surrounding the display, a higher resolution is needed in order to enhance the switching speed of the TFTs by scaling down the channel length to the submicrometer range. The question regarding the highest resolution achievable with a surface-energy pattern assisted printing was investigated by fabricating patterns with feature sizes down to 200 nm by means of electron-beam lithography. On this basis, dewetting of the ink from hydrophobic ribs was demonstrated, where droplets were split in two halves with submicrometer scale resolution [64]. Even if in this case, the need to predefine a pattern by lithography increases the process complexity and costs, the technique demonstrated an exceptional resolution capability achievable with conventional low-resolution ink-jet printing techniques.

9.3.3
Self-Aligned Printing

A technique that allowed to achieve a submicrometer resolution using ink-jet printing without the need of prepatterning the substrate would represent a more viable option in terms of manufacturing. This possibility was first demonstrated by defining channels in the sub-hundred-nanometer range through the self-alignment of PEDOT:PSS electrodes [65]. The fundamental mechanism, that allows to achieve such a high resolution with standard ink-jet printing equipment and without the need for any substrate prepatterning or precise alignment, still relies on the ink droplet motion induced by a contrast in surface energy on the substrate. This SAP technique can be summarized in the following three steps:

1) the first conductive pattern is ink-jet printed on the substrate;
2) the surface energy of the printed pattern is selectively lowered without altering the wettability of the substrate, so that a strong enough contrast in surface energy is induced between the pattern and the substrate;
3) a second conductive pattern is ink-jet printed on top of the first one with a small overlap, without requiring fine alignment.

The droplets deposited to print the second pattern experience a driving force that induces them to dewet from the first pattern. Depending on several parameters, such as the surface-energy contrast, the ink viscosity and its drying rate, and the flow of the second pattern can be controlled to create a very small gap between the two patterns (Figure 9.4a). No top-down step is involved in the fabrication of these narrow gaps and the process is compatible with the alignment resolution, which is available in standard printing systems.

Different techniques were employed to lower the surface energy of the PEDOT:PSS pattern without altering that of the substrate, such as a selective treatment of the surface or a segregation of surfactants. In the first case, the first

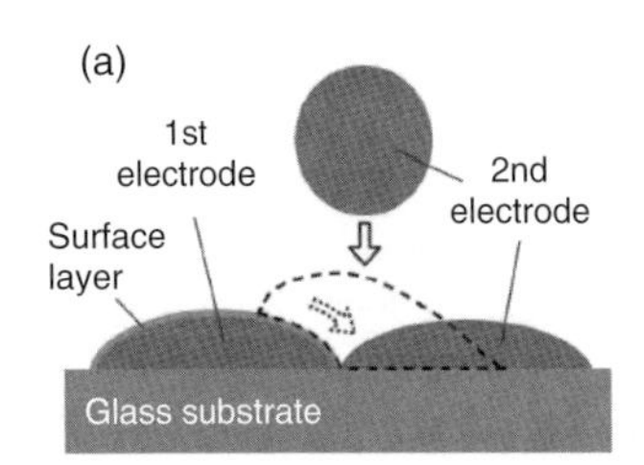

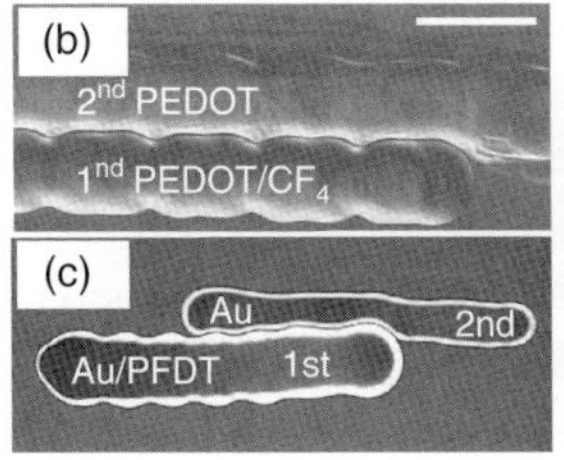

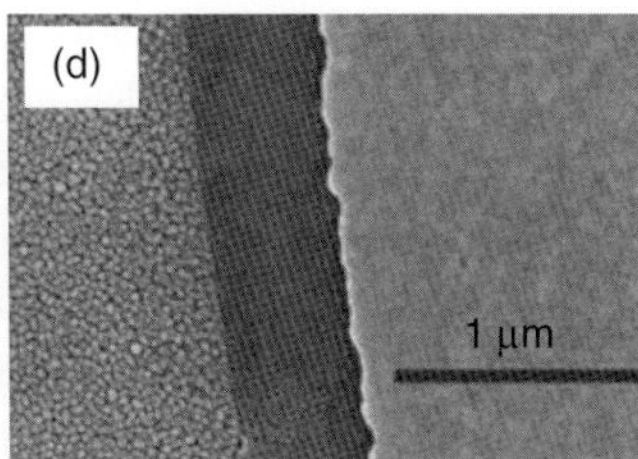

Figure 9.4 (a) Sketch of the SAP process: an ink droplet is printed in partial overlap (dashed line) with a previously printed, surface modified hydrophobic pattern; due to surface-energy contrast, the droplet experiences dewetting. (b) Optical micrograph of PEDOT:PSS self-aligned printed contacts; the surface energy of the first pattern was selectively modified using a CF_4 plasma. (a) and (b) reproduced with permission from [65]. (c) Optical micrograph of Au self-aligned printed contacts; the surface energy of the first pattern was selectively modified using a PFDT SAM. (d) SEM image of a 500 nm long Au SAP channel. (c) and (d) reproduced with permission from [66].

PEDOT:PSS pattern, printed on a glass substrate, was subjected to a postdeposition CF_4 plasma treatment. This treatment created a fluorinated surface layer on the PEDOT:PSS pattern, therefore, lowering its surface energy, while leaving the glass surface in a high energy state. This selective modification created the necessary surface-energy contrast that allowed the second pattern, printed in partial overlap with the first one, to dewet and to define a second electrically isolated contact at a submicrometer distance (Figure 9.4b). Typical gaps obtained with this method were shorter than 100 nm [65]. Among the parameters that had to be controlled in order to obtain a clean and complete dewetting, the receding contact angle of the ink flowing off the first pattern was found to be particularly critical. This had to be kept close to the advancing contact angle in order to avoid pinning of the contact line due to surface inhomogeneities. In an alternative approach, suitable surfactants, with polar head groups and nonpolar tail groups, were added to the PEDOT:PSS ink. The surfactants segregated to the surface of the water-based ink and created a low surface energy layer with the polar head group in contact with the ink and the nonpolar tail group at the interface with the air [67]. This approach gave results comparable to the plasma treatment described above. Nevertheless, it was found to show a much narrower process window when the second pattern was printed with an overlap, showing a higher sensitivity to droplet positioning. The optimum case was found by printing the second droplets apart from the first pattern, so that they could spread against it [1, 65].

The limited conductivity of polymer conductors such as PEDOT:PSS poses a severe limit to the currents obtainable in short-channel transistors. Therefore, highly conducting printable metal inks compatible with SAP were needed to extend the process applicability and to demonstrate the generality of the conceptual approach. Metal nanoparticles inks (Section 9.2.4), previously adopted in defining electrodes in low-resolution printing processes [68, 69], represented an obvious option. The first successful demonstration of highly conductive electrodes defined with the SAP technique was achieved with gold nanoparticles inks, where xylene and/or cyclohexylbenzene (CHB) were used as solvents, adopting a selective surface treatment approach [66]. After printing and sintering of the first gold pattern on glass or silicon/SiO_2, the substrate is subjected to a vapor deposition of a self-assembled SAM of 1H,1H,2H,2H-perfluoro-1-decanethiol (PFDT). The thiol, having a strong affinity with gold, grows selectively on the first printed pattern and lowers its surface energy. On the other hand, PFDT does not chemically bond to the substrate from which the physically deposited molecules can be easily washed away by solvent rinsing. Whereas the gold pattern surface energy is lowered, that of the substrate is unaffected by the SAM treatment and therefore a strong surface-energy contrast is achieved. When a second pattern is printed in partial overlap with the first pattern, the ink dewets from the first pattern and defines a narrow self-aligned channel (Figure 9.4c and d). The channel length is found to depend on fluid dynamical conditions, which were investigated by varying several parameters of the ink, such as the drying rate, the viscosity, and the surface tension, in a controlled experiment where the first gold pattern was defined by lithography. In particular, it is found that the channel length can be controlled from $\approx$50 nm to a few micrometers by adding an increasing amount of CHB to a xylene solution of the gold nanoparticles. CHB has higher surface tension and boiling point with respect to xylene, therefore a formulation with more CHB shows a higher repulsion from the hydrophobic SAM treated gold and a longer drying time, that allows the ink to dewet further before its contact line is pinned to the substrate because of solute precipitation due to solvent evaporation.

The SAP technique promises to have technological relevance since it shows to work, after a careful optimization of the parameters, in many different conditions. Self-aligned metallic contacts were defined both by printing the second pattern on a previously lithographically defined one [66] and, more interestingly, in an all-printed process, without any need for prepatterning (see next section). The SAP technique could be implemented with a polymeric conductive ink such as PEDOT:PSS, a custom-made gold nanoparticle ink, and by adopting a commercially available gold nanopaste, suitably diluted [70]. Moreover SAP patterns were printed with various single-nozzle piezo-driven commercial heads, such as MicroDrop and Microfab, and both with low (4 Hz) and high (up to 1 kHz) droplet ejection frequencies. The reliability and versatility of the process are promising features for its development and transfer to high-throughput multinozzle printing systems.

9.3.4 High Yield Printing of Single-Droplet Nanoscale Electrode Arrays

SAP is at present one of the only few examples [64, 65] of printing techniques that are capable of achieving sufficiently high resolution for definition of electrode gaps on a 100 nm scale. This resolution is needed for the probing of electronic properties of nanoscale materials such as carbon nanotubes, graphene, nanowires, or organic semiconductors, while offering in general a possibility to downscale organic electronic devices. Whether such high-resolution printing technique can become a widespread tool for applications in nanoelectronics on the 100 nm length scale, as an alternative and scalable patterning technique better suited to nonplanar, stretchable, and/or flexible substrates with respect to traditional electron-beam lithography and nanoimprinting, will depend much on what patterning yield and uniformity will be achievable.

The SAP technique was originally demonstrated in a configuration in which the two electrodes are linear shaped, printed from multiple droplets and run parallel to each other over several hundred micrometers. In this device configuration, it is generally found that the yield is moderate and a significant number of devices is shorted [64, 65]. Recently, an improved SAP configuration based on single-droplet contacts was reported, which allows achieving surprisingly high-device yield of 94–100 % on arrays with very low-leakage currents [71].

In Figure 9.5a and b, SEM images of the SAP gold contacts are shown. The first contact is printed on a glass substrate with a linear shape. The second contact is then printed at an angle of 90° with respect to the first one, realizing a characteristic "T-shape" (inset of Figure 9.5a). The length and shape of the two "arms" can be chosen to allow easy external electrical contacting and testing, but the channel

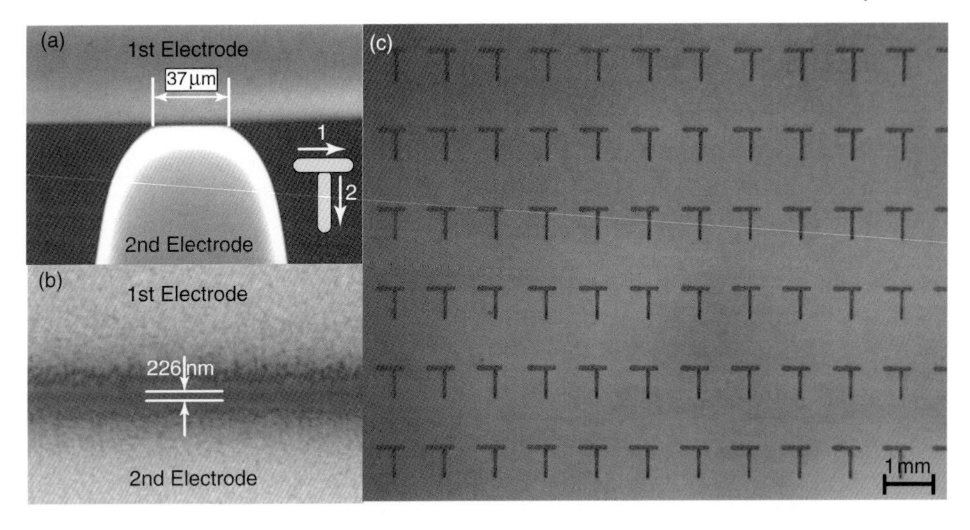

Figure 9.5 (a) SEM images of T-shape SAP gold electrodes, where the second contact is printed at an angle of 90° with respect to the first one and the channel width is defined by a single droplet. The naturally rounded shape of the droplet conforms to the edge of the first electrode and is deformed into a straight line along the edge of the first contact. In this region, a submicrometer electrically insulating gap is created: this is magnified in (b), where a clear gap of ≈ 200 nm can be observed. Optical microscope image (transmission) showing part of one of the 6 × 12 arrays of SAP contacts fabricated, where 6 rows and 11 columns can be seen.

width is defined by a single droplet only with diameter in the range of 30–80 μm. The first droplet of the second electrode pattern can be printed either partially on top of the first electrode or in close proximity to it: in the former case, the low wettability of the surface of the first electrode induced by PFDT coating leads to the complete dewetting of the droplet from the first pattern; in the latter case, the ink droplet is printed sufficiently close to the edge of the first electrode that the contact line of the spreading droplet comes in contact, but is repelled by the edge of the first electrode. In both cases, the droplet dries in close proximity to the edge of the first electrode and the naturally rounded shape of the droplet conforms to the edge of the first electrode and is deformed into a straight line. In this region, a submicrometer electrically insulating gap is created. A typical channel is magnified in Figure 9.5b, where a submicrometer gap can be observed. The adopted ink is a 1 : 3 volume dilution in xylene of a gold nanopaste manufactured by Harima Chemicals, Inc. and the printing is performed at room temperature with a jetting frequency of 1 kHz.

The T-shaped single-droplet electrode configuration has been found to be critical to achieve high yields for forming electrically insulating electrode gaps without shorts. The reason is that in the T-shaped architecture, the channel formation is determined by the fluid dynamics and dewetting of a single droplet only and allows easier optimization than for two parallel printed electrode lines. For the latter, dewetting of each droplet is also affected by the state of previously and subsequently deposited droplets. The effect of subsequently deposited droplets is, for example, to slow down the dewetting of a particular droplet because of it being pulled back by subsequent droplets that are being printed overlapping with the first electrode. In terms of failure analysis, there is the additional benefit that for a single-droplet channels, it is practical to observe the entire channel with SEM and correlate the occurrence of any electrical leakage current with specific defects along the edge of the channel. Arrays comprising 6 × 12 nano-spaced T-shaped contacts were realized (Figure 9.5c) in order to test the yield of channels formation. Typical leakage currents in ambient conditions can be below 1 pA at an applied bias voltage of 20 V (Figure 9.6a) and breakdown voltages are higher than $2\,\mathrm{MV\,cm^{-1}}$. As can be seen in Figure 9.6b, the typical yield achievable is higher than 90%, with only a very limited number of contacts showing higher leakage currents and no short circuits present. The origin of the higher leakage currents is typically related to the presence of small and localized conductive paths within the channels due to residual particle defects [71]. By improving the cleanliness of our process, these are likely to be eliminated. In fact, a number of arrays on which all 72 devices had a leakage current less than 2 pA at 10 V (Figure 9.6b) have been fabricated. The process window for positioning the first droplet of the second electrode with respect to the first electrode pattern was found to be larger than 20 μm, thus, requiring a droplet placement accuracy which is achievable with most commercial ink-jet printing systems [2].

It was found that the critical factor to successfully fabricate low-leakage SAP contacts is the drying time of the ink. By using high-boiling solvents for the ink formulation and allowing several minutes for slow drying at room temperature before the high-temperature sintering step, the ink of the second contact will

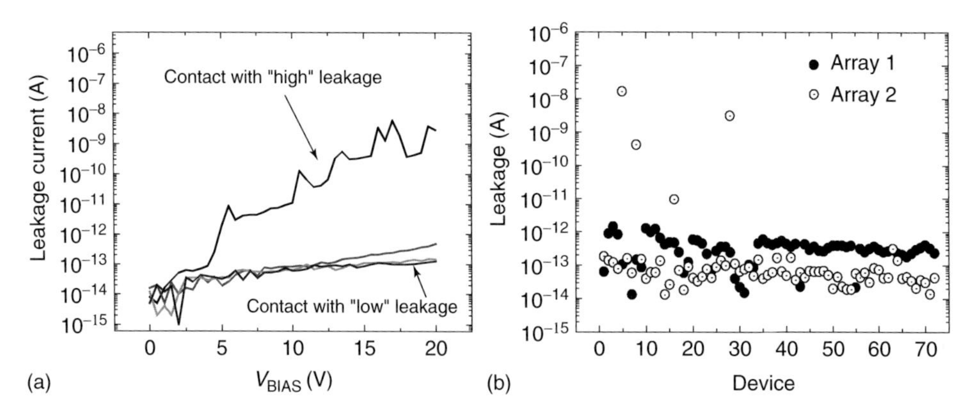

Figure 9.6 (a) Typical $I-V$ leakage curves of T-shaped SAP electrodes. (b) Yield measurement on 6 × 12 arrays of single-droplet SAP gold electrodes. The leakage current measured in air is plotted for each device. In a typical case (array 2), only 4 devices out of the 72 composing the array show a leakage higher than 1 pA and no short circuits are present ($V_{BIAS} = 20\,\mathrm{V}$). In the best case (array 1), all 72 devices have a leakage current less than 2 pA at 10 V.

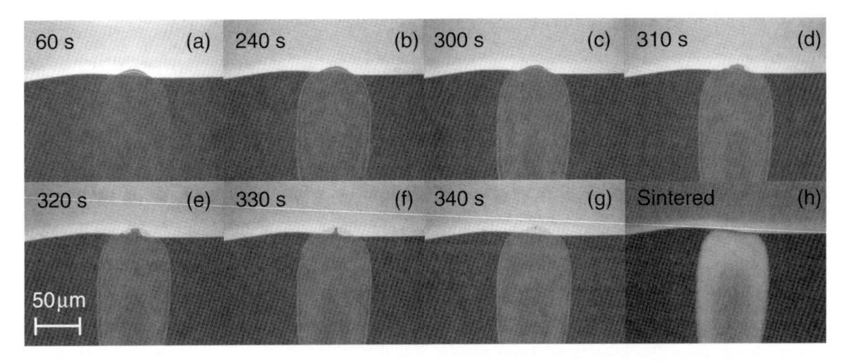

Figure 9.7 (a) Optical microscopic images illustrating the complete dewetting of a line, which shows overlap with the first pattern $t = 60$ s after printing. (b) For $t = 240$ s, substantially no variations can be observed, (c–g) while for $t = 300$ s, together with a variation of the ink color due to partial evaporation of the dispersing solvents, the contact line retracts and complete the dewetting process after 340 s. (h) Upon sintering of the second line, the SAP electrodes are reliably formed.

eventually dewet. A series of optical microscopy images are reported in Figure 9.7, where it is possible to follow the complete dewetting of the second line from the first one over a period of 6 min after printing. We observe that 60 s after printing the printed droplets still overlap significantly the first electrode and complete dewetting has not occurred yet (Figure 9.7a). No significant evolution can be observed over the next 3 min (Figure 9.7b). Subjecting the sample to the sintering temperature of 250 °C at this time would induce a fast drying of the ink and create short circuits. If the ink is left to dry at room temperature for longer time, complete dewetting from the first line occurs reliably on a characteristic time scale of 5–7 min for the ink formulation used here . It can be seen that after 310 s (Figure 9.7d) that the contact line starts to dewet from the surface of the first electrode in some regions, but remains pinned in others, presumably as a result of local defects on the surface of the first electrode. At 340 s (Figure 9.7e), the contact line has completely depinned from these defects. Dewetting is now complete and sintering at this stage results in reliable formation of a nanochannel (Figure 9.7h). We observe some color contrast in the optical microscopy images of the first electrode in the location where the droplets have dewetted (Figure 9.7g). This is either due to interactions of the ink solvent with the surface of the first electrode or to solvent/nanoparticle residues left after dewetting. However, this contrast disappears after sintering and does not appear to have any impact on the yield of nanochannel formation. We can follow the continuous drying of the ink through the observation of interference fringes and colors across the second printed line. This interference contrast changes markedly over the period of 6 min providing direct evidence that the driving force for the slow dewetting is the evaporation of solvent. This exerts an additional force on the contact line, which is needed to pull the pinned contact line back from the surface of the first electrode completely.

Single-droplet T-shaped contacts architecture allows the application of SAP for fabricating arrays of metal electrode gaps of the order of 200–500 nm with high yield and reliability. Furthermore, SAP can be realized with standard piezoelectric printing tools and it is therefore a potentially scalable patterning technique for submicrometer resolution metal contacts.

9.4 Printing of Downscaled Organic Thin Film Transistors

In order to become a fabrication tool enabling the embedding of electronic functionality in flexible substrates, any high-resolution printing technique would have to be integrated in a process capable to fully downscale the device while retaining the requirements for low-cost, large-area, flexible electronics manufacturing.

Some of the techniques to reduce the linewidth seen in the previous section and SAP have both shown to be promising tools for high-resolution patterning of electrodes. By applying such techniques, source and drain contacts for organic transistors could be fabricated in a bottom-up approach capable to overcome the poor resolution achievable with standard high-volume printing tools.

However, the downscaling of the channel length is not enough to achieve high-performing, fast-switching FETs. To fulfill the basic downscaling requirements, all the critical sizes governing the transistor functionality have to be correctly scaled.

This section is focused on downscaled OTFTs and reviews a few examples, where high-resolution printing techniques have been adopted to fabricate high-performance downscaled devices.

9.4.1 Downscaling Requirements

An organic TFT is a device that comprises two levels of contacts, one comprising the source and the drain electrodes, whose distance define the channel length L, and the other comprising the gate. The semiconducting active material and the dielectric complete the device. These layers can be arranged in different ways to produce various architectures. The so-called staggered architectures are particularly interesting since they allow to reduce contact resistance effects, thanks to current crowding [72], thus being particularly relevant for the downscaling of devices. A staggered TFT device can have a bottom-gate, top-contacts (BGTC) or a top-gate, bottom-contacts (TGBC) configuration: these are exemplified in the schematic sections of Figure 9.8. The choice between these two architectures is usually driven by the particular fabrication technology adopted.

When reducing L, it is necessary that the dielectric capacitance per unit area is correspondingly increased in order to keep the transverse electric field between the gate and the channel much larger than the longitudinal electric field between source and drain. Failing to do so would lead to severe short-channel effects [73]. This can be achieved, for example, by reducing the gate dielectric thickness d, while keeping the gate leakage current low. It is also important to minimize the overlap (x_{OV}) between the gate and the source and drain contacts (Figure 9.8),

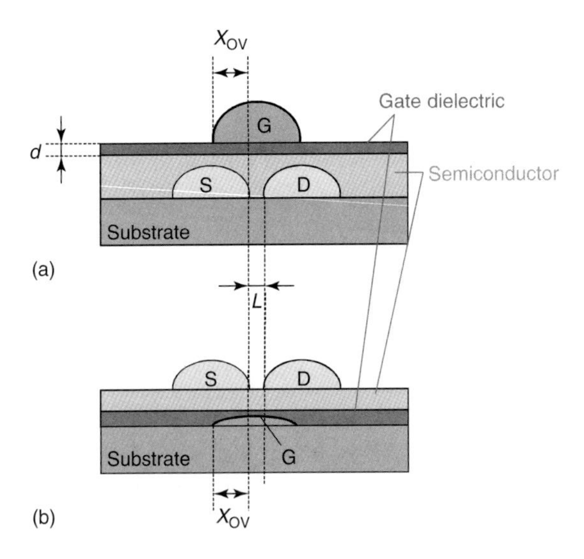

Figure 9.8 Schematic cross-sections of printed (a) top-gate, bottom-contacts and (b) bottom-gate, top-contacts transistors showing their main features. The printed electrodes are represented in light gray (S: source, D: drain), defining the channel length L. The printed gate contact is shown in gray (G: gate). d = dielectric thickness. x_{OV} = gate overlap with the source and drain electrodes, which is the main source of gate parasitic capacitance.

which has a strong influence on the transition frequency (f_T, see Figure 9.12) of the transistor. Moreover a compatible solution-processable semiconductor with a relatively high mobility ($\mu > 0.1\ cm^2\,V^{-1}\,s^{-1}$) is needed to reduce the charge carriers transit time between source and drain electrodes. To complete the device architecture, a printing process capable of defining the gate metal level on top of the dielectric in a BGTC architecture or the source/drain contacts on top of the semiconductor in the TGBC needs to be implemented.

9.4.2
Gate Dielectrics for Downscaled Organic TFTs

As discussed in Section 9.2.5.2, examples of printed gate dielectrics are very rare in the literature, given the difficulties of producing high-quality, pin-holes free printed layers from high-viscosity inks. However, the correct scaling of the specific capacitance is vital for the operation of downscaled devices and therefore we want here to review different approaches adopted to realize this.

The dielectric capacitance per unit area is expressed as

$$C' = \varepsilon_0 \frac{k}{d} \tag{9.2}$$

where ε_0 is permittivity of vacuum and k the dielectric constant. When downscaling a TFT channel, C' must be increased to retain ideal transistor behavior and, importantly, to reduce the operative voltages of the device through a reduction of its threshold voltage V_T and subthreshold swing S. In order to increase C', two different routes can be followed, one targeting a reduction of the thickness d of a given dielectric, and the other focusing on increasing k by developing alternative dielectric systems. In the following, a brief review of these two approaches for the development of high-capacitance gate dielectrics applied to organic TFTs is given, with a special attention to the adopted fabrication processes and their compatibility or possible integration with printing.

9.4.2.1 High-*k* Dielectrics

In the search for high-k dielectrics compatible with organic transistors, several approaches have been demonstrated. Various relied on the use of metal oxides, which can show in general much larger k than many polymer dielectrics. To enhance the device characteristics, these dielectrics are usually capped with apolar polymeric layers in order to produce a smoother interface and to avoid polaron localization at the gate insulator/semiconductor due to dipolar disorder [74, 75]. Examples are high-quality thin film of Al_2O_3 and TiO_2 [76] fabricated by means of anodization. A 7–8-nm thick TiO_2 layer in combination with a ≈10 nm-thick poly(α-methylstyrene) (PαMS) capping layer, show C' in the order of 200 nF cm^{-2} and lead to pentacene FETs with an operative voltage as low as 1 V and $S = 130$ mV dec^{-1}. Another example is the fabrication of nanocomposite gate dielectrics by combining sputtered Al_2O_3 or ZrO_2 with PαMS or poly(vinyl cinnamate) deposited by spin coating. These are found to show C' between 100 nF cm^{-2} and 200 nF cm^{-2} for a nanocomposite layers of around 50 nm, and when integrated in a pentacene BGTC FET lead to operative voltages of a few volts with $S = 100$ mV dec^{-1} [77]. Therefore, it is possible to fabricate very good quality metal oxides with high-k dielectric/organic semiconductor interfaces, but, even if compatible with flexible substrates, they are not yet compatible with a low-cost printing manufacturing process.

Instead of using inorganic oxides, solution-processable high-k polymer dielectrics, offering easier patterning, were also proposed, although the dielectric constant in this case is found to be not as high as for metal oxides. The highest k value demonstrated so far is 12.6 in the case of a cross-linked cyanoethylated poly vinyl alcohol (CR-V) [78]. Pentacene BGTC transistors adopting a 120 nm thick layer of cross-linked CR-V, with a leakage of 10^{-7} A cm^{-2} at 2 MV cm^{-1}, showed operative voltages of 3 V, $C' = 92.9$ nF cm^{-2} and $S = 185$ mV dec^{-1}. Higher k could be obtainable, for example, with ferroelectric polymers, which, however, inherently show hysteretic behaviors more suitable for memory devices than FETs [79].

Another interesting approach, which is receiving increasing attention, is the use of solution-processable solid electrolytes, which, due to their ionic nature, can show very-high capacitance values. It was demonstrated that polymer electrolytes, such as polyethylene oxide containing the salt $LiClO_4$, can gate organic TFTs with very high specific capacitances of the order of 10 μF cm^{-2} [80–82]. This is because of the fact that when, for example, a negative gate bias is applied, the mobile cations are attracted toward the gate electrode, while mobile anions move toward the semiconductor/dielectric interface. This leads to the formation at both interfaces of two sheets of charges, the ions and the corresponding compensation charges in the semiconductor and in the gate electrode, which are separated by only a few angstrom. However, the ionic conductivities of such solid electrolytes are low ($10^{-4} - 10^{-5}$ S cm^{-1} [83]), severely limiting the polarization speed of the dielectric and therefore the switching speed of the transistor, which cannot operate faster than a few hertz. Furthermore, displacement of the ions in this case could lead to electrochemical doping of the semiconductor.

An increase of the polarization speed for solid electrolytes can be obtained by adopting polyelectrolytes. These are polymers obtained by the repetition of a monomer bearing an electrolyte group. For example, the random copolymer of vinyl phosphonic acid and acrylic acid, P(VPA-AA), is a polyanionic electrolyte, where the phosphonic acid groups are the source of protons [84]. These under the effect of a bias can move, leaving behind anions in the form of deprotonated P(VPA-AA) chains, which are substantially immobile. In the case of a negative gate bias, protons depletion can occur rapidly and similarly the positively-doped channel formation, induced by the presence of anions in the proximity of the gate/semiconductor interface. While offering similar high capacitances (10 μF cm^{-2}), the polarization is faster than in the case of polymer electrolytes, avoiding electrochemical doping because of the anions being immobile. This solution-processable polyelectrolyte found a promising application in a downscaled SAP TFT based on P3HT with submicrometer channel length, where clear saturation at a gate voltage as low as −1 V and switching times of ≈50 μs could be obtained [85].

A new class of polymer electrolytes, namely, ion gels, was proposed as well. Here a block copolymer, typically poly(styrene-*block*-ethylene oxide-*block*-styrene) (PS-PEO-PS), is gelated in a room temperature ionic liquid, which usually comprises a cation with an imidazole skeleton and various anions, such as hexafluorophosphate $[PF_6]^-$ and bis(trifluoromethylsulfonyl)imide $[TFSI]^-$ [86]. These are characterized by a high-ionic conductivity, which can be above 10^{-3} S cm^{-1}. Since the polymer weight fraction at which gelation occurs is limited to a few percents for these systems, the gel ionic conductivity is not substantially affected. Therefore, reduced polarization response times with operative frequencies in the 1–10 kHz range, can be obtained together with very high capacitances of the order of 10–40 μF cm^{-2} [83, 87]. Ion gels are suitable for printing and all-printed TGBC P3HT TFTs arrays and inverters were successfully demonstrated by means of aerosol jet. In this case, water insoluble poly(styrene-*block*-methylmethacrylate-*block*-styrene) was adopted instead of PS-PEO-PS to avoid gel dissolution by the water dispersion of PEDOT:PSS adopted for the gate contact. Furthermore, given the high polarizability of ion gels, TFTs could be in principle fabricated with the gate contact displaced from the source and drain contacts, relaxing the requirement for the gate electrode printing registration [87]. Although fully downscaled devices have not been demonstrated so far, overall ion-gels dielectrics are very promising candidates for low cost, flexible electronics, the current limitations being represented by generally high-leakage currents likely due to gel impurities and to a high-displacement contribution.

9.4.2.2 **Ultra-Thin Dielectrics**

In order to reduce the dielectric thickness, the adoption of SAMs or multilayers represents a very promising approach, where a high-specific capacitance of around $1\ \mu F\ cm^{-2}$ can be obtained by downscaling d to a thickness comparable with the length of single molecules, in the range of a few nanometers. Molecular packing of the SAMs has a great influence on the leakage current and therefore is a key factor for their application in organic electronics [88]. Different kind of molecules where proposed, as carboxyl terminated [89] and phenoxy terminated [90] alkyltrichlorosilanes, together with multilayers composed of α,ω-difunctionalized hydrocarbon chains, highly polarizable stilbazolium molecules, and octachlorotrisiloxanes [91]. These systems were adopted to demonstrate organic TFTs on silicon, hence the choice of silane anchor groups. A step forward was made with the demonstration of very low leakage gate dielectrics fabricated by growing phosphonic acid functionalized alkyl chains on patterned aluminum gate contacts, where the surface is intentionally oxidized to provide more hydroxyl groups for a dense packing of the SAM. Even if aluminum is thermally evaporated, it offers the possibility to have a highly conductive gate contact that can be patterned at low temperature. A capacitance as high as $\approx 0.7\ \mu F\ cm^{-2}$ can be obtained with a 2.1-nm thick n-octadecylphosphonic acid SAM on the top of a 3.6-nm thick AlO_x, with an impressively low-leakage current in the range of 10^{-8}–$10^{-7}\ A\ cm^{-2}$ for a field higher than $3\ MV\ cm^{-1}$. BGTC TFTs, based on evaporated p and n-type small molecules, adopting this SAM dielectric, could be operated at voltages of a few volts, showing a subthreshold swing of $100\ mV\ dec^{-1}$, enabling the realization of complementary logic gates and ring oscillators. Adoption of anthryl-alkylphosphonic acid was shown to lead to similar device performances, with further reduction of leakage currents and with S as low as $85\ mV\ dec^{-1}$ [92]. Extensions of this approach to organic TFTs, where the semiconductor is solution processed were demonstrated [93], where a good coverage of the SAM, typically characterized by a low surface energy, is obtained by selecting semiconductor solutions with proper surface properties [94].

Although the use of SAMs required masks and thermal evaporation for the fabrication of the patterned Al/AlO_x layer, this is an approach in principle compatible with low-cost and flexible electronic devices. However, the yield and reliability of a SAM dielectric in a real fabrication process need to be verified.

Another promising approach for the realization of scaled dielectric thicknesses relies on ultra-thin cross-linked polymer layers [95]. Several groups reported cross-linkable systems as thin as 15 nm for BG organic transistors [96–98]. These examples are limited to the BG architectures because of the high cross-linking temperatures involved, which are typically higher than 150 °C. These temperatures would cause, in most cases, unacceptable damage to the underlying semiconductor layer in a TG architecture. Moreover, a common additional problem is the difficulty in finding suitable orthogonal solvents. These issues could be overcome with the selection of a suitable blend of a polymer dielectric and a cross-linking agent, essentially a trichlorosilyl-alkane, both soluble in the same orthogonal solvent [96]. The advantage of this approach, which extends the use of ultra-thin cross-linked polymer layers to TG architectures, is the spontaneous cross-linking that can develop at room temperature in the presence of moisture and oxygen. First examples were based on blends of poly(methylmethacrylate) (PMMA) or poly-4-vinylphenol (PVP) as the base polymer dielectric, 1,6-bis-(trichlorosilyl)hexane as the crosslinker and *n*-butyl acetate as the solvent (Figure 9.9a). When the blend is exposed to air, a physical cross-linking takes place, where the chlorosilane reacts with water to form a siloxane network through the polymer, therefore, reducing the free volume of the composite cross-linked material. Cross-linked PMMA (C-PMMA) was shown to have performances comparable to those obtainable with thermal SiO_2 grown on Si [99], with a $\approx$30–50 nm thick C-PMMA layer exhibiting a breakdown field higher than $3\,MV\,cm^{-1}$ and a leakage current as low as $10–100\,nA\,mm^{-2}$ at a field of $2\,MV\,cm^{-1}$ (Figure 9.9b). These systems were shown to be compatible with various high-mobility p-type polymeric semiconductors [100], but one major drawback is that their range of applicability does not cover, for example, some solution-processable small molecule semiconductors, given the nonperfect orthogonality of the *n*-butyl acetate, which is used to dissolve the PMMA. Recently, a more orthogonal cross-linked polymer system, based on an amorphous fluoropolymer gate dielectric, poly(perfluorobutenylvinylether) commercially known as *Cytop*, was demonstrated [101]. By adopting the same cross-linking agent, blend dielectric (C-Cytop) films as thin as 50 nm can be deposited by spin coating, allowing the fabrication of reliable p- and n-channel top-gate OFETs operating at very low voltages (<5 V). The most remarkable properties of this new C-Cytop gate dielectric is the excellent device yield ($\approx$100%) for thicknesses <100 nm and dramatically reduced sensitivity to the underlying semiconductor film morphology. Furthermore C-Cytop based TFTs exhibit reduced bias stress and better air stability with respect to other systems thanks to the inert perfluorinated chemical structure of this polymer.

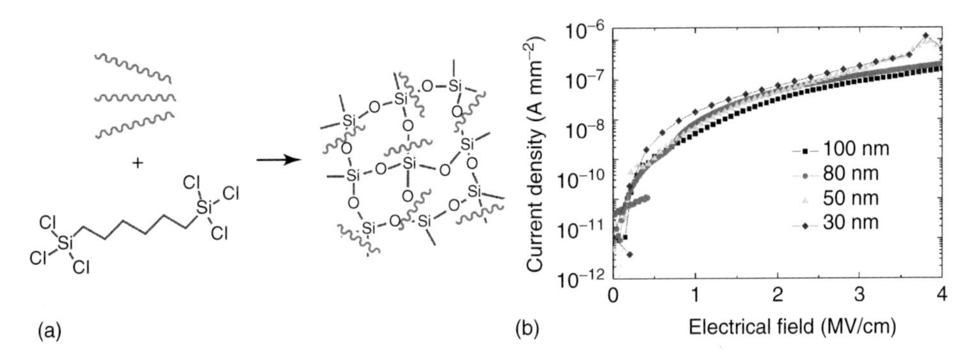

Figure 9.9 (a) Schematic representation of a cross-linked polymer system: 1,6-bis(trichlorosilyl)hexane (left) is mixed with a polymer solution and upon reacting with water it forms a siloxane network (right) throughout the polymer chains. (b) Leakage current density versus applied voltage of C-PMMA polymer films with thickness ranging between 30 and 100 nm sandwiched between gold electrodes. Adapted from [70].

9.4.3
Organic TFTs Printed with Subfemtoliter Printer

High-resolution printing of Ag contacts by means of subfemtoliter e-jet printing (Section 9.3), combined with a SAM dielectric layer (Section 9.4.2.2), allowed the fabrication of downscaled BGTC organic TFTs based on evaporated n-type and p-type small molecule semiconductors [7] (Figure 9.10a). The thin SAM dielectric is prepared, as previously described, by growing from solution a 2.1-nm thick *n*-octadecylphosphonic acid layer on a patterned Al gate contact, previously treated with oxygen plasma to produce a 3.6-nm thick AlO_x layer. Subsequently a patterned 30 nm thick layer of pentacene or $F_{16}CuPc$ is thermally evaporated through a shadow mask. The source and drain contacts are then printed directly onto the semiconductor with an Ag nanoparticles ink (Harima NPS-J-HP). Thanks to the high resolution of the adopted e-jet printing system, capable of ejecting subfemtoliter droplets, channel lengths as narrow as 1 μm (Figure 9.10b) could be obtained. The sintering temperature of the Ag lines printed with subfemtoliter accuracy is of 130 °C, low enough to be compatible with the presence of an organic semiconductor layer underneath. Moreover, since much of the solvent in these very small droplets, characterized by a high ratio between surface area and volume, evaporates before hitting the semiconductor, no damage is induced on the latter during the printing of the contacts. Pentacene TFTs showed very good linear

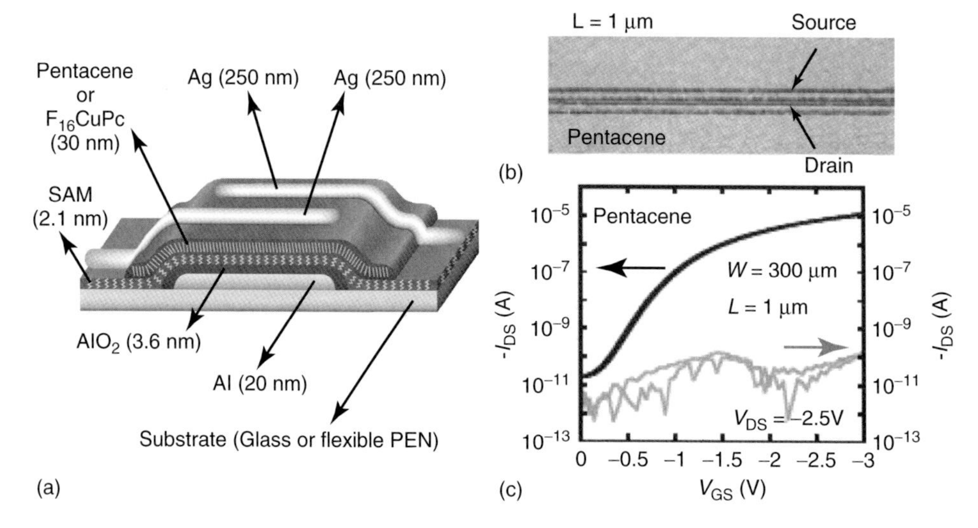

Figure 9.10 (a) Schematic cross-section of the OTFT with e-jet printed Ag nanoparticle source/drain contacts and ultra-thin SAM gate dielectric. (b) Optical microscope image of source and drain Ag nanoparticle contacts with a channel length of 1 μm printed on top of pentacene. (c) Transfer curves of a p-channel pentacene TFT with a $L = 1$ μm and $W = 300$ μm. Reproduced with permission from [7].

and saturation curves (Figure 9.10c) at voltages as low as 3 V, with a mobility as high as 0.3 $cm^2 V^{-1} s^{-1}$ and an on/off current ratio of $\approx 10^6$. The n-type $F_{16}CuPc$ TFTs similarly have nice curves, with an on/off ratio $>10^4$ and a mobility of 0.02 $cm^2 V^{-1} s^{-1}$, typical for this semiconductor. In these devices, the parasitic gate capacitance related to x_{OV} is limited without the need of any precise registration or self-aligned scheme (Section 9.12) because of the small linewidth (2 μm) for source and drain contacts.

This demonstration shows that subfemtoliter printing can be adopted for the fabrication of downscaled, high-performance BGTC organic transistors. The printing of the top contacts does not require any lithographic process or prepatterning, however, since this is a very low-throughput technique, in a production environment, it could be implemented only in the steps where critical features are actually involved, while relying on a standard high-throughput printing technique elsewhere. In the case of the work described above the patterning of the semiconductors and of the gate contact was realized by thermal evaporation through shadow masks. In principle, this step could be substituted with a printing technique, as long as it is demonstrated that a SAM dielectric of enough good quality can be grown on a printed electrode.

9.4.4 Mask-Free, All Solution Processed SAP TFTs

It was shown in Section 9.3 that with the SAP technique, it is possible to reach a submicrometer resolution in the definition of the gap between two metallic electrodes, while preserving the high throughput of picoliter piezoelectric printing. Downscaled BCTG TFTs, completely processed from solution, were fabricated by adopting the single-droplet Au contacts with 200 nm gaps as bottom source-drain electrodes [71]. The semiconducting polymer poly(9,9-dioctylfluorene-cobithiophene) (F8T2) was adopted as semiconductor and deposited on the contacts by spin coating. On top of this, a thin $\approx$50 nm C-PMMA layer (Section 9.4.2.2) was spun to realize the scaled dielectric needed to control the submicrometer channel. For these TFTs and generally for any integrated device structures that utilizes such SAP contacts, it is necessary to have at least two levels of printed metallization. For the top-level metallization (gate level) a silver ink based on an organic complex (TEC-IJ-010 InkTec Co., Ltd.) was adopted. This printable ink can be converted into a highly conducting electrode at low temperatures to avoid degradation of the underlying organic semiconductor and dielectric layers. After only 3 min at 130 °C in ambient air the lines show a resistance of a few ohms, corresponding to an estimated resistivity in the range of $5-10 \times 10^{-8}$ Ωm, close to the silver bulk conductivity. In Figure 9.11, a F8T2 FET with the T-shaped gold SAP source and drain contacts and a printed silver gate electrode can be seen. The device has a compact design and a small footprint, suitable for circuit integration, limited only by the printed electrode linewidth. In Figure 9.11b and c, the output characteristics of an all-printed F8T2 FET with $L \approx 200$ nm are reported. Clean saturation at voltages as low as a few volts is observed reflecting the correct scaling of the gate dielectric thickness [70].

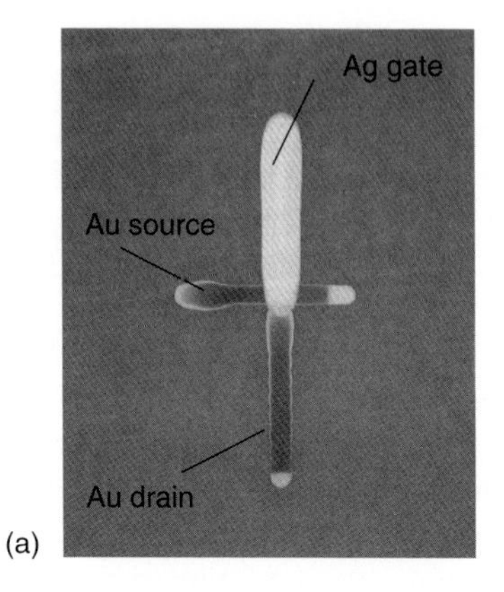

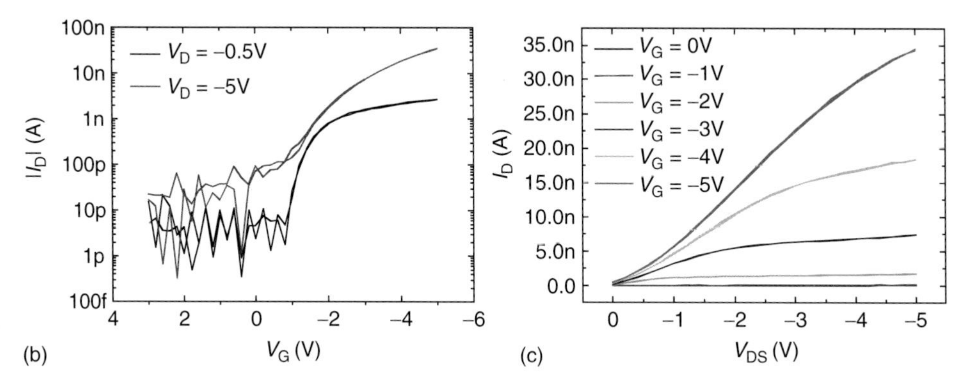

Figure 9.11 (a) Optical micrograph of an all-printed F8T2 self-aligned printed transistor with printed gold electrodes ($L \approx 200$ nm, $W \approx 40$ μm), a ≈50 nm thick C-PMMA gate dielectric and a silver printed gate contact. In (b) and (c), the transfer and output curves of the device are plotted, respectively.

The device operates at voltages lower than 5 V and has a field-effect mobility of $0.002\,\text{cm}^2\ \text{V}^{-1}\,\text{s}^{-1}$, which is typical for unaligned F8T2 transistors with PMMA gate dielectric.

This demonstrates the feasibility of the fabrication of downscaled organic TFTs based on the high-resolution SAP T-shape contacts with a completely bottom-up approach, without the need of any mask and where each component is processed from solution.

9.4.5
Self-Aligned Gate Contacts for Fast-Switching Transistors

The reduction of the channel length allows to extract more current from a transistor, but is not sufficient to obtain the maximum device operation speed if the parasitic gate capacitance is not minimized. A figure of merit of the highest operating frequency achievable in a FET is the transition frequency

f_T (or unity-gain frequency), corresponding to the crossover point at which an AC-modulated channel current in response to a gate voltage modulation becomes equal to the parasitic current flowing through the capacitance between the gate and source drain [70]. Equation (9.3) provides an expression for f_T based on the gradual channel approximation [73, 102]

$$f_T = \frac{g_m}{2\pi(C_{gd} + C_{gs})} \tag{9.3}$$

where g_m is the transistor transconductance, C_{gd} is the gate-drain capacitance and C_{gs} is the gate-source capacitance. Since $g_m \propto L^{-1}$, Eq. (9.3) shows that f_T can be increased proportionally to L^{-1} if C_{gd} and C_{gs} are constant. However if also the gate and source-drain capacitances could be scaled, limiting the total gate capacitance area to the geometrical channel area $L \times W$, f_T would scale with L^{-2}. Therefore, it is evident that, in order to boost the operational frequency of a SAP FET, the control of the gate overlap capacitance is crucial.

As an example, Figure 9.13b shows that f_T for a top-gate, bottom-contacts F8T2 transistor, with $L = 5$ μm, is around 1 kHz. When the channel length is reduced of an order of magnitude in a F8T2 SAP device, the drain current correspondingly increases, but if no attempt is made in order to control the overlap x_{OV}, this may result in an even larger parasitic gate current than in the long-channel case. This is due to the fact that the micrometer scale device shows an interdigitated contact pattern where the ratio of the channel area to the total area of the pattern is higher than in the case of the SAP device, where the width of the metal contacts is comparable, but the channel is much shorter. This severely limits the final f_T, thus, loosing the advantage of a scaled L. Even if a more controlled printing of the gate line was performed, the overlap between the gate and the source drain would, however, be associated with the printed gate linewidth, which is of the order of 50–100 μm. This would create an overlapping area approximately two orders of magnitude wider than the actual channel area, introducing an unacceptable parasitic capacitance.

This limitation extends to all the state-of-the-art volume printing techniques, not capable of dispensing small liquid volumes in order to significantly reduce the linewidths. Even if a considerable linewidth reduction could be achieved, for example, using a subfemtoliter printing system [7], this would have to be compatible with large-area manufacturing, and more importantly, a fine gate-channel alignment, which would be strongly beneficial in the case of a submicrometer channel TFT, would still be required. For these reasons, it is essential to pattern the gate with a self-aligned process.

To achieve this, different solutions were proposed in the literature. One approach used a self-aligned surface energy pattern, defined thanks to a topographical contrast defined during patterning of a lower layer [103] or thanks to a self-aligned photopatterning of a SAM [104]. A more general approach, fully compatible with large-area manufacturing of top-gate SAP FETs was proposed, in which the source and drain contacts were used as a photomask to develop a trench in a thick photoresist film deposited on top of the dielectric layer [70].

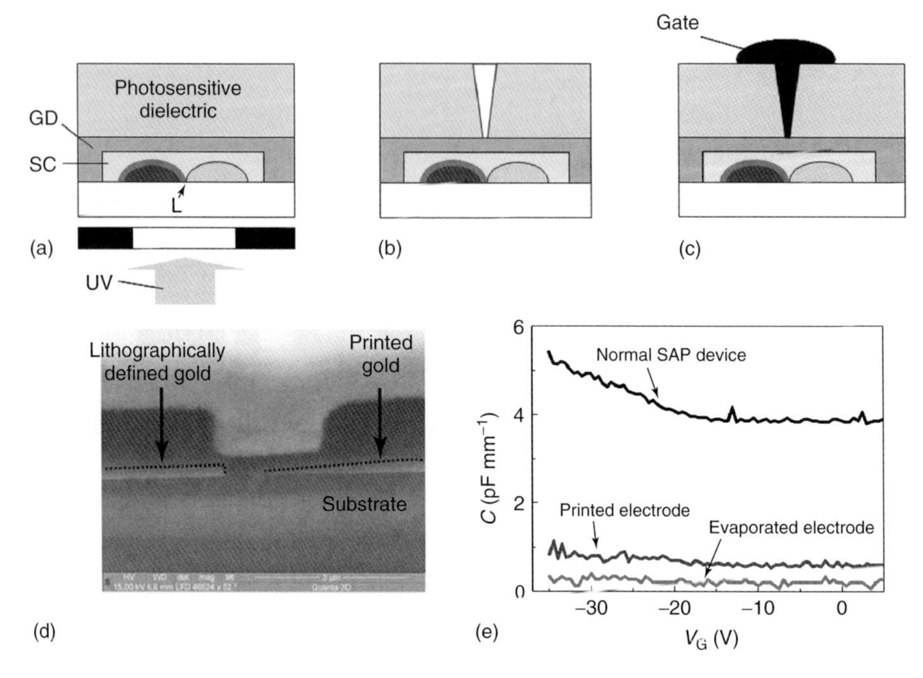

Figure 9.12 (a–c) Schematic representation of the self-aligned gate (SAG) process: (a) a thick layer of photoresist is spun on top of the thin dielectric layer (*L*: channel length, SC: semiconductor, GD: gate dielectric); (b) the sample is exposed from the bottom to UV light, using the SAP channel as a photomask; a trench, self-aligned to the channel, is developed; and (c) a gate contact is printed in the trench. (d) SEM cross-section of an SAP-SAG FET. (e) Capacitance–voltage characteristics of SAP FETs (semiconductor: F8T2; gate dielectric: C-PMMA) with self-aligned printed source/drain electrodes. In these devices, the first contact was defined by evaporation and photolithography (evaporated electrode) and the second was defined by SAP (printed electrode). The gate contact was made of ink-jet printed PEDOT:PSS. The black line shows the overlap capacitance between the gate and the source/drain electrodes of a SAP device without SAG. The dark gray line shows the overlap capacitance between the gate and the printed electrode in an SAG-SAP device. The light gray line shows the overlap capacitance between the gate and the evaporated electrode in an SAG-SAP device. Reproduced with permission from [70].

The process comprises the following steps:

1) a thick (1–2 μm) UV photosensitive dielectric layer is deposited on top of the thin dielectric layer;
2) the substrate is irradiated with light through the back, selectively exposing the channel region thanks to the masking action provided by SAP contacts (Figure 9.12a);
3) the photoresist is developed, obtaining a trench structure self-aligned to the edges of the source and drain contacts (Figure 9.12b);
4) a wide gate electrode is ink-jet printed with no need for fine alignment (Figure 9.12c).

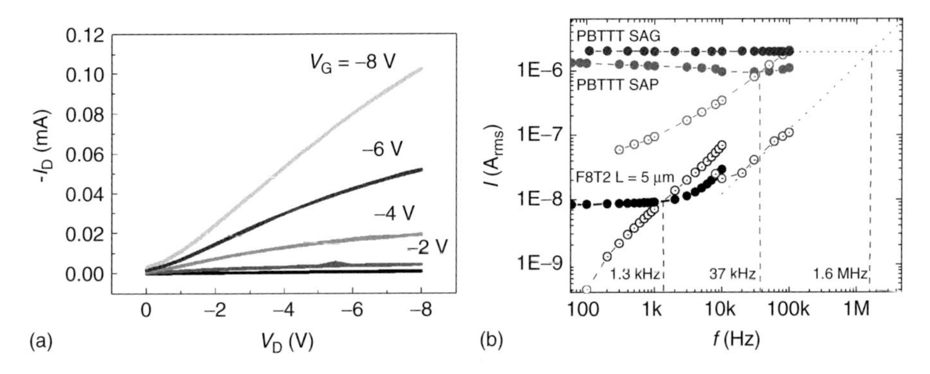

Figure 9.13 (a) Output characteristics of an SAG-SAP PBTTT-based FET ($L \approx 200$ nm, $W = 500$ μm, ≈50 nm). (b) f_T measurements for an SAG-SAP PBTTT FET (dark gray lines), a normal SAP PBTTT FET without SAG (light gray lines) and a reference long-channel F8T2 device (black lines). Reproduced with permission from [70].

Thanks to this self-aligned gate (SAG) process, the gate dielectric is thin only in correspondence of the channel, where this is required for correct gating of the device (Figure 9.12d). Everywhere else the photoresist increases the effective thickness of the gate dielectric and therefore strongly reduces the overlap capacitance, which is found to be as low as 1 pF mm^{-1}, up to an order of magnitude lower than in reference devices fabricated without the SAG structure (Figure 9.12e). Interestingly, the narrow trench region where the dielectric is thin contributes only for ≈10% of the overall overlap capacitance, which is still dominated by the surrounding regions, where the thick second dielectric is present. Therefore, a further reduction of the overlap capacitance is possible by employing a thicker resist layer.

The overall self-aligned fabrication process of the transistor, comprising the SAP and SAG processes, is compatible with well-known polymer semiconductors, such as F8T2 and high-mobility pBTTT [105]: no degradation was observed during UV exposure or standard development process. The adoption of pBTTT allowed the fabrication of high-performance transistors with field-effect mobility in the range of 0.1 $cm^2\,V^{-1}\,s^{-1}$, operating at 5–8 V (Figure 9.13a). The SAG process enabled full downscaling of the devices, which reached f_T values of around 1.6 MHz, with a strong improvement compared to a pBTTT transistor incorporating only SAP electrodes, which showed f_T limited to 40 kHz (Figure 9.13b).

The integration of SAP and SAG techniques offers a viable process for the fabrication of high-performance organic transistors, paving the way to the realization of fast-switching organic logic compatible with the manufacturing requirements of low-cost, large-area flexible circuits.

Although specifically developed in the field of organic transistors, this can be regarded as an example of how the control of surface energy and of wetting properties of the inks, as well as the development of novel architectures, can be effectively used to overcome resolution limitations of ink-jet printing and make this technique suitable for high-frequency electronic applications.

9.5
Conclusions and Outlook

In this work, we have reviewed the recent advancements in the ink-jet printing of downscaled organic electronic devices. We have introduced the basics of ink-jet printing, providing an overview of the available techniques and their most promising applications in the field of organic electronics. Emerging approaches toward the development of high-resolution tools and techniques, overcoming the limitations of standard mass production equipment, were reported in detail, and their application in the fabrication of downscaled, high-performance OTFTs was presented, analyzing their compatibility with low-cost and large-area flexible electronics manufacturing.

One of the first successful applications of piezoelectric DOD ink-jet printing in organic electronics has been in the fabrication of OLED displays, where the possibility to selectively deposit small amounts of functional materials on a substrate offers an obvious advantage. The same holds true for the patterning of arrays of detectors, where specific spatial distributions could be realized in a single step, with no need of postpatterning. Several printable functional materials, comprising organic semiconductors and highly conductive metal inks, have recently been developed, covering a wide range of applications. To overcome somc constraints on the viscosity of printable functional inks, other contact-less and mask-free printing tools are being developed as well, such as aerosol jet printing.

These techniques currently allow the fabrication of all the various functional layers needed to fabricate an organic electronic device. However, the low resolution offered by tools developed for the graphical arts limits the smallest feature size achievable. This limitation is reflected in low circuit switching speed despite the recent development of high-mobility organic semiconductor inks. This has represented a strong driving force toward the search for new techniques and tools allowing to overcome these limits. Self-aligned printing techniques and subfemtoliter e-jet printing are very good examples of how, through novel approaches and interdisciplinary efforts, high-resolution printing is achievable, opening up various opportunities for the downscaling of OTFTs. This represents a great opportunity for the development of new and smart device architectures, with a wide range of fundamental and interesting aspects, regarding structure-processing-property relationships of functional materials. An open question still holds for the adoption of these advanced techniques in a mass production environment, where the constraints on yield and uniformity are very stringent and not completely achievable with the current techniques for most applications. Recent developments in other fields, such as ink-jet printing adoption in the volume manufacturing of LCDs, and the demonstration of high-resolution patterning with high yield at laboratory level, show that these technological challenges can be overcome. However, more progress is needed to show that high-resolution printing of organic electronic circuits is viable and scalable. In particular, more efforts are needed in the development of new tools and tailored functional materials. Finally, a better scientific understanding of fundamentals properties of functionals inks and their deposition through jetting, especially in the case of high-resolution printing, need to be gained.

Acknowledgments

The authors gratefully acknowledge the fundamental work of various students and postdoctoral researchers at the Cavendish Laboratories in Cambridge (UK) that made possible part of the work reported here. In particular, we would like to thank Prof Y.-Y. Noh, Dr N. Zhao, Dr X. Cheng, and Mrs M. Lee that, at different stages of the research, contributed with great enthusiasm and invaluable skills in the development of the self-aligned printed transistors. The Engineering and Physical Sciences Research Council (EPSRC), the Cambridge Integrated Knowledge Center (CIKC), and Plastic Logic Ltd. are gratefully acknowledged for financial support and help in the development of the printing equipment.

References

1. Sirringhaus, H., Sele, C.W., von Werne, T., and Ramsdale, C. (2006) Manufacturing of organic transistor circuits by solution-based printing, *Organic Electronics, Materials, Manufacturing and Applications*, Wiley-VCH Verlag GmbH, pp. 294–322.
2. Creagh, L.T. and Mcdonald, M. (2003) Design and performance of inkjet print heads for non-graphic-arts applications. *MRS Bull.*, **28** (11), 807–811.
3. Rayleigh, F.R.S. (1878) On the instability of jets. *Proc. London Math. Soc.*, **10** (4), 4–13.
4. Le, H.P. (1998) Progress and trends in ink-jet printing technology. *J. Imaging Sci. Technol.*, **42** (1), 49–62.
5. Fujifilm Dimatix (2008) *Materials Printer & Cartridges dmp-2831 & dmc-11601/11610 - datasheet.* Datasheet http://www.dimatix.com. (accessed 2010).
6. Elrod, S.A., Hadimioglu, B., Yakub, K.B.T., Rawson, E.G., Richley, E., Quate, C.F., Mansour, N.N., and Lundgren, T.S. (1989) Nozzleless droplet formation with focused acoustic beams. *J. Appl. Phys.*, **65** (9), 3441–3447.
7. Sekitani, T., Noguchi, Y., Zschieschang, U., Klauk, H., and Someya, T. (2008) Organic transistors manufactured using inkjet technology with subfemtoliter accuracy. *Proc. Natl. Acad. Sci. U.S.A.*, **105** (13), 4976–4980.
8. Optomec. *Aerosol Jet 300 Datasheet*, Datasheet http://www.optomec.com. (accessed 2010).
9. Mette, A., Richter, P.L., Glunz, S.W., and Willeke, G. (2006) Novel metal jet printing technique for the front side metallization of highly efficient industrial silicon solar cells. 21st European Photovoltaic Solar Energy Conference, Barcelona, Spain.
10. Kahn, B.E. (2007) The m3d aerosol jet system, an alternative to inkjet printing for printed electronics. Organic and Printed Electronics, pp. 14–17.
11. Lindner, T.J. (2005) Printed electronics - hp's technology beyond ink on paper. Presentation at Printed Electronics USA.
12. Inc Microfab Technologies. *Technote 99-02: Fluid Properties Effects*, www.microfab.com. (accessed 2010).
13. Paul, K.E., Wong, W.S., Ready, S.E., and Street, R.A. (2003) Additive jet printing of polymer thin-film transistors. *Appl. Phys. Lett.*, **83** (10), 2070–2072.
14. Hayati, I., Bailey, A.I., and Tadros, Th. (1986) Mechanism of stable jet

formation in electrohydrodynamic atomization. *Nature*, **319** (6048), 41–43.
15. Suwan, N., Jayasinghe, A., Qureshi, N., and Eagles, P.A.M. (2006) Electrohydrodynamic jet processing: an advanced electric-field-driven jetting phenomenon for processing living cells13. *Small*, **2** (2), 216–219.
16. Park, J.U., Hardy, M., Kang, S.J., Barton, K., Adair, K., Mukhopadhyay, D.K., Lee, C.Y., Strano, M.S., Alleyne, A.G., Georgiadis, J.G., Ferreira, P.M., and Rogers, J.A. (2007) High-resolution electrohydrodynamic jet printing. *Nat. Mater.*, **6** (10), 782–789.
17. Murata, K., Matsumoto, J., Tezuka, A., Matsuba, Y., and Yokoyama, H. (2005) Super-fine ink-jet printing: toward the minimal manufacturing system. *Microsyst. Technol.*, **12** (1), 2–7.
18. Lee, D.Y., Shin, Y.S., Park, S.E., Yu, T.U., and Hwang, J. (2007) Electrohydrodynamic printing of silver nanoparticles by using a focused nanocolloid jet. *Appl. Phys. Lett.*, **90** (8), 081905+.
19. H.C. Starck GmbH *Clevios Conducting Polymers for Organic Thin Film Transistors.* Datasheet http://www.clevios.com.
20. Huang, D., Liao, F., Molesa, S., Redinger, D., and Subramanian, V. (2003) Plastic-compatible low resistance printable gold nanoparticle conductors for flexible electronics. *J. Electrochem. Soc.*, **150** (7), G412–G417.
21. Chung, J., Ko, S., Bieri, N.R., Grigoropoulos, C.P., and Poulikakos, D. (2004) Conductor microstructures by laser curing of printed gold nanoparticle ink. *Appl. Phys. Lett.*, **84** (5), 801–803.
22. Novacentrix. *Metalon ici-001 Datasheet*, Datasheet http://www.novacentrix.com. (accessed 2010).
23. Inktec Corporation. *Transparent Silver Ink-jet Inks.* Datasheet http://www.inktec.com. (accessed 2010).
24. Kao, C.Y. and Chou, K.S. (2007) Electroless copper plating onto printed lines of nanosized silver seeds. *Electrochem. Solid-State Lett.*, **10** (3), D32–D34.
25. Buffat, P. and Borel, J.P. (1976) Size effect on the melting temperature of gold particles. *Phys. Rev. A*, **13** (6), 2287+.
26. Ko, S.H., Pan, H., Grigoropoulos, C.P., Luscombe, C.K., Fréchet, J.M.J., and Poulikakos, D. (2007) All-inkjet-printed flexible electronics fabrication on a polymer substrate by low-temperature high-resolution selective laser sintering of metal nanoparticles. *Nanotechnology*, **18** (34), 1–8.
27. Perelaer, J., Klokkenburg, M., Hendriks, C.E., and Schubert, U.S. (2009) Microwave flash sintering of inkjet-printed silver tracks on polymer substrates. *Adv. Mater.*, **21** (47), 4830–4834.
28. Hong, C.M. and Wagner, S. (2000) Inkjet printed copper source/drain metallization for amorphous silicon thin-film transistors. *Electron. Device Lett. IEEE*, **21** (8), 384–386.
29. Park, B., Kim, D., Jeong, S., Moon, J., and Kim, J. (2007) Direct writing of copper conductive patterns by ink-jet printing. *Thin Solid Films*, **515** (19), 7706–7711.
30. Luechinger, N.A., Athanassiou, E.K., and Stark, W.J. (2008) Graphene-stabilized copper nanoparticles as an air-stable substitute for silver and gold in low-cost ink-jet printable electronics. *Nanotechnology*, **19** (44), 445201–445206.
31. Macdonald, W.A. (2004) Engineered films for display technologies. *J. Mater. Chem.*, **14** (1), 4–10.
32. Singh, M., Haverinen, H.M., Dhagat, P., and Jabbour, G.E. (2009) Inkjet printing — process and its applications. *Adv. Mater.*, **21**, 1–13.
33. Forrest, S.R. (2004) The path to ubiquitous and low-cost organic electronic appliances on plastic. *Nature*, **428** (6986), 911–918.
34. Tekin, E., Wijlaars, H., Holder, E., Egbe, D.A.M., and Schubert, U.S. (2006) Film thickness dependency of the emission colors of ppe-ppvs in inkjet printed libraries. *J. Mater. Chem.*, **16**, 4294–4298.
35. Hebner, T.R., Wu, C.C., Marcy, D., Lu, M.H. and Sturm, J.C. (1998) Ink-jet printing of doped polymers for organic

light emitting devices. *Appl. Phys. Lett.*, **72**, 519–521.

36. Chang, S.-C., Liu, J., Bharathan, J., Yang, Y., Onohara, J., and Kido, J. (1999) Multicolor organic light-emitting diodes processed by hybrid inkjet printing. *Adv. Mater.*, **11**, 734–737.

37. Shimoda, T., Morii, K., Seki, S., and Kiguchi, H. (2003) Inkjet printing of light-emitting polymer displays. *MRS Bull.*, **28** (11), 821–827.

38. Sirringhaus, H. (2005) Device physics of solution-processed organic field-effect transistors. *Adv. Mater.*, **17** (20), 2411–2425.

39. Yan, H., Chen, Z., Zheng, Y., Newman, C., Quinn, J.R., Dötz, F., Kastler, M., and Facchetti, A. (2009) A high-mobility electron-transporting polymer for printed transistors. *Nature*, **457**, 679–686.

40. Sirringhaus, H., Bürgi, L., Kawase, T., and Friend, R.H. (2003) Polymer transistor circuits fabricated by solution processing and direct printing, *Thin Film Transistors*, Marcell Dekker Inc, pp. 427–474.

41. Arias, A.C., Ready, S.E., Lujan, R., Wong, W.S., Paul, K.E., Salleo, A., Chabinyc, M.L., Apte, R., and Street, R.A. (2004) All jet-printed polymer thin-film transistor active-matrix backplanes. *Appl. Phys. Lett.*, **85** (15), 3304–3306.

42. Song, D.H., Choi, M.H., Kim, J.Y., Janga, J., and Kirchmeyer, S. (2007) Process optimization of organic thin-film transistor by ink-jet printing of dh4t on plastic. *Appl. Phys. Lett.*, **90**, 053504.

43. Choi, M.H., Han, S.H., Lee, S.H., Choo, D.J., Jang, J., and Kwon, S.K. (2009) Effect of active layer thickness on environmental stability of printed thin-film transistor. *Org. Electron.*, **10**, 421–425.

44. Lim, J.A., Lee, W.H., Lee, H.S., Lee, J.H., Park, Y.D., and Cho, K. (2008) Self-organization of ink-jet-printed triisopropylsilylethynyl pentacene via evaporation-induced flows in a drying droplet. *Adv. Funct. Mater.*, **18**, 229–234.

45. Ah Lim, J., Lee, W.H., Kwak, D., and Cho, K. (2009) Evaporation-induced self-organization of inkjet-printed organic semiconductors on surface-modified dielectrics for high-performance organic transistors. *Langmuir*, **25** (9), 5404–5410.

46. Ng, T.N., Sambandan, S., Lujan, R., Arias, A.C., Newman, C.R., Yan, H., and Facchetti, A. (2009) Electrical stability of inkjet-patterned organic complementary inverters measured in ambient conditions. *Appl. Phys. Lett.*, (994), 233307+.

47. Molesa, S.E., Volkman, S.K., Redinger, D.R., de la Fuente Vornbrock, A., and Subramanian, V. (2004) A high-performance all-inkjetted organic transistor technology. IEEE International Electron Devices Meeting 2004 - IEDM Technical Digest, San Francisco, USA, p. 1072.

48. Liu, Y., Varahramyan, K., and Cui, T. (2005) Low-voltage all-polymer field-effect transistor fabricated using an inkjet printing technique. *Macromol. Rapid Commun.*, **26**, 1955–1959.

49. Subramanian, V., Frechet, J.M.J., Chang, P.C., Huang, D.C., Lee, J.B., Molesa, S.E., Murphy, A.R., Redinger, D.R., and Volkman, S.K. (2005) Progress toward development of all-printed rfid tags: materials, processes, and devices. *Proc. IEEE*, **93** (7), 1330–1338.

50. Shaheen, S.E., Radspinner, R., Peyghambarian, N., and Jabbour, G.E. (2001) Fabrication of bulk heterojunction plastic solar cells by screen printing. *Appl. Phys. Lett.*, **79** (18), 2996.

51. Tuomikoski, M. and Kopola, P. (2006) Technologies for Polymer Electronics TPE 06, Rudolstadt, Germany.

52. Hoth, C.N., Choulis, S.A., Schilinsky, P., and Brabec, C.J. (2007) High photovoltaic performance of inkjet printed polymer: fullerene blends. *Adv. Mater.*, **19**, 3973–3978.

53. Hoth, C.N., Choulis, S.A., Schilinsky, P., and Brabec, C.J. (2009) On the effect of poly(3-hexylthiophene) regioregularity on inkjet printed organic solar cells. *J. Mater. Chem.*, **19**, 5398–5404.

54. Aernouts, T., Aleksandrov, T., Girotto, C., Genoe, J., and Poortmans, J. (2008) Polymer based organic solar cells using ink-jet printed active layers. *Appl. Phys. Lett.*, **92**, 033306.
55. Someya, T., Sakurai, T., Sekitani, T., and Noguchi, Y. (2007) Printed organic transistors for large-area electronics. *IEEE Polytronic Conference Proceedings*, Odaiba, Tokyo, pp. 6–11.
56. Böberl, M., Kovalenko, M.V., Gamerith, S., List, E.J.W., and Heiss, W. (2007) Inkjet-printed nanocrystal photodetectors operating up to 3 micron wavelengths. *Adv. Mater.*, **19**, 3574–3578.
57. Antognazza, M.R., Scherf, U., Monti, P., and Lanzani, G. (2007) Organic-based tristimuli colorimeter. *Appl. Phys. Lett.*, **90**, 163–509.
58. Scott, J.C. and Bozano, L.D. (2007) Nonvolatile memory elements based on organic materials. *Adv. Mater.*, **19**, 1452–1463.
59. Doggart, J., Wu, Y., and Zhu, S. (2009) Inkjet printing narrow electrodes with $< 50\,\mu m$ line width and channel length for organic thin-film transistors. *Appl. Phys. Lett.*, **94**, 163503.
60. van Osch, T.H.J., Perelaer, J., de Laat, A.W.M., and Schubert, U.S. (2008) Inkjet printing of narrow conductive tracks on untreated polymeric substrates. *Adv. Mater.*, **20**, 343–345.
61. Hendriks, C.E., Smith, P.J., Perelaer, J., van den Berg, A.M.J., and Schubert, U.S. (2008) "invisible" silver tracks produced by combining hot-embossing and inkjet printing. *Adv. Funct. Mater.*, **18**, 1031–1038.
62. Sirringhaus, H., Kawase, T., Friend, R.H., Shimoda, T., Inbasekaran, M., Wu, W., and Woo, E.P. (2000) High-resolution inkjet printing of all-polymer transistor circuits. *Science*, **290** (5499), 2123–2126.
63. Burns, S.E., Cain, P., Mills, J., Wang, J., and Sirringhaus, H. (2003) Inkjet printing of polymer thin-film transistor circuits. *MRS Bull.*, **28** (11), 829–833.
64. Wang, J.Z., Zheng, Z.H., Li, H.W., Huck, W.T.S., and Sirringhaus, H. (2004) Dewetting of conducting polymer inkjet droplets on patterned surfaces. *Nat. Mater.*, **3**, 171–176.
65. Sele, C.W., von Werne, T., Friend, R.H., and Sirringhaus, H. (2005) Lithography-free, self-aligned inkjet printing with sub-hundred-nanometer resolution. *Adv. Mater.*, **17** (8), 997–1001.
66. Zhao, N., Chiesa, M., Sirringhaus, H., Li, Y., Wu, Y., and Ong, B. (2007) Self-aligned inkjet printing of highly conducting gold electrodes with submicron resolution. *J. Appl. Phys.*, **101**, 064513-1–064513-6.
67. Porter, M.R. (1994) *Handbook of Surfactants*, Blackie Academic & Professional.
68. Li, Y., Wu, Y., and Ong, B.S. (2005) Facile synthesis of silver nanoparticles useful for fabrication of high-conductivity elements for printed electronics. *J. Am. Chem. Soc.*, **127** (10), 3266–3267.
69. Szczech, J.B., Megaridis, C.M., Gamota, D.R., and Zhang, J. (2002) Fine-line conductor manufacturing using drop-on demand pzt printing technology. *IEEE Trans. Electron. Packaging Manuf.*, **25**, 26–33.
70. Noh, Y.Y., Zhao, N., Caironi, M., and Sirringhaus, H. (2007) Downscaling of self-aligned, all-printed polymer thin-film transistors. *Nat. Nano*, **2** (12), 784–789.
71. Caironi, M., Gili, E., and Sirringhaus, H. (2010) High yield, single droplet electrode arrays for nanoscale printed electronics. *ACS Nano*, **4** (3), pp. 1451–1456.
72. Richards, T.J. and Sirringhaus, H. (2007) Analysis of the contact resistance in staggered, top-gate organic field-effect transistors. *J. Appl. Phys.*, **102**, 094510.
73. Sze, S.M. (1981) *Physics of Semiconductor Devices*, John Wiley & Sons, Inc.
74. Veres, J., Ogier, S., Lloyd, G., and de Leeuw, D. (2004) Gate insulators in organic field-effect transistors. *Chem. Mater.*, **16**, 4543–4555.

75. Zhao, N., Noh, Y.-Y., Chang, J.-F., Heeney, M., McCulloch, I., and Sirringhaus, H. (2009) Polaron localization at interfaces in high-mobility microcrystalline conjugated polymers. *Adv. Mater.*, **21**, 1–5.
76. Majewski, L.A., Schroeder, R., and Grell, M. (2005) One volt organic transistor. *Adv. Mater.*, **17**, 192–196.
77. Zirkl, M., Haase, A., Fian, A., Schön, H., Sommer, C., Jakopic, G., Leising, G., Stadlober, B., Graz, I., Gaar, N., Schwödiauer, R., Bauer-Gogonea, S., and Bauer, S. (2007) Low-voltage organic thin-film transistors with high-k nanocomposite gate dielectrics for flexible electronics and optothermal sensors. *Adv. Mater.*, **19**, 2241–2245.
78. Kim, S.H., Yang, S.Y., Shin, K., Jeon, H., Lee, J.W., Hong, K.P., and Park, C.E. (2006) Low-operating-voltage pentacene field-effect transistor with a high-dielectric-constant polymeric gate dielectric. *Appl. Phys. Lett.*, **89**, 183516.
79. Naber, R.C.G., Tanase, C., Blom, P.W.M., Gelinck, G.H., Marsman, A.W., Touwslager, F.J., Setayesh, S., and De Leeuw, D.M. (2005) High-performance solution-processed polymer ferroelectric field-effect transistors. *Nat. Mater.*, **4**, 243–248.
80. Panzer, M.J., Newman, C.R., and Frisbie, C.D. (2005) Low-voltage operation of a pentacene field-effect transistor with a polymer electrolyte gate dielectric. *Appl. Phys. Lett.*, **86**, 103503.
81. Panzer, M.J. and Frisbie, C.D. (2005) Polymer electrolyte gate dielectric reveals finite windows of high conductivity in organic thin film transistors at high charge carrier densities. *J. Am. Chem. Soc.*, **127**, 6960–6961.
82. Dhoot, A.S., Yuen, J.D., Heeney, M., McCulloch, I., Moses, D., and Heeger, A.J. (2006) Beyond the metal-insulator transition in polymer electrolyte gated polymer field-effect transistors. *Proc. Natl. Acad. Sci. U.S.A.*, **103**, 11834–11837.
83. Lee, J., Panzer, M.J., He, Y., Lodge, T.P., and Frisbie, C.D. (2007) Ion gel gated polymer thin-film transistors. *J. Am. Chem. Soc.*, **129**, 4532–4533.
84. Herlogsson, L., Crispin, X., Robinson, N.D., Sandberg, M., Hagel, O.-J., Gustafsson, G., and Berggren, M. (2007) Low-voltage polymer field-effect transistors gated via a proton conductor. *Adv. Mater.*, **19**, 97–101.
85. Herlogsson, L., Noh, Y.-Y., Zhao, N., Crispin, X., Sirringhaus, H., and Berggren, M. (2008) Downscaling of organic field-effect transistors with a polyelectrolyte gate insulator. *Adv. Mater.*, **20**, 4708–4713.
86. Cho, J.H., Lee, J., He, Y., Kim, B., Lodge, T.P., and Frisbie, C.D. (2008) High-capacitance ion gel gate dielectrics with faster polarization response times for organic thin film transistors. *Adv. Mater.*, **20**, 686–690.
87. Cho, J.H., Lee, J., Xia, Y., Kim, B., He, Y., Renn, M.J., Lodge, T.P., and Frisbie, C.D. (2008) Printable ion-gel gate dielectrics for low-voltage polymer thin-film transistors on plastic. *Nat. Mater.*, **7**, 900–906.
88. Boulas, C., Davidovits, J.V., Rondelez, F., and Vuillaume, D. (1996) Suppression of charge carrier tunneling through organic self-assembled monolayers. *Phys. Rev. Lett.*, **76**, 4797–4800.
89. Collet, J., Tharaud, O., Chapoton, A., and Vuillaume, D. (2000) Low-voltage, 30 nm channel length, organic transistors with a self-assembled monolayer as gate insulating films. *Appl. Phys. Lett.*, **76** (14), 1941–1943.
90. Halik, M., Klauk, H., Zschieschang, U., Schmid, G., Dehm, C., Schütz, M., Maisch, S., Effenberger, F., Brunnbauer, M., and Stellacci, F. (2004) Low-voltage organic transistors with an amorphous molecular gate dielectric. *Nature*, **431**, 963–966.
91. Yoon, M.-H., Facchetti, A., and Marks, T.J. (2005) $\sigma - \pi$ molecular dielectric multilayers for low-voltage organic thin-film transistors. *Proc. Natl. Acad. Sci. U.S.A.*, **102** (13), 4678–4682.
92. Ma, H., Acton, O., Ting, G., Ka, J.W., Yip, H.-L., Tucker, N., Schofield, R., and Jen, A.K.-Y. (2008) Low-voltage organic thin-film transistors with $\pi - \sigma$-phosphonic acid molecular dielectric monolayers. *Appl. Phys. Lett.*, **92**, 113303.

93. Park, Y.D., Kim, D.H., Jang, Y., Hwang, M., Lim, J.A., and Choa, K. (2005) Low-voltage polymer thin-film transistors with a self-assembled monolayer as the gate dielectric. *Appl. Phys. Lett.*, **87**, 243509.
94. Wöbkenberg, P.H., Ball, J., Kooistra, F.B., Hummelen, J.C., de Leeuw, D.M., Bradley, D.D.C., and Anthopoulos, T.D. (2008) Low-voltage organic transistors based on solution processed semiconductors and self-assembled monolayer gate dielectrics. *Appl. Phys. Lett.*, **93**, 013303.
95. Facchetti, A., Yoon, M.-H., and Marks, T.J. (2005) Gate dielectrics for organic field-effect transistors: new opportunities for organic electronics. *Adv. Mat.*, **17**, 1705–1725.
96. Yoon, M.H., Yan, H., Facchetti, A., and Marks, T.J. (2005) Low-voltage organic field-effect transistors and inverters enabled by ultrathin cross-linked polymers as gate dielectrics. *J. Am. Chem. Soc.*, **127** (29), 10388–10395.
97. Jang, Y., Kim, D.H., Park, Y.D., Cho, J.H., Hwang, M., and Cho, K. (2006) Low-voltage and high-field-effect mobility organic transistors with a polymer insulator. *Appl. Phys. Lett.*, **88**, 72101.
98. Yang, S.Y., Kim, S.H., Shin, K., Jeon, H., and Park, C.E. (2006) Low-voltage pentacene field-effect transistors with ultrathin polymer gate dielectrics. *Appl. Phys. Lett.*, **88**, 173507.
99. Frank, D.J., Dennard, R.H., Nowak, E., Solomon, P.M., Taur, Y., and Wong, H.-S.P. (2001) Device scaling limits of si mosfets and their application dependencies. *Proc. IEEE*, **89** (3), 259–288.
100. Noh, Y. and Sirringhaus, H. (2009) Ultra-thin polymer gate dielectrics for top-gate polymer field-effect transistors. *Org. Electron.*, **10** (1), 174–180.
101. Cheng, X., Caironi, M., Noh, Y.-Y., Wang, J., Newman, C., Yan, H., Facchetti, A., and Sirringhaus, H. (2010) Air stable crosslinked cytoptm ultrathin gate dielectric for high yields low-voltage top-gate organic field-effect transistors. *Chem. Mater.*, **22** (4), pp. 1559–1566.
102. Sedra, A.S. and Smith, K.C. (1998) *Microelectronic Circuits*, 4th edn, Oxford University Press.
103. Stutzmann, N., Friend, R.H., and Sirringhaus, H. (2003) Self-aligned, vertical-channel, polymer field-effect transistors. *Science*, **299** (5614), 1881–1884.
104. Ando, M., Kawasaki, M., Imazeki, S., Sasaki, H., and Kamata, T. (2004) Self-aligned self-assembly process for fabricating organic thin-film transistors. *Appl. Phys. Lett.*, **85** (10), 1849–1851.
105. Mcculloch, I., Heeney, M., Bailey, C., Genevicius, K., Macdonald, I., Shkunov, M., Sparrowe, D., Tierney, S., Wagner, R., Zhang, W., Chabinyc, M.L., Kline, R.J., Mcgehee, M.D., and Toney, M.F. (2006) Liquid-crystalline semiconducting polymers with high charge-carrier mobility. *Nat. Mater.*, **5**, 328–333.

10
Interplay between Processing, Structure, and Electronic Properties in Soluble Small-Molecule Organic Semiconductors

Oana D. Jurchescu, Devin A. Mourey, Yuanyuan Li, David J. Gundlach, and Thomas N. Jackson

10.1
Introduction

Organic field-effect transistors (OFETs) have experienced impressive improvements in electrical performance in recent years by careful design of novel materials and device structures. As a result, the mobility was improved 5 orders of magnitude in the last 15 years, and several polymeric and small-molecule semiconductors are presently rivaling and even surpassing α-Si:H, while offering unique advantages such as versatility and additional functionality [1–9]. For example, highly purified single crystals of pentacene and rubrene have allowed demonstrations of field-effect mobility as high as 20–40 cm^2 $(Vs)^{-1}$ [10–12], and metal-like behavior [13–16]. However, their solubility is limited, or does not preserve the excellent material properties. This represents a significant drawback, as the use of organic semiconductors in consumer applications implies requirements not only of high mobility, but also of less expensive processability that yields high performance. Solution-processed organic thin-film transistors (OTFTs) are of particular interest owing to their potential use in low-cost manufacturing techniques such as printing and roll-to-roll processing. Molecular design has allowed introduction and synthesis of a multitude of soluble organic semiconductors with remarkable electrical properties. Several functionalization strategies were proposed to improve the solubility of acenes and tune the solid state order. For instance, addition of side-groups promotes considerably better solubility and stability, and at the same time a face-to-face π-stacking molecular arrangement, as opposed to the edge-to-face herringbone arrangement present in molecules investigated earlier, for example, pentacene [17–19]. Different types and amounts of π-overlap can be controlled by the nature, size, and position of the substituent, leading to various stacking motifs, which reflect in the performance of field-effect transistors fabricated with these molecules. High mobility spin-cast OTFTs were reported for acene and thiophene derivatives with strong π-stacking interactions [17, 18, 20–22]. Fluorination of the molecular backbone was introduced as a method to further improve stability and accelerate film crystallization. This allowed demonstrations

Organic Electronics II: More Materials and Applications, First Edition. Edited by Hagen Klauk.

of mobilities as high as 1.5 cm^2 $(Vs)^{-1}$ in 2,8-difluoro-5,11-bis(triethylsilylethynyl) anthradithiophene (F-TES ADT) films [20, 23–25].

Nevertheless, these results are based on testing a large number of compounds, and a rational design of novel materials is hindered by the strong dependence of the transport properties on the processing parameters. The reason for this peculiar behavior is that with solution deposition, the organic semiconductor is dissolved in a solvent, and the solution is deposited on the substrate by using various techniques such as spin coating, dip coating, inkjet printing, drop casting, spray coating, and so on. The solvent evaporates, the solution becomes supersaturated, and crystals are formed on the substrate via a nucleation and growth mechanism. This is a very dynamic and complex process, driven by the interplay between solvent/substrate, solvent/vapor, solute/substrate, and solute/solvent interactions. The properties of the resulting film can vary considerably with the choice of deposition method, surface energy modifications, nature of dielectric, or post-processing. Consequently, the field-effect mobility can span several orders of magnitude in thin films of nominally the same material [25–27]. One of the causes of this large spread in performance is the different microstructure that results from various fabrication routes. This strongly affects charge transport, making it difficult to establish direct correlations between molecular structure and device performance. In this chapter, we will use OTFTs as experimental tools to measure carrier mobilities in organic semiconductors and relate it to the molecular packing and microstructure effects resulting from various processing methods. OTFTs are the basic building blocks of applications based on organic electronics and at the same time, from the fundamental standpoint, they represent excellent experimental platforms, allowing studies of charge transport with great control of the charge density achieved by electrostatic doping.

The chapter is organized as follows: Section 10.2 highlights several basic concepts relevant to charge transport in crystalline OTFTs. In Section 10.3 we describe the role of processing parameters on the performance of organic device. Specifically, we discuss the effects of deposition conditions (solvent, method, surface treatment) on the nucleation and growth of small-molecule organic semiconductor films, and show that they are critical parameters dictating structural and electronic properties of the film. Section 10.4 reviews several approaches which exploit the dependence of thin-film microstructure on processing parameters to achieve high-performance OTFTs.

10.2
Transport Limits in Crystalline Semiconductors

10.2.1
Crystallinity – Role of Structural Order

The intermolecular forces in organic molecular solids are weak, of van der Waals and electrostatic type; therefore the bonding energies are considerably lower than in covalent and inorganic semiconductors, leading to a fundamentally different mechanism of charge transport [28–30]. Here, charge carriers interact strongly with the lattice environment leading to polarization effects and tendency of charge-carrier localization. Owing to the weak van der Waals interactions, the electronic bandwidths are narrow, which results in strong interactions between free charge carriers and the lattice, and polaron formation. The physical quantity usually used to describe charge transport in these materials is *mobility* (μ), and this is discussed in detail in this chapter. The mobility of the charge carriers reflects the drift velocity of the charges in the lattice (v_d), and it is thus a function of the electric field E and all interactions that the charge encounters, following the expression:

$$\mu = \frac{v_d}{E} \tag{10.1}$$

The electronic properties of the solid are strongly influenced by the degree of order and the structural pattern of molecules in the crystal. At nanoscale level, short intermolecular distances and parallel, planar molecular orientations allow better overlap between the π orbitals of the neighboring molecules and are beneficial for charge transport. The long-range order of the $\pi-\pi$ stacked planes of the organic semiconductor molecules is often intimately related with charge transport in these materials. At the mesoscopic scale, the presence of defects, impurities, and phonons decreases the drift velocity of charges, and consequently the electronic mobility. The effect of the scattering events on the value of mobility can be easily understood by considering the following expression for mobility:

$$\mu = \frac{e\tau}{m^*} \tag{10.2}$$

where e represents the elementary charge, τ is the mean time between two consecutive scattering events, and m^* is the effective mass. This expression provides a direct relation between the morphology of the materials and their mobility. In a simple picture then, increased number of scattering sites reduces the value of mobility. The mechanism through which the defects create trapping sites in the bandgap of the organic semiconductors is still not characterized well, but it was empirically demonstrated that various types of defects are of critical importance [12, 16, 25, 29, 31–34]. Defects can be of intrinsic (molecular sliding, tilting, torsion, and vibrations [31, 32]) or extrinsic (impurities such as oxygen, water [35–39], or other organic molecules trapped in the host lattice [12, 16, 40–42]) nature. In this chapter we describe several intrinsic defects arising from various device fabrication methods.

10.2.2
Grain Boundaries

In Section 10.2.1, we have described the importance of the structural order, where the molecular orientation and crystal defects govern the transport. But as charge transport in OTFTs occurs over many molecular units, the long-range order and structural homogeneity also play significant roles. To address this concept, in the following section we describe mesoscopic structural features such as the size and distribution of the crystalline domains and the grain boundaries connecting them, which influence the operation of OTFTs. Recently, Teague *et al.* have used Scanning Kelvin probe measurements (SKPMs) and demonstrated that grain boundaries can be responsible for significant voltage drops in transistor channels [43]. Kelley and Frisbie have measured 1 order of magnitude larger resistance across the grain boundaries than within the grains using conducting probe atomic force microscopy (CP-AFM) [33]. These results demonstrate that charge transport at the grain boundaries can be significantly different than within the grains, and can greatly affect the macroscopic electrical properties of the films. The conclusion extracted from these studies was that the organic semiconductor film often consists of high-mobility regions within crystalline grains and low mobility regions connecting them, where the intra-grain mobility is an intrinsic property of the semiconductor. The effective field-effect mobility of the charge carriers in the organic film is a weighted sum of the two, and thus can be critically affected by the concentration of the grain boundaries, with larger grain size then being beneficial for charge transport. Indeed, Lee *et al.* have tuned the grain size in the channel of TES ADT (chemical structure presented in Scheme 10.1) TFTs 3 orders of magnitude, from 30 to 2700 μm, and have observed dramatic changes in the value of mobility, which allowed them to estimate the mobilities of the grains and grain boundaries, respectively [44]. Although no detailed description of the degree of order within the grain is given, this report presents a direct quantitative description of the effect of the grain boundaries on charge transport in organic thin films. Numerical simulations have been developed to model grain boundaries as defects responsible for charge-carrier trapping [45]. In the framework of these models, the presence of grain boundaries alters the mobility by inducing traps, which lead to band bending followed by creation of a potential well, E_B. For example, this barrier height was experimentally estimated to be an average of about 100 meV in sexithiophene (6T) [33]. The change in the grain size then induces variations in the trap densities, and consequently variations in thin-film mobility. The extent of

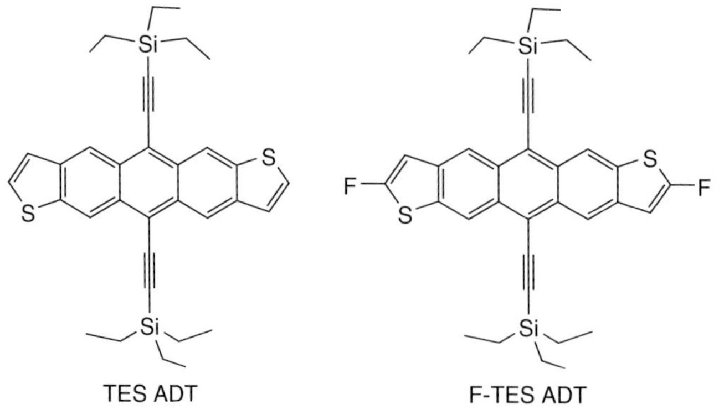

TES ADT

F-TES ADT

Scheme 10.1 Chemical formulae of triethylsilylethynyl anthradithiophene (TES ADT) and 2,8-difluoro-5,11-bis(triethylsilylethynyl) anthradithiophene (F-TES ADT).

charge trapping generated by grain boundaries is directly related to the thin-film fabrication conditions, as we demonstrate later, and it is very difficult to control in a reproducible manner. For this reason, the electrical properties of materials are often masked by the presence of various trap densities, distributions, and energies, and accessing the intrinsic properties is challenging.

10.2.3
Single Crystals – Model Systems to Study Intrinsic Properties of Organic Semiconductors

Although large-area electronic applications are based on thin-film devices, as we have pointed out earlier, with thin films the intrinsic properties may be altered by grain boundaries or other structural defects that localize charge carriers, and their performance is severely dependent on the processing conditions. A reliable insight into the mechanism of charge transport in organic molecular crystals, as well as on the interplay between the chemical structure, crystal structure, and charge transport can thus only be obtained when single crystals are investigated. Single crystals provide a well-defined structure, the ultimate molecular long-range order, and can serve as model systems, for which structure–properties correlations can be established and the limit of achievable electrical properties can be explored [3, 46, 47].

10.3
Structure–Processing–Properties Relationship in Small-Molecule Organic Thin-Film Transistors

10.3.1
Microstructure and Mobility

In Section 10.2 we described how the performance of OTFTs is dominated by the complex nanoscale pattern (molecular packing) and microstructure of the film (crystal orientation, grain boundaries), which in turn are dictated by the chemical structure and processing. We discussed that this represents a serious issue, and makes the investigation of novel organic semiconductors very challenging. In this section we illustrate these effects in several small-molecule organic semiconductors. For example, in Figure 10.1, we present the electrical properties of transistors fabricated on single crystals (a) and thin films (b) of fluorinated F-TES ADT (chemical formula presented in Scheme 10.1). When comparing the evolution of the drain current with the gate voltage (Figure 10.1c,d), it can be observed that the single-crystal devices (Figure 10.1c) show a very sharp turn on, large on/off ratio, and almost perfect linear behavior in the $\sqrt{I_D}$ versus V_{GS} plot, as expected from the expression governing the saturation regime in an FET:

$$I_D = \frac{1}{2}\frac{W}{L}C_i \cdot \mu \cdot (V_{GS} - V_T)^2 \tag{10.3}$$

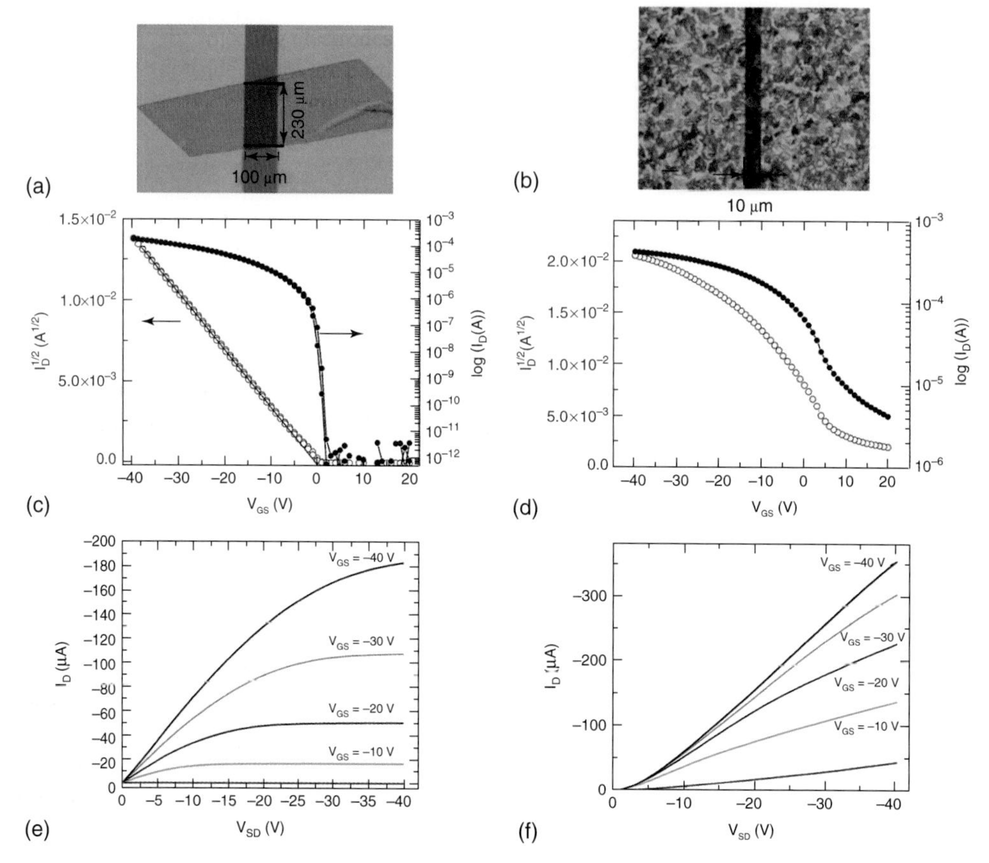

Figure 10.1 Field-effect transistors fabricated on organic single crystals (a,c,e) and thin films (b,d,f). Optical micrograph of the single-crystal device is shown in (a), and of a thin-film device in (b). (c) and (d) show the evolution of the drain current I_D with the gate voltage V_{GS} in the saturation regime ($V_{DS} = -40\,\text{V}$) for single-crystal devices (c) and TFTs (d). In (e) and (f) we show the transport characteristics for single-crystal (e) and thin-film (f) devices.

where W and L are the channel width and length, respectively, C_i is the gate oxide capacitance per unit area, μ represents the field-effect mobility, V_{GS} is the gate to source voltage, and V_{T} is the threshold voltage. On the contrary, thin-film devices show less sharp turn on and severe deviations from linearity in the $\sqrt{I_D}$ versus V_{GS} plot (Figure 10.1d). Similar effects can be observed in the evolution of the drain current with the drain voltage (Figure 10.1e,f). The electrical characteristics of single crystals (Figure 10.1e) show very well-defined transistor behavior, in which at small drain-source V_{SD} voltages the current depends linearly on the applied voltage and the device acts as a gate-voltage-controlled variable resistor (linear regime). At higher voltages the current becomes independent of the drain voltage (saturation regime) [20]. In thin-film devices (Figure 10.1f), the linear regime shows serious current compression and deviation from linearity, and the saturation regime is not reached [23, 25]. This behavior was described in terms of contact injection problems [48–50]. The hypothesis is supported by the severe voltage drop at the contacts measured using SKPM [43]. In spite of the excellent I–V characteristics in single-crystal devices, a voltage drop was also measured here [51]. Although this drop is considerably lower than in the case of thin-film devices, it demonstrates that even single-crystal devices are incapable of pointing out intrinsic material properties. Single crystals can reveal some limits for charge transport in organic crystalline semiconductors, although not always the maximum limits. The electrical properties measured in thin films are always inferior to the single-crystal properties (for this particular example $\mu_{\text{single crystal}} = 6\ \text{cm}^2\ (\text{Vs})^{-1}$ and $\mu_{\text{thin film}} = 0.2\ \text{cm}^2\ (\text{Vs})^{-1}$), but both single crystal and thin film results represent an underestimation of the intrinsic mobility of the material. Another factor limiting the transport in thin films, and possibly being responsible for the nonlinearities in I_D versus V_{SD} curves, is the fact that the molecular orientation on the gold and in its immediate vicinity may be different than in the bulk, which may promote the formation of different surface dipoles at metal surfaces and considerably change the energetics of carrier injection [52].

Simultaneously with these intriguing effects present at organic/metal interfaces, in the transistor channel the presence of the grain boundaries can seriously affect transport, as described in Section 10.2.2. The deviations from ideality present in thin-film transistors are also in agreement with the voltage drop measured along the grain boundaries using SKPM [43]. The different microstructure of the films usually results in a large variation in mobility with FET channel length. This effect was described in F-TES ADT, where mobility was tuned from 10^{-2} to $0.4\,\text{cm}^2\ (\text{Vs})^{-1}$ by controlling the crystalline grain size using chemical modifications of the contacts and gate dielectric [24, 25]. Here, the very different electrical properties were mirrored by changes detected in the drain current noise, and devices with smaller grains in the transistor channel showed not only lower mobility but also much larger noise values. The figure of merit used to quantitatively describe the noise measurements was the Hooge parameter. We describe in more detail later how 1/f flicker noise can be used to detect subtle changes in organic semiconductor thin-film microstructure.

Given the important role that molecular order plays in determining the performance of devices and its strong dependence on the fabrication conditions, careful attention must be given to details of the processing. For example, substrate temperature, solvent, deposition method, or surface chemical modifications are factors that strongly dictate the film microstructure and properties. Achieving large-area uniform microstructure and molecular ordering in organic semiconductors deposited in solution is challenging, but critical, as device performance is often strongly related to crystalline order. Therefore, processes which enhance the crystal structure of solution-based organic films during or after the film deposition are of particular interest.

10.3.2 Controlling Film Morphology by Surface Chemical Modifications

Self-assembled monolayers (SAMs) have been widely used to improve both the contact injection and organic/dielectric interface in organic thin-film transistors [53–56]. More recently, it was demonstrated that chemical modifications at the interfaces can also induce changes in microstructure of the organic films [24]. F-TES ADT showed an interesting differential microstructure on pentafluorobenzenethiol (PFBT) treated Au source/drain electrodes and on non-treated Au or gate dielectric area. The difference of mobility in the two microstructures is more than 2 orders of magnitude. Figure 10.2 shows optical micrographs of spin-cast F-TES ADT thin films from chlorobenzene solutions on untreated Au (a), and on PFBT-treated Au (b). It can be observed from this figure that the film morphology on untreated Au shows small grains, and the film on PFBT-treated Au electrode shows large crystallized structures that extend from the contact edge into SiO_2 area. F-TES ADT films deposited on a PFBT-treated Au surfaces present large grains, molecular terraces, and strong X-ray diffraction (XRD) peaks. The XRD results show good molecular ordering, with a vertical intermolecular spacing of 16.3 Å, similar to the d-spacing characteristic in single crystals [57]. On the contrary, the film on bare Au presents a distinct structure and shows the signature of different molecular orientations, with both edge-on and face-on being present. It was suggested that the differential microstructure on PFBT and non-PFBT-treated areas may be the result of interfacial effects such as surface energy, chemical effect (electron withdrawing) of self-assembled monolayer (PFBT layer) on F-TES ADT molecule, and molecular wiring between fluorine molecules in PFBT and F-TES ADT. The large platelets forming on PFBT-treated electrodes are a result of the presence of a cofacial molecular orientation within the grain (001), which is also the fast growth direction in the plane of the film. On untreated electrodes and the dielectric, the film is formed by (111) and (001) platelets, for which the fast growth axis is very different, competing along the growth front of the film, and as a result smaller grains are observed. For the PFBT-treated contacts, the nucleated crystals grow considerably, forming large grains, and for transistors with sufficiently narrow channels, these grains bridge the channel and provide a high-mobility path for the charge carriers. On the contrary, for untreated contacts and for the dielectric surface, the film presents small grains consisting of (111) and (001) platelets. The relative amount of the two molecular orientations, and thus the dynamics of the grain growth, can be further tuned with oxide dielectric treatment.

The differences in molecular ordering and orientation induce significant differences in electrical characteristics in the two regions. The $\pi-\pi$ stacking on the PFBT Au is beneficial for charge transport and leads to high mobilities (as high as 0.9 cm^2 $(Vs)^{-1}$, as can be observed in the graph presented in Figure 10.2, in blue). On the untreated contacts, the presence of the mixed molecular orientations result in 2 orders of magnitude lower mobility measured for the same material ($\mu = 0.01$ cm^2 $(Vs)^{-1}$ for untreated contacts; see the graph in Figure 10.2, in red). The electrical properties and 1/f low frequency noise of three different types of

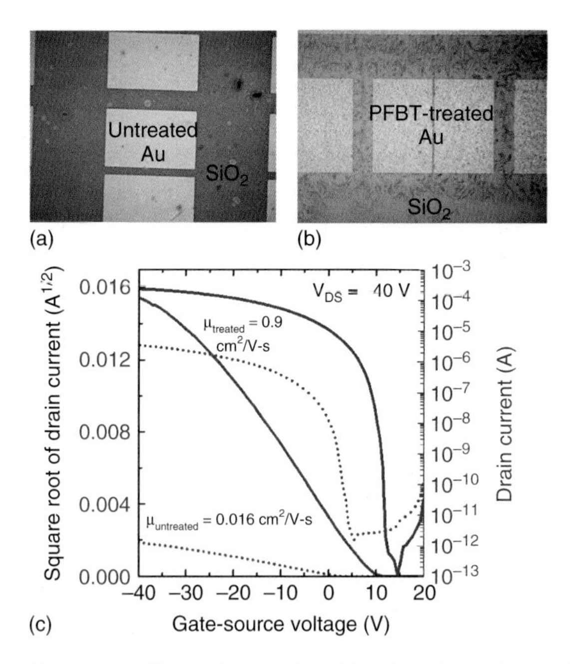

Figure 10.2 Electrical properties of F-TES ADT transistors with untreated (red) and PFBT-treated (blue) contacts. The optical micrographs show the different film morphology.

films of F-TES ADT with various microstructures can be observed in Figure 10.3 [25]. Here, black symbols correspond to films with untreated contacts and dielectric, containing very fine grains (optical micrograph presented in Figure 10.2). In red we represent the properties measured in films with PFBT-treated contacts, which exhibit large grains (approximately 15–25 μm) growing off the treated contacts into the channel (optical micrograph presented in Figure 10.2). Here, devices with channel lengths smaller than the grain size present high mobilities. When the channel length becomes larger than the grain diameter, smaller grains consisting of (111) and (001) platelets form in the middle of the channel, and the mobility decreases abruptly, which can be depicted as a change in slope in the mobility versus channel length plot (upper panel, Figure 10.3, in red). The blue symbols in Figure 10.3 correspond to films with even larger grain sizes (approximately 30–50 μm), obtained by using PFBT-treated contacts and hexamethyldisilazane (HMDS)-treated oxide dielectric. For this type of film, because of the larger grain size, the transition from purely (001) large grains to the mixture of (001) and (111) fine grains occurs at channel lengths $L = 40–50$ μm. In the lower panel of Figure 10.3, the Hooge parameter (α) for the same devices is plotted as a function of channel length. In parallel with mobility this parameter, obtained from 1/f noise measurements, is capable of indicating OTFT device properties and is very sensitive to microstructure changes in the film. The Hooge parameter α is related to the drain current noise spectral density (S_I/I^2), the total number of free charge

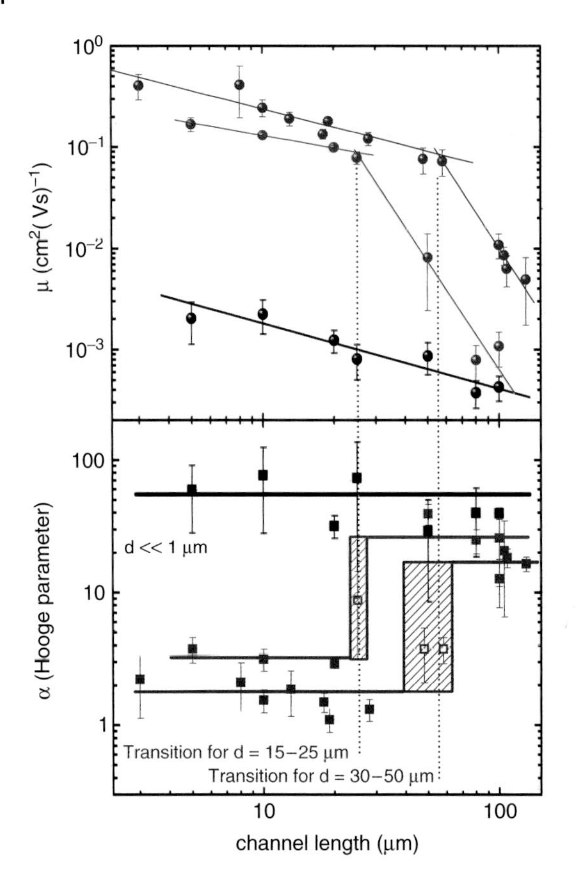

Figure 10.3 Upper panel: mobility versus channel length for large (blue), medium (red) and small (black) grains. Lower panel: Hooge parameter for the same devices.

carriers (N), and the frequency (f) via Hooge's empirical expression:

$$\frac{S_I}{I^2} = \frac{\alpha}{Nf} \tag{10.4}$$

The important information delivered by Figure 10.3 is that devices consisting of small grain films not only present a low mobility but also exhibit a very large Hooge parameter: 55.25 (±8.06), shown in black in the lower panel. As the grain size increases, the noise is considerably reduced, and α becomes 3.36 (±0.54) for PFBT-treated contact devices (red), at short channel lengths. The Hooge parameter is constant for the homogenous channels present for channel lengths 5 μm $\leq L \leq$ 25 μm, and consisting of 001-oriented molecules forming large grains. At larger channel lengths, $L \geq$ 25 μm, when the fine grains form in the middle of the channel, α increases abruptly, approaching the value measured in untreated contact devices, as the low mobility fine grain film regions dominate the transport. For films with the largest grain size (blue), the scenario is similar, but as expected, the Hooge parameter is even lower, 1.64 (±0.39) and the transition takes place at larger channel length, $L \geq 40\,\mu m$. Both mobility and 1/f noise measurements demonstrate that the properties of organic devices are very sensitive to film microstructure, which results from interactions at the interface between organic semiconductors and contacts or dielectrics, as well as SAMs deposited at their surface.

10.3.3 Processing Parameters Affecting Electrical Properties

10.3.3.1 Deposition Method

The method used for deposition of the organic semiconductor is another important parameter in thin-film processing, which often has a strong effect on the growth habit and morphology of the film. Film growth is constrained not only by solution/substrate and solution/air interactions but also by intermolecular interactions within the organic semiconductor crystallites. Strong intermolecular forces encourage a 2D π-stacked crystal, and also a more pronounced 2-D morphology of the film. With weaker 2D intermolecular interactions, a 2-D film typically exhibits small grains, and sometimes even poor film continuity. Strong 1D intermolecular interactions usually induce long needle-like crystals, with the long axis of the needle being the π-stacking direction.

The fabrication method can alter grain size, shape, and orientation, as well as molecular packing within the grains. For example, in Figure 10.4 we present the very distinct film morphologies obtained in F-TES ADT films deposited by drop casting (a), spin coating (b), inkjet printing (c), and spray deposition (d). For drop casting (Figure 10.4a), several solution drops are placed on the substrate, and the solvent slowly evaporates, forming large crystallites. The evaporation rate can be tuned by the choice of solvent and by varying the substrate temperature. With spin coating, the substrate rotates with a very high angular speed (1000–4000 rpm), and the solvent evaporates very rapidly. As a result, the grain size is considerably smaller (Figure 10.4b). The inkjet printing process is similar to drop casting, with the only exception that the drop size is considerably smaller. Here, for a typical transistor several drops are printed in a row across the device channel with a drop spacing of 1–10 μm. These closely spaced drops remain liquid sufficiently long to coalesce and form a larger drop. The large drops will then dry with the formation of large crystals that extend across transistor channel areas (Figure 10.4c). When the

Figure 10.4 Thin films of F-TES ADT fabricated by using various deposition techniques: drop casting (a), spin coating (b), inkjet printing (c), and spray deposition (d).

organic semiconductor layer is fabricated by spray deposition, the semiconductor solution is aerosolized under high pressure of inert gas, and the small droplets are collected on the device structure, positioned at a distance below the airbrush [58]. This distance between the sample and the spray nozzle can be controlled to balance film properties. For very short distances, the solution is blown away by the incoming flow, and for large distances, the solvent evaporates before reaching the substrate, owing to large solution–air interface. For intermediate distances, a highly crystalline film is obtained on large areas (Figure 10.4d).

The different crystalline environment created by the various evaporation rates results in distinguishable electrical properties for each sample. For F-TES ADT the mobilities obtained by using the above-described methods are $\mu_{\text{drop cast}} = 1\text{–}2\ \text{cm}^2\ (\text{Vs})^{-1}$, $\mu_{\text{spin coat}} = 0.2\text{–}1.5\ \text{cm}^2\ (\text{Vs})^{-1}$, $\mu_{\text{inkjet}} = 0.1\text{–}0.35\ \text{cm}^2\ (\text{Vs})^{-1}$, and $\mu_{\text{spray}} = 0.1\text{–}0.2\ \text{cm}^2\ (\text{Vs})^{-1}$. Drop casting allows the formation of highly crystalline domains and high-performance devices, but the films are often inhomogeneous in crystalline order and orientation, making them unsuitable for applications where large-area uniformity is important. The mobility measured in spin-coated devices is quite high. However, even if this method is extensively used for rapid screening of novel materials, its incompatibility with larger areas represents a serious drawback. On the other hand, inkjet printing is of increasing interest although it promotes a relatively low mobility because it allows non-contact patterning and digital pattern control, potentially useful for fabrication of low-cost products customized and produced in small quantities. Spray deposition is particularly attractive as it can sustain fast coating of large areas, with the use of very small quantities of organic semiconductors.

10.3.3.2 **Solvent**

It was experimentally observed that even when using the same fabrication method, deposition of crystalline organic semiconductors from solvents with different vapor pressures results in significant differences in measured field-effect mobilities. Solvent evaporation rate can drive the crystal formation environment, leading to different mesoscale morphology, which will be reflected in electronic properties [27]. Low vapor pressure solvents promote slower evaporation rate and thus longer crystal formation time, allowing for molecular arrangement prior to transition to solid phase. This usually results in larger grains and a higher degree of molecular order, beneficial for charge-carrier mobility. On the contrary, for films deposited from high vapor pressure solvents, faster solvent evaporation considerably reduces the ability of the semiconductor molecules to organize. These films usually show substantially smaller grains as a result of rapid film formation, inhibiting large crystal formation. Figure 10.5 presents the electrical properties of F-TES ADT TFTs, with the organic film being spun-coated from two different solvents (toluene and chlorobenzene). The use of higher vapor pressure solvents such as toluene (vapor pressure of toluene is $\sim$24 mmHg at 22 °C), results in the formation of substantially smaller grains compared with lower vapor pressure solvents such as chlorobenzene (vapor pressure of chlorobenzene is $\sim$10 mmHg at 22 °C). The AFM images obtained on the films spun-coated from toluene and chlorobenzene

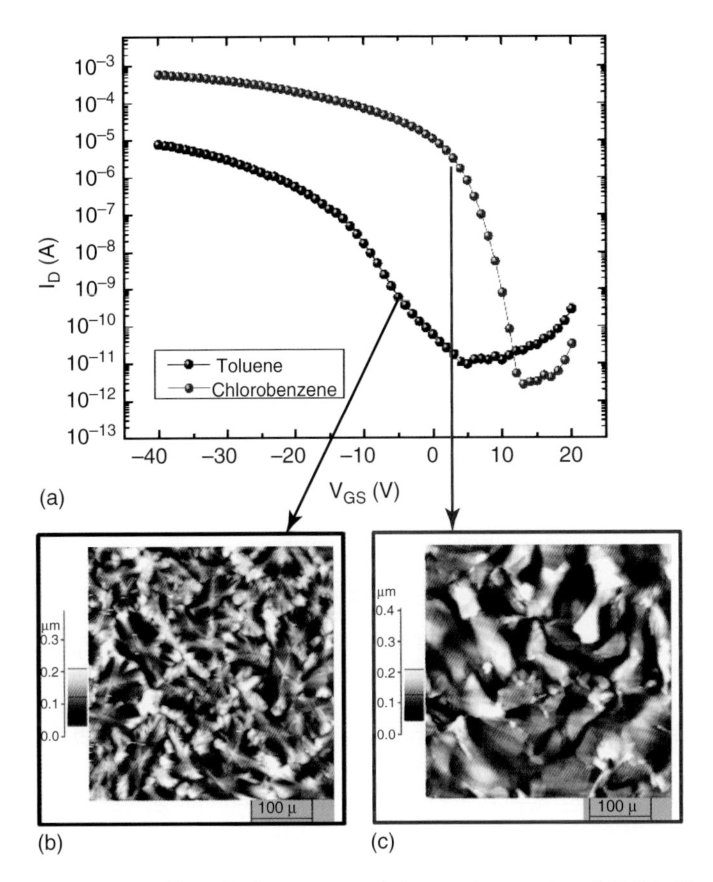

Figure 10.5 Effect of solvent on morphology and properties of OTFTs. The electrical properties and AFM images for the case of toluene (black) and chlorobenzene (black) are shown.

are presented in Figure 10.5. The small grain size and broader distribution of the orientation of the crystalline domains correlate well with a lower mobility measured in toluene, $\mu_{\text{toluene}} = 0.1\ \text{cm}^2\ (\text{Vs})^{-1}$, compared with $\mu_{\text{chlorobenzene}} = 0.4\ \text{cm}^2\ (\text{Vs})^{-1}$ obtained in films spun-coated from a chlorobenzene solution. The transistor transfer characteristics measured in the saturation regime ($V_{DS} = -40$ V) and plotted in Figure 10.5a demonstrate significant differences in device operation for the films spun-cast from toluene (black) and chlorobenzene (red). The smaller grains present in films deposited from toluene produce a larger number of grain boundaries, which act as scattering and trapping sites; thus the sub-threshold region is not as abrupt as in chlorobenzene devices. This figure also shows that the threshold voltage is affected by the grain size and possibly by the polarity of the solvent, with OTFTs deposited from chlorobenzene consistently showing positive shifts.

Table 10.1 summarizes the different mobilities measured in F-TES ADT TFTs when spin coating the organic semiconductor under the same conditions, from

Table 10.1 F-TES ADT mobility for thin films spun coated from various solvents. The boiling temperature and vapor pressures at 22 °C are also indicated.

Solvent	Boiling point (°C)	Vapor pressure at 22 °C (mmHg)	F-TES ADT field-effect mobility (cm^2 $(Vs)^{-1}$)
Chlorobenzene	132	10	0.7
Toluene	111	24	0.24
Mesitylene	165	2	0.65
Dichlorobenzene	180	1.5	0.012
Fluorobenzene	85	50	0.043
p-Xylene	138	7	0.32
Ethylbenzene	136	8	0.25

various solvents. It can be observed that the mobility can be varied by 2 orders of magnitude in part at least by controlling the drying dynamics.

A low vapor pressure solvent induces slow evaporation rate and can usually create large crystals and higher mobility (see Table 10.1). Unfortunately, the vapor pressure of the solvent is not sufficient to describe the different values of mobility measured. The film formation process is more complex and other factors such as surface energy of the substrate and wettability of the solvent play a crucial role. For example, in Table 10.1 the F-TES ADT film deposited from dichlorobenzene exhibits the lowest mobility ($\mu = 0.012$ cm^2 $(Vs)^{-1}$), in spite of the low vapor pressure of the solvent used. This effect is more likely due to the increased heterogeneity of the film, with poor connectivity between the crystalline domains, which inhibits coherent charge transport. Only small and non-uniform crystal domains are formed from dichlorobenzene solutions, and each drop shows a ring-like shape, which is referred to as *coffee stain* effect. Mixed-solvent solution is sometimes used to control the evaporation and obtain a uniform morphology, with large crystallites. Generally, there are two types of evaporation-induced flows for a droplet: convective flow toward the contact line and Marangoni flow induced by the surface tension gradient. Coffee stain behavior can be avoided by balancing Marangoni and convective flows.

10.4 Advanced Film Processing

10.4.1 How Sensibility to Processing Details Can Be Advantageous

We have shown in Sections 10.2 and 10.3 that processing conditions and interactions with different surfaces can cause large variations in results, making it challenging to characterize the intrinsic properties of organic semiconductors. In this section, we describe how exploring the dependence of film morphology on processing parameters can offer an opportunity to improve the electrical properties of devices, and to achieve high-mobility OTFTs by manipulating interactions at interfaces and in the bulk of the crystals. Organic semiconductors are weakly bonded systems, and thus are very sensitive to a variety of physical and/or chemical stimuli. We have described in previous sections the effects of processing details such as substrate chemistry and energy, as well as deposition method and solvent. These are important factors that may induce very different properties for nominally identical materials. Post-processing can also change dramatically, sometimes irreversibly, the properties of organic films. Temperature, pressure, presence of gas molecules, as well as solvent molecules can alter the film properties in various ways. Pentacene, for example, exhibits structural phase transitions driven by temperature or exposure to solvents [59]. The 15.4 Å polymorph can be converted to the 14.4 Å by exposure to solvents such as acetone, isopropyl alcohol, or ethanol. The presence of polymorphs is often encountered in molecular crystals and often different polymorphs exhibit very different electrical properties, and can contribute to poor device reproducibility [57]. Nevertheless, this property is not always detrimental or a source of instability in organic devices, but has been extensively used as a method for improving the quality of a large number of organic films.

10.4.2 Solvent Annealing

A variety of block copolymers, bulk heterojunction organic photovoltaics, and small-molecule organic TFTs have been improved by exposure of the film to solvent vapors, a process called *solvent annealing*. For example, the small-molecule organic semiconductor TES ADT was shown to form an amorphous film when spin-coated, as the solvent evaporation occurs faster than the crystallization [26]. OTFTs fabricated by using this procedure exhibited a poor field-effect mobility of $\mu = 0.002\ \text{cm}^2\ (\text{Vs})^{-1}$, and severe current hysteresis. Subsequent exposure of the films to a solvent (solvent annealing) substantially enhances crystallinity of the TES ADT film, improving the electrical response to $\mu = 0.11\ \text{cm}^2\ (\text{Vs})^{-1}$. Consistent shifts in the threshold voltages upon solvent exposure suggested that solvent molecules deeply penetrated the organic semiconductor film, inducing molecular rearrangements in the entire depth of the film, and reaching the interface with the dielectric, the surface measured in transistors. No correlations were observed between the effectiveness of the solvent annealing process and the polarity, boiling point, or vapor pressure of the solvent to which the organic film was exposed.

The structural rearrangement of the film upon solvent annealing is not encountered in all small-molecule organic semiconductors, and the effectiveness of this approach is directly related to the strength of the intermolecular interactions. For example, the F-TES ADT film is already highly ordered after spin casting (see Section 10.3), as a result of crystallization via F–F and F–S interactions, and post-growth solvent annealing has little to no effect on the film morphology, inducing only a modest improvement in device performance compared to normal spin-cast films. The solvent annealing method is an easy approach to substantially change the crystallinity in a variety of organic semiconductors, and can be easily implemented in solution-based deposition systems, but is most effective for very weakly bonded organic crystals.

10.4.3
Deposition under Solvent Vapors

In the following section, we describe a different type of interaction between organic semiconductor molecules and solvent vapors: solvent-vapor enhanced spin casting. The concept derives from the method presented in Section 10.4.2, with the exception that here the solvent vapor/organic semiconductor interactions occur *during* film deposition, as opposed to being encountered in a post-fabrication step. This modification of the processing conditions used for film deposition results in substantially larger crystalline grains in small-molecule organic semiconductors, and improved charge transport, compared to simply spin coating the active film. The solvent spin-casting process was designed to reproducibly enhance the grain growth in the organic semiconductor film. An inert gas is used to carry solvent vapor to the spinning sample enhancing the size of the crystalline domains and improving the wetting of the solution on the substrate. This process allows the spinning sample to be exposed to vapors of the solvent(s), with the vapor pressure controlled by the temperature of the solvent(s).

A schematic of the traditional spin-casting and solvent-enhanced spin-casting setups are shown in Figure 10.6a. While spin casting is performed in open atmosphere, solvent-enhanced spin casting takes place in a controlled solvent-vapor atmosphere, produced by bubbling N_2 through solvent. Figure 10.6b presents optical micrographs of TES ADT thin-film transistors fabricated by spin casting (left) and solvent spin cast by bubbling ~250 sccm of N_2 through toluene held at 55 °C (vapor pressure ~114 torr) (right). Note that this is the same material for which we have presented earlier the effect of post-deposition solvent annealing. Differential interference optical micrographs show uniform weakly ordered films with very small crystallites in the normal spin-cast devices, and large areas of ordered domains in the solvent spin-cast film. The solvent spin-cast technique thus results in a pronounced increase in film order.

A substantial change in device properties accompanies this effect. It can be observed in Figure 10.6c that the weakly ordered film obtained in normal spin-cast films exhibit very poor field-effect mobility ($\sim 10^{-4}$ cm^2 (Vs)$^{-1}$). The change in morphology obtained in solvent spin-cast films promotes a substantial increase in mobility to 0.25–0.3 cm^2 (Vs)$^{-1}$. This improvement in electrical characteristics is accompanied by a positive shift in the threshold voltage from ~2 to ~25 V. This may be the result of a change in the interface state density during the rapid crystallization and is possibly related to the polarity of the solvent as suggested by Dickey *et al.* [26]. In summary, TES ADT is very poorly ordered in normal spin-cast films, and the solvent-vapor enhanced spin casting improves its transistor performance considerably.

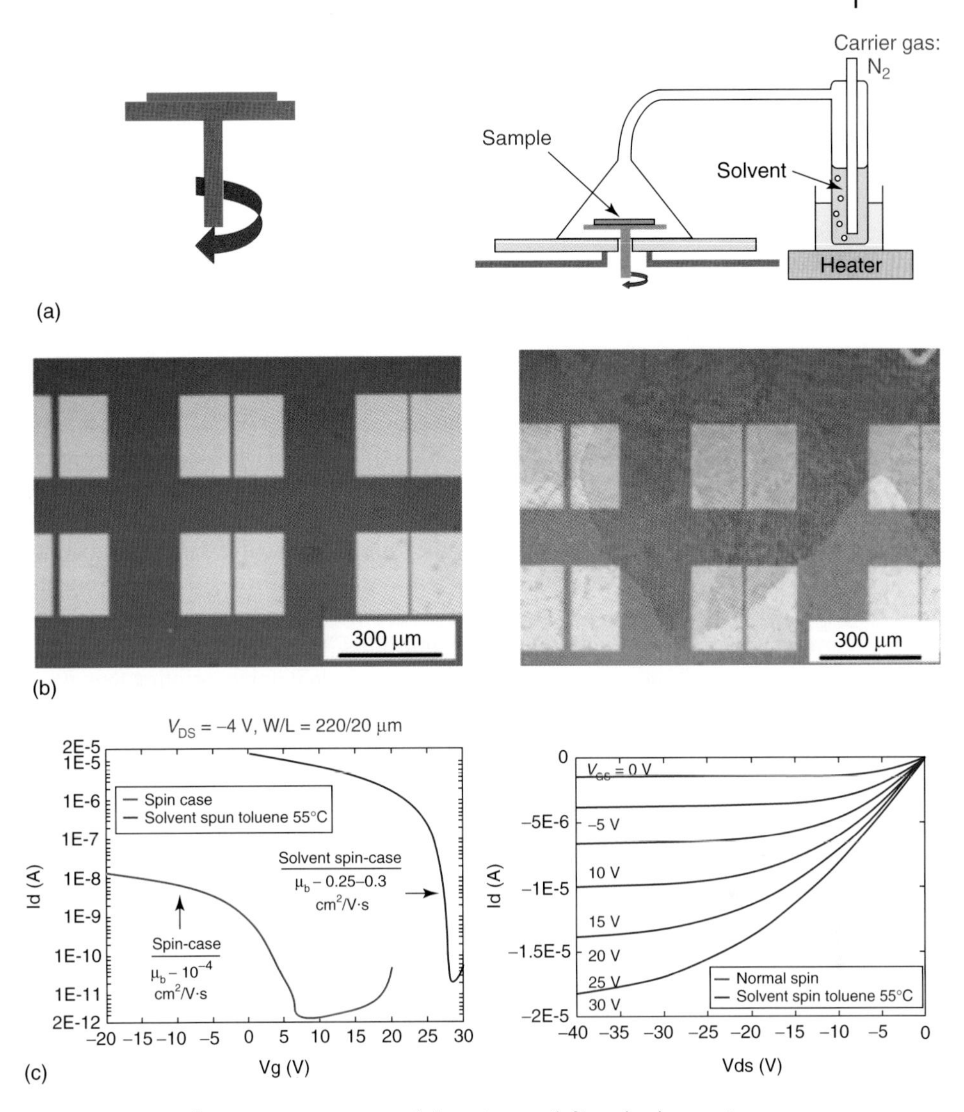

Figure 10.6 (a) Schematic representation of the spin-cast (left) and solvent spin-cast (right) deposition. (b) Optical micrographs of TES ADT devices fabricated by spin cast (left) and solvent spin cast (right). (c) Transfer (left) and transport (right) characteristics of spin-cast devices in red and solvent spin-cast devices in blue.

By further combining solvent spin casting with SAM surface treatments, a 3 orders of magnitude improvement in field-effect mobility of TES ADT was achieved ($\mu = 0.4\ \text{cm}^2\ (\text{Vs})^{-1}$). In comparison, when solvent spin casting was used to deposit F-TES ADT, the material that is highly ordered in normal spin-cast films on PFBT-treated electrodes, only minor differences in the film formation of these highly crystalline films were observed, primarily consisting in an increase in grain size. While the normal spin-cast film has a grain size of ~10–20 μm, for solvent spin-cast films the grains are typically ~30–50 μm. The solvent spin-casting process also substantially improved the wetting of the films: the normal spin-cast films treated were ~75% wet while the solvent spin-cast films were nearly 100% wet.

10.4.4
Patterning Organic Thin-Film Transistors

One critical step in OTFT fabrication is *patterning* the organic film to minimize the parasitic current paths, reduce cross talk between adjacent devices in integrated circuits, decrease the power consumption, and improve device performance. The ability to pattern organic thin films, the active layers in OTFTs, is of paramount importance for manufacturing organic electronic devices [60–62]. However, there are still no efficient patterning technologies for solution-deposited organic films. Photolithography techniques developed for inorganic semiconductors are not suited for organic materials [63, 64], as they yield considerable damage and device performance degradation. This limitation severely restricts the use of organic semiconductors in advanced device architectures. Several approaches were taken toward non-destructive patterning of organic films. They include direct evaporation of organic semiconductors through shadow masks [65], inkjet printing [66–70] or screen printing [71], lithographic stamping [72, 73], selective removal of organic layer by UV exposure [74], photochemical cross-linking [75], cold welding [73], microcutting [76], embossing [77], or micromolding [78]. Unfortunately, many of these methods consist of elaborate processing steps, which can be a barrier for low-cost mass production of organic electronics. Additionally, they present limited resolution and fail when scaled to large-area processing, allowing only restricted applicability. In this section, we describe unconventional, high-yield patterning techniques for OTFTs, which exploit unique film-forming properties giving rise to formation of differential microstructure on the contacts and dielectric, or SAMs deposited on their surface. The difference in mobility for the two microstructures can be more than 3 orders of magnitude. Given this significant difference, this method may be a viable approach to thin-film patterning technology for low-cost and simple organic electronics fabrication.

We illustrate the concept of patterning via differential microstructure using F-TES ADT organic semiconductor. We have presented in Sections 10.2 and 10.3 how the electrical properties of this material are highly sensitive to processing conditions. We recall here the results obtained in spin-coated devices. Spin-cast F-TES ADT field-effect transistors from 2 to 3 wt% toluene or chlorobenzene solutions on clean Au electrodes showed moderately low mobility between 10^{-3} and 10^{-2} $cm^2\ (Vs)^{-1}$. For comparison, the OTFTs fabricated from the same solutions of F-TES ADT spin cast on PFBT-treated Au electrodes typically have mobilities of 0.2–1.5 $cm^2\ (Vs)^{-1}$. The difference measured in mobilities of the same material deposited on different surfaces is related to the effect of the SAM treatment, which was described in greater detail in Section 10.3.2. A highly ordered film forms in spin-cast films on PFBT-treated Au electrodes, with laterally grown crystals extending onto the SiO_2 dielectric. Here, the molecules are preferentially oriented along the π-stacking direction, which is the high-mobility direction. On the contrary, small grains with mixed molecular orientations are obtained without treatment, yielding low mobility regions. The PFBT-treated devices thus present very different microstructure on PFBT-treated Au electrodes and on non-treated SiO_2 regions. For sufficiently narrow channels ($L < 30\,\mu m$) the crystallites growing off the treated Au contacts bridge the source and drain electrodes, giving high-mobility regions in the transistor channel, while high-resistivity regions surround the devices, inhibiting the formation of parasitic current paths, and thus in a sense automatically patterning the thin-film transistor (Figure 10.7c). As the formation of the differential microstructure responsible for patterning occurs during film formation, without additional processing steps involved, this patterning is referred to as *self-patterning* [24].

This intriguing behavior can allow the development of low-complexity and low-cost methods of patterning organic semiconductors. Practical control of this property and its implementation in complicated architectures require a deep understanding of the mechanism that drives its development. It is not fully described, but it is believed that the unique differential microstructure on PFBT- and non-PFBT-treated areas may be the result of interfacial effects such as surface energy, and chemical interactions between the sulfur and fluorine atoms present in the chemical structure of the organic semiconductor molecule, and the fluorine present in the SAM structure. In view of these hypotheses, formation of the high-ordered regions occurs then during thin-film deposition, via a chemical interaction between the halogen atom of the SAM used for the contact treatment (PFBT) and the halogen and/or sulfur atom in the organic semiconductor (see Figure 10.7a). Around the devices, PFBT is not present, thus no F–F and/or F–S interactions develop (Figure 10.7b), and for this reason thin-film mobility in that region is very low, consistent with the results presented in Section 10.2 on untreated contact devices and long channel lengths of treated devices (Figure 10.3). These results demonstrate that efficient organic thin-film patterning can be achieved at low cost, by connecting molecular design with device processing.

This high efficiency patterning technique was used to fabricate seven-stage ring-oscillator circuits. The reduced complexity of this procedure allowed demonstrations of ring-oscillator circuits not only on SiO_2 wafers but also on flexible substrates (Figure 10.8a).

The contact treatment induced microstructures, with large crystallites on the PFBT-treated electrodes and small grains on polyimide regions, as can be seen in Figures 10.8b,d. Figure 10.8d shows the OTFT pair (driver and load) forming the inverter stage of the ring oscillator. The poor mobility regions resulting in the field regions provide electrical isolation between adjacent inverters, and eliminates the need for further processing steps to etch the organic semiconductor film. For the high-resistivity regions surrounding the devices, we estimate that the sheet resistance is higher than $10^9\ \Omega$ per square. This level of insulation between adjacent devices is sufficient for a large number of circuit applications and electronic

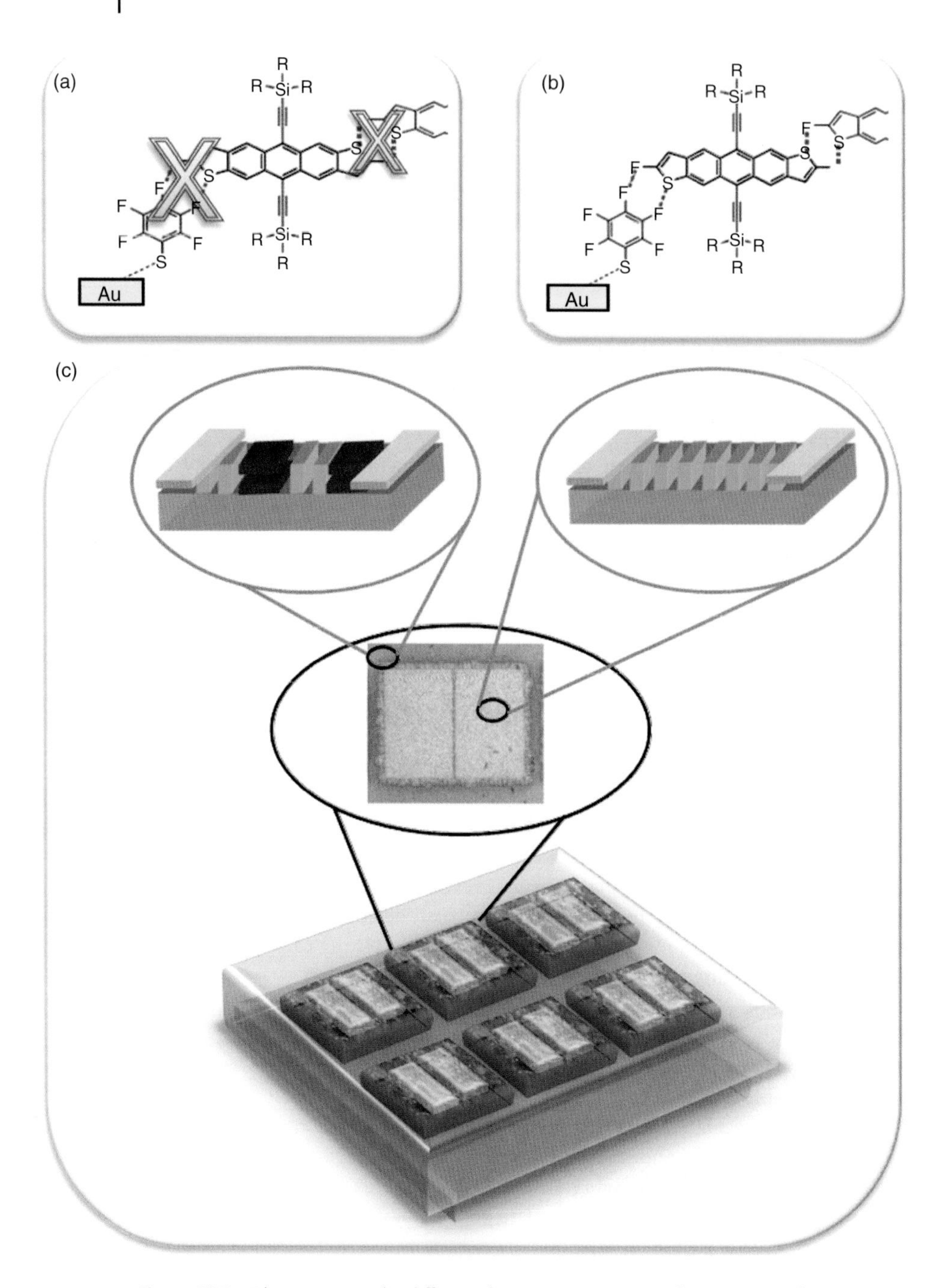

Figure 10.7 Film patterning by differential microstructure: mixed orientations form around the contacts yielding the high-resistivity regions, and highly ordered film areas form on the SAM-treated contact giving high-conductivity regions.

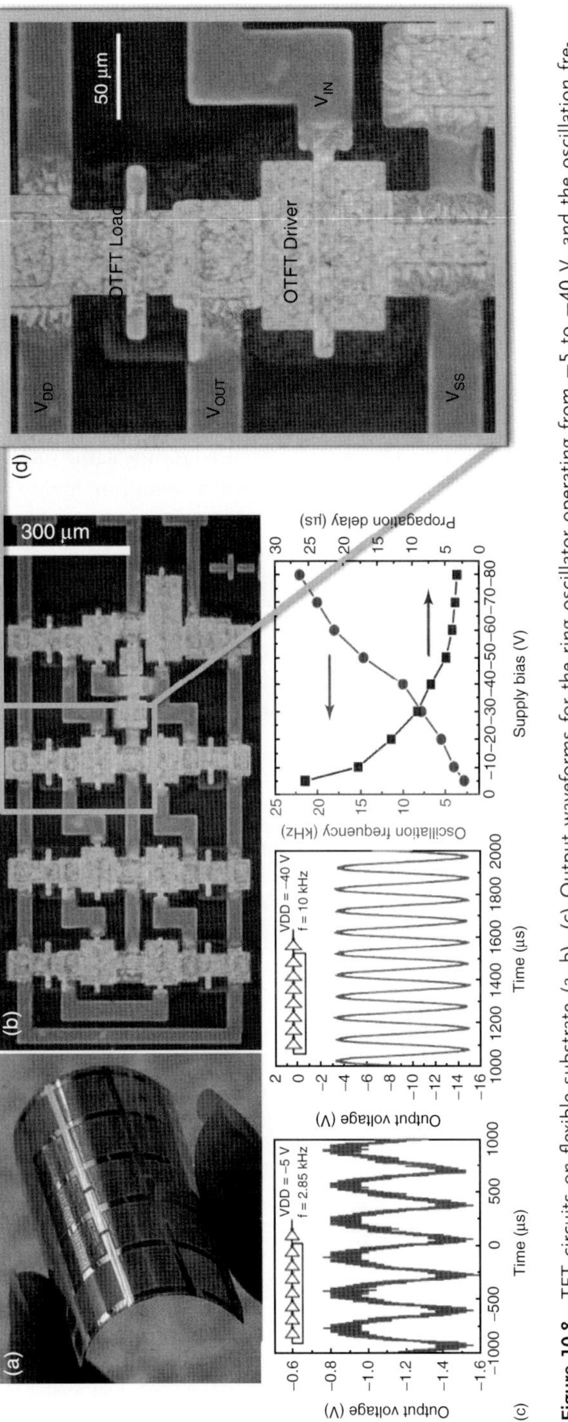

Figure 10.8 TFT circuits on flexible substrate (a, b). (c) Output waveforms for the ring oscillator operating from −5 to −40 V, and the oscillation frequency and per stage propagation delay as a function of supply voltage. (d) Driver and load forming the inverter stage.

paper. The properties of the above-mentioned devices are clearly superior to the unpatterned circuits reported in the literature (see Figure 10.8c plotting the output waveforms for the ring oscillator and the oscillation frequency and the single-stage propagation delay as a function of the supply voltage). This self-patterning approach has also been applied to hybrid organic–inorganic F-TES ADT circuits [79].

10.5 Summary

The driving force for the organic electronics field is the potential for low-cost and high-diversity applications fabricated by solution-based techniques. Although impressive achievements have been demonstrated, we have described in this chapter how processing details affect film formation and structure in OTFTs, giving rise to a wide range of electrical properties measured for the same material. Large, highly ordered crystalline grains are usually critical to achieving high-performance devices, but several other factors contribute to device performance such as contacts, grain boundaries, traps, and the molecular packing of the organic semiconductor. Various routes can be pursued for the optimization of these factors. The most common approaches consist of modifications in surface energies and chemistries by using SAMs, the choice of deposition method or solvent, or introduction of post-processing steps. In many cases these effects are interrelated and can make control and optimization challenging for solution-processed OTFTs.

References

1. Klauk, H. (2010) Organic thin-film transistors. *Chem. Soc. Rev.*, **39** (7), 2643–2666.
2. Alves, H., Molinari, A.S., Xie, H.X., and Morpurgo, A.F. (2008) Metallic conduction at organic charge-transfer interfaces. *Nat. Mater.*, **7** (7), 574–580.
3. Gershenson, M.E., Podzorov, V., and Morpurgo, A.F. (2006) Colloquium: electronic transport in single-crystal organic transistors. *Rev. Mod. Phys.*, **78** (3), 973–989.
4. Anthopoulos, T.D., Setayesh, S., Smits, E., Colle, M., Cantatore, E., de Boer, B., Blom, P.W.M., and de Leeuw, D.M. (2006) Air-stable complementary-like circuits based on organic ambipolar transistors. *Adv. Mater.*, **18** (14), 1900–1904.
5. Gelinck, G.H., Huitema, H.E.A., Van Veenendaal, E., Cantatore, E., Schrijnemakers, L., Van der Putten, J., Geuns, T.C.T., Beenhakkers, M., Giesbers, J.B., Huisman, B.H., Meijer, E.J., Benito, E.M., Touwslager, F.J., Marsman, A.W., Van Rens, B.J.E., and De Leeuw, D.M. (2004) Flexible active-matrix displays and shift registers based on solution-processed organic transistors. *Nat. Mater.*, **3** (2), 106–110.
6. Huitema, H.E.A., Gelinck, G.H., van der Putten, J., Kuijk, K.E., Hart, C.M., Cantatore, E., Herwig, P.T., van Breemen, A., and de Leeuw, D.M. (2001) Plastic transistors in active-matrix displays – The handling of grey levels by these large displays paves the way for electronic paper. *Nature*, **414** (6864), 599–599.
7. Meijer, E.J., De Leeuw, D.M., Setayesh, S., Van Veenendaal, E., Huisman, B.H., Blom, P.W.M., Hummelen, J.C., Scherf, U., and Klapwijk, T.M. (2003) Solution-processed ambipolar organic

field-effect transistors and inverters. *Nat. Mater.*, **2** (10), 678–682.

8. Smits, E.C.P., Mathijssen, S.G.J., van Hal, P.A., Setayesh, S., Geuns, T.C.T., Mutsaers, K., Cantatore, E., Wondergem, H.J., Werzer, O., Resel, R., Kemerink, M., Kirchmeyer, S., Muzafarov, A.M., Ponomarenko, S.A., de Boer, B., Blom, P.W.M., and de Leeuw, D.M. (2008) Bottom-up organic integrated circuits. *Nature*, **455** (7215), 956–959.
9. Arias, A.C., MacKenzie, J.D., McCulloch, I., Rivnay, J., and Salleo, A. (2010) Materials and applications for large area electronics: solution-based approaches. *Chem. Rev.*, **110** (1), 3–24.
10. Sundar, V.C., Zaumseil, J., Podzorov, V., Menard, E., Willett, R.L., Someya, T., Gershenson, M.E., and Rogers, J.A. (2004) Elastomeric transistor stamps: reversible probing of charge transport in organic crystals. *Science*, **303** (5664), 1644–1646.
11. Hulea, I.N., Fratini, S., Xie, H., Mulder, C.L., Iossad, N.N., Rastelli, G., Ciuchi, S., and Morpurgo, A.F. (2006) Tunable Frohlich polarons in organic single-crystal transistors. *Nat. Mater.*, **5** (12), 982–986.
12. Jurchescu, O.D., Popinciuc, M., van Wees, B.J., and Palstra, T.T.M. (2007) Interface-controlled, high-mobility organic transistors. *Adv. Mater.*, **19** (5), 688–692.
13. Podzorov, V., Menard, E., Borissov, A., Kiryukhin, V., Rogers, J.A., and Gershenson, M.E. (2004) Intrinsic charge transport on the surface of organic semiconductors. *Phys. Rev. Lett.*, **93** (8), 086602.
14. Podzorov, V., Menard, E., Rogers, J.A., and Gershenson, M.E. (2005) Hall effect in the accumulation layers on the surface of organic semiconductors. *Phys. Rev. Lett.*, **95** (22), 226601.
15. Takeya, J., Tsukagoshi, K., Aoyagi, Y., Takenobu, T., and Iwasa, Y. (2005) Hall effect of quasi-hole gas in organic single-crystal transistors. *Jpn. J. Appl. Phys. Part 2: Lett. Express Lett.*, **44** (46–49), L1393–L1396.
16. Jurchescu, O.D., Baas, J., and Palstra, T.T.M. (2004) Effect of impurities on the mobility of single crystal pentacene. *Appl. Phys. Lett.*, **84** (16), 3061–3063.
17. Park, S.K., Jackson, T.N., Anthony, J.E., and Mourey, D.A. (2007) High mobility solution processed 6,13-bis(triisopropyl-silylethynyl) pentacene organic thin film transistors. *Appl. Phys. Lett.*, **91** (6), 063514.
18. Payne, M.M., Parkin, S.R., Anthony, J.E., Kuo, C.C., and Jackson, T.N. (2005) Organic field-effect transistors from solution-deposited functionalized acenes with mobilities as high as 1 cm(2)/V-s. *J. Am. Chem. Soc.*, **127** (14), 4986–4987.
19. Sheraw, C.D., Jackson, T.N., Eaton, D.L., and Anthony, J.E. (2003) Functionalized pentacene active layer organic thin-film transistors. *Adv. Mater.*, **15** (23), 2009–2011.
20. Jurchescu, O.D., Subramanian, S., Kline, R.J., Hudson, S.D., Anthony, J.E., Jackson, T.N., and Gundlach, D.J. (2008) Organic single-crystal field-effect transistors of a soluble anthradithiophene. *Chem. Mater.*, **20** (21), 6733–6737.
21. Ostroverkhova, O., Cooke, D.G., Hegmann, F.A., Anthony, J.E., Podzorov, V., Gershenson, M.E., Jurchescu, O.D., and Palstra, T.T.M. (2006) Ultrafast carrier dynamics in pentacene, functionalized pentacene, tetracene, and rubrene single crystals. *Appl. Phys. Lett.*, **88** (16), 162101.
22. Park, S.K., Mourey, D.A., Subramanian, S., Anthony, J.E., and Jackson, T.N. (2008) High-mobility spin-cast organic thin film transistors. *Appl. Phys. Lett.*, **93** (4), 043301.
23. Subramanian, S., Park, S.K., Parkin, S.R., Podzorov, V., Jackson, T.N., and Anthony, J.E. (2008) Chromophore fluorination enhances crystallization and stability of soluble anthradithiophene semiconductors. *J. Am. Chem. Soc.*, **130** (9), 2706–2707.
24. Gundlach, D.J., Royer, J.E., Park, S.K., Subramanian, S., Jurchescu, O.D., Hamadani, B.H., Moad, A.J., Kline, R.J., Teague, L.C., Kirillov, O., Richter, C.A., Kushmerick, J.G., Richter, L.J., Parkin, S.R., Jackson, T.N., and Anthony, J.E. (2008) Contact-induced crystallinity for high-performance soluble acene-based

transistors and circuits. *Nat. Mater.*, **7** (3), 216–221.

25. Jurchescu, O.D., Hamadani, B.H., Xiong, H.D., Park, S.K., Subramanian, S., Zimmerman, N.M., Anthony, J.E., Jackson, T.N., and Gundlach, D.J. (2008) Correlation between microstructure, electronic properties and flicker noise in organic thin film transistors. *Appl. Phys. Lett.*, **92** (13), 132103.
26. Dickey, K.C., Anthony, J.E., and Loo, Y.L. (2006) Improving organic thin-film transistor performance through solvent-vapor annealing of solution-processable triethylsilylethynyl anthradithiophene. *Adv. Mater.*, **18** (13), 1721–1726.
27. Kim, C.S., Lee, S., Gomez, E.D., Anthony, J.E., and Loo, Y.L. (2008) Solvent-dependent electrical characteristics and stability of organic thin-film transistors with drop cast bis(triisopropylsilylethynyl) pentacene. *Appl. Phys. Lett.*, **93** (10), 103302.
28. Coropceanu, V., Cornil, J., da Silva, D.A., Olivier, Y., Silbey, R., and Bredas, J.L. (2007) Charge transport in organic semiconductors. *Chem. Rev.*, **107** (4), 926–952.
29. Troisi, A. and Orlandi, G. (2006) Charge-transport regime of crystalline organic semiconductors: diffusion limited by thermal off-diagonal electronic disorder. *Phys. Rev. Lett.*, **96** (8), 086601.
30. Hannewald, K. and Bobbert, P.A. (2004) Ab initio theory of charge-carrier conduction in ultrapure organic crystals. *Appl. Phys. Lett.*, **85** (9), 1535–1537.
31. Kang, J.H., da Silva, D., Bredas, J.L., and Zhu, X.Y. (2005) Shallow trap states in pentacene thin films from molecular sliding. *Appl. Phys. Lett.*, **86** (15), 152115.
32. Troisi, A. and Cheung, D.L. (2009) Transition from dynamic to static disorder in one-dimensional organic semiconductors. *J. Chem. Phys.*, **131** (1), 014703.
33. Kelley, T.W. and Frisbie, C.D. (2001) Gate voltage dependent resistance of a single organic semiconductor grain boundary. *J. Phys. Chem. B*, **105** (20), 4538–4540.
34. Xia, Y., Xie, W., Ruden, P.P., and Frisbie, C.D. (2010) Carrier localization on surfaces of organic semiconductors gated with electrolytes. *Phys. Rev. Lett.*, **105** (3), 036802.
35. Jurchescu, O.D., Baas, J., and Palstra, T.T.M. (2005) Electronic transport properties of pentacene single crystals upon exposure to air. *Appl. Phys. Lett.*, **87** (5), 052102.
36. Vollmer, A., Jurchescu, O.D., Arfaoui, I., Salzmann, I., Palstra, T.T.M., Rudolf, P., Niemax, J., Pflaum, J., Rabe, J.P., and Koch, N. (2005) The effect of oxygen exposure on pentacene electronic structure. *Eur. Phys. J. E*, **17** (3), 339–343.
37. Goldmann, C., Gundlach, D.J., and Batlogg, B. (2006) Evidence of water-related discrete trap state formation in pentacene single-crystal field-effect transistors. *Appl. Phys. Lett.*, **88** (6), 063501.
38. Krellner, C., Haas, S., Goldmann, C., Pernstich, K.P., Gundlach, D.J., and Batlogg, B. (2007) Density of bulk trap states in organic semiconductor crystals: discrete levels induced by oxygen in rubrene. *Phys. Rev. B*, **75** (24), 245115.
39. Pernstich, K.P., Oberhoff, D., Goldmann, C., and Batlogg, B. (2006) Modeling the water related trap state created in pentacene transistors. *Appl. Phys. Lett.*, **89** (21), 213509.
40. Conrad, B.R., Gomar-Nadal, E., Cullen, W.G., Pimpinelli, A., Einstein, T.L., and Williams, E.D. (2008) Effect of impurities on pentacene island nucleation. *Phys. Rev. B*, **77** (20), 205328.
41. Gomar-Nadal, E., Conrad, B.R., Cullen, W.G., and Willams, E.A. (2008) Effect of impurities on pentacene thin film growth for field-effect transistors. *J. Phys. Chem. C*, **112** (14), 5646–5650.
42. Pflaum, J., Niemax, J., and Tripathi, A.K. (2006) Chemical and structural effects on the electronic transport in organic single crystals. *Chem. Phys.*, **325** (1), 152–159.
43. Teague, L.C., Hamadani, B.H., Jurchescu, O.D., Subramanian, S., Anthony, J.E., Jackson, T.N., Richter, C.A., Gundlach, D.J., and Kushmerick, J.G. (2008) Surface potential imaging of solution processable acene-based thin film transistors. *Adv. Mater.*, **20** (23), 4513–4516.

44. Lee, S.S., Kim, C.S., Gomez, E.D., Purushothaman, B., Toney, M.F., Wang, C., Hexemer, A., Anthony, J.E., and Loo, Y.L. (2009) Controlling nucleation and crystallization in solution-processed organic semiconductors for thin-film transistors. *Adv. Mater.*, **21** (35), 3605–3609.
45. Di Carlo, A., Piacenza, F., Bolognesi, A., Stadlober, B., and Maresch, H. (2005) Influence of grain sizes on the mobility of organic thin-film transistors. *Appl. Phys. Lett.*, **86** (26), 263501.
46. de Boer, R.W.I., Gershenson, M.E., Morpurgo, A.F., and Podzorov, V. (2004) Organic single-crystal field-effect transistors. *Phys. Status Solidi A: Appl. Res.*, **201** (6), 1302–1331.
47. Briseno, A.L., Mannsfeld, S.C.B., Ling, M.M., Liu, S.H., Tseng, R.J., Reese, C., Roberts, M.E., Yang, Y., Wudl, F., and Bao, Z.N. (2006) Patterning organic single-crystal transistor arrays. *Nature*, **444** (7121), 913–917.
48. Gundlach, D.J., Zhou, L., Nichols, J.A., Jackson, T.N., Necliudov, P.V., and Shur, M.S. (2006) An experimental study of contact effects in organic thin film transistors. *J. Appl. Phys.*, **100** (2), 024509.
49. Reese, C. and Bao, Z. (2009) Detailed characterization of contact resistance, gate-bias-dependent field-effect mobility, and short-channel effects with microscale elastomeric single-crystal field-effect transistors. *Adv. Funct. Mater.*, **19** (5), 763–771.
50. Hamadani, B.H. and Natelson, D. (2005) Nonlinear charge injection in organic field-effect transistors. *J. Appl. Phys.*, **97** (6), 064508.
51. Teague, L.C., Jurchescu, O.D., Richter, C.A., Subramanian, S., Anthony, J.E., Jackson, T.N., Gundlach, D.J., and Kushmerick, J.G. (2010) Probing stress effects in single crystal organic transistors by scanning Kelvin probe microscopy. *Appl. Phys. Lett.*, **96** (20), 203305.
52. Duhm, S., Heimel, G., Salzmann, I., Glowatzki, H., Johnson, R.L., Vollmer, A., Rabe, J.P., and Koch, N. (2008) Orientation-dependent ionization energies and interface dipoles in ordered molecular assemblies. *Nat. Mater.*, **7** (4), 326–332.
53. Lin, Y.Y., Gundlach, D.J., Nelson, S.F., and Jackson, T.N. (1997) Stacked pentacene layer organic thin-film transistors with improved characteristics. *IEEE Electron Device Lett.*, **18** (12), 606–608.
54. Salleo, A., Chabinyc, M.L., Yang, M.S., and Street, R.A. (2002) Polymer thin-film transistors with chemically modified dielectric interfaces. *Appl. Phys. Lett.*, **81** (23), 4383–4385.
55. Gundlach, D.J., Jia, L.L., and Jackson, T.N. (2001) Pentacene TFT with improved linear region characteristics using chemically modified source and drain electrodes. *IEEE Electron Device Lett.*, **22** (12), 571–573.
56. de Boer, B., Hadipour, A., Mandoc, M.M., van Woudenbergh, T., and Blom, P.W.M. (2005) Tuning of metal work functions with self-assembled monolayers. *Adv. Mater.*, **17** (5), 621–625
57. Jurchescu, O.D., Mourey, D.A., Subramanian, S., Parkin, S.R., Vogel, B.M., Anthony, J.E., Jackson, T.N., and Gundlach, D.J. (2009) Effects of polymorphism on charge transport in organic semiconductors. *Phys. Rev. B*, **80** (8), 085201.
58. Azarova, N.A., Owen, J.W., McLellan, C.A., Grimminger, M.A., Chapman, E.K., Anthony J.E., and Jurchescu, O.D. (2010) Fabrication of organic thin-film transistors by spray-deposition for low-cost, large-area electronics *Org. Electron.*, **11** (12), 1960–1965.
59. Gundlach, D.J., Jackson, T.N., Schlom, D.G., and Nelson, S.F. (1999) Solvent-induced phase transition in thermally evaporated pentacene films. *Appl. Phys. Lett.*, **74** (22), 3302–3304.
60. Yoo, J.E., Lee, K.S., Garcia, A., Tarver, J., Gomez, E.D., Baldwin, K., Sun, Y.M., Meng, H., Nguyen, T.Q., and Loo, Y.L. (2010) Directly patternable, highly conducting polymers for broad applications in organic electronics. *Proc. Natl. Acad. Sci. U.S.A.*, **107** (13), 5712–5717.
61. Liu, S.H., Becerril, H.A., LeMieux, M.C., Wang, W.C.M., Oh, J.H., and Bao, Z.N. (2009) Direct patterning of organic-thin-film-transistor arrays via a

"dry-taping" approach. *Adv. Mater.*, **21** (12), 1266–1270.

62. Kim, S.J., Beveridge, H., Koberstein, J.T., and Kymissis, I. (2009) Isolation of organic field-effect transistors by surface patterning with an UV/ozone process. *J. Vac. Sci. Technol. B*, **27** (3), 1057–1059.

63. Kane, M.G., Campi, J., Hammond, M.S., Cuomo, F.P., Greening, B., Sheraw, C.D., Nichols, J.A., Gundlach, D.J., Huang, J.R., Kuo, C.C., Jia, L., Klauk, H., and Jackson, T.N. (2000) Analog and digital circuits using organic thin-film transistors on polyester substrates. *IEEE Electron Device Lett.*, **21** (11), 534–536.

64. Kuo, C.C. and Jackson, T.N. (2009) Direct lithographic top contacts for pentacene organic thin-film transistors. *Appl. Phys. Lett.*, **94** (5), 053304.

65. Chung, Y., Murmann, B., Selvarasah, S., Dokmeci, M.R., and Bao, Z.N. (2010) Low-voltage and short-channel pentacene field-effect transistors with top-contact geometry using parylene-C shadow masks. *Appl. Phys. Lett.*, **96** (13), 133306.

66. Sekitani, T., Nakajima, H., Maeda, H., Fukushima, T., Aida, T., Hata, K., and Someya, T. (2009) Stretchable active-matrix organic light-emitting diode display using printable elastic conductors. *Nat. Mater.*, **8** (6), 494–499.

67. Sekitani, T., Noguchi, Y., Zschieschang, U., Klauk, H., and Someya, T. (2008) Organic transistors manufactured using inkjet technology with subfemtoliter accuracy. *Proc. Natl. Acad. Sci. U.S.A.*, **105** (13), 4976–4980.

68. Cho, J.H., Lee, J., Xia, Y., Kim, B., He, Y.Y., Renn, M.J., Lodge, T.P., and Frisbie, C.D. (2008) Printable ion-gel gate dielectrics for low-voltage polymer thin-film transistors on plastic. *Nat. Mater.*, **7** (11), 900–906.

69. Xia, Y., Zhang, W., Ha, M.J., Cho, J.H., Renn, M.J., Kim, C.H., and Frisbie, C.D. (2010) Printed sub-2 V gel-electrolyte-gated polymer transistors and circuits. *Adv. Funct. Mater.*, **20** (4), 587–594.

70. Chang, S.C., Bharathan, J., Yang, Y., Helgeson, R., Wudl, F., Ramey, M.B., and Reynolds, J.R. (1998) Dual-color polymer light-emitting pixels processed by hybrid inkjet printing. *Appl. Phys. Lett.*, **73** (18), 2561–2563.

71. Bao, Z.N., Feng, Y., Dodabalapur, A., Raju, V.R., and Lovinger, A.J. (1997) High-performance plastic transistors fabricated by printing techniques. *Chem. Mater.*, **9** (6), 1299–1301.

72. Rogers, J.A., Bao, Z.N., Makhija, A., and Braun, P. (1999) Printing process suitable for reel-to-reel production of high-performance organic transistors and circuits. *Adv. Mater.*, **11** (9), 741–745.

73. Kim, C., Burrows, P.E., and Forrest, S.R. (2000) Micropatterning of organic electronic devices by cold-welding. *Science*, **288** (5467), 831–833.

74. Dickey, K.C., Subramanian, S., Anthony, J.E., Han, L.H., Chen, S., and Loo, Y.L. (2007) Large-area patterning of a solution-processable organic semiconductor to reduce parasitic leakage and off currents in thin-film transistors. *Appl. Phys. Lett.*, **90** (24), 244103.

75. Touwslager, F.J., Willard, N.P., and de Leeuw, D.M. (2002) I-Line lithography of poly-(3,4-ethylenedioxythiophene) electrodes and application in all-polymer integrated circuits. *Appl. Phys. Lett.*, **81** (24), 4556–4558.

76. Stutzmann, N., Tervoort, T.A., Bastiaansen, K., and Smith, P. (2000) Patterning of polymer-supported metal films by microcutting. *Nature*, **407** (6804), 613–616.

77. Stutzmann, N., Tervoort, T.A., Bastiaansen, C.W.M., Feldman, K., and Smith, P. (2000) Solid-state replication of relief structures in semicrystalline polymers. *Adv. Mater.*, **12** (8), 557–562.

78. Kim, E., Xia, Y.N., and Whitesides, G.M. (1995) Polymer microstructures formed by molding in capillaries. *Nature*, **376** (6541), 581–584.

79. Mourey, D.A., Park, S.K., Zhao, D.L.A., Sun, J., Li, Y.Y.V., Subramanian, S., Nelson, S.F., Levy, D.H., Anthony, J.E., and Jackson, T.N. (2009) Fast, simple ZnO/organic CMOS integrated circuits. *Org. Electron.*, **10** (8), 1632–1635.

Part III
Applications

Organic Electronics II: More Materials and Applications, First Edition. Edited by Hagen Klauk.

11
Light-Emitting Organic Transistors

Jana Zaumseil

11.1
Introduction

Over the past years, research in organic electronics has been aimed mainly at developing organic counterparts to existing basic inorganic electronic components such as light-emitting diodes (LEDs), field-effect transistors (FETs), and photovoltaic (PV) cells. The performance of organic semiconductors in these devices can now compete with that of some traditional inorganic materials, for example, amorphous silicon, while adding advantages such as low temperature or solution processing, flexibility, and low cost. Nevertheless, organic electronics are not likely to replace existing, mature technologies except for special applications where they offer a real advantage or new functionality.

One organic electronic device that exhibits a functionality not yet covered by traditional inorganic components is the light-emitting transistor, which integrates light emission and switching in a single device. This is particularly interesting for active matrix displays. The simplest design of a pixel in an active matrix organic light-emitting diode (AMOLED) display comprises an LED, a storage capacitor, and two FETs: one drive-FET and one switch-FET (Figure 11.1) [1, 2]. During a programming cycle, the switch-FET is turned on by the select line and a voltage according to the desired brightness level is stored by the capacitor. During the drive cycle, the switch-FET is off and the drive-FET turns on so that current can flow through the LED until the next programming cycle begins. In reality, a more complicated circuit design is often applied to avoid performance degradation due to threshold shifts of the transistors. This increases the amount of connections and space necessary for each pixel. A light-emitting transistor would replace the drive-FET and LED and thus save space, interconnects, and processing steps.

The idea of a light-emitting transistor was first proposed in the 1980s by Hack *et al.* after ambipolar charge transport was observed in silicon FETs [3, 4]. However, bulk silicon is not suitable for light emission because of its indirect bandgap and associated low radiative recombination rate. A solution to this problem is quantum confinement, that is, using ultrathin silicon [5] or silicon nanocrystals

Organic Electronics II: More Materials and Applications, First Edition. Edited by Hagen Klauk.

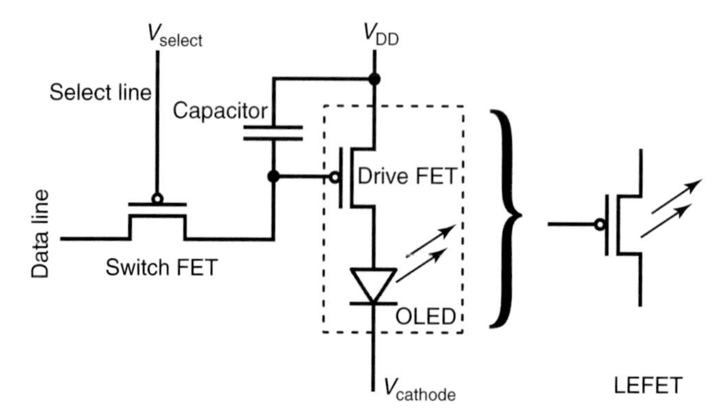

Figure 11.1 Schematic illustration of an AMOLED pixel and a light-emitting field-effect transistor (LEFET). OLED, organic light-emitting diode.

[6, 7]. The emission wavelength of silicon nanocrystals depends on their size and they can exhibit quantum efficiencies of tens of percent. The limiting factor for electroluminescence from silicon nanocrystal devices is inefficient charge injection. In one possible device architecture, the silicon nanocrystals were embedded in the gate dielectric of a metal-oxide-semiconductor field-effect transistor (MOSFET). When holes and electrons were injected sequentially under AC-bias conditions they emitted near-infrared light, although with low overall power conversion efficiencies (<0.1%) [8, 9].

A different approach to inorganic light-emitting transistors was recently demonstrated using direct bandgap InGaP/GaAs and AlGaInP/InGaP heterojunctions. These heterojunction bipolar transistors (HBTs) emit near-infrared and visible light, respectively [10, 11]. This approach, however, relies on carefully epitaxially grown layers of expensive III–V alloys, which is not feasible for large area applications such as displays.

Organic semiconductors, on the other hand, offer several advantages as materials for light-emitting transistors. They are suitable for large area processing on flexible substrates at low temperatures and low cost. Many of them exhibit reasonably good charge transport properties for switching and often have high photoluminescence efficiencies covering a wide range of emission colors across the visible spectrum.

In addition to the proposed use of organic light-emitting transistors in active matrix displays, they have spurred hopes to finally realize the elusive electrically pumped organic laser. Many organic semiconductors show stimulated emission with high gains when optically pumped [12], but attempts to fabricate organic laser diodes have so far failed because of a combination of limited maximum current density, Joule heating, singlet–triplet quenching, polaron quenching, and optical losses caused by the nearness of the anode and cathode in organic LEDs [13, 14]. Light-emitting FETs are attractive alternatives because they can sustain higher current densities than organic LEDs without breaking down. In addition, light emission can be spatially removed from metal electrodes that are responsible for optical losses, and feedback structures can be integrated easily.

In this chapter, we will discuss the two main groups of organic light-emitting field-effect transistors (LEFETs); those that emit light in the unipolar charge transport regime and, more importantly, those that operate in the ambipolar regime. Furthermore, we will briefly introduce light-emitting device concepts that differ from classic planar FETs but achieve similar switching and emission properties.

11.2 Unipolar Light-Emitting FETs

The first reported organic light-emitting transistor was a unipolar FET based on a thin film of vacuum-deposited tetracene demonstrated by Hepp *et al.* in 2003, which emitted green light at highly negative gate (V_g) and source–drain (V_{ds}) voltages [15]. The term "*unipolar*" in this context means that the current–voltage characteristics of the transistor indicate that only one polarity of charge carriers is present in the channel, which in this case was holes. The high work function of the gold electrodes (~5 eV), ambient measurement conditions and electron traps at the dielectric surface (SiO_2) made electron injection and electron transport rather unlikely in this device because of the small electron affinity of tetracene (lowest unoccupied molecular orbital (LUMO) level ~2.4 eV). However, in order to produce excitons by charge recombination, electrons had to be somehow injected into the tetracene. Micrographs of these and other unipolar light-emitting transistors [16–18] revealed that emission was exclusively localized at the edges of the drain electrodes (Figure 11.2). The external quantum efficiency of these LEFETs increased

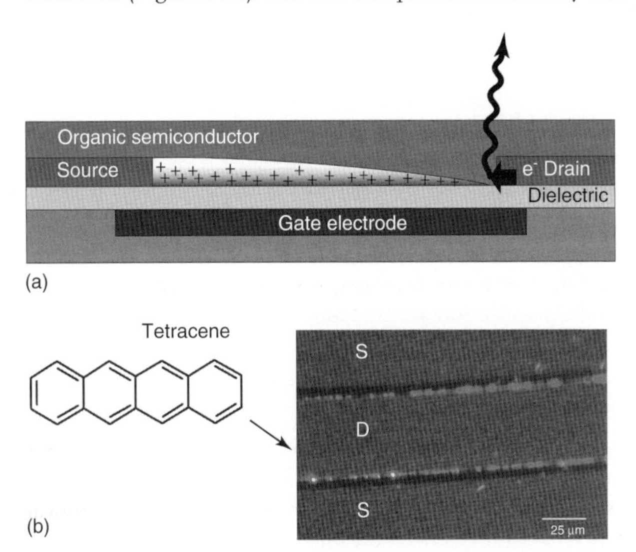

Figure 11.2 (a) Schematic illustration of light emission from a unipolar LEFET. (b) Optical micrograph of electroluminescence from a tetracene LEFET. Emission only occurs at the edge of the drain electrode (D). Reprinted with permission from Ref. [15]. Copyright 2003 by the American Physical Society.

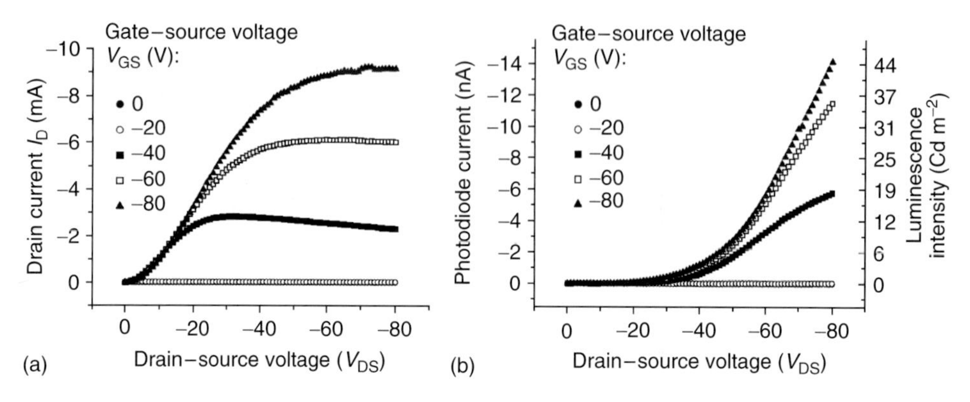

Figure 11.3 (a) Output characteristics of a tetracene LEFET for different gate–source voltages. (b) Optical output characteristics. Reprinted with permission from Ref. [15]. Copyright 2003 by the American Physical Society.

exponentially with V_{ds} while higher gate voltages enhanced the hole current and brightness but not efficiency (Figure 11.3). This indicates that electrons can be injected from gold into the LUMO level of tetracene and other low electron affinity semiconductors when the voltage drop created at the metal–semiconductor interface at the drain electrode is large enough to cause a distortion of the HOMO (highest occupied molecular orbital) and LUMO levels, enabling electron injection by tunneling as suggested by Santato *et al.* [19]. The probability of tunneling of electrons through the barrier and recombination with holes increases exponentially with V_{ds}. The external quantum efficiency for these devices is very small (10^{-6} photons/charge carrier). Apart from the low photoluminescence efficiency of tetracene, the limited injection of electrons and thus large excess of holes is responsible for the low photon yield.

In a p-channel, unipolar LEFET electron injection and thus recombination and quantum efficiency greatly increase when using a low work-function metal such as aluminum, magnesium, or calcium instead of gold for the drain electrode. A number of groups have used this approach to produce LEFETs with a range of materials from conjugated polymers to organic single crystals [20–28]. Despite the improved electron injection, no electron accumulation layer was formed and the emission remained localized at the drain in all cases. This was probably the result of electron traps within the semiconductor and at the dielectric surface preventing electron transport within the channel region.

Regardless, unipolar LEFETs with asymmetric injection electrodes can achieve remarkable brightness and reasonable efficiencies. For example, Namdas *et al.* reported a peak brightness of 310 cd m^{-2} and an external quantum efficiency of 0.5% for a unipolar LEFET based on the soluble poly(phenylene vinylene) (PPV) derivative "SuperYellow" with a gold source and calcium drain electrode [27]. In order to improve the hole mobility of this device and thus the overall source–drain current (I_{ds}), the same group introduced a charge transport layer of poly(2,5-bis(3-tetradecylthiopen-2-yl)thieno[3,2-b]thiophene) between the gate dielectric and

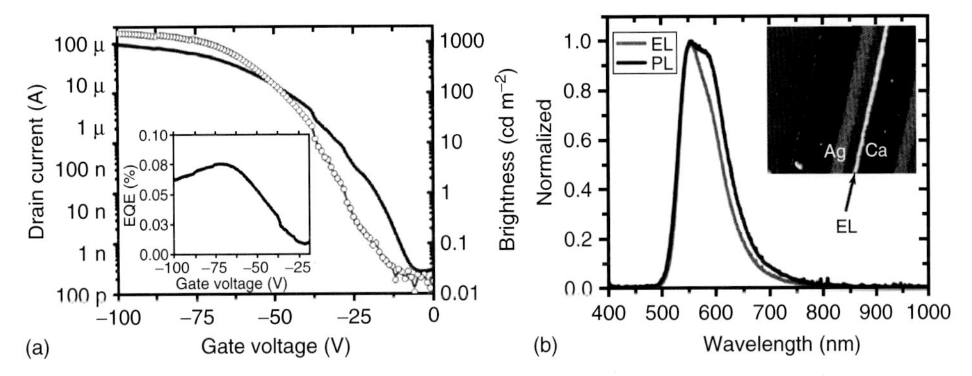

Figure 11.4 (a) Electrical and optical transfer characteristics of an LEFET with asymmetric source–drain electrodes and double layer of high hole mobility polymer poly(2,5-bis(3-tetradecylthiophen-2-yl) thieno[3,2-b] thiophene) and emissive poly(phenylene-vinylene) derivative (SuperYellow). (b) Emission spectrum and optical micrograph of emission close to the calcium drain electrode. Reprinted with permission from Ref. [26]. Copyright 2008, American Institute of Physics.

the emitting polymer that had a field-effect mobility 100 times higher than "SuperYellow." Although the quantum efficiency did not improve compared to the devices without the hole-transport layer, the increased source–drain current resulted in a maximum brightness of 2500 cd m^{-2} (Figure 11.4) [26].

Shortening the transistor channel length increases the brightness of unipolar LEFETs because of the much larger source–drain current, without significantly decreasing efficiency. Oyamada *et al.* demonstrated a peak luminance of 200 cd m^{-2} for an LEFET with gold electrodes on Si/SiO_2 and a channel length of 400 nm using tetraphenylpyrene doped with rubrene as the active layer, which has a photoluminescence efficiency of 99% and hole mobility of about 10^{-5} cm^2 V^{-1} s^{-1}. Its electroluminescence efficiency of 0.2% was similar to that of long channel LEFETs, which showed orders of magnitude lower levels of brightness. Further improvements of brightness and efficiency were achieved by employing double-layer source–drain electrodes of Mg/Au alloy and Au [29]. Oyamada *et al.* successfully applied the same approach to a range of materials covering the visible range from blue to red, and even some near-UV to deep blue emitting 9,9-diarylfluorene oligomers, which have very low lasing thresholds when optically pumped [22, 30].

Although the measures described above lessen the disadvantage of incomplete recombination of holes and electrons in a unipolar LEFET, another significant loss channel remains. When recombination and emission take place very close to a metal electrode, interference and quenching occur, thus diminishing the amount of light that is eventually coupled out of the LEFET. Gehlhaar *et al.* investigated the dependence of outcoupled light on the device structure of an LEFET using the finite difference time domain (FDTD) method [31]. They found that the outcoupling efficiency of light generated next to a gold electrode on Si/SiO_2 was only about 3%. Introducing reflecting layers and changing the electrode shape and material could increase outcoupling to up to 20%. Moving the emission zone away from the injecting electrodes by 100 nm nearly tripled the outcoupling efficiency. These considerations are particularly important when trying to optimize LEFET structures for external brightness or attempting to introduce feedback structures.

Despite these challenges in achieving high recombination and outcoupling efficiencies in LEFETs operating in the unipolar regime, in addition to their high operating voltages and often insufficient environmental stability, they have the potential to be implemented in a pixel of an active matrix display. Their brightness increases continuously with the applied gate voltage at a fixed source–drain bias, whereas emission characteristics of ambipolar LEFETs are much more complex as we will see in the next section.

11.3 Ambipolar Light-Emitting FETs

11.3.1 Ambipolar Device Characteristics

An ambipolar (sometimes also called *bipolar field-effect transistor*) is characterized by its ability to accumulate both holes and electrons depending on the applied voltages and so, in a sense, it behaves like a combination of an n-channel and a p-channel transistor. In addition to unipolar hole accumulation for negative gate voltages and unipolar electron accumulation for positive gate voltages, ambipolar FETs have a third transport regime. In this ambipolar regime, both hole and electron accumulation zones exist within the channel either in series or in parallel depending on the device configuration (single semiconductor, bilayer heterojunction, or blend/bulk heterojunction). The current–voltage characteristics of many of these different ambipolar transistors are remarkably similar. Ambipolar transfer characteristics (I_{ds} vs V_g for constant V_{ds}) are characterized by a distinct V-shape and the output characteristics (I_{ds} vs V_{ds} for different constant V_g) show quadratic current increases at low $|V_g|$ and high $|V_{ds}|$ (Figure 11.5).

These current–voltage characteristics can be easily understood considering the voltages applied to the source, the drain, and the gate electrode and the resulting local potentials within the transistor channel. Let us assume an ambipolar FET with thresholds $V_{Th,h}$ and $V_{Th,e}$ for hole and electron transport, respectively. The source potential is zero ($V_s = 0$), that is, the source is grounded, and the drain potential V_d is positive. Applying a positive gate voltage that is larger than the electron accumulation threshold ($V_g > V_{Th,e}$) leads to the injection of electrons from the source that drift through the channel toward the drain as in any n-channel unipolar FET. For the case that the gate voltage is smaller than the drain potential ($V_g < V_d$), the gate is effectively negative with respect to the drain. In a unipolar electron channel FET this would have no effect, but in an ambipolar FET the drain could now act as a hole source with an applied negative gate voltage of $V_g - V_d$. As this potential difference becomes more negative than the hole-accumulation threshold ($V_g - V_d < V_{Th,h}$), holes are injected from the drain into the channel resulting in a current flow even when V_g is smaller than $V_{Th,e}$ and an n-channel FET would be in its off-state. Since the difference between the drain potential and the gate potential depends directly on V_{ds}, the magnitude of the hole current changes with it leading to distinct ambipolar current–voltage characteristics for a set of source–drain voltages (Figure 11.5). For negative V_g and V_{ds}, the source injects holes and the drain injects electrons. It would be more accurate to describe the electrodes in an ambipolar FET as hole source and electron source, but by convention we will continue to call the grounded electrode the source.

Overall, one usually finds three distinct regimes in an ambipolar transistor: unipolar electron accumulation for $V_g \geq V_{Th,e}$ and $V_g - V_d > V_{Th,h}$, ambipolar transport for $V_g \geq V_{Th,e}$ and $V_g - V_d \leq V_{Th,h}$, and unipolar hole transport for $V_g < V_{Th,e}$ and $V_g - V_d \leq V_{Th,h}$. For light-emitting transistors, the ambipolar regime is the most interesting because both holes and electrons are present in the channel and have a high probability to meet and recombine radiatively. How and where this takes place exactly depends on the device structures that are described next.

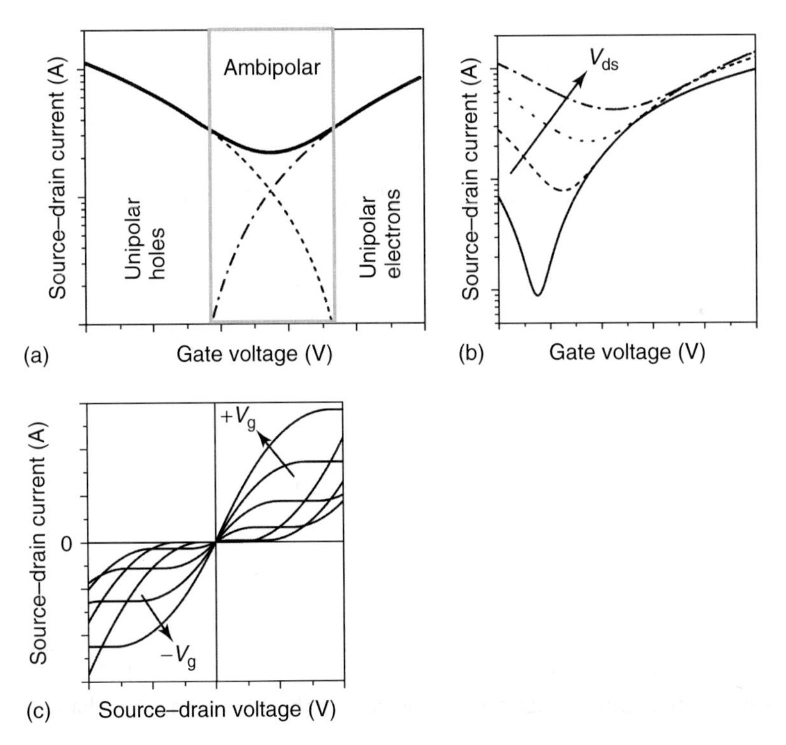

Figure 11.5 (a) Idealized transfer characteristics of ambipolar field-effect transistor (solid line) and unipolar currents of holes and electrons (dashed lines). The gray box indicates the ambipolar transport regime. (b) Dependence of ambipolar transfer characteristics on source–drain voltage. (c) Ambipolar output characteristics for positive and negative source–drain voltages.

11.3.2
Ambipolar Blends with Bulk Heterojunctions

A convenient way of fabricating an ambipolar FET is blending organic semiconductors with high electron affinities, for example, C_{60}, [6,6]-phenyl-C_{61}-butyric acid methyl ester (PCBM), N,N′-ditridecylperylene-3,4,9,10-tetracarboxylic diimide (PTCDI-$C_{13}H_{27}$), with those that have low electron affinities, for example, OC_1C_{10}-PPV, poly(3-hexylthiophene) (P3HT), and α-quinquethiophene (α-T5) [32–37]. Although they are not doped, these semiconductors are often referred to as *n-type* and *p-type*, respectively, because in standard FET structures with gold electrodes and a silicon dioxide gate dielectric they exhibit only electron or only hole accumulation. Blending these two components allows for injection and transport of charge carriers of both polarities in a single device. Depending on the position of the particular HOMO and LUMO levels, holes and electrons remain confined within the p-type and n-type material, respectively. Consequently, charges have to follow a percolation path across the length of the channel. The equivalent circuit for a blended ambipolar FET would be an n-type and a p-type FET in parallel but with reduced effective channel widths and increased channel lengths compared to the actual device dimensions. Blending organic semiconductors significantly lowers the effective field-effect mobility for one or both components, often by up to an order of magnitude, compared to values for the pure semiconductors [32, 34–36]. At the interface between the two materials, forming essentially a bulk heterojunction, holes and electrons can recombine and lead to light emission similar to blend polymer LEDs.

Rost *et al.* first used a blend approach to produce an ambipolar LEFET based on coevaporated α-T5 for hole transport and PTCDI-$C_{13}H_{27}$ for electron transport and light emission (Figure 11.6a) [35]. They found relatively similar hole and electron mobilities of 10^{-4} and 10^{-3} $cm^2\ V^{-1}s^{-1}$, respectively, and light emission at large negative source–drain and small negative gate voltages when electron accumulation is expected to dominate as shown in Figure 11.6b. Testing a range of composition ratios revealed that these FETs emitted light both for balanced ratios of α-T5 and PTCDI-$C_{13}H_{27}$, which led to ambipolar transport characteristics as well as for excess amounts of PTCDI-$C_{13}H_{27}$ that resulted in unipolar current–voltage characteristics. In contrast, excess amounts of α-T5 caused ambipolar transport without any light emission [38], probably due to increased exciton transfer to and nonradiative decay in H-type aggregates of α-T5 instead of emission from PTCDI-$C_{13}H_{27}$. Emission intensities appeared to be very low and in neither case was a spatial resolution of the emitted light reported. The bulk heterojunctions where charge recombination and emission are expected to take place could also be a source of quenching because of exciton dissociation at the interface. While detrimental for light emission, exciton dissociation in ambipolar FETs with bulk heterojunctions can be used to detect incident light and thus enables a different type of multifunctional FET [39, 40].

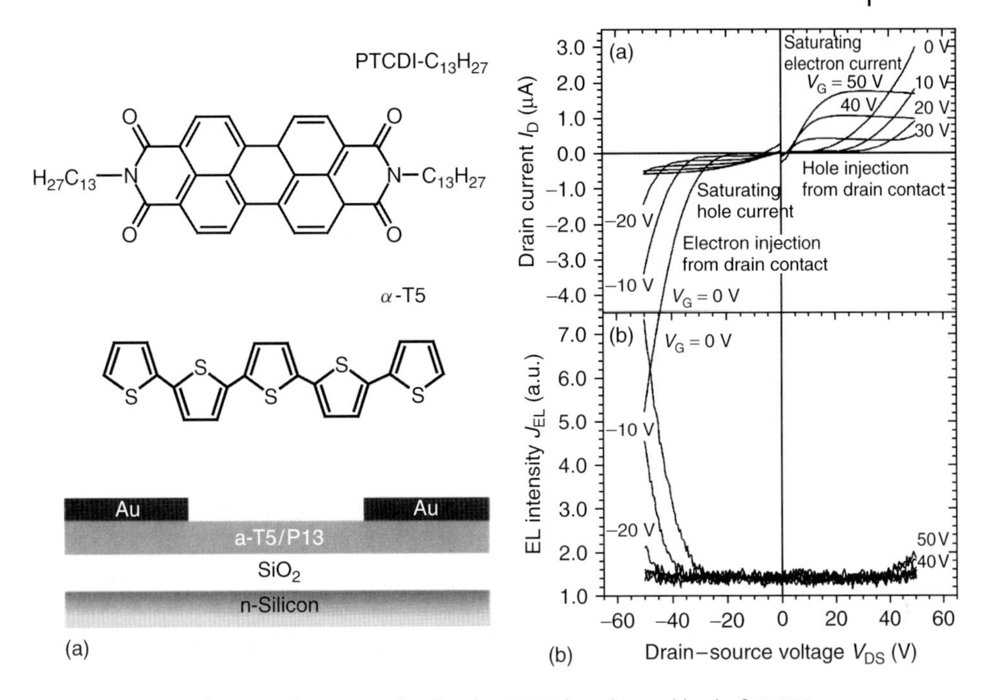

Figure 11.6 (a) Schematic illustration of ambipolar LEFET based on a blend of PTCDI-$C_{13}H_{27}$ and α-T5. (b) Electrical and optical output characteristics. Reprinted with permission from Ref. [35]. Copyright 2004, American Institute of Physics.

11.3.3
Double Layers and Lateral Heterojunctions

Semiconductor bilayers and spatially defined heterojunctions give rise to ambipolar device characteristics and light emission, while avoiding the problem of percolation paths. Dinelli *et al.* demonstrated light emission from an FET with a bilayer of PTCDI-$C_{13}H_{27}$ for electron transport and α,ω-dihexyl-quaterthiophene (DHT4) for hole transport [41]. When PTCDI-$C_{13}H_{27}$ was deposited on top of DH4T in an FET using top gold electrodes for injection hole and electron mobilities remained high (both 0.03 cm^2 V^{-1} s^{-1}). For the inverted layer structure, however, the hole mobility dropped by an order of magnitude probably due to growth compatibility issues. The hole and electron channels of these devices are in parallel, and the interaction area of holes and electrons is greatly reduced compared to bulk heterojunction FETs. Light emission was detected for both bilayer systems, although only in the unipolar electron accumulation regime when PTCDI-$C_{13}H_{27}$ was in direct contact with the dielectric and in the unipolar hole-accumulation regime when DH4T was at the bottom. Spatial resolution of the emission zone was not attempted but one could assume that emission occurred at the bilayer interface close to the drain electrode and thus forming a type of bilayer LED structure in series with an FET similar to the unipolar LEFET demonstrated by Namdas *et al.* [26]. Similar behavior was found for pentacene/PTCDI-$C_{13}H_{27}$ bilayer FETs [42].

In order to move the emission zone a significant distance away from the drain electrodes, de Vusser *et al.* developed a technique to sequentially deposit PTCDI-$C_{13}H_{27}$ for electron transport and a hole-transporting oligo(phenylene vinylene) derivative (1,4-bis(octyloxy)-2,5-bis[(E)-4-(E)-styrylstyryl]benzene, O-octyl-OPV5) at an angle so that they only overlapped in a certain area several micrometers away from the electrodes (Figure 11.7a) [43]. The HOMO and LUMO levels of the two semiconductors are such that there is only a small barrier for holes to cross over from the O-octyl-OPV5 to PTCDI-$C_{13}H_{27}$, while there is a large barrier for electrons to do the reverse. Bottom contact gold electrodes served as hole and electron sources, respectively. Figure 11.7b shows the current–voltage characteristics of this heterojunction FET (the electron source being the drain electrode with V_d while the hole source was grounded $V_s = 0$). For small negative V_d and large negative V_g only, holes accumulated along the channel with a low mobility within the PTCDI-$C_{13}H_{27}$ and thus low source–drain current. When the difference between V_g and V_d surpassed the electron accumulation threshold, the current increased quadratically with $-V_d$ as expected for an n-type FET under these bias conditions. When V_d decreased

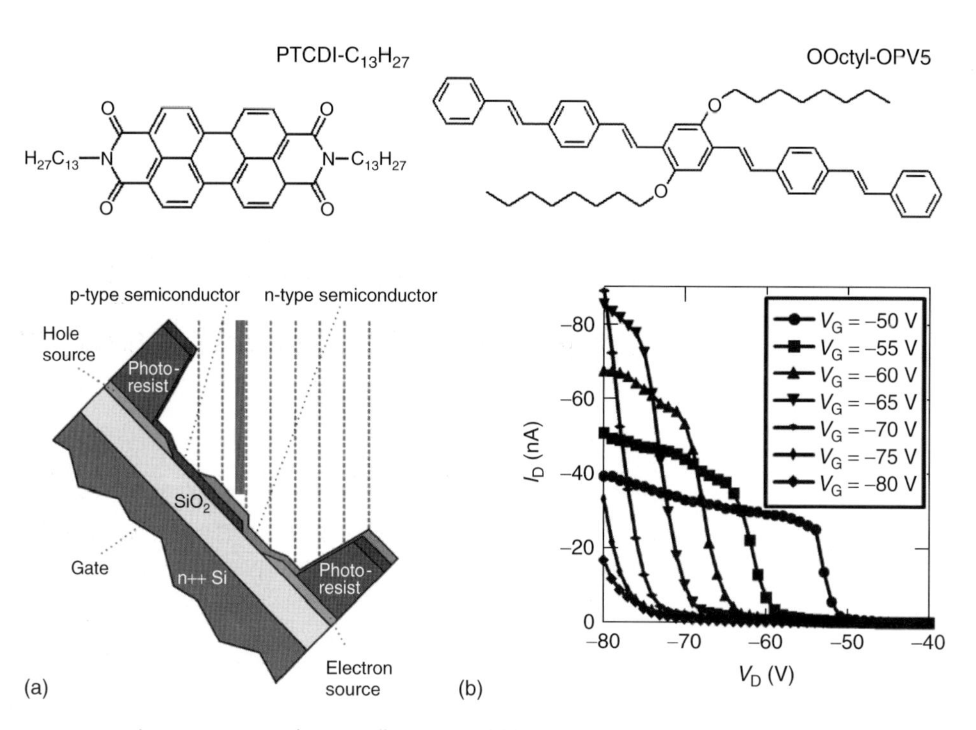

Figure 11.7 (a) Schematic illustration of fabrication process of lateral heterojunction LEFET by vacuum deposition of n-type PTCDI-$C_{13}H_{27}$ and p-type OOctyl-OPV5. (b) Output characteristics. Reprinted with permission from Ref. [43]. Copyright 2006, American Institute of Physics.

further for a constant V_g the electron current started to equal the maximum hole current and thus the FET went into saturation. Charge recombination and light emission exclusively occurred within the PTCDI-$C_{13}H_{27}$ layer and the emission position could be modeled to range from close to the drain electrode for small negative V_d to the pn-heterojunction at large negative V_d and thus far away from the injecting electrodes. The emission intensity increased with increasing current. In this heterojunction device, the hole and electron channels were in series because of the lateral separation of the two semiconductors except where the two layers overlapped.

11.3.4
Single Semiconductor Ambipolar FETs

In an ambipolar FET with only one semiconducting material, the hole- and electron-accumulation layers extending from the source and drain electrodes, respectively, meet at a zone within the channel where the local potential equals the gate voltage. The effective gate voltage is zero and thus the charge density is close to zero as well. Holes and electrons recombine within this recombination zone resulting in light emission. Unlike in unipolar or heterojunction LEFETs, this emission zone can be moved through the entire channel and its position is determined by the gate and source–drain voltage (Figure 11.8). It is very unlikely that charge carriers penetrate an accumulation layer of opposite polarity carriers for more than a few hundred

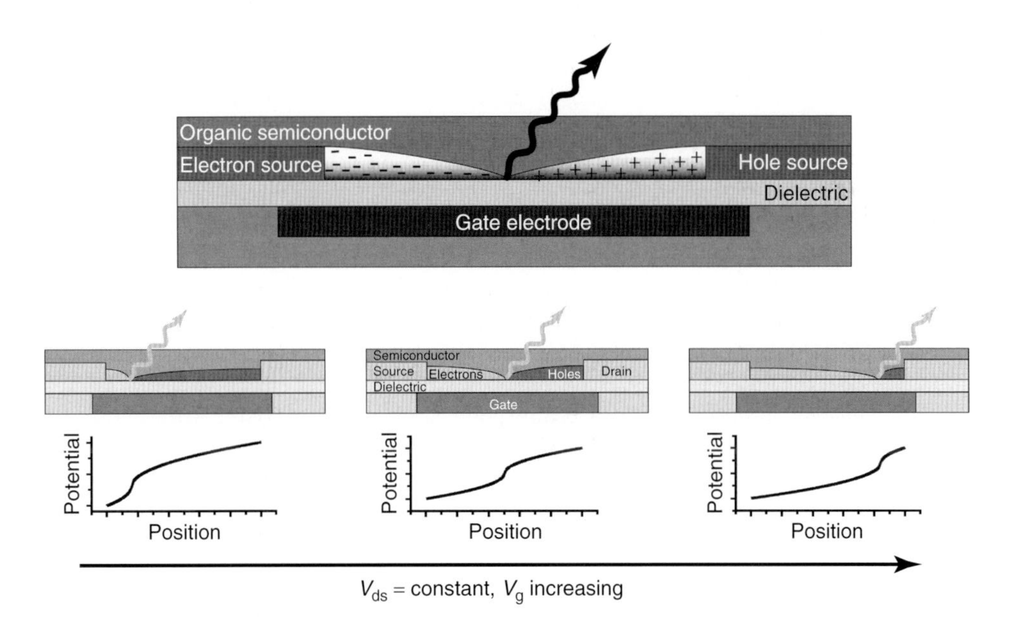

Figure 11.8 Schematic illustration of charge distribution and recombination zone within an ambipolar LEFET, including dependence of the position of recombination and emission zone on gate voltage.

nanometers without recombining, so that for a sufficiently long channel length we can assume that all injected charges recombine in the ambipolar regime. Hence, hole and electron currents are always equal and the two pinched-off channels behave as connected in series. The absolute number of formed electron/hole pairs is given directly by I_{ds}. Consequently, the electroluminescence efficiency of an ambipolar LEFET should be determined solely by the photoluminescence efficiency of the material, the singlet/triplet ratio and light outcoupling efficiency.

11.3.4.1 Intrinsic Ambipolar Transport in Organic Semiconductors

In order to achieve ambipolar transport in an FET with a single organic semiconductor, a few conditions have to be fulfilled. First, the semiconductor must be capable of conducting both holes and electrons. This is, in principle, the case for almost all organic semiconductors. Theoretical studies investigating charge transfer rates between small conjugated molecules as well as conjugated polymers found no significant difference between hole and electron transfer integrals [44–46]. That raises the question, why do most organic semiconductors apparently only exhibit one type of charge transport in time-of-flight (TOF) experiments, FETs, and diodes? Impurities and trap states are the major causes of that. Depending on the HOMO/LUMO levels of a molecule or polymer, impurities can act as hole or electron traps, and hence high purity of the organic semiconductor is crucial in order to observe intrinsic behavior especially in low charge carrier density measurements such as TOF or space charge-limited diodes. Trapping of electrons is the most common issue because of the small electron affinity of many organic semiconductors. Materials with electron affinities <4 eV are unstable toward reduction of O_2/H_2O [47]. These materials would only exhibit electron transport under inert conditions, for example, in vacuum or in a dry nitrogen glovebox but not in ambient environment.

In addition to the semiconductor, the gate dielectrics used in FETs obviously have a large influence on the transport of charges. Charge carrier densities in FETs are several orders of magnitude higher than in diode structures [48] and thus many intrinsic charge traps become saturated. However, the charge accumulation layer extends only a few nanometres away from the dielectric–semiconductor interface, and electron traps at the dielectric surface severely inhibit electron transport. Chua *et al.* found that replacing gate dielectrics that have surface hydroxyl groups, as does the widely used SiO_2, with dielectrics devoid of these electron trapping groups enables observation of electron transport in a wide range of organic semiconductors previously assumed to be exclusively p-type [49].

Given that the first condition for ambipolar transport is fulfilled, charges still need to be injected into the semiconductor before they can accumulate and drift. Gold electrodes are often used in organic FETs because of their environmental stability. Their work-function between 4.8 and 5.0 eV is reasonably well aligned with the HOMO level of many organic semiconductors such as pentacene, P3HT, or rubrene and thus allows for efficient injection of holes and low contact resistances. Owing to their large bandgap, the LUMO levels of these and many other organic semiconductors are more than 2 eV away from the Fermi level of gold, so that electron injection is strongly inhibited and can only occur by tunneling at high fields as described for unipolar LEFETs. Semiconductors with large electron affinities (>3.5 eV) such as C_{60}, PTCDI-$C_{13}H_{27}$ or perfluoropentacene allow for injection of electrons from gold into the LUMO level, but their HOMO levels are too low for hole injection, and so they appear to be n-type.

11.3.4.2 **Ambipolar FETs with Asymmetric Electrodes**

Several approaches are feasible in solving the injection problem and achieving ambipolar transport and associated light emission in single semiconductor organic FETs. One option is employing asymmetric work-function source/drain electrodes as already demonstrated for unipolar LEFETs. For example, by using gold for hole injection and low work-function metals such as calcium or magnesium for electron injection, in addition to electron trap-free gate dielectrics under inert conditions, it is possible to realize ambipolar transport in organic semiconductors with large bandgaps (2–3 eV) that emit light in the visible. The first ambipolar LEFETs based on this concept were demonstrated by groups in Cambridge and Santa Barbara using poly(phenylene-vinylene) derivatives for transport and emission [50, 51]. Both could spatially resolve a narrow emission zone (Figure 11.9) that could be moved from the source electrode through the channel to the drain electrode and back again by changing the applied voltages. The same approach was later successfully applied to single crystals of tetracene [52] and rubrene [53] as well as thin films of α,ω-bis(biphenyl-4-yl)-terthiophene [54] and pentacene [55].

The visual observation of the recombination and emission zone proved experimentally that the hole and electron channels were in series and the narrow width of the emission zone corroborated the assumption of complete charge recombination. Furthermore, the demonstration of ambipolar transport in high-purity single crystals of organic semiconductors confirmed that organic semiconductors are indeed intrinsically ambipolar.

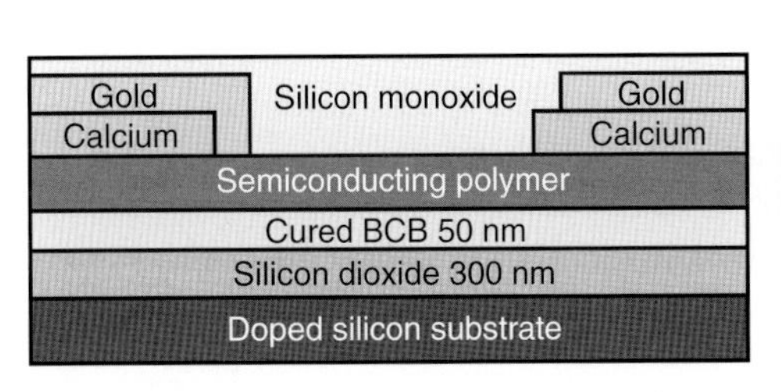

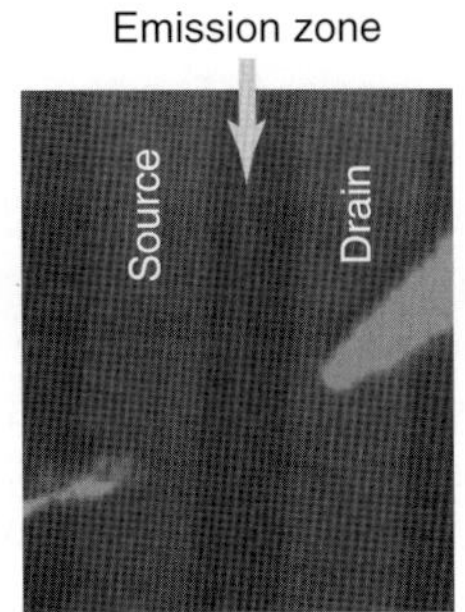

Figure 11.9 Device structure and optical micrograph of emission zone of ambipolar poly(2-methoxy-5-(3,7-dimethyloctoxy)-p-phenylene-vinylene) (OC_1C_{10}-PPV) transistor with asymmetric injection electrodes and a dielectric buffer layer (BCB) to remove electron traps at the SiO_2 surface.

11.3.4.3 **Ambipolar FETs with Narrow Bandgap Semiconductors**

Although low work-function metals enable injection of electrons into many organic semiconductors, they are by themselves very air sensitive, so that it is not possible to pattern them by standard photolithography, and some degrade quickly even under inert conditions. Circumventing this problem, Smits *et al.* proposed to use gold electrodes and organic semiconductors with bandgaps of <1.7 eV for ambipolar transistors [56]. When the Fermi level of gold lies approximately in the middle of the bandgap injection barriers are reasonably small for both holes and electrons. This also means that these semiconductors absorb and emit light in the near-infrared, which could be useful for sensors and optical communication systems. One example of a small bandgap semiconductor is the solution processable laser dye SQ1 (2,4-di-3-guaiazulenyl-1,3-dihydroxycyclobutenediyliumdixydroxide-bis inner salt, see inset in Figure 11.10a), which shows peak emission at 940 nm and hole and electron mobilities of 10^{-4} $cm^2\ V^{-1}\ s^{-1}$. It can be used to fabricate a near-infrared LEFET (Figure 11.10) with nicely ambipolar transfer characteristics. The concurrent emission characteristics show a distinct change of intensity with gate voltage that is very different from unipolar LEFETs. There are two emission maxima that do not coincide with the maximum current. The electroluminescence efficiency, which is proportional to the ratio of light-output (here measured as current of a photodiode) to source–drain current, is highest for gate voltages within the ambipolar regime when emission takes place away from the injecting electrodes and recombination should be complete [57].

Building on this concept Bürgi *et al.* developed a small bandgap and high-mobility bithiophene 2,5-dihydropyrrolo[3,4-c]pyrrole-1,4-dione copolymer that emitted near-infrared light around 860 nm. Depending on device structure and gate dielectric, this copolymer exhibited electron mobilities of up to 0.09 $cm^2\ V^{-1}\ s^{-1}$ and hole mobilities of up to 0.11 $cm^2\ V^{-1}\ s^{-1}$. For this LEFET, the emission zone was spatially resolved and could be moved through the channel region by adjusting the applied voltages [58]. Unfortunately, small bandgap organic semiconductors tend to have very low photoluminescence efficiencies and the external quantum efficiencies for these near-infrared LEFETs were <0.005%.

11.3.4.4 **Ambipolar FETs with Bottom Contact/Top Gate Electrodes**

As mentioned above, the large bandgap of most organic semiconductors leads to a high injection barrier for at least one polarity of charge carriers and thus very high contact resistance. Contact resistance, however, does not only depend on the energy level alignment of the materials but also on device structure and processing. For example, in bottom contact/top gate transistor structures a significant overlap of gate electrode and injecting electrodes decreases the effective contact resistance substantially. Aided by the gate field, more charges are injected from the relatively large electrode surface than from just the edge of the electrode as in a bottom gate/bottom contact FET. This principle was applied by the Cambridge group to realize ambipolar transport and light emission from a bottom contact/top gate FET using poly(9,9-dioctylfluorene-co-benzothiadiazole) (F8BT). This green-emitting conjugated polymer with a bandgap of 2.6 eV had

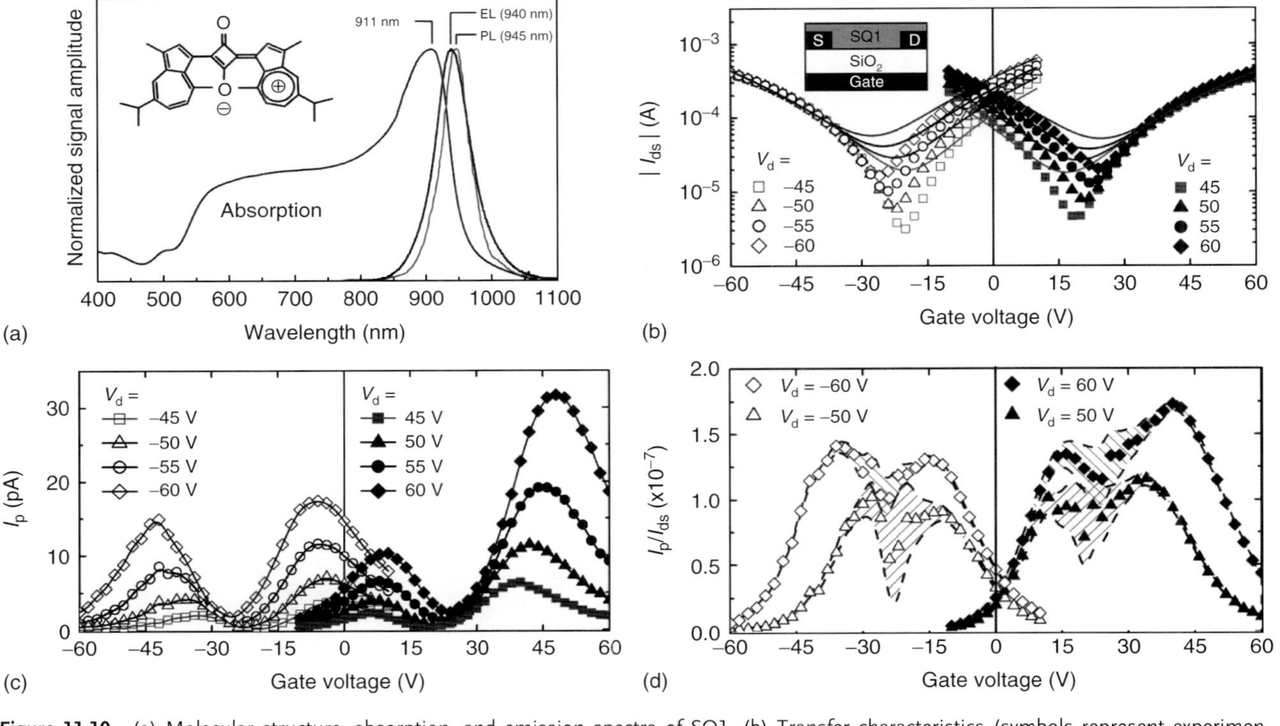

Figure 11.10 (a) Molecular structure, absorption, and emission spectra of SQ1. (b) Transfer characteristics (symbols represent experimental data and solid lines theoretical fits), (c) light output (as current of photodiode I_p) and (d) external efficiency of ambipolar SQ1-FET that is proportional to I_p/I_{ds}. Reprinted with permission from Ref. [57]. Copyright 2007, Wiley-VCH, Weinheim.

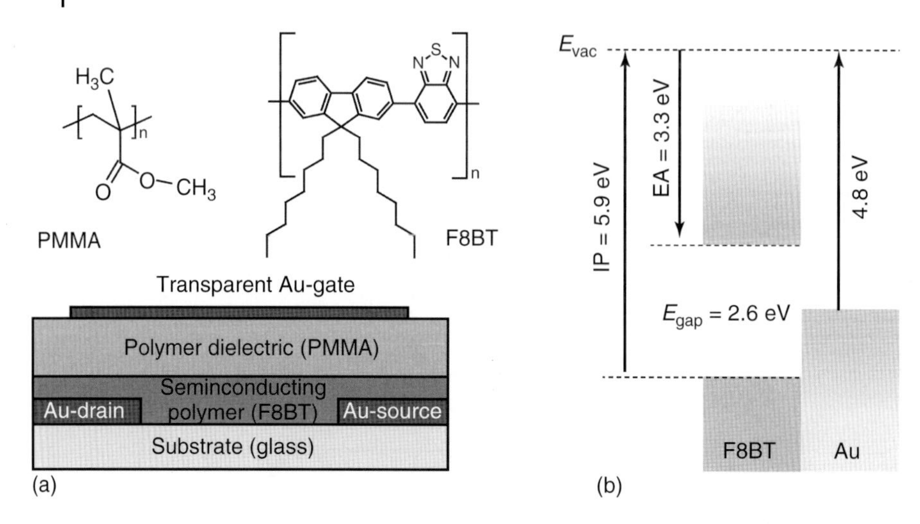

Figure 11.11 (a) Molecular structure of PMMA and F8BT and schematic device structure of bottom contact/top gate FET. (b) Energy level diagram of F8BT and gold (E_{vac} – vacuum level, IP – ionization potential, and EA – electron affinity).

been previously used in PV cells and LEDs as the electron transport layer. The HOMO and LUMO levels of F8BT are such that the Fermi level of gold lies almost exactly in the middle of the bandgap. Nevertheless, the injection barriers for holes and electrons (~1.3 eV) are still very high. The device structure shown in Figure 11.11 consists of photolithographically defined interdigitated gold electrodes on a glass substrate with spincoated and annealed F8BT and spincast poly(methyl methacrylate) (PMMA) as a gate dielectric without electron traps. A shadow mask evaporated gold gate electrode that overlapped completely with the source and drain electrodes completed the device.

The resulting transistors showed ambipolar transport characteristics with balanced hole and electron mobilities of around 8×10^{-4} cm^2 V^{-1} s^{-1} [59]. As shown in Figure 11.12a, the transfer characteristics of F8BT transistors exhibited the expected evolution of I_{ds} with V_{ds}. The hole and electron thresholds were stable and little current hysteresis occurred. The output characteristics show a strong suppression of current for low V_{ds}, revealing large non-ohmic contact resistance for both hole and electron transport. Figure 11.13 shows the emission zone of an F8BT transistor with interdigitated source and drain electrodes and a channel length of 20 μm for different gate voltages, while the source–drain voltage was kept constant. At low gate voltages, the emission zone appears at the source electrode and the channel region is filled with holes. With increasing V_g, the electron accumulation layer extends into the channel and the emission zone moves through the channel toward the drain. At high V_g, the channel is filled with electrons, and continued injection of holes from the drain electrode leads to light emission there. By reversing the voltage sweep, the emission zone movement is reversed as well. The observed width of the emission zone is constant throughout the channel and about 2 μm. This width does not necessarily represent the overlap of hole and

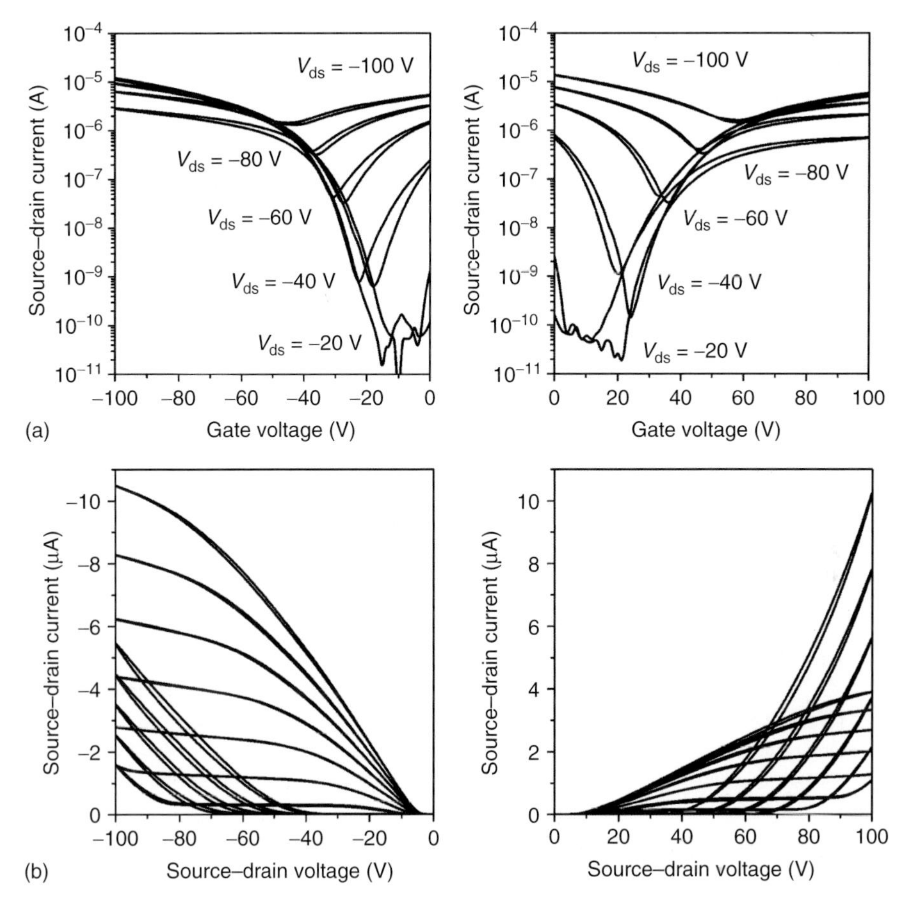

Figure 11.12 (a) F8BT-FET transfer characteristics for different V_{ds}. (b) Output characteristics for V_g from 0 to −100 V (step −10 V) and from 0 to 100 V (step 10 V).

electron accumulation in the channel where excitons are formed, but could be broadened by optical resolution limitations.

The position of the emission zone changes in a characteristic manner with gate voltage, which is symmetrical with respect to the lowest current point of the transfer curve. At that point, the emission zone is located in the middle of the channel (Figure 11.13). This symmetrical shape of the position versus gate voltage curve is due to the balanced hole and electron mobilities of F8BT. The ratio of hole to electron mobility, threshold voltages, and contact resistance has a strong influence. A substantial imbalance of hole and electron mobilities leads to an asymmetric curve as shown for F8TBT (a red-emitting alternating copolymer of thiophene, fluorene, and benzothiadiazole). The hole mobility in F8TBT is 10 times higher than its electron mobility (Figure 11.14) [60]. In this case, the emission zone of F8TBT moves very quickly with decreasing gate voltage away from the hole source

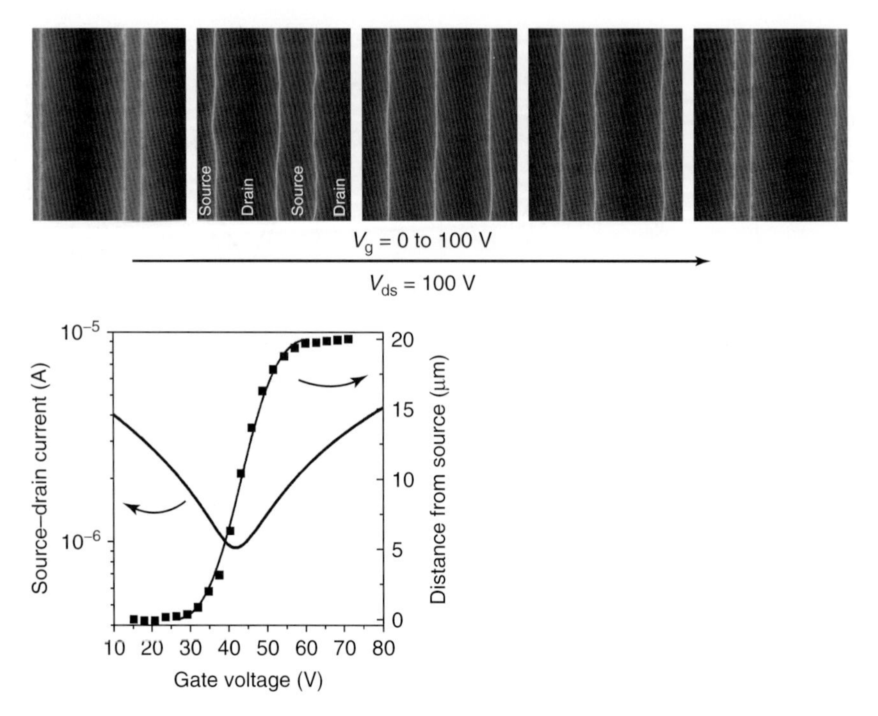

Figure 11.13 (a) Optical micrographs of emission zone in ambipolar F8BT-FET with interdigitated source–drain electrodes at different gate voltages and constant source–drain voltage. (b) Transfer characteristics (solid line) and concurrent emission zone position at $V_{ds} = 100$ V (solid squares).

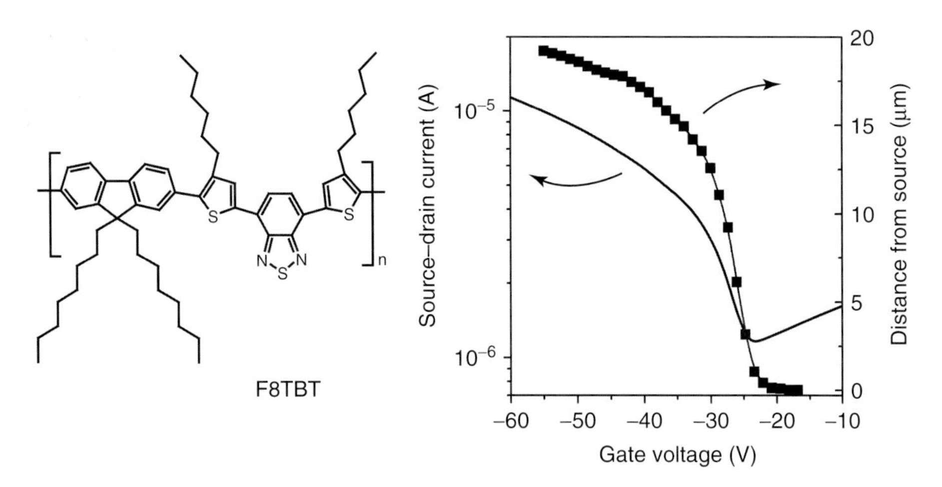

Figure 11.14 Transfer characteristics (solid line) and position (solid squares) versus gate voltage ($V_{ds} = -120$ V) for an ambipolar LEFET based on F8TBT with hole and electron mobilities of 5×10^{-3} cm^2 V^{-1} s^{-1} and 3×10^{-4} cm^2 V^{-1} s^{-1}, respectively. Adapted from Ref. [60]. Copyright 2008, American Institute of Physics.

and through the larger part of the channel before slowly approaching the electron source. Hole and electron currents must be equal under ambipolar conditions, but the conductivity of the electron channel is much lower than that of the hole channel. Therefore, a larger voltage drop across the electron channel or a shorter length is necessary to balance hole and electron currents. This also means that the voltage window for observing the emission zone within the channel region decreases with the increasing imbalance of hole and electron mobilities. For FETs with hole mobilities orders of magnitude higher than their electron mobilities, the emission zone will be almost exclusively positioned close to the drain, as in a unipolar LEFET. Furthermore, large threshold voltages and voltage drops at the electrodes (i.e., high contact resistance) increase the steepness of the position versus the gate voltage curve. Unequal injection barriers for holes and electrons cause an overall V_g-shift [61].

The luminance versus gate voltage characteristics of ambipolar LEFETs are quite different from unipolar LEFETs. As shown for F8BT in Figure 11.15 and for the

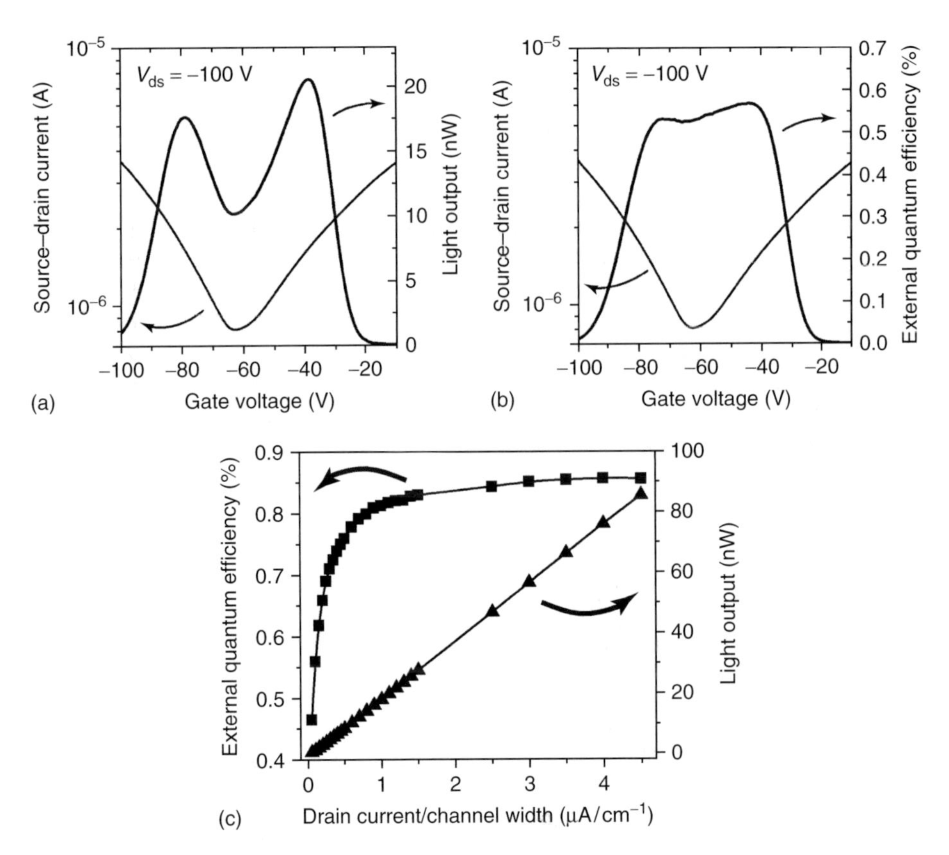

Figure 11.15 (a) Transfer and light-output characteristics for F8BT transistor with $L = 20\ \mu m$, $W = 1$ cm. (b) Concurrent external quantum efficiency versus gate voltage. (c) Light output and external quantum efficiency versus normalized drain current.

near-infrared LEFET in Figure 11.10, the overall light output during a transfer scan has two maxima and a minimum centered at the source–drain current minimum. When calculating the quantum efficiency, we find that it reaches a plateau around the current minimum when the emission zone is located in the middle of the channel. The efficiency drops quickly when the emission zone approaches the source or drain electrodes. This is easily explained by the ambipolar nature of charge transport. As long as the emission zone, which is clearly much narrower than the channel, is located away from the electrodes complete recombination of holes and electrons must take place. The hole and electron currents are equal ($I_{holes} = I_{electrons} = I_{ds}$) and the external quantum efficiency in the ambipolar regime depends on the ratio of singlets to triplets (usually assumed to be 1 : 3), the fraction of singlets decaying radiatively, and the outcoupling efficiency. At very high and very low V_g when the emission zone is close to the injecting electrodes, charges can escape into the electrodes without recombining, and hole and electron currents become unequal leading to lower emission efficiency. In addition, outcoupling is reduced and exciton quenching increases near the metal electrodes.

According to the above, the maximum quantum efficiency should be independent of the source–drain voltage. However, for both F8BT and SQ1 near-infrared LEFETs, experimental data show an overall increase of quantum efficiency with V_{ds}. Furthermore, the efficiency plateau is not completely flat but shows a small dip at the minimum current gate voltage. This behavior can be explained when plotting quantum efficiency versus source–drain current. Electroluminescence efficiency actually increases with current density before saturating around 0.8% for the F8BT transistor (Figure 11.15c). This current density dependence might be caused by a limited number of exciton quenching centers at the semiconductor–dielectric interface, which become saturated at high current densities. For very high current densities, the quantum efficiency of F8BT-LEFETs decreases again, which is likely a result of singlet quenching by polarons or long-lived triplets [62].

The external quantum efficiencies of ambipolar LEFET are generally higher than those of unipolar LEFETs, but that does not make them necessarily more attractive for applications in displays. As shown above, the emission intensity of ambipolar LEFETs does not simply increase with gate voltage, as it does for unipolar LEFETs. On top of that, the position of emission moves depending on the applied voltages. Although this could be interesting for optoelectronic switches involving waveguides, it is not very useful for displays. In a pixel, a certain voltage should result in a preset brightness level. A different biasing regime would be necessary to achieve that with ambipolar LEFETs. For example, grounding the gate electrode instead of the source and applying a drain voltage V_d and source voltage V_s that equals $-V_d$ will keep the transistor in the ambipolar regime once the threshold voltages are overcome. This way, while increasing the source–drain bias ($V_{ds} = V_d - V_s = 2V_d$) and thus increasing I_{ds} and brightness proportionally, the emission zone remains in the middle of the channel [62].

Apart from possible practical applications, ambipolar light-emitting transistors are an interesting platform to study recombination and emission physics of organic semiconductors because of their planar structure in contrast to organic LEDs.

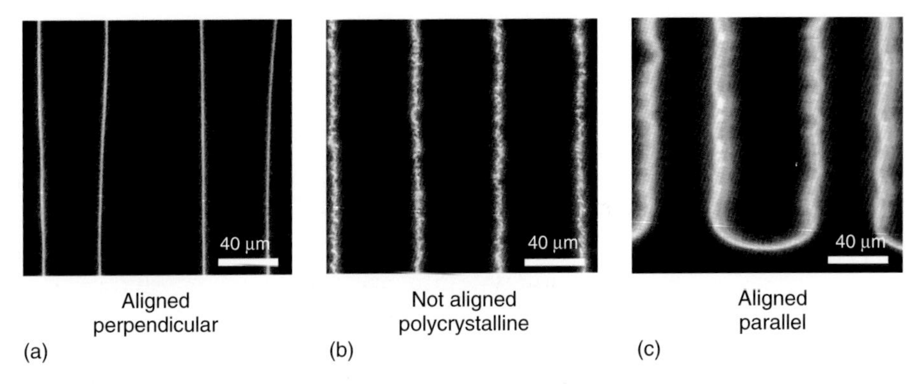

Figure 11.16 Optical micrographs of emission zones of light-emitting FETs with F8BT thermotropically aligned perpendicular or parallel to the charge transport direction, and with unaligned, polycrystalline F8BT. Adapted with permission from Ref. [65]. Copyright 2008, Wiley-VCH, Weinheim.

The recombination and emission zone is directly accessible by scanning probe techniques such as scanning Kelvin probe microscopy [63] and scanning optical microscopy [64]. Even simple optical microscopy can reveal interesting phenomena. For example, light-emitting LEFETs based on uniaxially aligned polymer films show a strong dependence of emission zone width on alignment direction, as shown in Figure 11.16. When the polymer chains are aligned parallel to the charge transport direction, the emission zone can be as wide as 10 μm and asymmetric, while in the opposite case the emission zone is even narrower (<2 μm) than for unaligned polymer films [65]. For a polycrystalline polymer FET (Figure 11.16b), the emission zone appears grainy, with darker and brighter areas, while photoluminescence from the same area is uniform. Since emission is proportional to the product of hole and electron densities, the emission profile in this polycrystalline film, where mobilities vary depending on molecule orientation and grain boundaries, directly visualizes the current density distribution without the help of scanning probe techniques [66].

11.3.5 Device Modeling

Being able to model the current–voltage and light emission characteristics of ambipolar FETs is an important step toward understanding the underlying physical processes. Several levels of approximation for a single semiconductor FETs are possible. In first approximation, the recombination rate of holes and electrons can be set as infinite and the width of the recombination zone and charge density at that point to zero. In the simplest case, one can now use standard FET equations within the gradual channel approximation [67] and take advantage of the fact that the lengths of the hole and electron channels add up to the overall channel length $L = L_{\text{holes}} + L_{\text{electrons}}$ with $I_{\text{holes}} = I_{\text{electrons}} = I_{\text{ds}}$. Both channels behave as if they are in saturation, that is, they are pinched off at the point of recombination. The source–drain current in the ambipolar regime then is

$$|I_{ds}| = \frac{WC_i}{2L} \cdot \{\mu_e \cdot (V_g - V_{Th,e})^2 + \mu_h (V_{ds} - [V_g - V_{Th,h}])^2\} \quad (11.1)$$

with gate dielectric capacitance C_i, channel width W, channel length L, hole mobility μ_h, and electron mobility μ_c, assuming that $V_{Th,e} \neq V_{Th,h}$. The position of the emission zone (x_0) defined as distance from the source can now be derived as

$$x_0 = \frac{L \cdot (V_g - V_{Th,e})^2}{(V_g - V_{Th,e})^2 + \frac{\mu_h}{\mu_e}\left(V_{ds} - (V_g - V_{Th,h})\right)^2} \quad (11.2)$$

Although this model is very crude and ignores the charge density dependence of mobility as well as contact resistance, it can qualitatively reproduce experimentally observed trends. The transfer and output characteristics in Figure 11.5 were calculated using these equations.

A more sophisticated model of charge transport is based on variable range hopping in exponential densities of states and adequately describes charge transport in disordered materials, such as semiconducting polymers [68]. Here, the effective mobility μ_{eff} depends on the local charge carrier density and thus on the effective gate potential ($V_{eff} = V(x) - V_g + V_{Th}$) at position x with $\mu_{eff} = f_0 V_{eff}^{\beta-2}$ where f_0 is a prefactor and $\beta = 2T_0/T.T_0$ is a measure of the width of the exponential density of states and T is the absolute temperature [63]. Still assuming an infinite recombination rate, the source–drain current in the ambipolar regime can be derived as

$$I_{ds} = \frac{WC_i}{L}\left[\frac{f_{o,B}}{(\beta_e - 1)\beta_e}(V_g - V_{Th,e})^{\beta_e} + \frac{f_{o,h}}{(\beta_h - 1)\beta_h}(V_{ds} + (V_g - V_{Th,h})^{\beta_h}\right] \quad (11.3)$$

For $\beta = 2(T_0 = T)$ and $f_0 = \mu$, this leads to the standard transistor equations again. This transport model reproduces the gate voltage dependence of mobility and fits current–voltage and local potential data from ambipolar transistors based on SQ1 (Figure 11.10) and nickel-dithiolene very well [56, 57, 63]. Reported values for β range between 2.3 and 4 depending on the organic semiconductor.

In order to make an estimate of the recombination and emission zone width, the bimolecular recombination rate as the product of hole and electron densities and the recombination constant has to be taken into account. For disordered organic semiconductors, the rate limiting step is assumed to be the diffusion of holes and electrons in their mutual Coulomb field (Langevin recombination). The recombination rate constant B is derived to be $\frac{q_e}{\varepsilon \cdot \varepsilon_0} \cdot (\mu_h + \mu_\varepsilon)$, where q_e is the elementary charge, ε_0 is the permittivity in vacuum and ε is the dielectric constant [69]. The recombination rate can be included in the continuity equation for holes and electrons and solved numerically or analytically with certain assumptions. A very simple relation for the recombination zone width was found analytically with $w = \sqrt{4.34 \cdot d \cdot \delta \cdot}$ [70]. Here the width w (defined as width at 1/e of the maximum recombination rate) only depends on the gate dielectric thickness d and height of the accumulation zone δ(1–10 nm). The values obtained this way are very similar to those of numerical calculations and are corroborated by scanning Kelvin probe measurements of ambipolar nickel-dithiolene FETs [63].

The maximum recombination zone width was estimated to be 500 nm, which is much narrower than optically observed emission zones. This discrepancy between theory and experiment could be accounted for by limited optical resolution in many cases. However, emission zones of several micrometer width as observed for aligned polymers (Figure 11.16) and single organic crystals cannot be explained this way and their emission profile must be a representation of the recombination rate profile within the transistor channel [52, 65]. To produce such broad recombination zones, the recombination rate constants must be much smaller than expected for Langevin-type recombination. Analytical and numerical calculations show that decreasing the recombination rate by a factor of 100 compared to the expected Langevin values widens the emission zone approximately by a factor of 10 [65, 70]. Observation of these very wide emission zones suggests that for semiconductors in which transport is anisotropic and mobilities are highly dependent on molecular orientation as, for example, in some organic single crystals and uniaxially aligned liquid crystalline polymers, the Langevin equation is not applicable and a more detailed analysis of recombination rate constants is necessary [71].

11.3.6
Toward Electrically Pumped Organic Lasers

Light-emitting transistors are potential candidates for the realization of electrically pumped organic lasers. One of the fundamental problems of this endeavor is the need for an organic semiconductor with very high charge carrier mobilities to achieve sufficiently high current and exciton densities, while minimizing polaron quenching. Simultaneously, photoluminescence yield and optical gain must be maximized. Organic semiconductors with very high mobilities (1–10 $cm^2\ V^{-1}\ s^{-1}$) have a high degree of molecular order as, for example, single crystals and polycrystalline thin films of small molecules or polymers. This causes strong concentration quenching, that is, excitons diffuse rapidly to low energy or trap sites and decay nonradiatively. For example, rubrene, which has the highest hole mobilities reported for organic semiconductors so far (up to 30 $cm^2\ V^{-1}\ s^{-1}$ [72]) is a highly luminescent molecule in low concentrations but in single crystals its photoluminescence yield is below 1% [22, 53, 73]. Many conjugated polymers have high photoluminescence yields even in thin films, but they are restricted in their charge transport properties with field-effect mobilities usually much lower than 0.01 $cm^2\ V^{-1}\ s^{-1}$. Recent reports suggest that it is possible to optimize both charge carrier mobility and photoluminescence yield simultaneously through synthesis [74], but for most common organic semiconductors a compromise between the two is inevitable.

Milestones on the way to accomplishing electrically pumped organic lasers with a LEFET are the integration of feedback structures and reduction of optical losses. Integration of a distributed feedback (DFB) structure in an ambipolar polymer

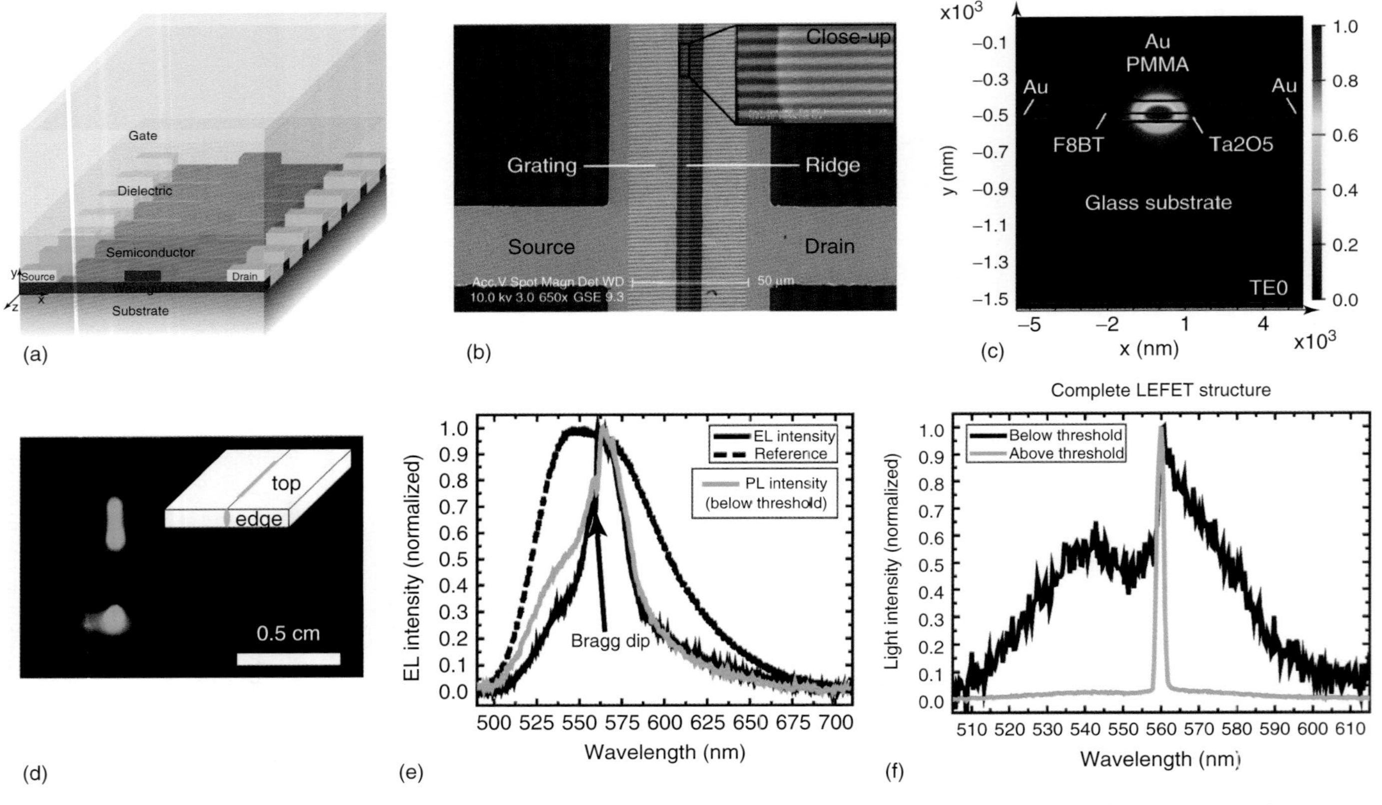

Figure 11.17 (a) Schematic illustration of bottom contact/top gate LEFET with integrated rib waveguide. (b) Scanning electron micrograph. (c) Visualization of ground-mode (TE_0). (d) Top and edge emission from of F8BT-based LEFET with rib waveguide. (e) Electroluminescence spectrum of LEFET with waveguide (solid line) compared to a reference device on bare glass (dashed line) and photoluminescence spectrum of the same device below threshold under optical excitation (green line). (f) Emission spectra of optically excited device structure including all three transistor electrodes and optimized gate architecture below and above threshold. Reprinted with permission from Ref. [75]. Copyright 2009, Wiley-VCH, Weinheim.

LEFET was recently demonstrated by Gwinner *et al.* [75]. They adopted the previously described bottom contact/top gate FET structure with F8BT as the emitting semiconductor and added a Ta_2O_5 ridge waveguide with a one-dimensional DFB grating etched into it as shown in Figure 11.17. Owing to the high refractive index of Ta_2O_5, light is efficiently coupled and confined in the waveguide when the emission zone is positioned on top of it. Consequently, top and edge emission are observed. Optimizing the periodicity of the DFB grating, the thickness of the semiconductor and polymer dielectric layer, as well as the gate electrode material led to a complete LEFET structure, which showed the same low lasing thresholds and high slope efficiencies when optically pumped as DFB samples without any electrodes (Figure 11.17f). Although F8BT LEFETs could not possibly achieve lasing because the estimated maximum exciton density (10^9 cm^{-2} for hole and electron mobilities of 0.01 cm^2 V^{-1} s^{-1} and an emission zone width of 2 μm) was 4 orders of magnitude smaller than necessary even for optically pumped lasing ($\sim 10^{12}$ cm^{-2}) [62], it demonstrated that ambipolar LEFETs are generally suitable to reduce optical losses and efficiently integrate feedback structures in organic light-emitting devices.

Organic single crystals have charge carrier mobilities that could enable sufficiently high exciton densities for lasing. Takenobu *et al.* demonstrated light emission from ambipolar tetracene and rubrene single crystal LEFETs at high current densities using gold and calcium source–drain electrodes [53]. The hole and electron mobilities were reasonably high, but the external quantum efficiencies were very low with 0.03 and 0.015%, respectively, because of the low radiative decay yield in these single crystals (<1%). A promising material to mitigate this problem is α,ω-bis(biphenylyl) terthiophene (BP3T), whose single crystals exhibit self-waveguided spectrally narrowed emission with internal efficiencies of up to 80% [76]. Bisri *et al.* realized ambipolar LEFETs with BP3T single crystals and found high mobilities (hole mobility: 1.64 cm^2 V^{-1} s^{-1}, electron mobility: 0.17 cm^2 V^{-1} s^{-1}) and strong, polarized edge emission (Figure 11.18). Similarly, the Adachi group recently found that single crystals of bis-styrylbenzene derivatives have low amplified stimulated emission thresholds (25 μJ cm^{-1}), high photoluminescence yield ($\sim$89%) and reasonably high hole and electron mobilities (0.1 and 0.009 cm^2 V^{-1} s^{-1}, respectively) [77]. These examples show that high carrier mobilities and high luminescence yield are not always mutually exclusive in organic semiconductors and high exciton densities are possible.

Despite these promising results, another fundamental obstacle to the realization of electrically pumped organic lasers remains. At high current densities, the density of long-lived triplets, which are formed three times more often than singlets, increases drastically. The associated singlet–triplet quenching could push the threshold for lasing well beyond reach in these devices [78] and thus indirect electrical pumping schemes, for example, a polymer feedback structure pumped with an InGaN LED [79], might be more feasible in the end.

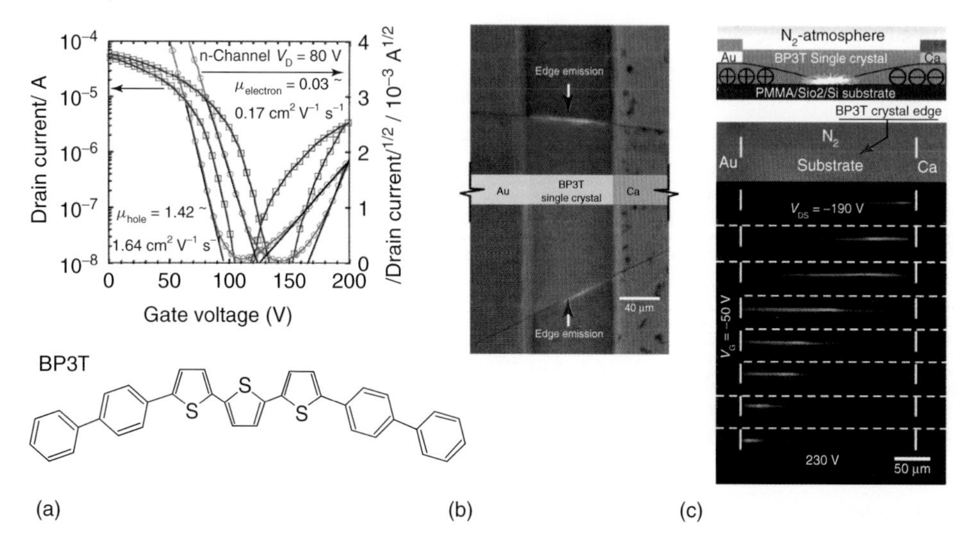

Figure 11.18 (a) Transfer characteristics of ambipolar FET based on single crystal BP3T using asymmetric gold and calcium electrodes for hole and electron injection. (b) Emission was observed only from the edges of the BP3T crystal and the emission zone moved with applied voltages from source to drain (c). Reprinted with permission from Ref. [73]. Copyright 2009, Wiley-VCH, Weinheim.

11.4 Other Field-Effect-Based Light-Emitting Devices

11.4.1 Vertical Light-Emitting Transistors

For the application of light-emitting transistors in active matrix displays, it is important to achieve low operating voltages and a high aperture ratio, that is, the ratio of light-emitting area to pixel size. Lateral LEFETs usually require high drive voltages because of their long channel lengths and low mobilities of organic semiconductors. Using a vertical transistor structure integrated with an organic LED layer enables high drive currents for small applied voltages because of the short vertical channel length as well as high aperture ratios.

In a vertical LEFET structure proposed by Xu *et al.* [80], the gate electrode is separated from the source electrode, which also acts as the electron injecting cathode, by a thin insulating layer. Applying a positive bias changes the injection barrier for electrons into the light-emitting polymer and thus modulates current and brightness. Kudo *et al.* followed a different approach and combined a static induction transistor (SIT) with an organic LED structure [81]. They inserted a grid-type Al-gate electrode into the hole-transporting layer that was sandwiched between the indium tin oxide (ITO) anode (source), an electron-transporting and emissive layer and the cathode (drain). The Al-gate blocked current by forming a double Schottky barrier and controlled the flow of holes toward the drain by adjusting the potential barrier near the gate electrode fingers. The device operated like a depletion-FET with the drain current and thus luminance decreased when a positive gate voltage was applied.

The most promising vertical LEFET structure, a metal-insulator-semiconductor-type organic light-emitting transistor (MIS-OLET), was recently employed in a flexible 16 × 16 pixel active matrix display by Nakamura *et al.* [82, 83]. As shown in Figure 11.19a, the gate electrode was separated from a slit source electrode by an insulating dielectric. A hole-transporting semiconductor such as pentacene was deposited on top so that holes would be injected and accumulated at negative gate voltages. Above the source electrode fingers, an insulating charge restriction layer was deposited in order to diffuse the holes through the hole-transport layer above and prevent leakage. The cathode (drain) electrode injected electrons at negative drain voltages into the electron transport and emission layer. The gate electrode modulated the hole density in the pentacene layer and thus drain current and brightness (up to 150 cd m^{-2}) of the MIS-OLET with sufficiently high on/off ratios for application in an active matrix display (Figure 11.19b).

11.4.2
Field-Effect Enhanced LEDs

One objective of light-emitting FETs is to spatially remove the emission zone from the injecting electrodes to reduce optical losses and enable integration of waveguide structures. Schols *et al.* proposed and demonstrated a novel LED structure that does exactly that without introducing an extra gate electrode [84, 85]. This hybrid structure combines a standard LED with an ITO anode, hole-transporting, emission, and electron-transporting layer with lateral field-effect transport above an additional insulating layer (SiO_2), as shown in Figure 11.20a. The cathode (LiF/Al) is placed above the insulator and is displaced from the emission zone by several micrometers so that the anode acts like a gate electrode short-circuited with the drain electrode/anode. The inset in Figure 11.20b shows the equivalent circuit diagram. When the device is in forward bias, electrons are injected from the cathode into the electron-transporting layer (e.g., PTCDI-$C_{13}H_{27}$). They accumulate and move along the interface with the emission layer (e.g., tris(8-hydroxyquinoline) aluminum (Alq_3) doped with 4-(Dicyanomethylene)-2-methyl-6(julolindin-4-yl-vinyl)-4H-pyran (DCM_2)). At the edge of the insulator, the higher vertical electrical field leads to injection of electrons into the emission layer, while holes are injected from the hole-transporting layer (e.g., poly(triarylamine) PTAA), resulting in light emission at this edge with an emission zone width of 2 μm. Decreasing the distance between cathode and insulator edge increases current density and brightness. The observed electroluminescence efficiencies (0.02–0.25% depending on the electron transport layer) are similar to comparable LEDs but break down voltages (~30 V) and maximum current densities are much higher (13 A cm^{-2}). Also, efficiencies do not roll off, as often found in high brightness organic LEDs [14, 86, 87]. These advantageous properties could lead to the application of this novel device as waveguide organic LEDs or even open another possible route to electrically pumped organic lasers.

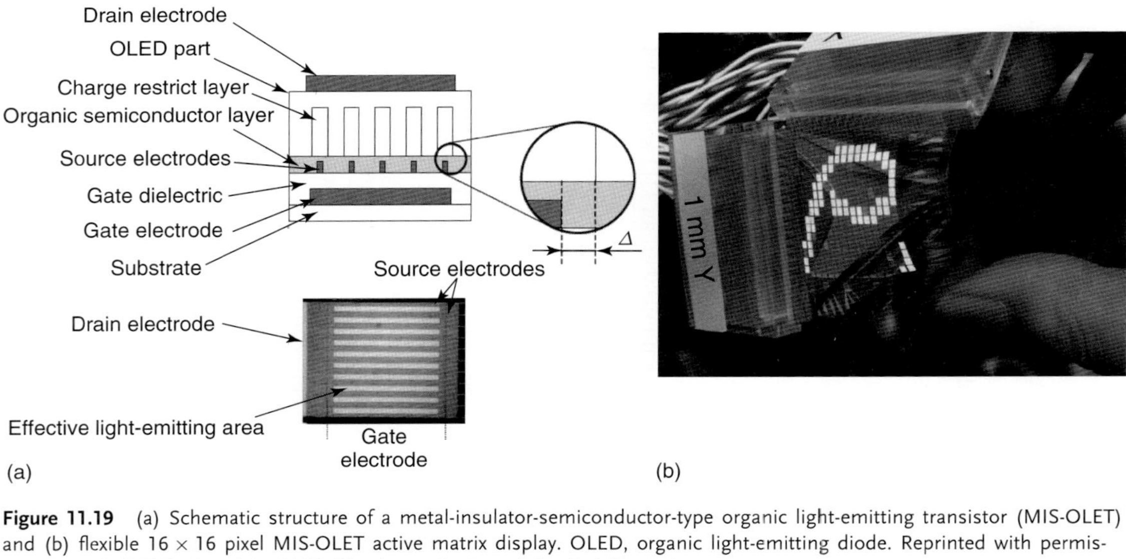

Figure 11.19 (a) Schematic structure of a metal-insulator-semiconductor-type organic light-emitting transistor (MIS-OLET) and (b) flexible 16 × 16 pixel MIS-OLET active matrix display. OLED, organic light-emitting diode. Reprinted with permission from Ref. [83]. Copyright 2008 (The Japan Society of Applied Physics).

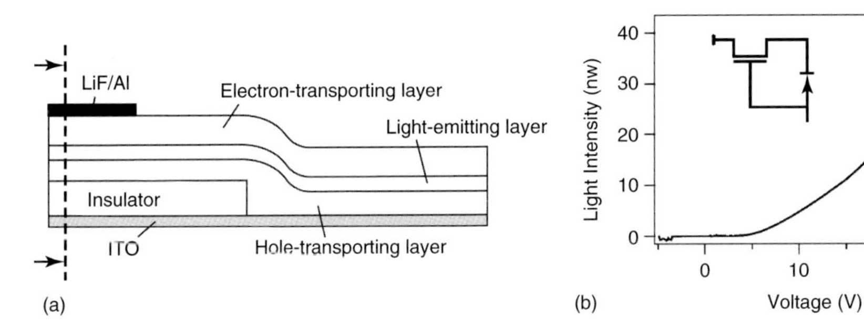

Figure 11.20 (a) Schematic device structure of LED with lateral field-effect and (b) light-intensity versus applied bias characteristics for a device with a width of 1 mm and distance between top electrode and insulator edge of 3.5 μm. Inset: equivalent circuit. Reprinted with permission from Refs. [84, 85]. Copyright 2008, Wiley-VCH, Weinheim.

11.5 Conclusions

Within a few years the field of organic light-emitting transistors as multifunctional optoelectronic devices has seen tremendous progress. Various device structures and materials were proposed and light emission over the entire visible spectrum and in the near-infrared has been demonstrated. Some of them have the potential to be applied in active matrix displays and other optoelectronic circuits. The development of new materials, which combine high-mobility charge transport with high photoluminescence yield, and continuous improvement of device architectures and environmental stability will make them more attractive for technological applications in the future. Scientifically, their planar architecture makes them a valuable tool to investigate charge injection and transport as well as recombination and emission physics of organic semiconductors and thus could help improving other types of organic electronic devices.

Acknowledgments

It is my pleasure to acknowledge the people who have made the work presented in this chapter possible, namely, Professor Henning Sirringhaus, Professor Richard H. Friend, and the Optoelectronics Group at the Cavendish Laboratory, University of Cambridge.

References

1. Sirringhaus, H., Tessler, N., and Friend, R.H. (1998) *Science*, **280**, 1741–1744.
2. Zhou, L.S., Wanga, A., Wu, S.C., Sun, J., Park, S., and Jackson, T.N. (2006) *Appl. Phys. Lett.*, **88**, 083502-1–083502-3.
3. Hack, M., Shur, M., and Czubatyj, W. (1986) *Appl. Phys. Lett.*, **48**, 1386–1388.
4. Pfleiderer, H. and Kusian, W. (1986) *Solid-State Electron.*, **29**, 317–319.

5. Saito, S., Hisamoto, D., Shimizu, H., Hamamura, H., Tsuchiya, R., Matsui, Y., Mine, T., Arai, T., Sugii, N., Torii, K., Kimura, S., and Onai, T. (2006) *Appl. Phys. Lett.*, **89**, 163504-1–163504-3.
6. Valenta, J., Juhasz, R., and Linnros, J. (2002) *Appl. Phys. Lett.*, **80**, 1070–1072.
7. Walters, R.J., Kalkman, J., Polman, A., Atwater, H.A., and de Dood, M.J.A. (2006) *Phys. Rev. B*, **73**, 132302-1–132302-4.
8. Walters, R.J., Bourianoff, G.I., and Atwater, H.A. (2005) *Nat. Mater.*, **4**, 143–146.
9. Peralvarez, M., Carreras, J., Barreto, J., Morales, A., Dominguez, C., and Garrido, B. (2008) *Appl. Phys. Lett.*, **92**, 241104-1–241104-3.
10. Feng, M., Holonyak, N., and Hafez, W. (2004) *Appl. Phys. Lett.*, **84**, 151–153.
11. Dixon, F., Chan, R., Walter, G., Holonyak, N., Feng, M., Zhang, X.B., Ryou, J.H., and Dupuis, R.D. (2006) *Appl. Phys. Lett.*, **88**, 012108-1–012108-3.
12. Samuel, I.D.W. and Turnbull, G.A. (2007) *Chem. Rev.*, **107**, 1272–1295.
13. Baldo, M.A., Holmes, R.J., and Forrest, S.R. (2002) *Phys. Rev. B*, **66**, 035321-1–035321-16.
14. Gartner, C., Karnutsch, C., Lemmer, U., and Pflumm, C. (2007) *J. Appl. Phys.*, **101**, 023107-1–023107-9.
15. Hepp, A., Heil, H., Weise, W., Ahles, M., Schmechel, R., and von Seggern, H. (2003) *Phys. Rev. Lett.*, **91**, 157406.
16. Ahles, M., Hepp, A., Schmechel, R., and von Seggern, H. (2004) *Appl. Phys. Lett.*, **84**, 428–430.
17. Cicoira, F., Santato, C., Dadvand, A., Harnagea, C., Pignolet, A., Bellutti, P., Xiang, Z., Rosei, F., Meng, H., and Perepichka, D.F. (2008) *J. Mater. Chem.*, **18**, 158–161.
18. Cicoira, F., Santato, C., Melucci, M., Favaretto, L., Gazzano, M., Muccini, M., and Barbarella, G. (2006) *Adv. Mater.*, **18**, 169–174.
19. Santato, C., Capelli, R., Loi, M.A., Murgia, M., Cicoira, F., Roy, V.A.L., Stallinga, P., Zamboni, R., Rost, C., Karg, S.E., and Muccini, M. (2004) *Synth. Met.*, **146**, 329–334.
20. Sakanoue, T., Fujiwara, E., Yamada, R., and Tada, H. (2004) *Appl. Phys. Lett.*, **84**, 3037–3039.
21. Nakamura, K., Ichikawa, M., Fushiki, R., Kamikawa, T., Inoue, M., Koyama, T., and Taniguchi, Y. (2005) *Jpn. J. Appl. Phys. Part 2: Lett. Express Lett.*, **44**, 1367–1369.
22. Oyamada, T., Uchiuzou, H., Sasabe, H., and Adachi, C. (2005) *J. Soc. Inf. Display*, **13**, 869–873.
23. Reynaert, J., Cheyns, D., Janssen, D., Muller, R., Arkhipov, V.I., Genoe, J., Borghs, G., and Heremans, P. (2005) *J. Appl. Phys.*, **97**, 114501.
24. Sakanoue, T., Fujiwara, E., Yamada, R., and Tada, H. (2005) *Chem. Lett.*, **34**, 494–495.
25. Swensen, J., Moses, D., and Heeger, A.J. (2005) *Synth. Met.*, **153**, 53–56.
26. Namdas, E.B., Ledochowitsch, P., Yuen, J.D., Moses, D., and Heeger, A.J. (2008) *Appl. Phys. Lett.*, **92**, 183304-1–183304-3.
27. Namdas, E.B., Swensen, J.S., Ledochowitsch, P., Yuen, J.D., Moses, D., and Heeger, A.J. (2008) *Adv. Mater.*, **20**, 1321–1324.
28. Sakanoue, T., Yahiro, M., Adachi, C., Burroughes, J.H., Oku, Y., Shimoji, N., Takahashi, T., and Toshimitsu, A. (2008) *Appl. Phys. Lett.*, **92**, 053505-1–053505-3.
29. Oyamada, T., Uchiuzou, H., Akiyama, S., Oku, Y., Shimoji, N., Matsushige, K., Sasabe, H., and Adachi, C. (2005) *J. Appl. Phys.*, **98**, 074506-1–074506-7.
30. Oyamada, T., Chang, C.H., Chao, T.C., Fang, F.C., Wu, C.C., Wong, K.T., Sasabe, H., and Adachi, C. (2007) *J. Phys. Chem. C*, **111**, 108–115.
31. Gehlhaar, R., Yahiro, M., and Adachi, C. (2008) *J. Appl. Phys.*, **104**, 033116-1–033116-5.
32. Meijer, E.J., de Leeuw, D.M., Setayesh, S., van Veenendaal, E., Huisman, B.H., Blom, P.W.M., Hummelen, J.C., Scherf, U., and Klapwijk, T.M. (2003) *Nat. Mater.*, **2**, 678–682.
33. Babel, A., Wind, J.D., and Jenekhe, S.A. (2004) *Adv. Funct. Mater.*, **14**, 891–898.
34. Shkunov, M., Simms, R., Heeney, M., Tierney, S., and McCulloch, I. (2005) *Adv. Mater.*, **17**, 2608–2612.

35. Rost, C., Karg, S., Riess, W., Loi, M.A., Murgia, M., and Muccini, M. (2004) *Appl. Phys. Lett.*, **85**, 1613–1615.
36. Opitz, A., Bronner, M., and Brutting, W. (2007) *J. Appl. Phys.*, **101**, 063709-1–063709-9.
37. Cho, S.N., Yuen, J., Kim, J.Y., Lee, K., and Heeger, A.J. (2006) *Appl. Phys. Lett.*, **89**, 153505-1–153505-3.
38. Loi, M.A., Rost-Bietsch, C., Murgia, M., Karg, S., Riess, W., and Muccini, M. (2006) *Adv. Funct. Mater.*, **16**, 41–47.
39. Cho, S., Yuen, J., Kim, J.Y., Lee, K., and Heeger, A.J. (2007) *Appl. Phys. Lett.*, **90**, 063511-1–063511-3.
40. Anthopoulos, T.D. (2007) *Appl. Phys. Lett.*, **91**, 113513-1–113513-3.
41. Dinelli, F., Capelli, R., Loi, M.A., Murgia, M., Muccini, M., Facchetti, A., and Marks, T.J. (2006) *Adv. Mater.*, **18**, 1416–1420.
42. Capelli, R., Dinelli, F., Loi, M.A., Murgia, M., Zamboni, R., and Muccini, M. (2006) *J. Phys.: Condes. Matter*, **18**, S2127–S2138.
43. de Vusser, S., Schols, S., Steudel, S., Verlaak, S., Genoe, J., Oosterbaan, W.D., Lutsen, L.J., Vanderzande, D.J.M., and Heremans, P. (2006) *Appl. Phys. Lett.*, **89**, 223504, 1–3.
44. Cornil, J., Gueli, I., Dkhissi, A., Sancho-Garcia, J.C., Hennebicq, E., Calbert, J.P., Lemaur, V., Beljonne, D., and Bredas, J.L. (2003) *J. Chem. Phys.*, **118**, 6615–6623.
45. Cornil, J., Bredas, J.L., Zaumseil, J., and Sirringhaus, H. (2007) *Adv. Mater.*, **19**, 1791–1799.
46. Coropceanu, V., Cornil, J., da Silva Filho, D.A., Olivier, Y., Silbey, R., and Bredas, J.L. (2007) *Chem. Rev.*, **107**, 926–952.
47. Anthopoulos, T.D., Anyfantis, G.C., Papavassiliou, G.C., and de Leeuw, D.M. (2007) *Appl. Phys. Lett.*, **90**, 122105-1–122105-3.
48. Tanase, C., Meijer, E.J., Blom, P.W.M., and de Leeuw, D.M. (2003) *Phys. Rev. Lett.*, **91**, 216601-1–216601-4.
49. Chua, L.L., Zaumseil, J., Chang, J.F., Ou, E.C.W., Ho, P.K.H., Sirringhaus, H., and Friend, R.H. (2005) *Nature*, **434**, 194–199.
50. Zaumseil, J., Friend, R.H., and Sirringhaus, H. (2006) *Nat. Mater.*, **5**, 69–74.
51. Swensen, J.S., Soci, C., and Heeger, A.J. (2005) *Appl. Phys. Lett.*, **87**, 253511-1–253511-3.
52. Takahashi, T., Takenobu, T., Takeya, J., and Iwasa, Y. (2007) *Adv. Funct. Mater.*, **17**, 1623–1628.
53. Takenobu, T., Bisri, S.Z., Takahashi, T., Yahiro, M., Adachi, C., and Iwasa, Y. (2008) *Phys. Rev. Lett.*, **100**, 066601-1–066601-4.
54. Yamane, K., Yanagi, H., Sawamoto, A., and Hotta, S. (2007) *Appl. Phys. Lett.*, **90**, 162108-1–162108-3.
55. Schidleja, M., Melzer, C., and von Seggern, H. (2009) *Appl. Phys. Lett.*, **94**, 123307-1–123307-3.
56. Smits, E.C.P., Anthopoulos, T.D., Setayesh, S., van Veenendaal, E., Coehoorn, R., Blom, P.W.M., de Boer, B., and de Leeuw, D.M. (2006) *Phys. Rev. B*, **73**, 205316-1–205316-9.
57. Smits, E.C.P., Setayesh, S., Anthopoulos, T.D., Buechel, M., Nijssen, W., Coehoorn, R., Blom, P.W.M., de Boer, B., and de Leeuw, D.M. (2007) *Adv. Mater.*, **19**, 734–738.
58. Burgi, L., Turbiez, M., Pfeiffer, R., Bienewald, F., Kirner, H.J., and Winnewisser, C. (2008) *Adv. Mater.*, **20**, 2217–2224.
59. Zaumseil, J., Donley, C.L., Kim, J.S., Friend, R.H., and Sirringhaus, H. (2006) *Adv. Mater.*, **18**, 2708–2712.
60. Zaumseil, J., McNeill, C.R., Bird, M., Smith, D.L., Ruden, P.P., Roberts, M., McKiernan, M.J., Friend, R.H., and Sirringhaus, H. (2008) *J. Appl. Phys.*, **103**, 064517-1–064517-10.
61. Schidleja, M., Melzer, C., and von Seggern, H. (2009) *Adv. Mater.*, **21**, 1172–1176.
62. Naber, R.C.G., Bird, M., and Sirringhaus, H. (2008) *Appl. Phys. Lett.*, **93**, 023301-1–023301-3.
63. Smits, E.C.P., Mathijssen, S.G.J., Colle, M., Mank, A.J.G., Bobbert, P.A., Blom, P.W.M., de Boer, B., and de Leeuw, D.M. (2007) *Phys. Rev. B*, **76**, 125202-1–125202-6.

64. Swensen, J.S., Yuen, J., Gargas, D., Buratto, S.K., and Heeger, A.J. (2007) *J. Appl. Phys.*, **102**, 013103-1–013103-5.
65. Zaumseil, J., Groves, C., Winfield, J.M., Greenham, N.C., and Sirringhaus, H. (2008) *Adv. Funct. Mater.*, **18**, 3630–3637.
66. Zaumseil, J., Kline, R.J., and Sirringhaus, H. (2008) *Appl. Phys. Lett.*, **92**, 073304-1–073304-3.
67. Schmechel, R., Ahles, M., and von Seggern, H. (2005) *J. Appl. Phys.*, **98**, 084511-1–084511-6.
68. Vissenberg, M. and Matters, M. (1998) *Phys. Rev. B*, **57**, 12964–12967.
69. Blom, P.W.M., de Jong, M.J.M., and Breedijk, S. (1997) *Appl. Phys. Lett.*, **71**, 930–932.
70. Kemerink, M., Charrier, D.S.H., Smits, E.C.P., Mathijssen, S.G.J., de Leeuw, D.M., and Janssen, R.A.J. (2008) *Appl. Phys. Lett.*, **93**, 033312-1–033312-3.
71. Groves, C. and Greenham, N.C. (2008) *Phys. Rev. B*, **78**, 155205-1–155205-8.
72. Hulea, I.N., Fratini, S., Xie, H., Mulder, C.L., Iossad, N.N., Rastelli, G., Ciuchi, S., and Morpurgo, A.F. (2006) *Nat. Mater.*, **5**, 982–986.
73. Bisri, S.Z., Takenobu, T., Yomogida, Y., Shimotani, H., Yamao, T., Hotta, S., and Iwasa, Y. (2009) *Adv. Funct. Mater.*, **19**, 1728–1735.
74. Yap, B.K., Xia, R.D., Campoy-Quiles, M., Stavrinou, P.N., and Bradley, D.D.C. (2008) *Nat. Mater.*, **7**, 376–380.
75. Gwinner, M.C., Khodabakhsh, S., Song, M.H., Schweizer, H., Giessen, H., and Sirringhaus, H. (2009) *Adv. Funct. Mater.*, **19**, 1360–1370.
76. Kanazawa, S., Ichikawa, M., Koyama, T., and Taniguchi, Y. (2006) *ChemPhysChem*, **7**, 1881–1884.
77. Kabe, R., Nakanotani, H., Sakanoue, T., Yahiro, M., and Adachi, C. (2009) *Adv. Mater.*, **21**, 4034–4038.
78. Giebink, N.C. and Forrest, S.R. (2009) *Phys. Rev. B*, **79**, 073302-1–073302-4.
79. Yang, Y., Turnbull, G.A., and Samuel, I.D.W. (2008) *Appl. Phys. Lett.*, **92**, 163306-1–163306-3.
80. Xu, Z., Li, S.H., Ma, L., Li, G., and Yang, Y. (2007) *Appl. Phys. Lett.*, **91**, 092911-1–092911-3.
81. Kudo, K. (2005) *Curr. Appl. Phys.*, **5**, 337–340.
82. Nakamura, K., Hata, T., Yoshizawa, A., Obata, K., Endo, H., and Kudo, K. (2006) *Appl. Phys. Lett.*, **89**, 103525-1–103525-3.
83. Nakamura, K., Hata, T., Yoshizawa, A., Obata, K., Endo, H., and Kudo, K. (2008) *Jpn. J. Appl. Phys.*, **47**, 1889–1893.
84. Schols, S., Verlaak, S., Rolin, C., Cheyns, D., Genoe, J., and Heremans, P. (2008) *Adv. Funct. Mater.*, **18**, 136–144.
85. Schols, S., McClatchey, C., Rolin, C., Bode, D., Genoe, J., Heremans, P., and Facchetti, A. (2008) *Adv. Funct. Mater.*, **18**, 3645–3652.
86. Baldo, M.A., Adachi, C., and Forrest, S.R. (2000) *Phys. Rev. B*, **62**, 10967–10977.
87. Nakanotani, H., Sasabe, H., and Adachi, C. (2005) *Appl. Phys. Lett.*, **86**, 213506-1–213506-3.

12
Design Methodologies for Organic RFID Tags and Sensor Readout on Foil

Kris Myny, Hagen Marien, Soeren Steudel, Peter Vicca, Monique J. Beenhakkers, Nick A.J.M. van Aerle, Gerwin H. Gelinck, Jan Genoe, Wim Dehaene, Michiel Steyaert, Paul Heremans, and Eugenio Cantatore

12.1
Introduction

A major scientific challenge in recent years is the quest to realize complex electronics functions based on organic thin-film transistors (OTFTs). The technology and modeling of these devices is still very much in development. Because of this, integrated circuit designers have to cope, on the one hand, with technology imperfections such as large statistical parameter variations, defects, and parameter instability under bias and, on the other hand, with poor electronic design automation (EDA) systems that are not capable of supporting design with accurate simulations and predictions of the experimental results.

Nevertheless, the applications of organic transistors are of broad industrial interest and span the domain of low-cost, large-area electronics. In this chapter, the most circuit design intensive applications of OTFTs are explored : (large area) OTFT-based sensor systems and radio frequency identification (RFID) tags.

The former application is based on the idea that organic transistors can interface large-area sensors to produce foils where the sensing function is distributed on a (relatively) large area. Example of this kind sensor foils are spoilage detectors for high-value packaged goods (from expensive food to pharmaceuticals), artificial tactile skins, intelligent surfaces that can sense the presence of electronic objects and provide wireless energy to recharge them, intelligent walls and floors for tracking and communication, and so on.

RFIDs, on the other hand, are an application that concentrates on low cost and integration of RFID electronics, antenna, and goods wrapping to enable identification solutions for each item in a retail shop. This kind of technology would enable far-reaching advancements both in the retail logistics and in the experience of users, who could checkout their shopping basket without any human intervention, could ask for an automatic inventory of their fridge, or leave their microwave oven choose the best recipe for the food they bought.

Organic Electronics II: More Materials and Applications, First Edition. Edited by Hagen Klauk.

In Section 12.2, the state of the art of RFID technology based on OTFTs is thoroughly investigated, discussing in detail the architectures used in RFID electronics and showing that nowadays organic RFID tags can be compliant with simple standards developed for mainstream Si CMOS solutions.

In the third part of this chapter, a detailed circuit-level analysis of state-of-the-art organic integrated electronics is presented. The discussion starts presenting several alternative approaches to the design of basic digital building blocks, which are vital to enable any possible OTFT circuit application: design trade-offs and the advantages and disadvantages of the different solutions are analyzed. Later on the state of the art of organic analog and mixed-signal OTFT circuit design is discussed, putting special attention on design techniques that can used to mitigate the adverse effects of technology imperfections. These circuits are paramount to enable large-area sensor and communication applications for OTFTs.

12.2 Organic RFID Tags

A typical application of organic circuits is RFID tags, that is, electronic circuits that are coupled to an antenna which enables a two-way electromagnetic link to a base station, also called *reader* (Figure 12.1). The reader sends a radio carrier to the tag, which is used on the tag to generate the DC voltage needed to power the electronics function. In its simplest embodiment, the RFID is capable of sending a digital message, an identification code stored in its memory, to the reader using the same carrier used for the power transmission. More sophisticated RFIDs can also provide a data link between the reader and the tag and employ rather complex communication protocols.

If one could build RFIDs tags on flexible substrates, these electronic labels could be used to identify items in the retail distribution, replacing the bar codes. To make this vision true, it is paramount to enable:

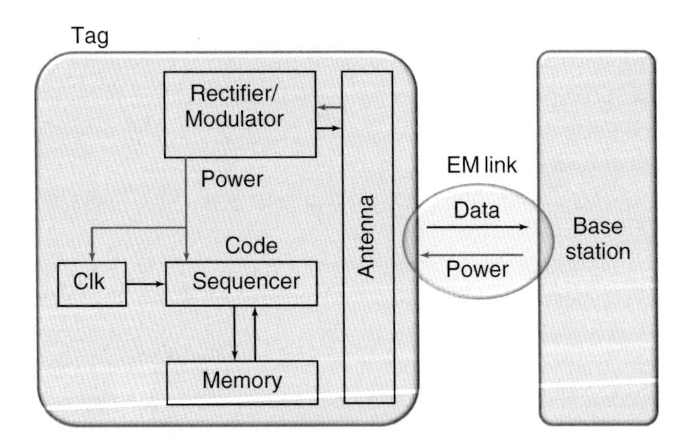

Figure 12.1 Schematic diagram of an RFID system.

- Ultra low cost (<10 dollarcent per tag, typically).
- Develop a simple and cheap technology flow that allows integration of the RFID tags with the wrapping of the retail articles.

Organic electronics, which can be manufactured at very low temperature on basically any kind of substrate and using high-throughput manufacturing processes such as industrial printing, holds the promise to enable both these characteristics: this is the basic motivation for the work on organic-based RFID tags.

To achieve low cost, not only the production of the electronic functions itself has to be extremely cheap but also inexpensive ways to make antennas and to couple the antenna to the electronics must be devised.

In the rest of this section, we discuss the state of the art of organic RFIDs, showing that organic technologies are capable to comply with the basic performance criteria used in standards for silicon RFID implementations. However, please note that the given circuits are realized using photolithography and that further steps are needed to enable a low-cost manufacturing technology for organic RFID tags. More work has to be done, however, both in the implementation of these electronic functions with lower cost using high-throughput printing-based manufacturing processes and in the implementation of low-cost antennas, including the development of the coupling process with the electronics.

12.2.1 Capacitively Coupled RFID Tags

The electromagnetic coupling between RFID reader and tag is mostly realized in the reactive near-field region (when the communication distance is $<0.16\,\lambda$); thus, the coupling can be realized with a mainly electric or mainly magnetic field. In the first case, the antennas will look like facing conductive plates, that is, like a capacitor, and we often speak of capacitive coupling. In the second case, the antennas are loops and the coupling can be though as a loosely coupled air-core transformer. In this case, we speak of inductive coupling.

As it will be shown in next subsection, inductive coupling between the RFID tag and the reader can be realized with cheap antennas only at relatively high frequency, (typically starting at the so-called HF RFID: 13.56 MHz), when a relatively high coil resistance can be tolerated and thus the thickness (and cost) of the antenna material can be kept low.

A schematic of capacitive coupling is shown in Figure 12.2, where the intentional capacitive coupling is shown with the capacitance C_c, the parasitic capacitance to ground is modeled by the capacitances C_{Rgnd} and C_{Tgnd}, the output impedance of the reader is Z_o, and the input impedance of the RFID tag is Z_{in}. The voltage generated on the reader side is V_{read}, and the input voltage on the tag is V_{in}. The coupling capacitance is inversely proportional to the distance between the reader and the tag. The impedance levels typically found in the divider network shown in the figure together with the maximum electric field that can be generated from the reader and the minimum voltage on the tag (the latter value depends on the minimum voltage that can be rectified and on the minimum voltage needed for the

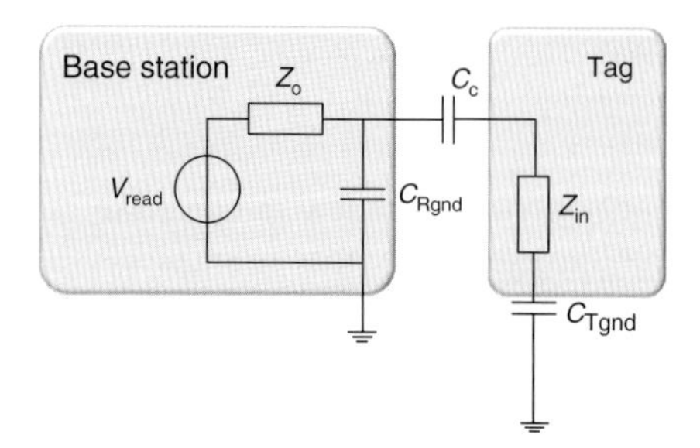

Figure 12.2 Schematic diagram of a capacitively coupled RFID.

tag electronics) are such that, for capacitive tags having the dimension of a credit card, the maximum reading distance is just a few centimeters [1].

Reduced reading distance is the main disadvantage of capacitive coupling. On the other hand, with this approach, the resistivity of the tag antenna flap is not important, because of the large cross-section of this antenna and to the high impedance level of the capacitive divider network. This makes it possible to build functional capacitive antennas with very cheap printed conductive layers based on carbon inks, as it has been demonstrated already by Rasul and Olson [2].

12.2.2 Inductively Coupled RFID Tags

A schematic of an inductively coupled RFID tag is shown in Figure 12.3. Such RFID tags use inductive antennas to communicate with the reader. These antennas are composed of an LC network, which resonates at the base carrier frequency used to transmit the power to the tag. The efficiency of this LC network is characterized by the quality (Q) factor. The Q factor of an antenna coil is proportional to the frequency and its inductance value and inversely proportional to the series resistance of the antenna (Eq. (12.1)).

$$Q = \frac{\omega L}{R}. \tag{12.1}$$

Inductive coupling is preferred as long as the Q factor can be made high ($Q > 10$). The series resistance of the antenna, or the thickness of the metal wire, is one of the determining factors of the cost of such an antenna. Starting at 13.56 MHz, sufficient vales of the quality factor can be obtained with low-cost antennas and thus inductive-coupled RFID systems are preferred. The efficiency of the antenna strongly influences the received voltage. As RFID systems are often limited by the amplitude of the voltage generated at the antenna output, the LC resonant network in the inductively coupled scheme is beneficial with its large impedance, and reading distances up to 1 m can be reached with inductively coupled tags for vicinity readers (ISO 15963 standard).

Our 13.56 MHz inductive-coupled RFID tag comprises three different building blocks: the antenna, the rectifier, and the transponder (Figure 12.3). The antenna

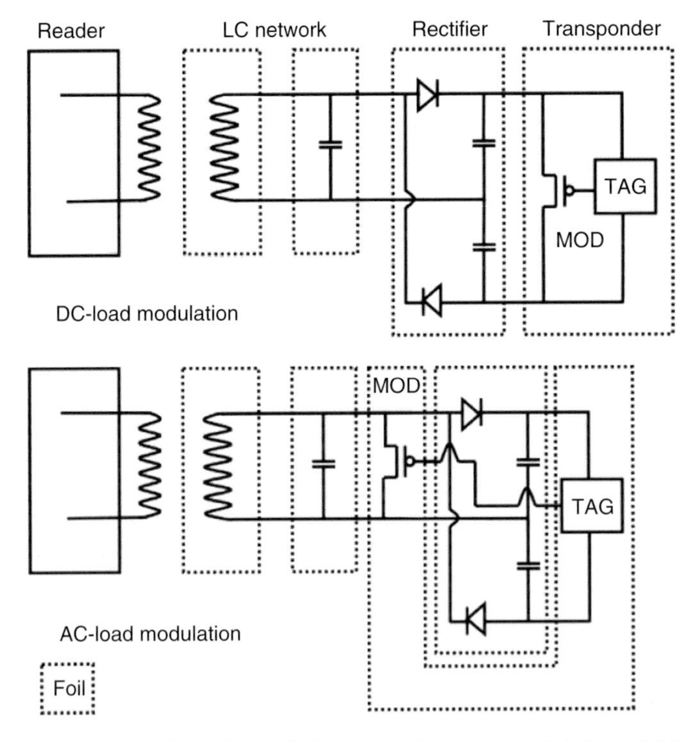

Figure 12.3 Inductively coupled organic RFID tags using DC (a) and AC (b) load modulation.

is composed of an inductor coil and a capacitor. The coil is made from etched copper on foil and was manufactured by Hueck Folien GmbH. The HF capacitor consists of a metal-insulator-metal (MIM) stack, processed on a 200 μm thick flexible polyethylene naphthalate (PEN) foil (Teonex Q65A, Dupont Teijin Films). The insulator material used for the capacitor is Parylene diX SR.

12.2.2.1 **Transponder Chip**

Organic transponder chips are made on a 25 μm thin plastic substrate using organic bottom-gate thin-film transistors. The organic electronics technology that is used was developed by Polymer Vision for commercialization in rollable active matrix displays and is described elsewhere [3]. The insulator layers and the semiconductor layers are organic materials processed from solution. The transistors, with a typical channel length of 5 μm, exhibit an average saturation mobility of $0.15\,\mathrm{cm^2\,V^{-1}\,s^{-1}}$.

In this section, we describe two transponder chips, a 64 bit and a 128 bit including extra functionality [4]. The block-level schematic of the 64 bit code generator is depicted in Figure 12.4. When powered, a clock signal is generated by a 19 stage ring oscillator. This signal is used to clock the output register, the 3 bit binary counter, and the 8 bit line select. The 8 bit line select has an internal 3 bit binary

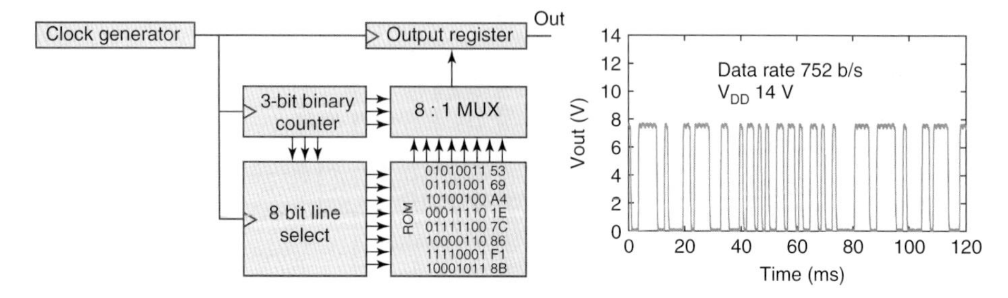

Figure 12.4 Block-level schematic of the digital logic of the 64 bit transponder chip and its output code.

counter and a 3–8 decoder. This block selects a row of 8 bits in the code memory. The 3 bit binary counter drives the 8 : 1 multiplexer, selecting a column of 8 bits in the code matrix. The data bit at the crossing of the active row and column is transported via a 8 : 1 multiplexer to the output register, which sends this bit on the rising edge of the clock to the modulation transistor. The outputs of the 3 bit binary counter are also used in the 8 bit line select block for selecting a new row after all 8 bits in a row are transmitted.

The 64 bit transponder foil comprises only 414 OTFTs. At 14 V supply voltage, it generates the correct code at a data rate of 752 b s^{-1}, as shown in Figure 12.4. This foil has been used to realize an integrated, 64 bit, organic RFID tag.

The schematic overview of the more advanced transponder chip is shown in Figure 12.5. This chip comprises a basic ALOHA anticollision protocol and Manchester encoding of the data. Also, the number of bits has been doubled to 128. This design employs 1286 OTFTs.

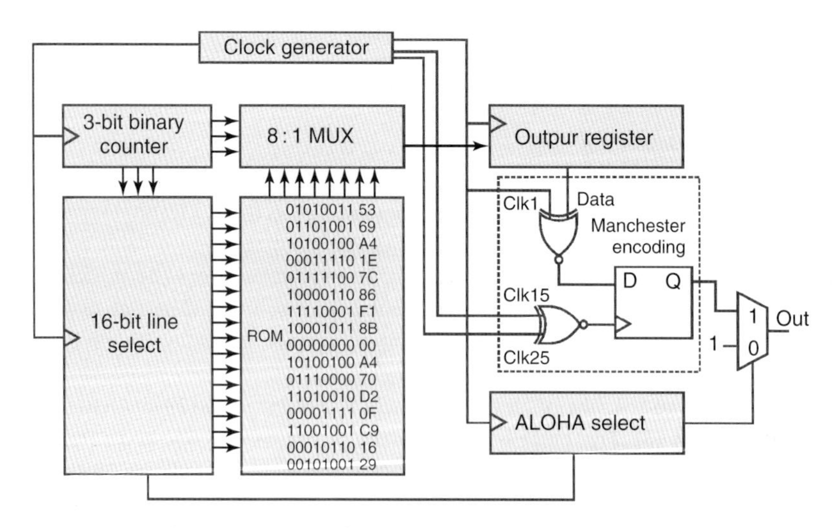

Figure 12.5 Schematic overview of the digital logic of the 128 bit transponder chip.

In this design, a 33 stage ring oscillator is used to generate the clock signal, determined by a larger critical time path as a consequence of the added complexity. The difference between the 128 and 64 bit transponder chip is the complexity of the line select, which is now a 16 bit line select. As in the previous transponder, the code is hard wired in the chip.

Data encoding is added to this transponder chip, more specifically Manchester encoding. A detail of the digital logic used to implement the Manchester encoding block is depicted on the right of Figure 12.5. Manchester encoded data requires, besides the normal bit transitions, also a transition in the middle of the bit. A transition from 0 to 1 corresponds indeed to a logic "0," while the opposite transition encodes a logic "1." In our design, every transition needs a rising edge of the clock. To include Manchester encoding to this scheme without losing data rate, thus, a clock having double frequency is generated making an EXNOR of the outputs of stages 15 and 25 of the 33 stage ring oscillator, as shown in Figure 12.5. The data is subsequently encoded using another EXNOR.

To enable the readout of multiple organic RFID tags at once, a basic anticollision protocol is added to the plastic RFID transponder chip. The anticollision protocol used is a basic version of ALOHA, which is a "tag talks first" protocol. A tag sends its code after which a silent period follows. The code is then retransmitted. During the silent period, another tag can be read out. If a tag transmits its code during the transmitting time of another tag, a collision occurs and the code is consequently not valid. A full ALOHA protocol should also allow the reader to acknowledge the successful detection of the code, after which the tag remains silent. This has not been implemented here as this transponder does not realize any communication from the reader to the tag.

In the implementation of this ALOHA protocol, a 12 state modulo-counter is used to select whether the data, coming from the Manchester encoder, should be sent out (value of the counter is 0000) or the supply voltage should be connected to the load modulator (all other values of the counter). The clock for this counter is generated by the 3 bit binary counter and the 16 bit line select. In this way, the silent period takes 12 times the time necessary to stream out all data bits. Figure 12.6 depicts the measurement results of the 128 bit transponder chip, including Manchester encoding and the ALOHA protocol. This chip was powered with a supply voltage of 24 V and employs 1286 OTFTs. The 128 bits can be read out in 83.7 ms, that is, at a bit rate of 1529 Hz.

12.2.2.2 **Integrated RFID Transponder**

The complete tag is realized by properly interconnecting the contacts of four foils (antenna, HF capacitor, rectifier, and transponder), which we achieved in an experimental setup where we plug the individual foils into sockets. Alternatively, we have also manufactured tags by lamination of the foils, whereby electrically conductive glue is used to interconnect the different contacts of the individual foils.

The reader setup conforms to the ECMA-356 standard for "RF Interface Test Methods." It comprises a field generating antenna and two parallel sense coils,

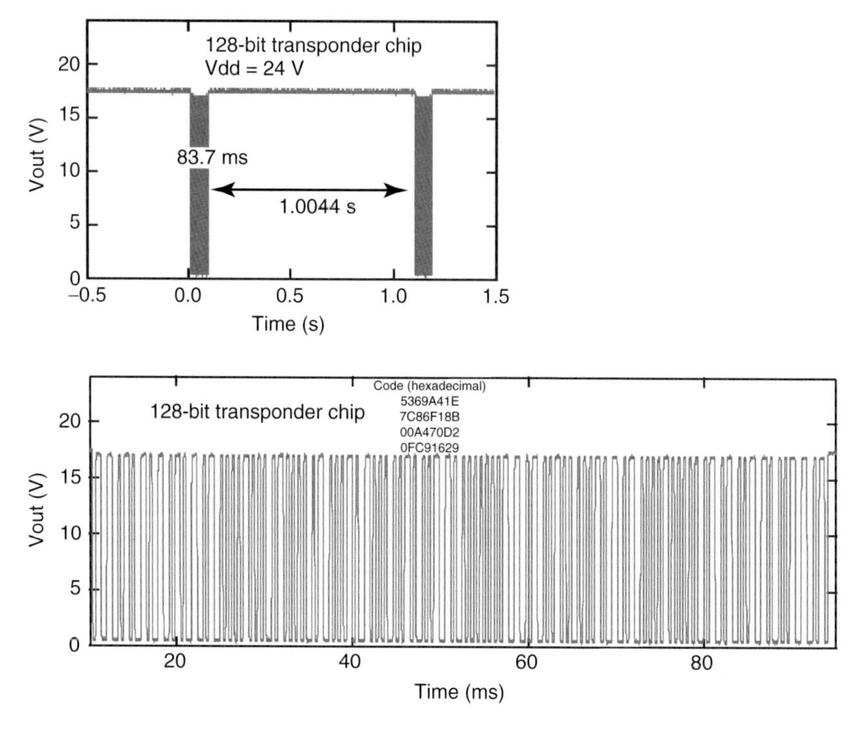

Figure 12.6 (a) Measured signal of the 128 bit organic transponder chip with a power supply of 24 V. (b) Zoom of one period of Figure 12.6a where the full code of the 128 bit organic transponder chip is shown.

which are matched to cancel the emitted field [4]. By this method, only the signal sent by the RFID tag is read out at the reader side. The detected signal is then demodulated by a simple envelope detector (inset of Figure 12.8), that is, a diode followed by a capacitor and a resistor, and shown on an oscilloscope.

The basic schematic of the organic RFID tag presented here is depicted in Figure 12.3. The coil and the HF capacitor form an LC tank resonating at the HF resonance frequency of 13.56 MHz, which provides the energy for the organic rectifier with an AC voltage at 13.56 MHz. The rectifier generates the DC supply voltage for the 64 bit organic transponder chip, which drives the modulation transistor between the on- and off-state with a 64 bit code sequence. Load modulation can be obtained in two different modes, depending on the position of the load modulation transistor in the RFID circuit, as shown in Figure 12.3. AC load modulation, whereby the modulation transistor is placed in front of the rectifier, sets demanding requirements to the OTFT, since it has to be able to operate at HF frequency. This is not obvious, as a consequence of the limited charge-carrier mobility of the OTFT, which is $0.1–1\ cm^2\ V^{-1}\ s^{-1}$ for the flavor of pentacene we use as organic semiconductor. Therefore, load modulation at the output of the rectifier (DC-load modulation) has been preferred in our organic RFID tags. In latter mode, the OTFT does not require to operate at HF frequency.

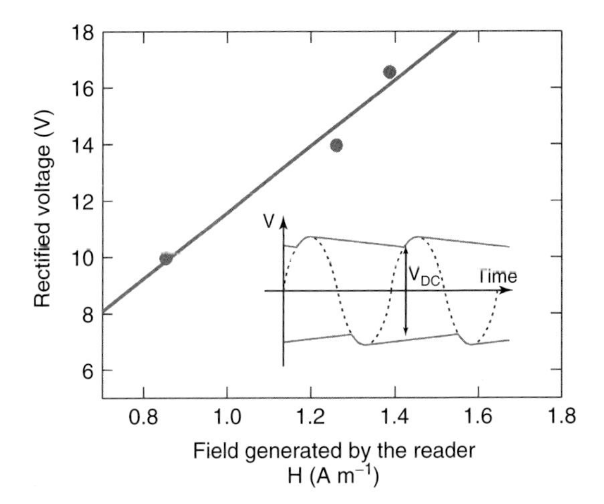

Figure 12.7 Output DC voltage generated by a double half-wave rectifier in an organic RFID tag plotted versus the strength of the 13.56 MHz magnetic field generated by the reader. Inset, schematic operation of a double half-wave rectifier.

The rectifier has to be able to rectify at the carrier frequency of 13.56 MHz and to deliver a minimal DC supply voltage of 14 V, sufficient to drive the 64 bit transponder. In order to achieve this, a more complex rectifier circuit has been implemented, more specifically a double half-wave rectifier [5]. A double half-wave rectifier circuit consists of two single half-wave rectifiers connected between the same nodes, with diodes connected as shown in Figure 12.3. Both single half-wave rectifiers rectify the AC input voltage: one rectifies the upper cycles of the AC input voltage and the other rectifies the lower cycles of the input voltage. The power and the ground voltage for the digital logic of the RFID tag are taken between both rectified signals, as schematically depicted in the inset of Figure 12.7. This figure also plots the rectified voltage obtained in DC-load modulation versus the field generated by the reader. As can be seen in the graph, a 14 V rectified voltage is obtained under the loading conditions found in the tag at a 13.56 MHz electromagnetic field of 1.26 $A\,m^{-1}$. The ISO 14443 standard states that RFID tags should be operational at a minimum required RF magnetic field strength of 1.5 $A\,m^{-1}$. The double half-wave rectifier circuit presented here, therefore, satisfies this ISO norm. After extrapolation of the measurement data, a DC voltage of 17.4 V can be obtained at a field of 1.5 A/m. If a single half-wave rectifier was used, then the rectified voltage would be limited to 8–9 V, which is too low for current organic digital circuits.

The DC voltage from the rectifier drives the transponder chip, which sends the code to the modulation transistor. The signal sent from the fully integrated, plastic tag is received by the reader and subsequently visualized using a simple envelope detector (see inset Figure 12.8) without amplification, as described above. The signal measured at the reader side is depicted in Figure 12.8. This shows the

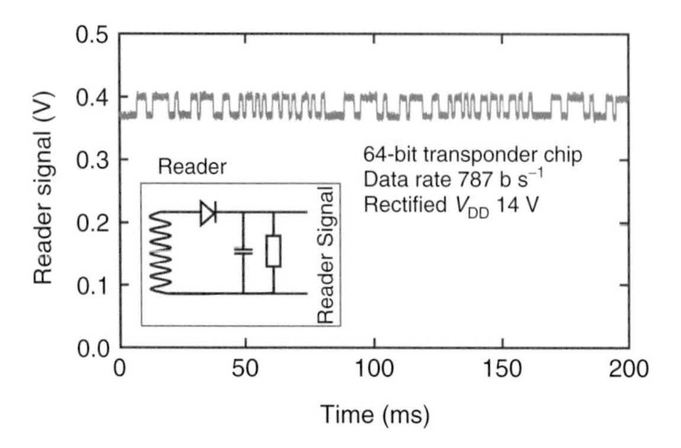

Figure 12.8 Signal of the 64 bit RFID tag measured on the reader (unamplified reader signal). The envelope detector of the reader is depicted in the inset.

fully functional, 64 bit RFID tag using an inductively coupled 13.56 MHz RFID configuration with a data rate of 787 b s^{-1}.

Two of the reader standards at 13.56 MHz base carrier frequency are the proximity (ISO 14443) and vicinity readers (ISO 15693). The main difference between those standards is the coil radius, being 7.5 cm for the proximity reader and 55 cm for the vicinity reader. This results in a maximum readout distance of typically 10 cm for the proximity and 1 m for the vicinity reader. As mentioned earlier, the former standard (ISO 14443) states also that the tag should be operational at an RF magnetic field of 1.5 A/m, which is significantly lower than the maximum allowed RF magnetic field of 7.5 A/m. One can calculate the required magnetic field at the reader antenna center in order to obtain the required field to operate the tag. In the demonstrators reported on here, the required electromagnetic field to operate

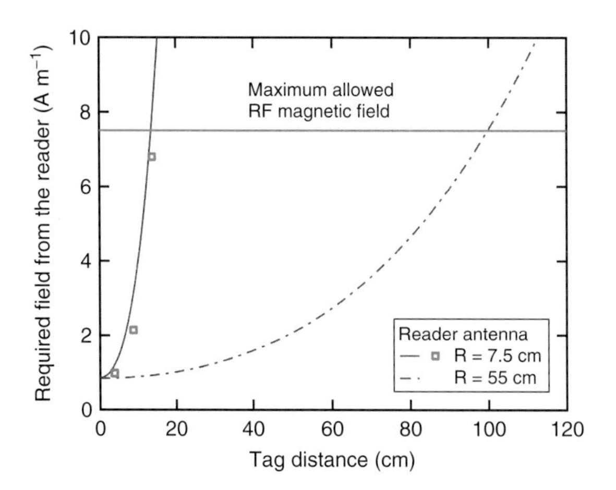

Figure 12.9 Calculation and experimentally obtained data of the required RF magnetic field at the reader side as a function of the tag distance in order to generate the required RF magnetic field to operate the tag.

an 8 bit organic RFID tag was 0.97 A/m (measured at the tag side). It is obvious that the corresponding RF field emitted by the reader must increase as the distance reader–tag increases. This is depicted in Figure 12.9. The dots in this graph show the experimental data at distances of 3.75, 8.75, and 13.75 cm with respect to the field generating antenna. This graph shows that it is possible to energize the 8 bit organic RFID tag at maximum readout distances for proximity readers below the maximum allowed RF magnetic field.

12.3 Transistor-Level Design with Organic Transistors

OTFTs may find application in basically two types of integrated circuits. Digital circuits process digital information. An example is a transponder chip for an identification tag, where Boolean information from a memory is read out and prepared for RF transmission to a base station. The second type of circuits is the so-called analog circuits, typically used for conditioning, filtering, and amplifying analog signals coming from sensors or used in (RF) communication. A special category of circuits, finally, is employed to convert analog signals into a stream of Boolean digits (mixed-signal circuits). Design considerations for both digital and analog/mixed-signal organic circuits will be discussed in the next two subsections.

12.3.1 Design Considerations for Digital Organic Circuits

Digital circuits process digital information, represented in most cases by two voltage levels, one near the supply voltage (logic "1") and one near the ground voltage (logic "0"). Building blocks not only can perform simple Boolean functions as an inverter or a NAND but can also provide more complex functions like a full adder or include memory of the previous inputs like a flipflop, and so on. In this subsection, we will focus our discussion on an inverter, a circuit that inverts the input signal. From a Boolean perspective, if the input is "0," the output becomes "1" and vice versa.

The design of an inverter can be easily extended to a NAND, and based on inverters and NANDs, any digital function can be, in principle, built, both with and without memory.

The inverter design in current Silicon MOSFET technologies is based on a CMOS inverter, whereby the pull-up stage is a *p*-type transistor and an *n*-type transistor is used for the pull down (Figure 12.10a). This design requires an approximate matching between *n*- and *p*-type transistors and therefore a technology that is more complex than a unipolar technology (a process where only one type of transistor is available). Current mature organic technologies offer only *p*-type transistors, resulting in *p*-type-only building blocks and circuits, although substantial progress is currently made toward organic complementary technologies. The design of

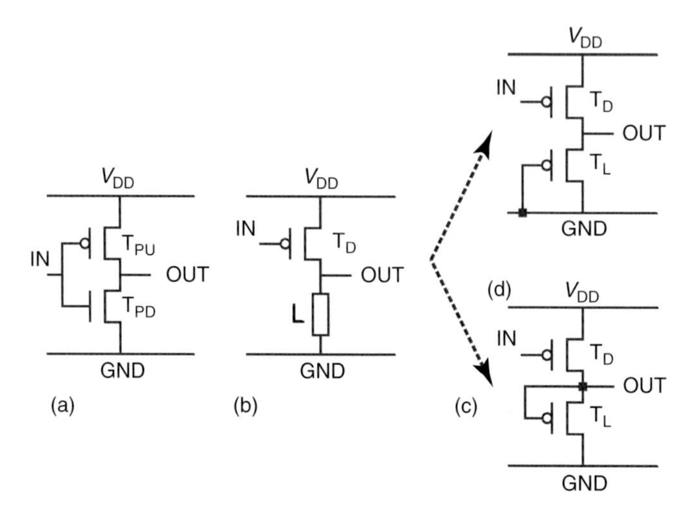

Figure 12.10 Different implementations for a digital inverter in (a) complementary logic and (b) unipolar *p*-type logic having a load L. This load is replaced by a *p*-type transistor that can be (c) diode-load or (d) zero-vgs-load connected.

a unipolar *p*-type inverter is shown in Figure 12.10b, whereby the load can be implemented as a resistor, for example, using a conductive material such as PEDOT:PSS [6]. However, to avoid the integration of such extra material in the thin film transistor (TFT) process, the load is most often implemented as a load transistor T_L (Figure 12.10c,d). This is similar to the silicon *n*-MOS logic used in the 1970s, except that, in the silicon technology, one could use the doping of the semiconductor to define a different threshold voltage for the driver and load transistors ("dual-V_T" technology), while in OTFT technology, no reproducible method is known yet to control the threshold voltage of two adjacent transistors independently. Therefore, OTFT logic is mostly unipolar *p*-type and single-V_T.

12.3.1.1 **Unipolar Logic**

The gate of the load transistor can be connected to its source (i.e., the output node of the inverter, see Figure 12.10d) or to the drain (i.e., the ground of the inverter, see Figure 12.10c). The former is called "*zero-vgs-load*," or "*depletion-load*" logic, as it requires that the load transistor would be a "depletion" transistor, that is, have a positive threshold voltage. The architecture of Figure 12.10c is called instead "*diode-load*" logic, because a transistor with shorted gate drain has diodelike current–voltage characteristics. In contrast to the zero-vgs-load architecture of Figure 12.10d, it is compatible with enhancement mode transistors and mildly depletion mode transistors.

Figure 12.11 depicts some figure of merit of an inverter stage, more precisely the trippoint (V_{trip}), the maximum gain, and noise margin. The noise margin [7] measures the immunity of the inverter against inevitable noise picked up by the circuit and adding to the input voltage. There are many definitions of the noise margin, but a convenient one is the "maximum equal criterion." According to

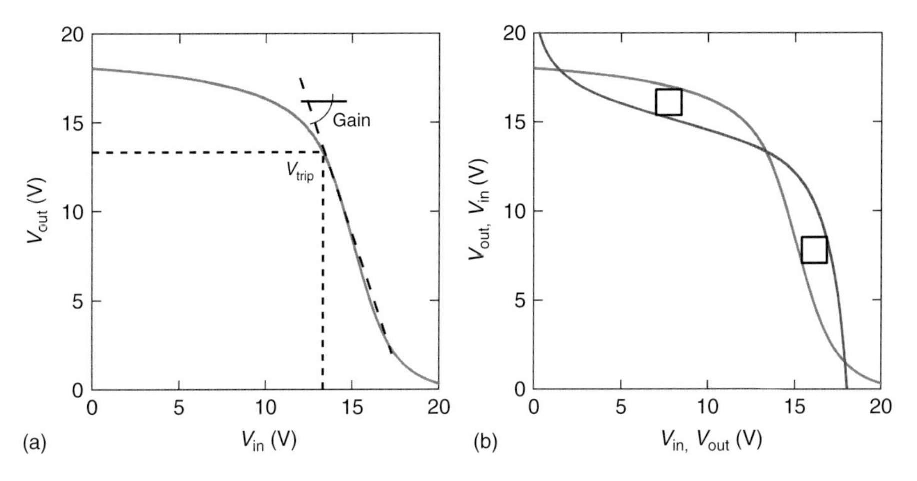

Figure 12.11 (a) Output voltage as a function of input voltage of the inverter circuit, with definition of trip voltage and gain. (b) Determination of the noise margin of the inverter as the size of the largest square fitting between inverter curve and its mirrored curve.

this criterion, the noise margin is found by mirroring input and output voltages in the "V_{out} versus V_{in}" graph and determining the size of the maximum square that fits between the original and mirrored curves, as shown in Figure 12.11b. The noise margin of a basic inverter allows predicting the size of a circuit that can be integrated with a certain yield [7].

The noise margin of unipolar organic inverters is limited by the poor gain of unipolar logic aggravated by the overall poor control and reproducibility of the threshold voltage of organic transistors [8], and by the difficulty to place the trip voltage at middle rail ($V_{DD}/2$) in a unipolar, single-V_T technology. The noise margin improves with increasing V_{DD}, which explains why many organic circuits with a significant amount of transistors require substantial V_{DD} to operate properly. It also improves when the gate dielectric can be made very thin, culminating in gate dielectrics formed as a self-assembled monolayer on the gate electrode [9].

The noise margin is a useful figure of merit to determine the yield of complex logic circuits integrating a large number of logic gates (or inverter stages) in the presence of parameter spread (like V_T and mobility spread). As an example, we show in Figure 12.12 the expected calculated yield as a function of the number of concatenated inverter stages, with as parameter the variability in the threshold voltage. Yield is calculated as the joint probability that all inverter stages will have a noise margin >0. The right panel shows that the number of stages increases with V_{DD} thanks to the increasing average noise margin.

12.3.1.2 **Complementary Logic**

Complementary technology is a powerful option to increase the noise margin of inverters and other logic gates and therefore increase yield and provide higher robustness against parameter variations compared to unipolar, single-V_T circuits.

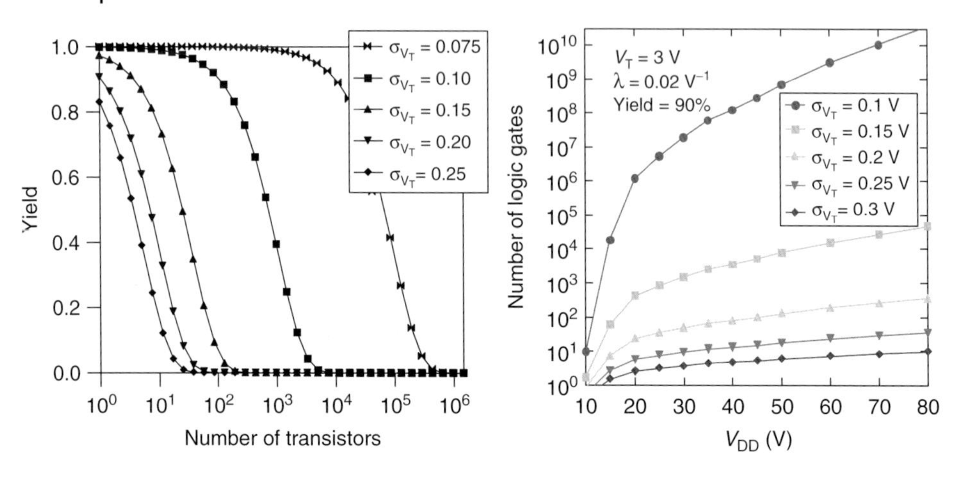

Figure 12.12 (a) Worst case calculation of the circuit yield as a function of the number of inverters, calculated for $V_{DD} = 15$ V and an average V_T of 3 V. (b) The average noise margin increases for increasing V_{DD}, such that the number of logic gates that can be integrated also increases (other parameters kept the same as for (a)).

The schematic of a complementary inverter is depicted in Figure 12.10a. It consists of a *p*-type pull-up transistor and an *n*-type pull-down transistor. In contrary to unipolar inverters, both transistors actively drive the output node. At a low input voltage, the pull-up transistor is switched on, while the pull-down transistor is (almost) off. This pulls the output node to V_{DD}. When the input voltage is high, the pull-down transistor pulls the output to 0 V while the pull-up transistor is (almost) off. A proper complementary technology will consequently yield a voltage swing from 0 to V_{DD}, called *rail-to-rail* and will avoid static power consumption in the inverter. Another advantage is a much larger gain than for unipolar inverters. The resulting noise margin is much larger (for a given V_{DD}) than for unipolar logic; therefore, the complexity of circuits that can be integrated with complementary logic is significantly larger than that of (single-V_T) unipolar logic [10]. This is quantified in Figure 12.13. Several research groups have demonstrated some complex integrated complementary circuits, more specifically a 4 bit organic RFID tag [11] and a 2 V organic CMOS decoder used in an electromagnetic interference (EMI) detection sheet [12].

12.3.1.3 Dual-V_T Logic

As already stated above, organic complementary technology requires reasonable matching between the performance of p- and n-type materials, both in terms of threshold and mobility. This has been for long time a difficult goal to reach with OTFT technology. A simpler alternative route to improve the robustness of unipolar organic circuits is the dual-V_T approach. If the threshold voltage of the load transistor in unipolar logic can be controlled independently of the threshold voltage of the drive transistor, the trip point of unipolar inverters (both zero-vgs-load and diode-load) can be designed to be at its ideal value $V_{DD}/2$ and the noise margin can

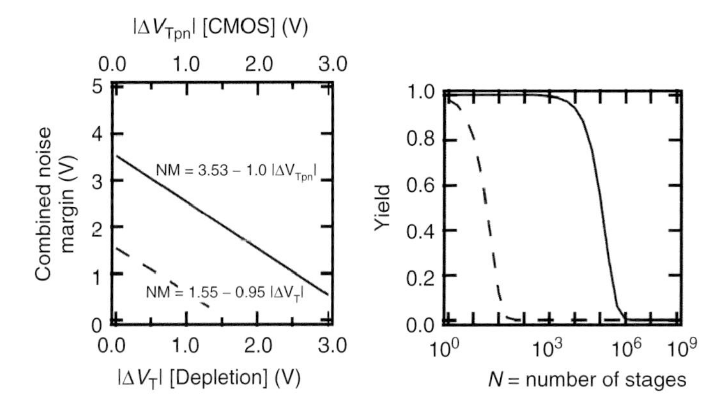

Figure 12.13 (a) Calculated noise margin for depletion-load *p*-logic inverter (Figure 12.10d) and complementary inverter (Figure 12.10a) and (b) corresponding to the larger noise margin of complementary logic, the number of inverter stages that can be integrated in complementary logic (full line) is much larger than for *p*-channel logic (dashed line) for a given yield. The parameters for the calculation are $V_T = 2.5$ V, $\sigma V_T = 0.5$ V, and $V_{DD} = 15$ V.

be significantly improved. A practical way to achieve a second threshold voltage is based on *p*-type organic transistors equipped with a double gate [13, 14]. The extra gate on the back side of the channel (backgate) is used to vary the V_T measured at the front gate using electrostatic control on the channel, which effectively leads to a dual-V_T *p*-type technology. Figure 12.14 depicts the measured transfer curve of a dual-gate, zero-vgs-load *p*-type-only inverter. When increasing the applied voltage to the backgate of the drive transistor ($V_{BG,D}$, see Figure 12.5), the threshold voltage of this transistor becomes increasingly more negative, and as a consequence, the trip point moves toward $V_{DD}/2$, yielding a higher noise margin and consequently improved robustness. This threshold voltage control enabled integration into a 64 bit RFID transponder chip, designed in both zero-vgs-load and diode-load topologies [15]. It emerged from Ref. [15] that dual-gate control yields speed for the diode-load logic, whereas mainly robustness at low-power voltage is obtained for zero-vgs-load logic. Several groups have demonstrated circuits using the effect of a dual-gate transistor, realizing dual-gate unipolar inverters [16, 17], dual-gate complementary inverters [18], dual-gate organic SRAMs [19], a dual-gate (zero-vgs load and diode load) 64 bit RFID transponder chip [15], and an analog-to-digital converter (ADC) [20].

12.3.2
Design Considerations for Analog and Mixed-Signal Organic Circuits

Technologies for organic electronics are still very young, compared to the mature monocrystalline silicon technologies. They still suffer from large process and parameter variations that influence transistor behavior and accordingly affect circuit accuracy and reliability. Analog circuits are very sensitive to all kinds of

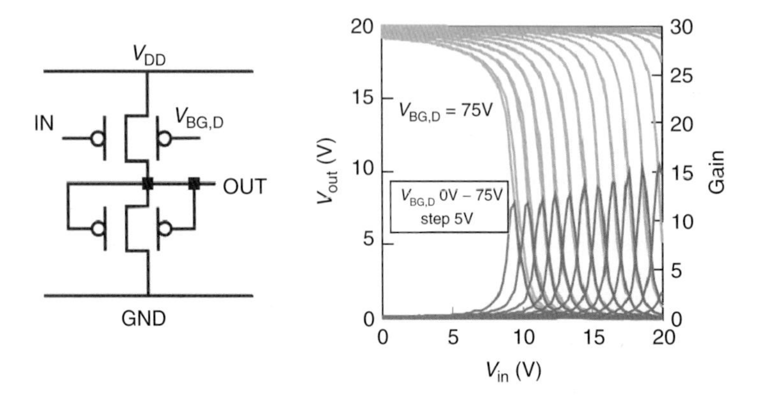

Figure 12.14 Zero-vgs-load topology whereby each transistor is replaced by a dual-gate transistor. The extra gate or backgate controls the threshold voltage. The measured transfer curves of this inverter are plotted to the right for a supply voltage of 20 V, as a function of the voltage on the V_T-control gate of the drive transistor.

parametric variations present in the technology. Therefore, the design of analog and mixed-signal circuits in organic technologies is very challenging. In order to overcome these challenges, several analog techniques can be applied to reduce the sensitivity of the analog building blocks to the variation of specific parameters.

12.3.2.1 **Technology Challenges**

In this subsection, an overview is given of the technological challenges that hamper analog circuit design. In the next subsection, circuit design techniques are suggested to mitigate the effect of these technology imperfections.

Probably the most important drawback at the moment is the evolutionary unbalance between *p*-type and *n*-type organic transistors. Until now, *p*-type transistors are applied in a vast majority of the organic technologies, whereas *n*-type transistors are still in their infancy. Although complementary circuits have already been published, such as in [21], the most literature on organic circuits concentrates on *p*-type-only solution. In the latter approach, the lack of *n*-type devices forms a very important drawback. This means that analog circuit blocks such as current mirrors, where the source contact of the transistor is connected to one of the supply voltages, can only be used when the *p*-OTFTs are connected to the positive supply voltage. Additionally, OTFTs suffer from a low intrinsic gain. This limits the gain in amplifiers and lowers the gain in feedback loops. A third important challenge for analog design is the considerable variation of the physical parameters of the transistors, especially of the threshold voltage V_t. Variations of the V_t induce a large and nonlinear current variation in the transistor current. Moreover, bias stress degradation and degradation caused by the O_2 and H_2O in the environment penetrating in the organic semiconductor result in V_t shifts [22–24] that are difficult to counteract. During the following sections, transistors with equal sizing and biasing will be assumed to undergo the same V_t variations. Following the discussion here above, it is very important to reduce V_t sensitivity when designing analog circuit blocks.

12.3.2.2 Circuit Design Techniques for Analog Building Blocks

As explained in the previous section, when designing a circuit with a given functionality, that is, an amplifier, design solutions must be found to reduce the sensitivity of the circuit behavior to V_t variations without hampering the desired functionality of the circuit, that is, the gain. Several analog techniques exist that can be applied to this purpose. In this section, a range of these techniques are discussed and compared, taking as example a single-stage amplifier on foil.

The search for a good amplifier starts with a single-ended amplifier, shown in Figure 12.15a, consisting of one OTFT and an ideal resistor. The gain of this amplifier is $g_m{}^*(r_{load}//r_0)$, being r_0 the output resistance of the transistor. This gain is typically maximized choosing a small gate overdrive $V_{sg} - V_t$. In this topology, a variation of V_t has an identical effect toward the output voltage as an inverse variation of the input voltage. This follows assuming a transistor current which is a quadratic function of $V_{sg} - V_t$. In order to quantify the sensitivity to V_t of a circuit, the V_t suppression ratio (VTSR) is defined in Eq. (12.2) as the voltage gain of the amplifier divided by the undesired influence of the input transistor V_t on the output voltage. For this single-ended circuit, VTSR = 1 = 0 dB.

$$\mathrm{VTSR} = \left| \frac{\dfrac{\partial V_{out}}{\partial V_{in}}}{\dfrac{\partial V_{out}}{\partial V_{t,in}}} \right| \tag{12.2}$$

A very well-known technique to separate desired and undesired signals in an amplifier is to use a differential amplifier, shown in Figure 12.15b. In such a differential amplifier, V_t has only a DC common-mode effect on the output voltage. Depending on the output resistance of the tail current source, the VTSR can theoretically go up to infinity. Nevertheless, the current source is not ideal and simulations based on suitable models of our transistors show a VTSR of 4.5 dB in this circuit. The sensitivity of this differential amplifier can be reduced by applying common-mode feedback (CMFB). The simplest kind of CMFB is the linear CMFB, shown in Figure 12.15c. This technique employs two transistors biased in the linear region, which is good for low V_t sensitivity (as the gate overdrive will be large). On the other hand, this topology results in a very low gain in the CMFB loop; therefore, it only has a small effect on the VTSR. The loop gain is slightly improved by adding a cascode transistor to the circuit, as shown in Figure 12.15d. The simulated VTSR is now 11.5 dB.

A solution must also be found to replace the ideal load resistors with transistors. This part suffers from the lack of *n*-type transistors. For technologies where *n*-type transistors are available, this paragraph is obsolete. A *p*-type transistor as load can, in the simplest form, be connected in two different ways, either with the gate connected to the drain, the so-called diode-connected load shown in Figure 12.16a, or with the gate connected to the source, the so-called zero-vgs load, shown in Figure 12.16b. For the diode-connected load, the transistor is in saturation, with high $V_{sg} - V_t$, which is interesting for a reduced V_t sensitivity. However, the drawback of this load topology is the very low resistance realized, namely $1/g_m$. The

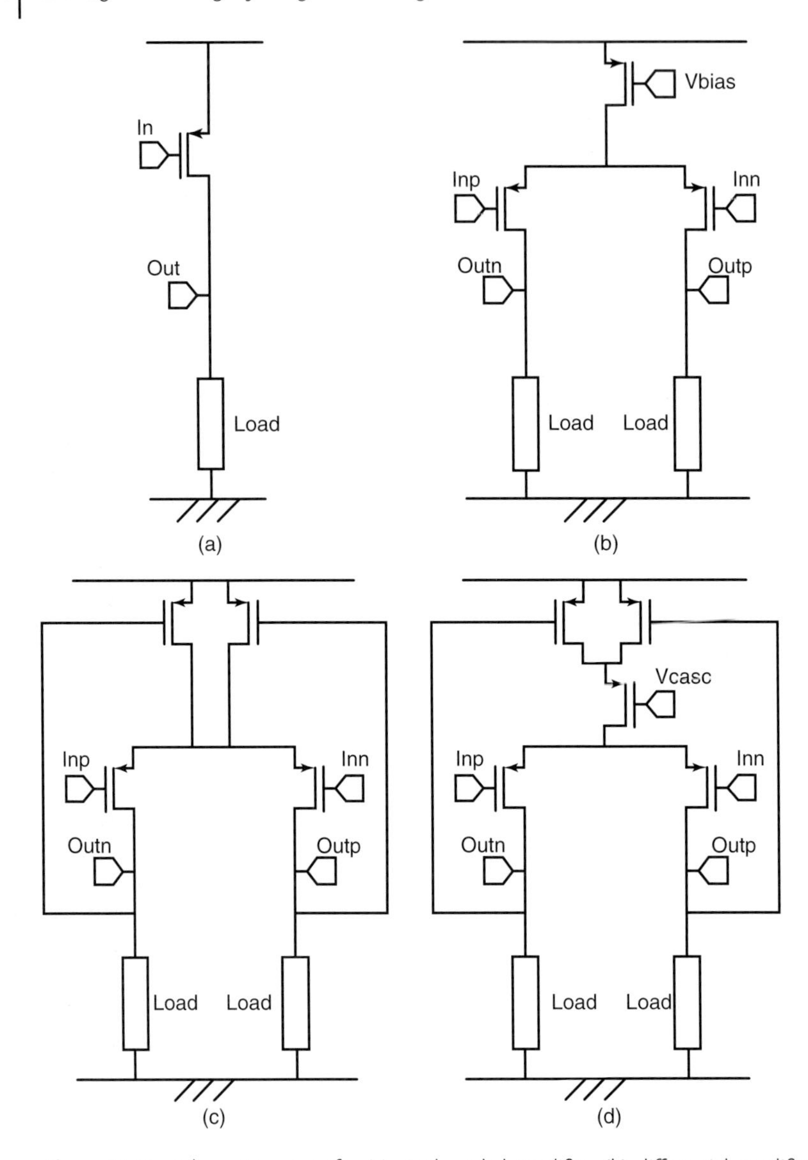

Figure 12.15 Schematic views of a (a) single-ended amplifier, (b) differential amplifier, (c) differential amplifier with CMFB, and (d) differential amplifier with CMFB and cascode transistor.

low-load resistance is the main reason for not choosing this topology. The zero-vgs load, on the other hand, has a high output resistance, which is good and preferable for an amplifier. However, this topology is much more sensitive to V_t than the diode-load topology, as the gate overdrive is minimal. Since both topologies have an important advantage over the other one, it would be interesting to apply a technique that combines both advantages in one topology. This is made possible with the

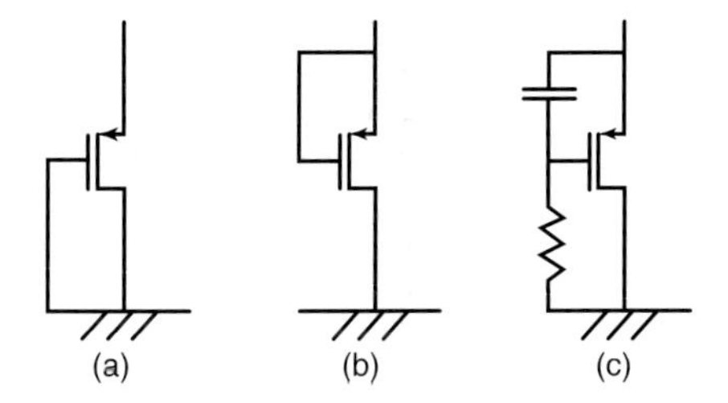

Figure 12.16 Three different *p*-OTFT load topologies: (a) diode-coupled load, (b) zero-vgs load, and (c) bootstrapped load.

topology shown in Figure 12.16c. This topology connects the source with the gate through a high-pass filter. Therefore, in DC, the circuit will behave as the diode configuration, enjoying the advantage of low V_t sensitivity, whereas, in AC, the circuit has a high output resistance r_0, which is beneficial for gain. Of course, also this topology needs to be carefully designed since the location of the filter cutoff frequency may interfere with the circuit behavior. The high-pass pole should be at the lowest frequency possible, which is, however, limited by the parasitic parallel resistance of the capacitor.

Simulations are performed to compare the behavior of the different load topologies. Figure 12.17 shows the behavior of each amplifier for the case where a V_t mismatch ΔV_t is present in one of the load transistors. In Figure 12.17a, the DC output offset voltage is plotted versus ΔV_t, whereas, Figure 12.17b, the corresponding circuit gain is visualized. Concerning the offset, both the diode and the bootstrapped loads, which behave equally in DC, are less sensitive than the zero-vgs load, which actually scores pretty bad in this case. The gain of the zero-vgs load is also very sensitive to mismatch, whereas for the bootstrapped load, the gain is high and insensitive to mismatch. The diode load shows a stable but low gain

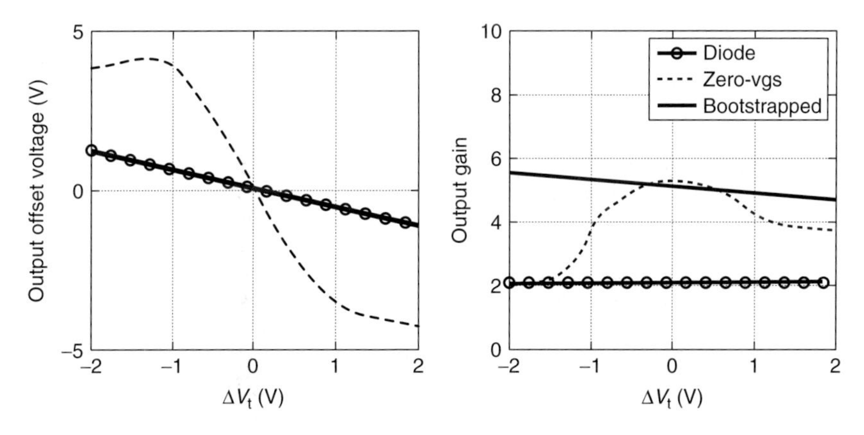

Figure 12.17 Simulated output offset and gain of three differential amplifiers, with different loads, when V_t mismatch ΔV_t is present in the load transistor.

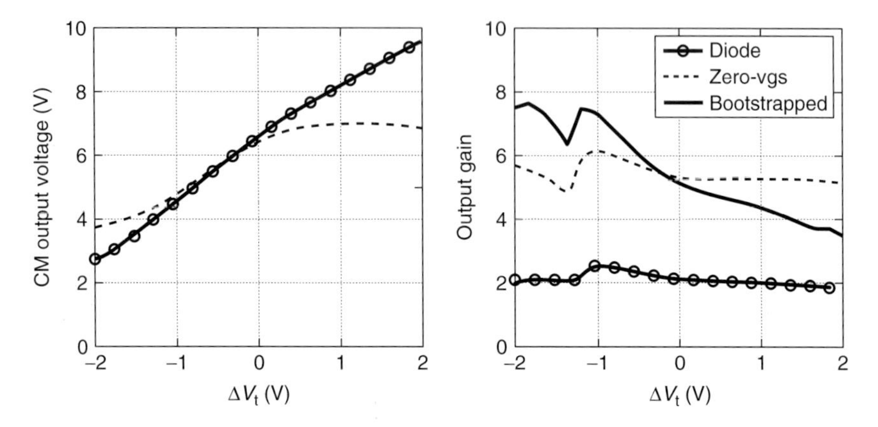

Figure 12.18 Simulated common-mode output voltage and gain of three differential amplifiers, with different loads, when a ΔV_t shift is present in the transistors.

as was expected. Figure 12.18 shows the behavior of the amplifiers with different loads for the case of a shift of the V_t common to all transistors. In the left panel, the common-mode DC output voltage is plotted. As expected, the DC behavior of the bootstrapped load is identical to the behavior of the diode load. The pretty large voltage swing of both loads is caused by the V_t sensitivity of the input pair. The swing for the zero-vgs load is smaller, since the load shows a similar V_t sensitivity to the input pair. This load performs well under this disturbance, especially for a high V_t, where the zero-vgs load is not in the cutoff region anymore. In the right panel, the AC behavior is visualized. Both the zero-vgs and the bootstrapped load perform with high gain, whereas the diode load has only a very low gain. The zero-vgs load has a constant gain, since the V_{gs} for both load and input transistors remains constant, whereas the bootstrapped load undergoes a considerable V_{gs} swing and hence a large variation of its gain. From the comparison in both figures, conclusions about the preferable loads can be drawn. Since the gain for the diode load is too low, this load topology is not preferable. The choice between the zero-vgs load and the bootstrapped load depends on the situation. The gain for zero-vgs load is more sensitive to mismatch, whereas the gain for bootstrapped load is more sensitive to an overall V_t shift. Moreover, as discussed in the previous sections, CMFB can be applied that also counteracts common-mode effects caused by the V_t shift. When V_t mismatch is small in a technology, the zero-vgs load can be suitable for analog design; however, the bootstrapped load is the more conservative kind of load available.

An additional interesting technique for analog circuit is backgate steering [13]. This technique applies a backgate on the top of the bottom-gate OTFT device. The technique has a twofold effect on the transistor behavior. It forms a distinct backgate transistor, with source, drain, and backgate pins, which draws its own drain current. This secondary transistor, however, turns out to have small influence on the overall drain current, due to the rather large thickness of the backgate dielectric. The

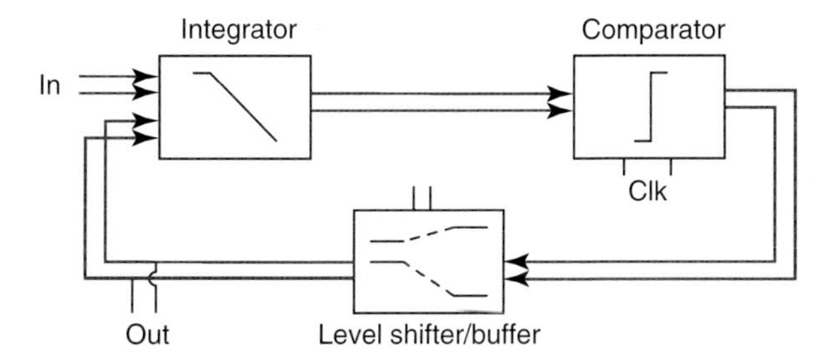

Figure 12.19 Schematic system level representation of the CT $\Delta\Sigma$ ADC and its subblocks, the integrator, and the comparator.

second observed behavior is that the backgate voltage influences the V_t of the initial transistor (with source, drain, and front gate). The effect is rather linear and results in a ~80% increase of g_m when gate and backgate are connected together.

The differential amplifier with CMFB, cascode, bootstrapped load, and backgate steering is processed and measured on foil. The circuit performs with 15 dB gain and has a gain-bandwidth (GBW) of 10 kHz. The measured CMRR, which corresponds to the VTSR, is 12 dB.

12.3.2.3 **Design of a Sigma-Delta Analog-to-Digital Converter**

The design considerations formulated in the previous paragraph are exploited to realize a more complex mixed-signal organic circuit on foil, a continuous-time (CT) delta-sigma ($\Delta\Sigma$) ADC [20]. In Figure 12.19, the schematic of this ADC is shown.

The sigma-delta modulator consists of an analog filter, a comparator, and a level shifter and counts a total of 129 transistors. The analog filter is a differential first-order opamp-C (operational amplifier) integrator, built with a three-stage opamp. Each one of the three stages of the opamp is a sized differential amplifier, built according to the considerations described in the previous section. In order to assure the stability in the unity feedback configuration, Miller capacitors and feedforward blocking resistors are added to the opamp. The comparator is built with multiple differential amplifier stages and two latches. Figure 12.20 shows the measured output spectrum of the $\Delta\Sigma$ ADC. The measured signal-to-noise ratio (SNR) is 26.5 dB for an oversampling ratio (OSR) of 16, a clock frequency of 500 Hz, and an input signal of 10 Hz. The circuit draws 100 μA from a 15 V supply voltage. The chip photograph of the integrated $\Delta\Sigma$ ADC is shown in Figure 12.21. A summary of the measurement results is given in Table 12.1. A more detailed analysis of this design can be found in [20].

12.4 Conclusions

In this chapter, we have shown how complex designs, both digital designs as analog designs, can be realized in *p*-type-only organic technologies. Organic RFID tags

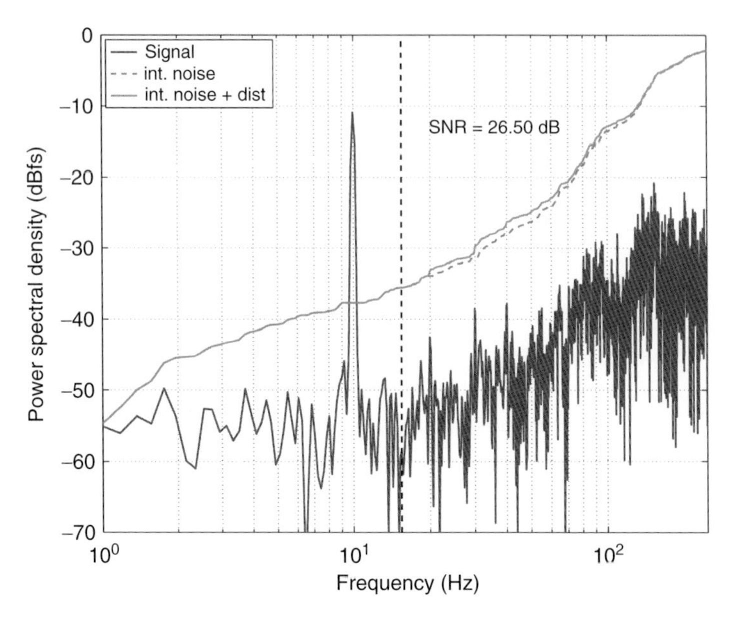

Figure 12.20 Measured output spectrum and integrated noise of the ΔΣ ADC for a 10 Hz sinusoidal input and a 500 Hz clock frequency.

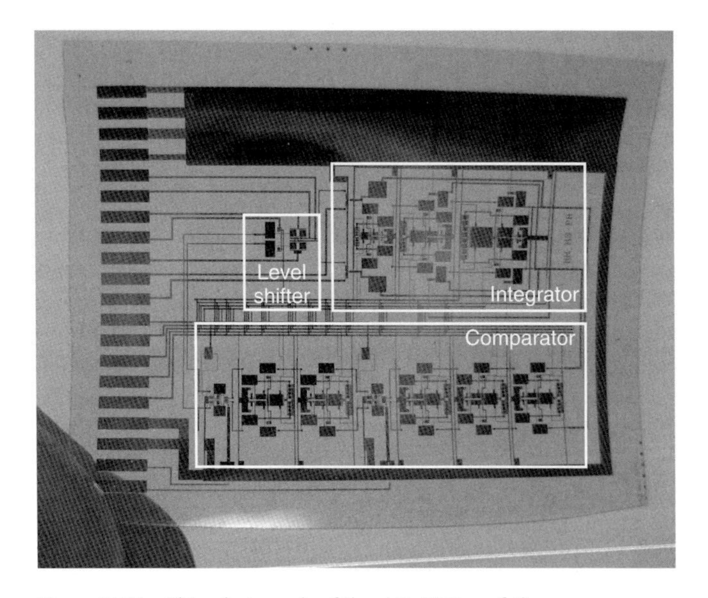

Figure 12.21 Chip photograph of the ΔΣ ADC on foil.

Table 12.1 Summary of the measurement results of the organic ΔΣ ADC.

Parameter	Value
Minimal width	25 μm
Minimal length	5 μm
Typical intrinsic gain	5
f_T for minimal transistor	20 kHz
Specification	Value
Current consumption	100 μA
Power supply	15 V
Clock frequency	500 Hz
SNR	26.5 dB
SNDR	24.5 dB
OSR	16
BW	17 Hz
Active area	$13 \times 20\ mm^2$

have been used as a template application for digital design in organic technologies, and the organic sigma-delta ADC shows the capabilities of current technologies in the field of analog and mixed-signal technologies. It emerges that threshold voltage control is crucial for organic technologies in both application fields. The use of dual-gate technologies enabled additional control over the threshold voltage and hence results in substantial improvements in yield and performance. This work shows that ubiquitous sensors and tags produced in a low-cost organic technology are certainly possible in future. However, as the results shown in this chapter have been obtained using classical technologies (such as lithography), further steps are still needed toward upscaling and low-cost production.

Acknowledgments

This work was performed in collaboration between IMEC and TNO in the frame of the HOLST Centre. Part of it has been supported by the EC-funded IP POLYAPPLY (IST #507143). The authors also thank Klaus Schmidegg of Hueck Folien GmbH for the antenna foils.

References

1. Cantatore, E., Geuns, T.C.T., Gelinck, G.H., van Veenendaal, E., Gruijthuijsen, A.F.A., Schrijnemakers, L., Drews, S., and de Leeuw, D.M. (2007) A 13.56-MHz RFID system based on organic transponders. *IEEE J. Solid-State Circuits*, **42**, 84–92.
2. Rasul, J. and Olson, W. (2002) Flip chip on paper assembly utilizing anisotropic conductive adhesive. Proceedings of the 2002 Electronic Components and Technology Conference, pp. 90–94.
3. Huitema, H.E.A. (2008) Rollable displays: the start of a new mobile device

generation. 7th Annual Flexible Electronics & Displays Conference USDC, Phoenix, Arizona, January 2008.

4. (a) Myny, K., Steudel, S, Vicca, P, Beenhakkers, M.J, van Aerle, N.A.J.M, Gelinck, G.H, Genoe, J, Dehaene, W, and Heremans, P. (2009) Plastic Circuits and tags for 13.56 MHz radio-frequency communication. *Solid State Electron.*, **53** (12), 1220–1226; (b) Myny, K, Van Winckel, S, Steudel, S, Vicca, P, De Jonge, S, Beenhakkers, M.J, Sele, C.W, van Aerle, N.A.J.M, Gelinck, G.H, Genoe, J, and Heremans, P. (2008) An inductive coupled 64 bit organic RFID tag operating at 13.56 MHz with a data rate of 787 b/s. *International Solid-State Circuits Conference Digest Technical Papers*, February 2008, pp. 282–283; (c) Myny, K, Beenhakkers, M.J, van Aerle, N.A.J.M, Gelinck, G.H, Genoe, J, Dehaene, W, and Heremans, P. (2009) A 128 bit organic RFID transponder chip, including Manchester encoding and ALOHA anti-collision protocol, operating with a data rate of 1529b/s. *International Solid-State Circuits Conference Digest Technical Papers*, February 2009, pp. 206–207.

5. Myny, K., Steudel, S., Vicca, P., Genoe, J., and Heremans, P. (2008) An integrated double half-wave organic Schottky diode rectifier on foil operating at 13.56 MHz. *Appl. Phys. Lett.*, **93**, 093305.

6. Sirringhaus, H., Kawase, T., Friend, R.H., Shimoda, T., Inbasekaran, M., Wu, W., and Woo, E.P. (2000) High-resolution inkjet printing of all-polymer transistor circuits. *Science*, **290** (5499), 2123–2126.

7. Lohstroh, J., Seevinck, E., and de Groot, J. (1983) Worst-case static noise margin criteria for logic circuits and their mathematical equivalence. *IEEE J. Solid-State Circuits.*, **SC-18** (6), 803–807.

8. De Vusser, S., Genoe, J., and Heremans, P. (2006) Influence of transistor parameters on the noise margin of organic digital circuits. *IEEE Trans. Electron Devices*, **53** (4), 601–610.

9. Halik, M., Klauk, H., Zschieschang, U., Schmid, G., Dehm, C., Schütz, M., Maisch, S., Effenberger, F., Brunnbauer, M., and Stellacci, F. (2004) Low-voltage organic transistors with an amorphous molecular gate dielectric. *Nature*, **431**, 963–966.

10. Bode, D., Rolin, C., Schols, S., Debucquoy, M., Steudel, S., Gelinck, G.H., Genoe, J., and Heremans, P. (2010) Noise margin analysis for organic thin film complementary technology. *IEEE Trans. Electron Devices*, **57** (1), 201–208.

11. Blache, R., Krumm, J., and Fix, W. (2009) Organic CMOS circuits for RFID applications. IEEE International Solid-State Circuits Conference–Digest of Technical Papers, pp. 208–209.

12. Ishida, K., Masunaga, N., Zhou, Z., Yasufuku, T., Sekitani, T., Zschieschang, U., Klauk, H., Takamiya, M., Someya, T., and Sakurai, T. (2009) A stretchable EMI measurement sheet with 8×8 coil array, 2V organic CMOS decoder, and -70dBm EMI detection circuits in 0.18 μm CMOS. IEEE International Solid-State Circuits Conference–Digest of Technical Papers, February 2009, pp. 472–473.

13. Gelinck, G.H., van Veenendaal, E., and Coehoorn, R. (2005) Dual-gate organic thin-film transistors. *Appl. Phys. Lett.*, **87**, 073508.

14. Iba, S., Sekitani, T., Kato, Y., Someya, T., Kawaguchi, H., Takamiya, M., Sakurai, T., and Takagi, S. (2005) Control of threshold voltage of organic field-effect transistors with double-gate structures. *Appl. Phys. Lett.*, **87**, 023509.

15. Myny, K., Beenhakkers, M.J., van Aerle, N.A.J.M., Gelinck, G.H., Genoe, J., Dehaene, W., and Heremans, P. (2011) Unipolar organic transistor circuits made robust by dual-gate technology. *IEEE J. Solid-State Circuits*, **46**, 1223–1230.

16. Koo, J.B., Ku, C.H., Lim, J.W., and Kim, S.H. (2007) Novel organic inverters with dual-gate pentacene thin-film transistors. *Org. Electron.*, **8**, 552–558.

17. Spijkman, M., Smits, E.C.P., Blom, P.W.M., de Leeuw, D.M., Bon Saint Côme, Y., Setayesh, S., and Cantatore, E. (2008) Increasing the noise margin in organic circuits using dual

gate field-effect transistors. *Appl. Phys. Lett.*, **92**, 143304.

18. Hizu, K., Sekitani, T., and Someya, T. (2007) Reduction in operation voltage of complementary organic thin-film transistor inverter circuits using double-gate structures. *Appl. Phys. Lett.*, **90** (9), 93504 -1–93504-3.

19. Takamiya, M., Sekitani, T., Kato, Y., Kawaguchi, H., Someya, T., and Sakurai, T. (2006) An organic FET SRAM for braille sheet display with back gate to increase static noise margin. IEEE International Solid-State Circuits Conference–Digest of Technical Papers, February, 2006, pp. 276–277.

20. (a) Marien, H., Steyaert, M.S.J., van Veenendaal, E., and Heremans, P. (2011) A fully integrated ΔΣ ADC in organic thin-film transistor technology on flexible plastic foil. IEEE J. Solid-State Circuits, **46**, 276–284; (b) Marien, H., Steyaert, M., van Aerle, N., and Heremans, P. (2010) An analog organic first-order CT ΔΣ ADC on a flexible plastic substrate with 26.5dB precision. International Solid-State Circuits Conference Digest Technical Papers, February 2010, pp. 136–137.

21. Xiong, W., Zschieschang, U., Klauk, H., and Murmann, B. (2010) A 3V 6b successive-approximation ADC using complementary organic thin-film transistors on glass. IEEE International Solid-State Circuits Conference–Digest of Technical Papers, February 2010, pp. 134–135.

22. Nga Ng, T., Chabinyc, M.L., Street, R.A., and Salleo, A. (2007) Bias stress effects in organic thin film transistors. Reliability Physics Symposium, 2007, Proceedings 45th Annual, IEEE International, pp. 243–247.

23. Genoe, J., Steudel, S., De Vusser, S., Verlaak, S., Janssen, D., and Heremans, P. (2004) Bias stress in pentacene transistors measured by four probe transistor structures. ESSDERC, 2004, Proceeding of the 34th European, pp. 413–416.

24. Pernstich, K.P., Oberhoff, D., Goldmann, C., and Batlogg, B. (2006) Modeling the water related trap state created in pentacene transistors. *Appl. Phys. Lett.*, **89**, p. 213509.

Index

Organic Electronics II: More Materials and Applications, First Edition. Edited by Hagen Klauk.